NONLINEAR MODELS IN MATHEMATICAL FINANCE: NEW RESEARCH TRENDS IN OPTION PRICING

Nonlinear Models in Mathematical Finance: New Research Trends in Option Pricing

Matthias Ehrhardt
Editor

Nova Science Publishers, Inc.
New York

For permission to use material from this book please contact us:
Telephone 631-231-7269; Fax 631-231-8175
Web Site: http://www.novapublishers.com

LIBRARY OF CONGRESS CATALOGING-IN-PUBLICATION DATA

Available upon request

ISBN 978-1-60456-931-5

Published by Nova Science Publishers, Inc. ✢ New York

CONTENTS

PREFACE
NONLINEAR MODELS IN OPTION PRICING

Matthias Ehrhardt*
Weierstrass Institute for Applied Analysis
and Stochastics, Berlin, Germany

ABSTRACT

Subject: Nonlinear Black-Scholes equations have been increasingly attracting interest over the last two decades, since they provide more accurate values than the classical linear model by taking into account more realistic assumptions, such as transaction costs, risks from an unprotected portfolio, large investor's preferences or illiquid markets, which may have an impact on the stock price, the volatility, the drift and the option price itself.

This book is concerned with several models from the most relevant class of nonlinear Black-Scholes equations for European and American options with a volatility depending on different factors, such as the stock price, the time, the option price and its derivatives. We will present both analytical techniques and numerical methods to solve adequately the arising nonlinear equations.

Purpose: The purpose of this book is to give an overview on the current state-of-the-art research on nonlinear option pricing. Nonlinear models are becoming more and more important since they take into account many effects that are not included in the linear model. However, in practice (i.e. in banks) still linear models are used giving rise to large errors in computing the fair price of options. Hence there exists a noticeable need for nonlinear modeling of financial products. This book should help to foster the usuage of nonlinear Black-Scholes models in practice.

The book is designed to be consisting of a collection of contributed chapters. Outstanding experts working successfully in this challenging research area will be invited to contribute each a chapter of roughly 30-40 pages to this volume.

Academic Level: The principal audience is graduate and Ph.D. students of mathematical finance, lecture of mathematical (computational) finance and people working in banks and stock markets that are interested in new tools for option pricing.

* E-mail address: ehrhardt@wias-berlin.de

INTRODUCTION

Nonlinear Models in Option Pricing - an Introduction, (20 pages)

by Matthias Ehrhardt, Weierstrass Institute for Applied Analysis and Stochastics, Berlin, Germany.

PART I: NONLINEAR BLACK-SCHOLES MODELS

Option Pricing and Hedging in the Presence of Transaction Costs and Nonlinear Partial Differential Equations (45 pages)

by Valeri Zakamouline, Faculty of Economics, University of Agder, Kristiansand, Norway.

Abstract: The celebrated Black and Scholes model for option pricing assumes perfect capital market where it is possible to replicate the payoff of an option by constructing a self-financing dynamic trading strategy. As a consequence, the absence of arbitrage dictates that the option price be equal to the cost of setting up the replicating portfolio. In the presence of transaction costs in capital markets the continuous replication policy is not feasible. Consequently, perfect hedging is impossible and the absence of arbitrage argument is no longer valid.

In this chapter we consider some approaches to option pricing and hedging in the presence of transaction costs. The distinguishing feature of all these approaches is that the solution for the option price and hedging strategy is given by a nonlinear partial differential equation. We start with a review of the classical Leland approach which yields a nonlinear parabolic partial differential equation for the option price, one of the first such in finance. Since the Leland's approach to option pricing has been criticized on different grounds, we present a justification of this approach and show how the performance of the Leland's hedging strategy can be improved. We also present the extensions of the Leland's approach for pricing and hedging of commodity futures contracts, path-dependent and basket options.

Then we proceed to the review of the most successful approach to option hedging with transaction costs, namely, the utility-based approach pioneered by Hodges and Neuberger in 1989. Judging against the best possible tradeoff between the risk and the costs of a hedging strategy, this approach seems to achieve excellent empirical performance. The asymptotic analysis of the option pricing and hedging in this approach reveals that the solution is also given by a nonlinear differential equation. However, this approach has one major drawback that prevents the broad application of this approach in practice, namely, the lack of a closed-form solution. The numerical computations are cumbersome to implement and the calculations of the optimal hedging strategy are time consuming.

Using the results of asymptotic analysis we suggest a simplified parameterized functional form of the optimal hedging strategy for either a single option or a portfolio of options and methods for finding the optimal parameters. Finally we provide an empirical testing of our optimized hedging strategies against some alternative strategies and illustrate that our strategies outperforms all the others.

Utility Indifference Pricing with Market Incompletness (35 pages)

by Michael Monoyios, Mathematical Institute, University of Oxford, Oxford, United Kingdom.

Abstract: Utility indifference pricing and hedging theory is presented, showing how it leads to linear or to non-linear pricing rules for contingent claims. Convex duality is first used to derive probabilistic representations for exponential utility-based prices, in a general setting with locally bounded semi-martingale price processes. The indifference price for a finite number of claims gives a non-linear pricing rule, which reduces to a linear pricing rule as the number of claims tends to zero, resulting in the so-called marginal utility-based price of the claim. Applications to basis risk models with lognormal price processes, under full and partial information scenarios are then worked out in detail. In the full information case, a claim on a non-traded asset is priced and hedged using a correlated traded asset. The resulting hedge requires knowledge of the drift parameters of the asset price processes, which are very difficult to estimate with any precision. This leads naturally to a further application, a partial information problem, with the drift parameters assumed to be random variables whose values are revealed to the hedger in a Bayesian fashion via a filtering algorithm. The indifference price is given by the solution to a non-linear PDE, reducing to a linear PDE for the marginal price when the number of claims becomes infinitesimally small.

PART II: ANALYTIC SOLUTIONS

Pricing Options in Illiquid Markets: Symmetry Reductions and Exact Solutions (29 pages)

by Ljudmila A. Bordag, Halmstad University, Halmstad, Sweden
and Rüdiger Frey, Department of Mathematics, Universität Leipzig, Postfach 10 09 20, 04009 Leipzig, Germany.

Abstract: This chapter is concerned with nonlinear Black-Scholes equations arising in certain option pricing models with a large trader and/or transaction costs. In the first part we give an overview of existing option pricing models with frictions. While the financial setup differs between models, it turns out that in many of these models derivative prices can be characterized by fully nonlinear versions of the standard parabolic Black-Scholes PDE. In the second part of this chapter we study a typical nonlinear Black-Scholes equation using methods from Lie group analysis. The equation possesses a rich symmetry group. By introducing invariant variables, invariant solutions can therefore be characterized in terms of solutions to ordinary differential equations. Finally we discuss properties and applications of these solutions.

Distributional Solutions to an Integro-differential Parabolic Problem Arising on Financial Mathematics (16 pages)

by Maria C. Mariani and Michael Eydenberg, New Mexico State University, USA.

Abstract: In this chapter we study solutions to a parabolic integro-differential operator based on the Black-Scholes model of option pricing in financial mathematics. We obtain solutions under suitable conditions using the method of upper and lower solutions in conjunction with standard methods for solving nonlinear parabolic differential equations. We also consider solutions for a constant-coefficient divergence-form operator when the integration term appears in the form of a convolution. This result is applied to obtain distribution-valued solutions in a simple setting.

PART III: NUMERICAL TREATMENT OF NONLINEAR BLACK-SCHOLES EQUATIONS

A Semidiscretization Method for Solving Nonlinear Black-Scholes Equations: Numerical Analysis and Computing (24 pages)

by Lucas Jódar, Rafael Company and José Ramón Pintos, Instituto de Matemática Multidisciplinar, Universidad Politécnica de Valencia, Valencia. Spain.

Abstract: This chapter deals with the numerical solution of nonlinear Black-Scholes equations modeling option pricing problems in markets with transaction costs. Our approach is based on the semidiscretization technique with respect to the underlying asset variable, allowing several discretization schemes and it is applicable to distinct models.

The proposed method is purely nonlinear in the sense that it does not involve any linearization step, neither in the computation nor in the numerical analysis of the solution.

We pay special attention to two representative models. The first, is the one introduced by Hoggard, Whalley and Wilmott and developed by Parás and Avellaneda where the adjusted volatility depends on the sign of the Gamma of the option price. The second selected model is due to Barles and Soner which involves a nonlinear correction of volatility function depending on the Gamma of the option price in a more complex way apart from the time and the asset variable.

Consistency and stability of the proposed numerical solution are studied and illustrative numerical examples are included.

Transformation Methods for Evaluating Approximations to the Optimal Exercise Boundary for a Linear and Nonlinear Black-Scholes Equation (50 pages)

by Daniel Ševčovič, Department of Applied Mathematics and Statistics, Division of Applied Mathematics, Comenius University, Slovakia.

Abstract: The purpose of this survey chapter is to present a transformation technique that can be used in analysis and numerical computation of the early exercise boundary for an American style of vanilla options that can be modelled by class of generalized Black-Scholes equations. We analyze qualitatively and quantitatively the early exercise boundary for a linear as well as a class of nonlinear Black-Scholes equations with a volatility coefficient which can be a nonlinear function of the second derivative of the option price itself. A motivation for studying the nonlinear Black-Scholes equation with a nonlinear volatility arises from option pricing models taking into account e.g. nontrivial transaction costs, investor's preferences, feedback and illiquid markets effects and risk from a volatile (unprotected) portfolio. We present a method how to transform the free boundary problem for the early exercise boundary position into a solution of a time depending nonlinear nonlocal parabolic equation defined on a fixed domain. We furthermore propose an iterative numerical scheme that can be used in order to find an approximation of the free boundary. In the case of a linear Black-Scholes equation we are able to derive a nonlinear integral equation for the position of the free boundary. We present results of numerical approximation of the early exercise boundary for various types of linear and nonlinear Black-Scholes equations and we discuss dependence of the free boundary on model parameters. Finally, we discuss an application of the transformation method for the pricing equation for American type of Asian options.

Global in Space Numerical Computation for the Nonlinear Black-Scholes Equation (25 pages)

by Naoyuki Ishimura, Graduate School of Economics, Hitotsubashi University, Kunitachi, Tokyo, Japan
and Hitoshi Imai, Institute of Technology and Science, University of Tokushima, Tokushima, Japan.

Abstract: In this chapter the numerical treatment of the nonlinear Black-Scholes equation, which is global in the space variable, is discussed. Exploiting suitable transformation of variables, we do not make any truncation nor approximation of unbounded domains of the equation. Numerical implementation shows that our method and scheme are well formulated. As model problems we deal with the Black-Scholes equation with the effect of transaction costs, which is known to become nonlinear. Moreover, we apply our machinery to other singular nonlinear partial differential equation arising in the optimal investment problem.

Fixed Domain Transformations and High Order Compact Schemes for Nonlinear Black-Scholes Equations for American Options (30 pages)

by Julia Ankudinova and Matthias Ehrhardt, Institute for Mathematics, TU Berlin, Germany.

Abstract: Due to transaction costs, illiquid markets, large investors or risks from an unprotected portfolio the assumptions in the classical Black-Scholes model become unrealistic and the model results in strongly or fully nonlinear, possibly degenerate, parabolic diffusion-convection equations, where the stock price, volatility, trend and option price may depend on the time, the stock price or the option price itself.

In this chapter we will be concerned with several models from the most relevant class of nonlinear Black-Scholes equations for American options with a volatility depending on different factors, such as the stock price, the time, the option price and its derivatives. We will analytically approach the option price by following the ideas proposed by Ševčovič and transforming the free boundary problem into a fully nonlinear nonlocal parabolic equation defined on a fixed, but unbounded domain.

Finally, we will present the results of different compact finite difference schemes for various volatility models including the Leland model, the Barles and Soner model and the Risk adjusted pricing methodology model.

Pricing Hydroelectric Power Plants with/without Operational Restrictions: a Stochastic Control Approach (31 pages)

by Zhuliang Chen and Peter Forsyth, Cheriton School of Computer Science, University of Waterloo, Waterloo, Canada.

Abstract: In this chapter, we price the value of a hydroelectric power plant under a stochastic control framework, taking into consideration the implication of operational constraints such as the ramping and minimum flow rate constraints for the purpose of environmental protection. The power plant valuation problem under the operational constraints is characterized as a bounded stochastic control problem, resulting in a Hamilton-Jacobi-Bellman (HJB) partial integrodifferential equation (PIDE). The valuation problem without the operational restrictions is characterized as an unbounded stochastic control problem; we propose an impulse control formulation, resulting in an HJB variational inequality, for the valuation problem in this scenario. We develop a consistent numerical scheme for solving both the HJB PIDE for the bounded control problem and the HJB variational inequality for the unbounded control problem. We prove the convergence of the numerical scheme to the viscosity solution of each pricing equation, provided a strong comparison result holds. Numerical results indicate that failing to consider operational constraints may considerably underestimate the value of hydroelectric power plants.

Numerical Solutions of Certain Nonlinear Models in European Options on a Distributed Computing Environment (17 pages)

by Choi-Hong Lai, School of Computing and Mathematical Sciences, University of Greenwich, United Kingdom.

Abstract: Financial modelling in the area of option pricing involves the understanding of the correlations between asset and movements of buy/sell in order to reduce risk in investment. Such activities depend on financial analysis tools being available to the trader with which he can make rapid and systematic evaluation of buy/sell contracts. In turn, analysis tools rely on fast numerical algorithms for the solution of financial mathematical models. There are many different financial activities apart from shares buy/sell activities. However it is not the intention of this chapter to discuss various financial activities. The main aim of this chapter is to discuss a distributed algorithm for the numerical solution of a European option. Both linear and non-linear cases are considered.

The algorithm is based on the concept of the Laplace transform and its numerical inverse. The scalability of the algorithm is examined. Numerical tests are used to demonstrate the effectiveness of the algorithm for financial analysis. Time dependent functions for volatility and interest rates are also discussed. Applications of the algorithm to non-linear Black-Scholes equation where the volatility and the interest rate are functions of the option value are included. Some qualitative results of the convergence behaviour of the algorithm is examined. This work also examines the various computational issues of the Laplace transformation method in terms of distributed computing. The idea of using a two-level temporal mesh in order to achieve distributed computation along the temporal axis is introduced. Finally the chapter ends with some conclusions.

PART IV: PARAMETER IDENTIFICATION (INVERSE PROBLEMS)

Calibration Problems in Option Pricing (31 pages)

by Bertram Düring, Institute for Analysis and Scientific Computing, Vienna University of Technology, Vienna, Austria.

Abstract: We review different approaches to calibrate volatility functions in Black-Scholes type models to market data. We focus on an optimal control approach, where a regularized cost functional is minimized over a suitable set of admissible volatilities. The cost functional measures the deviations of option prices obtained from a pricing model to the given market data. We discuss the Black-Scholes case as well as the extension to pricing models in markets with frictions, e.g. models for option pricing in the presence of transaction costs.

In: Nonlinear Models in Mathematical Finance...
Editor: Matthias Ehrhardt, pp. 1-19
ISBN 978-1-60456-931-5

NONLINEAR MODELS IN OPTION PRICING – AN INTRODUCTION

Matthias Ehrhardt
Weierstraß–Institut für Angewandte Analysis und Stochastik,
Mohrenstr. 39, 10117 Berlin, Germany

Abstract

Nonlinear Black–Scholes equations have been increasingly attracting interest over the last two decades, since they provide more accurate values by taking into account more realistic assumptions, such as transaction costs, risks from an unprotected portfolio, large investor's preferences or illiquid markets, which may have an impact on the stock price, the volatility, the drift and the option price itself.

This book consists of a collection of contributed chapters of well–known outstanding scientists working successfully in this challenging research area. It discusses concisely several models from the most relevant class of nonlinear Black–Scholes equations for European and American options with a volatility depending on different factors, such as the stock price, the time, the option price and its derivatives. We will present in this book both analytical techniques and numerical methods to solve adequately the arising nonlinear equations.

The purpose of this book is to give an overview on the current state-of-the-art research on nonlinear option pricing. The intended audience is on the one hand graduate and Ph.D. students of (mathematical) finance and on the other hand lecturer of mathematical finance and and people working in banks and stock markets that are interested in new tools for option pricing.

1. Introduction

Nonlinear models im mathematical finance are becoming more and more important since they take into account effects like the presence of transaction costs, feedback and illiquid market effects due to large traders choosing given stock-trading strategies, imperfect replication and investor's preferences and risk from unprotected portfolios.

Due to transaction costs, illiquid markets, large investors or risks from an unprotected portfolio the assumptions in the classical Black–Scholes model become unrealistic and the model results in strongly or fully nonlinear, possibly degenerate, parabolic diffusion–convection equations, where the stock price, volatility, trend and option price may depend on the time, the stock price or the option price itself.

In this chapter we will be concerned with several models from the most relevant class of nonlinear Black–Scholes equations for European and American options with a volatility depending on different factors, such as the stock price, the time, the option price and its derivatives, where the nonlinearity results from the presence of transaction costs.

In the following sections we will give a short introduction to option pricing.

2. Financial Derivatives

The interest in pricing financial derivatives – among them in pricing options – arises from the fact that financial derivatives, also called contingent claims, can be used to minimize losses caused by price fluctuations of the underlying assets. This process of protection is called *hedging*. There is a variety of financial products on the market, such as futures, forwards, swaps and options. In this introductory chapter we will focus on European and American Call and Put options.

Definition 2.1 *A* European Call option *is a contract where at a prescribed time in the future, known as the expiry or expiration date T ($t = 0$ means 'today'), the holder of the option may purchase a prescribed asset, known as the underlying asset or the underlying $S(t)$, for a prescribed amount, known as the exercise or strike price K. The opposite party, or the writer, has the obligation to sell the asset if the holder chooses to buy it.*

At the final time T the holder of the European Call option will check the current price of the underlying asset $S := S(T)$. If the price of the asset is greater than the strike price, $S \geq K$, then the holder will exercise the Call and buy the stock for the strike price K. Afterwards, the holder will immediately sell the asset for the price S and make a profit of $V = S - K$. In this case the *cash flow*, or the difference of the money received and spent, is positive and the option is said to be *in-the-money*. If $S = K$, the cash flow resulting from an immediate exercise of the option is zero and the option is said to be *at-the-money*. In case $S \leq K$, the cash flow is negative and the option is said to be *out-of-the-money*. In the last two cases the holder will not exercise the Call option, since the asset S can be purchased on the market for K or less than K, which makes the Call option worthless. Therefore, the value of the European Call option at expiry, known as the *pay-off function*, is

$$V(S,T) = (S - K)^+,$$

with the notation $f^+ = \max(f, 0)$.

Definition 2.2 *Reciprocally, a* European Put option *is the right to sell the underlying asset $S(t)$ at the expiry date T for the strike price K. The holder of the Put may exercise this option, the writer has the obligation to buy it in case the holder chooses to sell it.*

The Put is in-the-money if $K \geq S$, at-the-money if $K = S$ and out-of-the-money if $K \leq S$. The pay-off function for a European Put option is therefore

$$V(S,T) = (K - S)^+.$$

The pay-off functions for the European Call and Put option are plotted in Fig, 1 from the perspective of the holder. This perspective is called the *long position.* The perspective of the writer, or the *short position*, is reversed and can be seen when the pay-off functions in Fig. 1 are multiplied by -1. That means that the writer of a European Call option is taking the risk of a potentially unlimited loss and must carefully design a strategy to compensate for this risk [27].

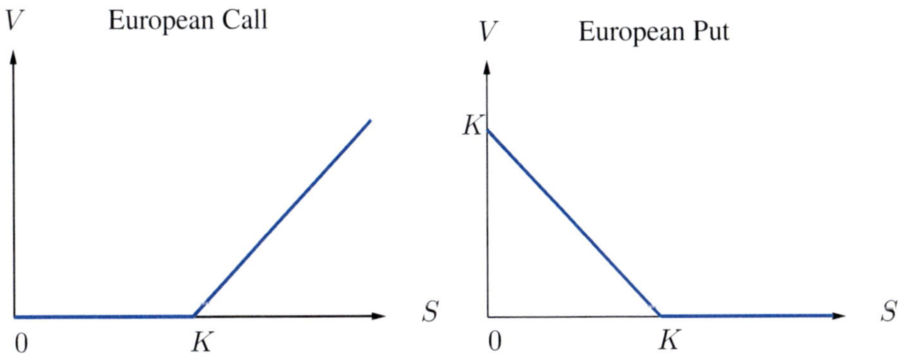

Figure 1. Pay-off functions for European options with a strike price K.

While European options can only be exercised at the expiry date T, *American options* can be exercised at any time until the expiration. Since an American option includes at least the same rights as the corresponding European option, the value of an American option V^{am} can never be smaller than the value of a European option V^{eur}, i.e.

$$V^{\mathrm{am}} \geq V^{\mathrm{eur}}.$$

Whether the values are equal depends on the *dividend yield* q, which describes the percentage rate of the returns on the underlying asset. Assuming that the underlying stock S pays no dividends, the values of a European and an American Call option are equal if all the other parameters remain the same (for details see [13, 34]). In case of an American Put option without dividend payments it can often be advantageous to exercise it before expiry, so that the values of a European and an American Put can differ substantially.

In the presence of a continuous dividend payment the fair price $V(S,0)$ of both an American Call and Put option is greater than the value of a European Call or Put, see Fig. 2.

Furthermore, it should be mentioned that the value of a Call option on an underlying without a dividend payment is always greater than the value of a Call option on an underlying with a dividend payment for both European and American options. For European and American Put options on an underlying without a dividend payment the value is less than on an underlying with a dividend payment. The influence of a dividend payment is summarized in Fig. 3.

Options, whose pay-offs only depend on the final value of the underlying asset, are called *vanilla* options. Options, whose pay-offs depend on the path of the underlying asset, are called *exotic* or *path-dependent* options. Examples are *Asian*, *Barrier* and *lookback* options. In this chapter, we will be solely concerned with plain vanilla European and American options.

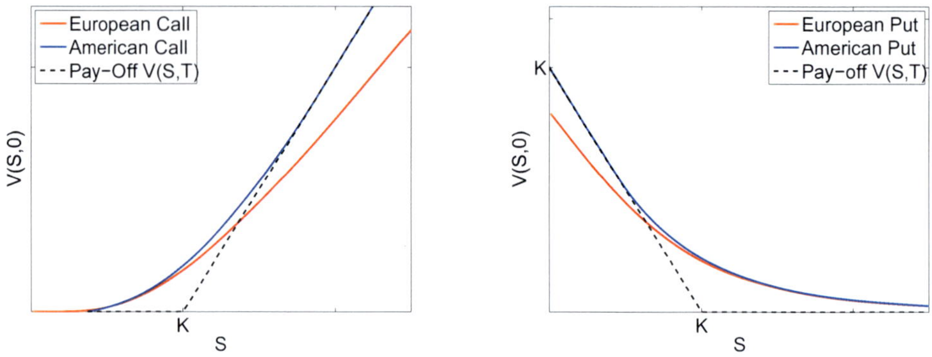

(a) American vs. European Call with dividends.

(b) American vs. European Put option.

Figure 2. Schematical values of American vs. European options at $t = 0$.

3. Linear Black–Scholes Equations

Option pricing theory has made a great leap forward since the development of the Black–Scholes option pricing model by Fischer Black and Myron Scholes in [3] in 1973 and previously by Robert Merton in [25]. The solution of the famous *(linear) Black–Scholes equation* [9]

$$0 = V_t + \frac{1}{2}\sigma^2 S^2 V_{SS} + rSV_S - rV, \tag{1}$$

where $S := S(t) > 0$ and $t \in (0, T)$, provides both an option pricing formula for a European option and a hedging portfolio that replicates the contingent claim assuming that [27]:

- The price of the asset price or underlying asset S follows a Geometric Brownian motion, meaning that if $W := W(t)$ is a standard Brownian motion (see Appendix A.6.), then S satisfies the following stochastic differential equation (SDE):

$$dS = \mu S dt + \sigma S dW.$$

- The *trend* or *drift* μ (measures the average rate of growth of the asset price), the *volatility* σ (measures the standard deviation of the returns) and the riskless interest rate r are constant for $0 \le t \le T$ and no dividends are paid in that time period.

- The market is *frictionless*, thus there are no transaction costs (fees or taxes), the interest rates for borrowing and lending money are equal, all parties have immediate access to any information, and all securities and credits are available at any time and any size. That is, all variables are perfectly divisible and may take any real number. Moreover, individual trading will not influence the price.

- There are no *arbitrage* opportunities, meaning that there are no opportunities of instantly making a risk-free profit ("There is no such thing as free lunch").

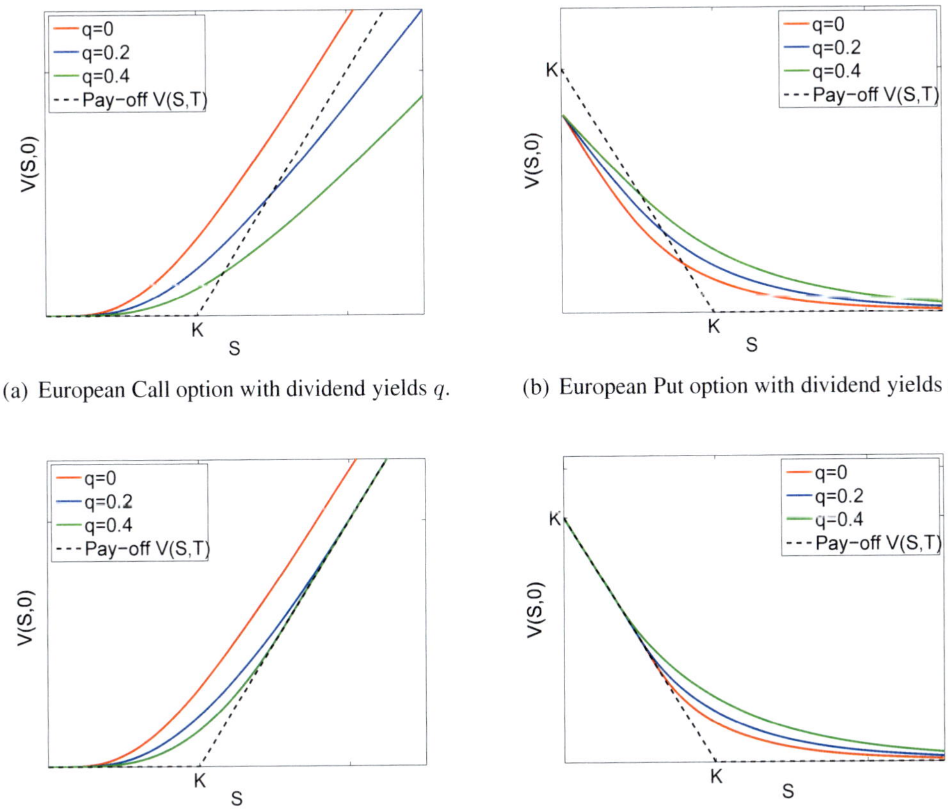

(a) European Call option with dividend yields q.

(b) European Put option with dividend yields q.

(c) American Call option with dividend yields q.

(d) American Put option with dividend yields q.

Figure 3. The influence of a dividend yield.

Under these assumptions the market is *complete*, which means that any derivative and any asset can be replicated or hedged with a portfolio of other assets in the market (see [31]). Then, it is well-known that the linear Black–Scholes equation (1) can be transformed into the heat equation and analytically solved to price the option [33]. The derivation of the solution can be found in [27], the formulae for the European Call and Put options are attached in Appendix B..

For American options, in general, analytic valuation formulae are not available, except for a few special types, which we are not going to address in this chapter. Those types are Calls on an asset that pays discrete dividends and *perpetual* Calls and Puts – meaning Calls and Puts with an infinite time to expiry [23].

4. Nonlinear Black–Scholes Equations

It is quite easy to imagine that the restrictive assumptions mentioned in the previous Section 3. are never fulfilled in reality. Due to transaction costs (cf. [2, 5, 24]), large investor preferences (cf. [11, 12, 26]) and incomplete markets (cf. [30]) these assumptions are likely to become unrealistic and the classical model results in strongly or fully nonlinear, possibly degenerate, parabolic convection–diffusion equations, where both the volatility σ and the drift μ can depend on the time t, the stock price S or the derivatives of the option price V itself.

In this chapter we will focus on several transaction cost models from the most relevant class of nonlinear Black–Scholes equations for European and American options with a constant drift μ and a nonconstant *modified volatility function*

$$\widetilde{\sigma}^2 := \widetilde{\sigma}^2(t, S, V_S, V_{SS}).$$

Under these circumstances (1) becomes the following *nonlinear Black–Scholes equation*, which we will consider for European options:

$$0 = V_t + \frac{1}{2}\widetilde{\sigma}^2(t, S, V_S, V_{SS})S^2 V_{SS} + rSV_S - rV, \tag{2}$$

where $dS = \mu S dt + \widetilde{\sigma} S dW$, $S > 0$ and $t \in (0, T)$.

Studying the linear Black–Scholes equation (1) for an American Call option would be redundant, since the value of an American Call option equals the value of a European Call option if no dividends are paid and the volatility is constant.

In order to make the model more realistic, we will consider a modification of the nonlinear Black–Scholes equation (2) for American options, where S pays out a *continuous dividend* $qSdt$ in a time step dt:

$$0 = V_t + \frac{1}{2}\widetilde{\sigma}^2(t, S, V_S, V_{SS})S^2 V_{SS} + (r - q)SV_S - rV, \tag{3}$$

where S follows the dynamics $dS = (\mu - q)Sdt + \widetilde{\sigma} SdW$, $S > 0$, $t \in (0, T)$ and the dividend yield q is constant.

In the mathematical sense the nonlinear Black–Scholes equations (2) and (3) are called convection–diffusion equations. The second-order term $\frac{1}{2}\widetilde{\sigma}^2(t, S, V_S, V_{SS})S^2 V_{SS}$ is responsible for the *diffusion*, the first-order term rSV_S or $(r - q)SV_S$ is called the *convection* term and $-rV$ can be interpreted as the *reaction* term (see [27, 32]).

In the financial sense, the partial derivatives indicate the sensitivity of the option price V to the corresponding parameter and are called *Greeks*. The option delta is denoted by $\Delta = V_S$, the option gamma by $\Gamma = V_{SS}$ and the option theta by $\theta = V_t$. For a detailed discussion of this issue we refer to [19].

5. Terminal and Boundary Conditions

In order to find a unique solution for the equation (2) we need to complete the problem by stating the terminal and boundary conditions for both the European Call and Put option.

Since American options can be exercised at any time before expiry, we need to find the optimal time t of exercise, known as the *optimal exercise time*. At this time, which mathematically is a *stopping time* (see Appendix A.5.), the asset price reaches the *optimal exercise price* or *optimal exercise boundary* $S_f(t)$. This leads to the formulation of the problem for American options by dividing the domain $[0, \infty[\times[0, T]$ of (3) into two parts along the curve $S_f(t)$ and analyzing each of them (see Fig. 4). Since $S_f(t)$ is not known in

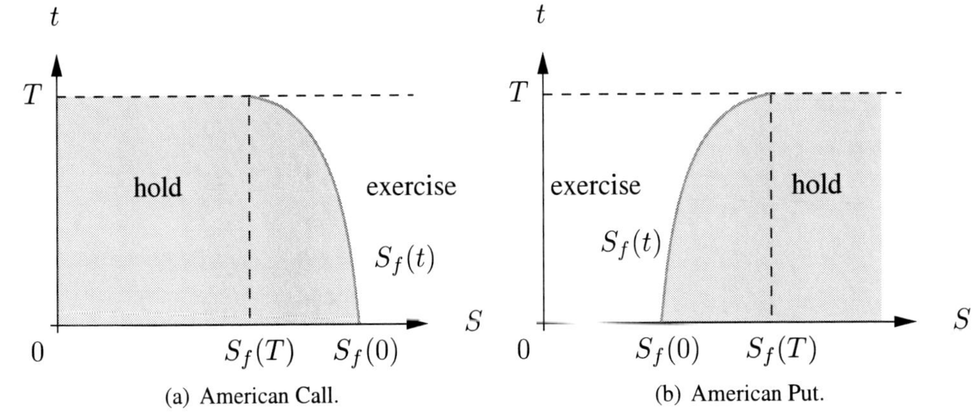

Figure 4. Exercising and holding regions for American options.

advance but has to be determined in the process of the solution, the problem is called *free boundary value problem* [34].

For different numerical approaches, the free boundary problem for American options can be reformulated into a *linear complementary problem* (LCP), a *variational inequality* and a *minimization problem* [13]. The most simple treatment is the formulation as a free boundary problem [8, 14].

Even though we will focus on Call options in this chapter, we state the conditions for Put options for the sake of completeness.

5.1. European Call Option

The value $V(S,t)$ of the European Call option is the solution to (2) on $0 \leq S < \infty$, $0 \leq t \leq T$ with the following terminal and boundary conditions:

$$\begin{aligned} V(S,T) &= (S-K)^+ && \text{for } 0 \leq S < \infty \\ V(0,t) &= 0 && \text{for } 0 \leq t \leq T \\ V(S,t) &\sim S - Ke^{-r(T-t)} && \text{as } S \to \infty. \end{aligned} \tag{4}$$

5.2. European Put Option

Reciprocally, the value $V(S,t)$ of the European Put option is the solution to (2) on $0 \leq S < \infty$, $0 \leq t \leq T$ with the pay–off function for the Put as the terminal condition and the boundary conditions:

$$\begin{aligned} V(S,T) &= (K-S)^+ && \text{for } 0 \leq S < \infty \\ V(0,t) &= Ke^{-r(T-t)} && \text{for } 0 \leq t \leq T \\ V(S,t) &\to 0 && \text{as } S \to \infty. \end{aligned} \tag{5}$$

5.3. American Call Option

For the American Call option the *spatial* domain is divided into two regions by the *free boundary* $S_f(t)$, the *stopping region* $S_f(t) < S < \infty$, $0 \leq t \leq T$, where the option is exercised or dead with $V(S,t) = S - K$ and the *continuation region* $0 \leq S \leq S_f(t)$, $0 \leq t \leq T$, where the option is held or stays alive and (3) is valid under the following terminal and boundary conditions (see Fig. 4(a)):

$$\begin{aligned}
V(S,T) &= (S-K)^+ && \text{for } 0 \leq S \leq S_f(T) \\
V(0,t) &= 0 && \text{for } 0 \leq t \leq T \\
V(S_f(t),t) &= S_f(t) - K && \text{for } 0 \leq t \leq T \\
V_S(S_f(t),t) &= 1 && \text{for } 0 \leq t \leq T \\
S_f(T) &= \max(K, rK/q). &&
\end{aligned} \tag{6}$$

For the sake of simplicity we will assume $r > q$ in this chapter, and therefore we have $S_f(T) = rK/q$ for the American Call.

The structure of the value of an American Call can be seen Fig. 5(a), where we notice that the free boundary $S_f(t)$ determines the position of the exercise. The exercising and holding regions are illustrated in Fig. 4(a).

5.4. American Put Option

The American Put option is exercised in the stopping region $0 \leq S < S_f(t)$, $0 \leq t \leq T$ where it has the value $V(S,t) = K - S$ (see Fig. 4(b)). In the continuation region $S_f(t) \leq S < \infty$, $0 \leq t \leq T$ the Put option stays alive and (3) is valid under the following terminal and boundary conditions:

$$\begin{aligned}
V(S,T) &= (K-S)^+ && \text{for } S_f(T) \leq S < \infty \\
\lim_{S \to \infty} V(S,t) &= 0 && \text{for } 0 \leq t \leq T \\
V(S_f(t),t) &= K - S_f(t) && \text{for } 0 \leq t \leq T \\
V_S(S_f(t),t) &= -1 && \text{for } 0 \leq t \leq T \\
S_f(T) &= \min(K, rK/q). &&
\end{aligned} \tag{7}$$

Since we assumed that $r > q$, we have $S_f(T) = K$ for the American Put. In Fig. 5(b) one can see how the free boundary $S_f(t)$ determines the structure of an American Put.

6. Volatility Models

The essential parameter of the standard Black–Scholes model, that is not directly observable and is assumed to be constant, is the volatility σ. There have been many approaches to improve the model by treating the volatility in different ways and using a *modified volatility function* $\widetilde{\sigma}(\cdot)$ to model the effects of transaction costs, illiquid markets and large traders, which is the reason for the nonlinearity of (2) and (3). We will first give a brief overview of several volatility models and then focus on the volatility models of transaction costs.

- The constant volatility σ in the standard Black–Scholes model can be replaced by the estimated volatility from the former values of the underlying. This volatility is known as the *historical volatility* [13].

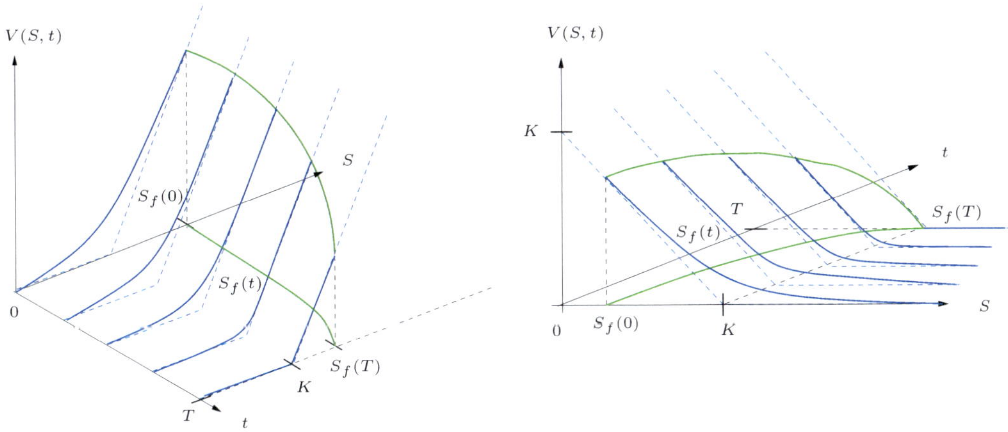

(a) American Call. (b) American Put.

Figure 5. Schematical values $V(S, t)$ of American options.

- If the price of the option and the other parameters are known, which is e.g. the case for the European Call and Put options (see Appendix B.), then the *implied volatility* can be calculated from those Black–Scholes formulae. The implied volatility is the value σ, for which (24) or (25) is true compared to the real market data. It can be calculated implicitly via the difference between the observed option price V (from the market data) and the Black–Scholes formulae (24) or (25), where all the parameters – except for the implied volatility σ – are taken from the market data (the stock price S, the time t, the expiration date T, the strike price K, the interest rate r the dividend rate q).

  Considering options with different strike prices K but otherwise identical parameters, we see that the implicit volatility changes depending on the strike price. If the implicit volatility for a certain strike price K is less than the implicit volatility for both the strike price greater and less than K, this effect is called *volatility smile* [22].

- Replacing the constant volatility with the observed implicit volatilities at each stock price and time leads to the term of the *local volatility* $\widetilde{\sigma} := \widetilde{\sigma}(S, t)$. Dupire [7] examined the dependencies and expressed the local volatility as a function of implicit volatilities.

- Hull and White [18] and Heston [15] developed a model, in which the volatility follows the dynamics of a stochastic process. This is known as the *stochastic volatility*.

- The assumption, that each security is available at any time and any size, or that individual trading will not influence the price, is not always true. Therefore, illiquid markets and large trader effects have been modeled by several authors. In [11] Frey and Stremme and later Frey and Patie [12] considered these effects on the price and come up with the result

$$\widetilde{\sigma} = \frac{\sigma}{1 - \rho\lambda(S) S V_{SS}}, \tag{8}$$

where σ the historical volatility, ρ constant, $\lambda(S)$ strictly convex function, $\lambda(S) \geq 1$. The function $\lambda(S)$ depends on the pay-off function of the financial derivative. For the European Call option, Frey and Patie show that $\lambda(S)$ is a smooth, slightly increasing function for $S \geq K$. Bordag and Chmakova [4] assumed that $\lambda(S)$ is constant and solve the problem (2) with the modified volatility (8) explicitly using Lie-group theory (see also [6]).

As the main scope of this general overview chapter, we draw our attention in the sequel to a more detailed description of several transaction cost models.

6.1. Transaction Costs

The Black–Scholes model requires a continuous portfolio adjustment in order to hedge the position without any risk. In the presence of transaction costs it is likely that this adjustment easily becomes expensive, since an infinite number of transactions is needed [23]. Thus, the hedger needs to find the balance between the transaction costs that are required to rebalance the portfolio and the implied costs of hedging errors. As a result to this "imperfect" hedging, the option might be over- or underpriced up to the extent where the riskless profit obtained by the arbitrageur is offset by the transaction costs, so that there is no single equilibrium price but a range of feasible prices.

It has been shown that in a market with transaction costs there is no replicating portfolio for the European Call option and the portfolio is required to dominate rather than replicate the value of the option (see [2]). Soner, Shreve and Cvitanič [29] proved that the minimal hedging portfolio that dominates a European Call is the trivial one (hence holding one share of the stock that the Call is written on), so that efforts have been made to find an alternate relaxation of the hedging conditions to better replicate the pay-offs of derivative securities.

6.2. The Model of Leland

Leland's idea [24] of relaxing the hedging conditions is to trade at discrete times, which promises to reduce the expenses of the portfolio adjustment. He assumes that the transaction cost $\kappa|\Delta|S/2$, where κ denotes the round trip transaction cost per unit dollar of the transaction and Δ the number of assets bought ($\Delta > 0$) or sold ($\Delta < 0$) at price S, is proportional to the monetary value of the assets bought or sold. Now consider a replicating portfolio with Δ units of the underlying and the *bond* B (a certificate of debt issued by a government or a corporation guaranteeing payment B plus interest by a specified future date):

$$\Pi = \Delta S + B.$$

After a small change in time of the size δt the change in the portfolio becomes

$$\delta\Pi = \Delta\delta S + rB\delta t - \frac{\kappa}{2}|\delta\Delta|S, \tag{9}$$

where δS is the change in price S, so that the first term represents the change in value, the second term represents the bond growth in δt time and $\delta\Delta$ represents the change in the number of assets, so that the last term becomes the transaction cost due to portfolio change.

We apply Itô's lemma (see 23 in Appendix A.7.) to the value of the option $V := V(S,t)$ and get

$$\delta V = V_S \delta S + (V_t + \frac{\sigma^2}{2} S^2 V_{SS}) \delta t. \tag{10}$$

Assuming that the option V is replicated by the portfolio Π, their values have to match at all times and there can be no risk-free profit. With this no-arbitrage argument we get

$$\delta \Pi = \delta V.$$

Matching the terms in (9) and (10) we get $\Delta - V_S$ and

$$rB\delta t - \frac{\kappa}{2} |\delta \Delta| S = (V_t + \frac{\sigma^2}{2} S^2 V_{SS}) \delta t. \tag{11}$$

Leland shows that

$$\frac{\kappa}{2} |\delta \Delta| S = \frac{\sigma^2}{2} \, \mathrm{Le} \, S^2 |V_{SS}| \delta t, \tag{12}$$

where Le denotes the *Leland number*, which is given by

$$\mathrm{Le} = \sqrt{\frac{2}{\pi}} \left(\frac{\kappa}{\sigma \sqrt{\delta t}} \right), \tag{13}$$

with δt being the transaction frequency (interval between successive revisions of the portfolio) and κ the round trip transaction cost per unit dollar of the transaction. Plugging (12) and $B = \Pi - \Delta S = V - SV_S$ into the equation (11) becomes

$$rV - rSV_S - \frac{\sigma^2}{2} \, \mathrm{Le} \, S^2 |V_{SS}| = V_t + \frac{\sigma^2}{2} S^2 V_{SS}. \tag{14}$$

Therefore, Leland deduces that the option price is the solution of the nonlinear Black–Scholes equation

$$0 = V_t + \frac{1}{2} \tilde{\sigma}^2 S^2 V_{SS} + rSV_S - rV,$$

with the *modified volatility*

$$\tilde{\sigma}^2 = \sigma^2 \Big(1 + \mathrm{Le} \, \mathrm{sign}(V_{SS}) \Big), \tag{15}$$

where σ represents the historical volatility and Le the Leland number. It follows from the definition of the Leland number (13) that the more frequent the rebalancing (δt smaller), the higher the transaction cost and the greater the value of V.

It is known that $V_{SS} > 0$ for European Puts and Calls in the absence of transaction costs. Assuming the same behavior in the presence of transaction costs, equation (2) becomes linear with an adjusted constant volatility $\tilde{\sigma}^2 = \sigma^2(1 + \mathrm{Le}) > \sigma^2$.

Leland's model has played a significant role in financial mathematics, even though it has been partly criticized by e.g. Kabanov and Safarian in [21], who prove that Leland's result has a hedging error. The restriction of his model is the convexity of the resulting option price V (hence $V_{SS} > 0$) and the possibility to only consider one option in the portfolio. Hoggard, Whalley and Wilmott studied equation (2) with the modified volatility (15) for several underlyings in [17]. An extension to this approach to general pay-offs is obtained by Avellaneda and Parás [1].

6.3. Barles and Soner

In [2] Barles and Soner derived a more complicated model by following the above utility function approach of Hodges and Neuberger [16]. Consider the process of bonds owned $X(s)$ and the process of shares owned $Y(s)$. Let the trading strategy $\big(L(s), M(s)\big)$ be a pair of nondecreasing processes with $L(t) = M(t) = 0$, which are interpreted as the cumulative transfers, measured in shares of stock. $L(s)$ is measured in shares from bond to stock and $M(s)$ is measured in shares from stock to bond. Let $\kappa \in (0,1)$ be the proportional transaction cost. The processes $X(s)$ and $Y(s)$ start with the initial values x and y, $s \in [t,T]$ and evolve according to

$$X(s) = x - \int_t^s S(\tau)(1+\kappa)\,dL(\tau) + \int_t^s S(\tau)(1-\kappa)\,dM(\tau) \tag{16}$$

and

$$Y(s) = y + L(s) - M(s). \tag{17}$$

The first integral in (16) represents buying shares of stock at a price increased by the proportional transaction cost, the second integral represents selling stock at a reduced price of the transaction cost. In (17) we add the amount of the stocks bought and subtract the amount for the stocks sold to the initial amount of stocks owned.

According to the utility maximization approach of Hodges and Neuberger [16], the price of a European Call option can be obtained as the difference between the maximum utility of the terminal wealth when there is no option liability and when there is such a liability. Following this approach, Barles and Soner considered two optimization problems. Let the *exponential utility function* be

$$U(\xi) = 1 - e^{-\gamma\xi}, \quad \xi \in \mathbb{R},$$

where $\gamma > 0$ is the *risk aversion factor*. The first value function is the expected utility from the final wealth without any option liabilities taken over the transfer processes

$$V_1(x,y,S(t),t) := \sup_{L(\cdot),M(\cdot)} E\big[U\big(X(T) + Y(T)S(T)\big)\big],$$

the second one is the expected utility from the final wealth assuming that we have sold N European Call options taken over the transfer processes

$$V_2(x,y,S(t),t) := \sup_{L(\cdot),M(\cdot)} E\big[U\big(X(T) + Y(T)S(T) - N(S(T)-K)^+\big)\big].$$

Hodges and Neuberger postulate that the price of each option is equal to the maximal solution Λ of the algebraic equation

$$\begin{aligned} V_2(x+N\Lambda,y,S(t),t) &= \sup_{L(\cdot),M(\cdot)} E\big[U\big(X(T) + N\Lambda + Y(T)S(T) \\ &\qquad - N(S(T)-K)^+\big)\big] \\ &= \sup_{L(\cdot),M(\cdot)} E\big[U\big(X(T) + Y(T)S(T)\big)\big] \\ &= V_1(x,y,S(t),t), \end{aligned}$$

which means that the option price Λ equals the increment of the initial capital at time t that is needed to cope with the option liabilities arising at T. By a linearity argument selling N options with risk aversion factor of γ yields the same price as selling one option with risk aversion factor γN. This leads to performing an asymptotic analysis as $\gamma N \to \infty$. Hence, we consider

$$U(\xi) = 1 - e^{-\gamma N \xi}$$

and

$$\varepsilon = \frac{1}{\gamma N}.$$

Then, we have

$$U_\varepsilon(\xi) = 1 - e^{-\frac{\xi}{\varepsilon}}, \quad \xi \in \mathbb{R}.$$

Our optimization problems become

$$V_1(x, y, S(t), t) = 1 - \inf_{L(\ ),M(\cdot)} E[e^{-\frac{1}{\varepsilon}(X(T)+Y(T)S(T))}]$$

and

$$V_2(x, y, S(t), t) = 1 - \inf_{L(\cdot),M(\cdot)} E[e^{-\frac{1}{\varepsilon}(X(T)+Y(T)S(T)-(S(T)-K)^+)}].$$

For analysis simplification Barles and Soner define $z_{1,2} : \mathbb{R} \times (0, \infty) \times (0, T) \to \mathbb{R}$ by

$$V_1(x, y, S(t), t) = 1 - e^{-\frac{1}{\varepsilon}\left(x+yS(t)-z_1(y,S(t),t)\right)}$$

and

$$V_2(x, y, S(t), t) = 1 - e^{-\frac{1}{\varepsilon}\left(x+yS(t)-z_2(y,S(t),t)\right)}.$$

Then

$$z_1(y, S(t), T) = 0 \quad \text{and} \quad z_2(y, S(t), T) = (S(T) - K)^+$$

and the option price

$$\Lambda(x, y, S(t), t; \frac{1}{\varepsilon}, 1) = z_2(y, S(t), t) - z_1(y, S(t), t).$$

By the theory of stochastic optimal control [10], Barles and Soner state that the value functions V_1 and V_2 are the unique solutions of the dynamic programming equation

$$\min\{-V_t + \frac{1}{2}\sigma^2 S^2 V_{SS} - rSV_S, -V_y + S(1+\kappa)V_x, V_y - S(1-\kappa)V_x\} = 0,$$

which leads to a dynamic programming equation for z_1 and z_2, which are independent of the variable x.

Supposing that the proportional transaction cost κ is equal to $a\sqrt{\varepsilon}$ for some constant $a > 0$, they prove that as $\varepsilon \to 0$ and $\kappa \to 0$

$$z_1 \to 0 \quad \text{and} \quad z_2 \to V,$$

where V is the unique (viscosity) solution of the nonlinear Black–Scholes equation

$$0 = V_t + \frac{1}{2}\tilde{\sigma}^2 S^2 V_{SS}^2 + rSV_S - rV,$$

where

$$\widetilde{\sigma}^2 = \sigma^2\Big(1 + \Psi(e^{r(T-t)}a^2S^2V_{SS})\Big). \tag{18}$$

Here σ denotes the historical volatility, $a = \kappa/\sqrt{\varepsilon}$ and $\Psi(x)$ is the solution to the following nonlinear ordinary differential equation (ODE)

$$\Psi'(x) = \frac{\Psi(x)+1}{2\sqrt{x\Psi(x)} - x}, \; x \neq 0, \tag{19a}$$

with the initial condition

$$\Psi(0) = 0. \tag{19b}$$

The analysis of this ODE (19) by Barles and Soner in [2] implies that

$$\lim_{x\to\infty} \frac{\Psi(x)}{x} = 1 \quad \text{and} \quad \lim_{x\to-\infty} \Psi(x) = -1. \tag{20}$$

The property (20) encourages to treat the function $\Psi(\cdot)$ as the identity for large arguments and therefore to simplify the calculations. In this case the volatility becomes

$$\widetilde{\sigma}^2 = \sigma^2(1 + e^{r(T-t)}a^2S^2V_{SS}). \tag{21}$$

The existence of a viscosity solution to (2) for European options with the volatility given by (18) is proved by Barles and Soner in [2] and their numerical results indicate an economically significant price difference between the standard Black–Scholes model and the nonlinear model with transaction costs.

6.4. Risk Adjusted Pricing Methodology

In this model, proposed by Kratka in [22] and improved by Jandačka and Ševčovič in [20], the optimal time-lag δt between the transactions is found to minimize the sum of the rate of the transaction costs and the rate of the risk from an unprotected portfolio. That way the portfolio is still well protected with the Risk Adjusted Pricing Methodology (RAPM) and the *modified volatility* is now of the form

$$\widetilde{\sigma}^2 = \sigma^2\Big(1 + 3(\frac{C^2M}{2\pi}SV_{SS})^{\frac{1}{3}}\Big), \tag{22}$$

where $M \geq 0$ is the transaction cost measure and $C \geq 0$ the risk premium measure.

It is worth mentioning that these nonlinear transaction cost models that are described above are all consistent with the linear model if the additional parameters for transaction costs are equal to zero and vanish (Le, $\Psi(\cdot)$, M).

Conclusion

In this chapter we provided a profound overview over nonlinear Black–Scholes equations for European and American options.

We introduced the reader to the financial terminology and to Black–Scholes equations and presented several reasons for their nonlinearity and focused on the nonlinearity resulting from a modified volatility function due to transaction costs. Here we focused on several transaction cost models, including Leland' model, Barles' and Soner's model, the identity model and the Risk Adjusted Pricing Methodology.

Acknowledgements

The author acknowledges very fruitful discussions with Daniel Ševčovič in the framework of a bilateral German–Slovakian DAAD project *Fin-Diff-Fin: Finite differences for Financial derivative models.*

Appendix

A. Stochastics

In this chapter, we used several terms and concepts of probability theory and stochastics. Thus, we recall some definitions (see e.g. [13, 27, 28] and the references therein).

A.1. Probability Space

Let Ω be a *sample space* representing all possible scenarios (e.g. all possible paths for the stock price over time). A subset of Ω is an *event* and $\omega \in \Omega$ a *sample point.*

Definition 1.1 *Let Ω be a nonempty set and $\mathcal{F}$ be a collection of subsets of Ω. $\mathcal{F}$ is called a σ-*algebra *(not related to the volatility σ), if*

i) $\Omega \in \mathcal{F}$,

ii) whenever a set A belongs to $\mathcal{F}$, its complement A^c also belongs to $\mathcal{F}$ and

iii) whenever a sequence of sets A_n, $n \in \mathbb{N}$ belongs to $\mathcal{F}$, their union $\bigcup_{n=1}^{\infty} A_n$ also belongs to $\mathcal{F}$.

In our financial scenario, $\mathcal{F}$ represents the space of events that are observable in the market and therefore, all the information available until the time t can be regarded as a σ-algebra $\mathcal{F}_t$. It is logical that $\mathcal{F}_t \subseteq \mathcal{F}_s$ for $t < s$, since the information that has been available t is still available at s.

Definition 1.2 *Let Ω be a nonempty set and $\mathcal{F}$ be a σ-algebra of subsets of Ω.* A probability measure *P is a function that assigns a number in $[0, 1]$ to every set $A \in \mathcal{F}$. The number is called the* probability *of A and is written $P(A)$. We require:*

- $P(\Omega) = 1$ *and*
- *whenever a sequence of disjoint sets A_n, $n \in \mathbb{N}$ belongs to $\mathcal{F}$, then*

$$P\left(\bigcup_{n=1}^{\infty} A_n\right) = \sum_{n=1}^{\infty} P(A_n).$$

The tripel $(\Omega, \mathcal{F}, P)$ is called a *probability space.*

A.2. Random Variable

Definition 1.3 *A real-valued function X on Ω is called a* random variable *if the sets*

$$\{X \leq x\} := \{\omega \in \Omega : X(\omega) \leq x\} = X^{-1}(]-\infty, x])$$

are measurable for all $x \in \mathbb{R}$. That is, $\{X \leq x\} \in \mathcal{F}$.

A.3. Stochastic Process

Definition 1.4 *A (continuous)* stochastic process $X(t) = X(\cdot, t)$*, $t \in [0, \infty[$, is a family of random variables $X : \Omega \times [0, \infty[\to \mathbb{R}$ with $t \mapsto X(\omega, t)$ continuous for all $\omega \in \Omega$.*

A.4. Itô Process

Definition 1.5 *An* Itô process *is a stochastic process of the form*

$$dX = a(X, t)dt + b(X, t)dW,$$

which is equivalent to

$$X(t) = X(0) + \int_0^t a(X, s)ds + \int_0^t b(X, s)dW,$$

where $X(0)$ is nonrandom, $W(t)$ is a standard Wiener process, $a(\cdot)$ and $b(\cdot)$ are sufficiently regular functions and the integrals are Itô integrals.

A.5. Stopping Time

Definition 1.6 *A* stopping time *t is a random variable taking values in $[0, \infty]$ and satisfying*

$$\{t \leq s\} \in \mathcal{F}_s \quad \forall s \geq 0.$$

A.6. Brownian Motion

Definition 1.7 *A* Brownian motion *or* Wiener process *is a time-continuous stochastic process $W(t)$ with the properties:*

- $W(0) = 0$.
- $W(t) \sim \mathcal{N}(0, t)$ *for all $t \geq 0$. That is, for each t the random variable $W(t)$ is normally distributed with mean* $\mathrm{E}[W(t)] = 0$ *and variance* $\mathrm{Var}[W(t)] = \mathrm{E}[W^2(t)] = t$.
- *All increments $\Delta W(t) := W(t + \Delta t) - W(t)$ on non-overlapping time intervals are independent. That is, $W(t_2) - W(t_1)$ and $W(t_4) - W(t_3)$ are independent for all $0 \leq t_1 < t_2 \leq t_3 < t_4$.*
- *$W(t)$ depends continuously on t.*

A.7. Itô's Lemma

Theorem 1.8 *Consider a function* $V(S,t) : \mathbb{R} \times [0,\infty[\to \mathbb{R}$ *with* $V \in C^{2,1}(\mathbb{R} \times [0,\infty[)$ *and suppose that* $S(t)$ *follows the Itô process*

$$dS = a(S,t)dt + b(S,t)dW,$$

where $W(t)$ *is a standard Wiener process. Then* V *follows an Itô process with the same Wiener process* $W(t)$*:*

$$dV = (aV_S + \frac{1}{2}b^2 V_{SS} + V_t)dt + bV_S dW, \tag{23}$$

where $a := a(S,t)$ *and* $b := b(S,t)$.

If we consider a special case, where $a(S,t) = \mu S$ and $b(S,t) = \sigma S$, then $S(t)$ follows the Geometric Brownian motion, where $W(t)$ is a standard Wiener process, and we have

$$dS = \mu S dt + \sigma S dW.$$

Then, Itô's Lemma yields

$$\begin{aligned} dV &= (\mu S V_S + \frac{1}{2}\sigma^2 S^2 V_{SS} + V_t)dt + \sigma S V_S dW \\ &= (\frac{1}{2}\sigma^2 S^2 V_{SS} + V_t)dt + V_S dS. \end{aligned}$$

B. Pricing Formulae

Theorem 2.1 *The solution to the linear Black–Scholes equation* (1) *with the terminal and boundary conditions* (4)*, or the value of the* European Call *option, is given by*

$$V(S,t) = Se^{-q(T-t)}\mathcal{N}(d_1) - Ke^{-r(T-t)}\mathcal{N}(d_2), \tag{24}$$

where

$$d_1 := \frac{\ln\left(\frac{S}{K}\right) + (r - q + \frac{\sigma^2}{2})(T-t)}{\sigma\sqrt{T-t}}$$

$$d_2 := \frac{\ln\left(\frac{S}{K}\right) + (r - q - \frac{\sigma^2}{2})(T-t)}{\sigma\sqrt{T-t}}$$

and $\mathcal{N}(x)$ *is the standard normal cumulative distribution function*

$$\mathcal{N}(x) = \frac{1}{\sqrt{2\pi}} \int_{-\infty}^{x} e^{-\frac{y^2}{2}}\, dy, \qquad x \in \mathbb{R}.$$

Respectively, the value of the European Put *option is the solution to the linear Black–Scholes equation* (1) *with the terminal and boundary conditions* (5) *and is given by*

$$V(S,t) = Se^{-q(T-t)}\mathcal{N}(d_1) - Ke^{-r(T-t)}\mathcal{N}(d_2). \tag{25}$$

References

[1] M. Avellaneda and A. Parás, *Dynamic hedging portfolios for derivative securities in the presence of large transaction costs, Appl. Math. Finance* **1** (1994), 165–193.

[2] G. Barles and H.M. Soner, Option pricing with transaction costs and a nonlinear Black–Scholes equation, *Finance Stoch.* **2** (1998), 369–397.

[3] F. Black and M. Scholes, The pricing of options and corporate liabilities, *J. Polit. Econ.* **81** (1973), 637–659.

[4] L. Bordag and A. Chmakova, Explicit solutions for a nonlinear model of financial derivatives, *Int. J. Theor. Appl. Finance* **10** (2007), 1–21.

[5] P. Boyle and T. Vorst, Option Replication in Discrete Time with Transaction Costs, *J. Finance* **47** (1992), 271–293.

[6] A. Chmakova, *Symmetriereduktionen und explizite Lösungen für ein nichtlineares Modell eines Preisbildungsprozesses in illiquiden Märkten*, Dissertation, Technische Universität Cottbus, 2005 (in german).

[7] B. Dupire, Pricing with a Smile, *Risk* **7** (1994), 18–20.

[8] M. Ehrhardt and R.E. Mickens, A fast, stable and accurate numerical method for the Black–Scholes equation of American options, accepted: *Int. J. Theoret. Appl. Finance.*

[9] G. Ferreyra, The Mathematics Behind the 1997 Nobel Prize in Economics, *e-MATH, What's New in Mathematics*, (electr. magazine of the Amer. Math. Soc.), 1997.

[10] W.H. Fleming and H.M. Soner, *Controlled Markov Processes and Viscosity Solutions*, Springer, New York, 1993.

[11] R. Frey and A. Stremme, Market Volatility and Feedback Effects from Dynamic Hedging, *Math. Finance* **4** (1997), 351–374.

[12] R. Frey and P. Patie, *Risk Management for Derivatives in Illiquid Markets: A Simulation Study*, Springer, Berlin, 2002.

[13] M. Günther and A. Jüngel, *Finanzderivate mit MATLAB*, Vieweg, 2003 (in german).

[14] H. Han and X. Wu, A fast numerical method for the Black–Scholes equation of American options, *SIAM J. Numer. Anal.* **41** (2003), 2081–2095.

[15] S. Heston, A Closed–Form Solution for Options with Stochastic Volatility with Applications to Bond and Currency Options, *Rev. Fin. Studies* **6** (1993), 327–343.

[16] S. Hodges and A. Neuberger, Optimal replication of contingent claims under transaction costs, *Review of Futures Markets* **8** (1989), 222–239.

[17] T. Hoggard and A. E. Whalley and P. Wilmott, Hedging option portfolios in the presence of transaction costs, *Adv. Fut. Opt. Res.* **7** (1994), 21–35.

[18] J. Hull and A. White, The Pricing of Options on Assets with Stochastic Volatilities, *J. Finance* **42** (1987), 281–300.

[19] J. Hull, *Options, Futures and Other Derivatives*, 5th Edition, Prentice Hall, 2002.

[20] M. Jandačka and D. Ševčovič, On the risk-adjusted pricing-methodology-based valuation of vanilla options and explanation of the volatility smile, *J. Appl. Math.* **3** (2005), 235–258.

[21] Y. Kabanov and M. Safarian, On Leland's Strategy of Option Pricing with Transactions Costs, *Finance Stoch.* **1** (1997), 239–250.

[22] M. Kratka, No Mystery Behind the Smile, *Risk* **9** (1998), 67–71.

[23] Y.-K. Kwok, *Mathematical Models of Financial Derivatives*, Springer Finance, 1998.

[24] H.E. Leland, Option pricing and replication with transactions costs, *J. Finance* **40** (1985), 1283–1301.

[25] R.C. Merton, Theory of rational option pricing, *Bell J. Econ. Manag. Sci.* **4** (1973), 141–183.

[26] P. Schönbucher and P. Wilmott, The feedback–effect of hedging in illiquid markets, *SIAM J. Appl.Math.* **61** (2000), 232–272.

[27] R. Seydel, *Tools for Computational Finance*, Second ed., Springer, 2004.

[28] S.E. Shreve, *Stochastic Calculus for Finance II: Continuous–Time Models*, Springer Finance, New York, 2004.

[29] H.M. Soner, S.E. Shreve and J. Cvitanič, There is No Nontrivial Hedging Portfolio for Option Pricing with Transaction Costs, *Ann. Appl. Probab.* **5** (1995), 327–355.

[30] H.M. Soner and N. Touzi, Superreplication Under Gamma Constraints, *SIAM J. Contr. Optim.* **39** (2001), 73–96.

[31] D. Tavella and C. Randall, *Pricing financial instruments: The finite difference method*, John Wiley & Sons, 2000.

[32] P. Wilmott, J. Dewynne and S. Howison, *Option Pricing: Mathematical Models and Computation*, Oxford, Financial Press, 1993.

[33] P. Wilmott, S. Howison, and J. Dewynne, *The Mathematics of Financial Derivatives, A Student Introduction,* Cambridge University Press, 2002.

[34] Y. Zhu, X. Wu and I. Chern, *Derivative securities and difference methods*, Springer Finance, New York, 2004.

Part I.
Nonlinear Black-Scholes Models

In: Nonlinear Models in Mathematical Finance... ISBN 978-1-60456-931-5
Editor: Matthias Ehrhardt, pp. 23-65

Chapter 1

OPTION PRICING AND HEDGING IN THE PRESENCE OF TRANSACTION COSTS AND NONLINEAR PARTIAL DIFFERENTIAL EQUATIONS

Valeri Zakamouline
University of Agder, Faculty of Economics,
Service Box 422, 4604 Kristiansand, Norway

Abstract

In the presence of transaction costs the perfect option replication is impossible which invalidates the celebrated Black and Scholes model. In this chapter we consider some approaches to option pricing and hedging in the presence of transaction costs. The distinguishing feature of all these approaches is that the solution for the option price and hedging strategy is given by a nonlinear partial differential equation (PDE).

We start with a review of the classical Leland approach which yields a nonlinear parabolic PDE for the option price, one of the first such in finance. Since the Leland's approach to option pricing has been criticized on different grounds, we present a justification of this approach and show how the performance of the Leland's hedging strategy can be improved. We extend the Leland's approach to cover the pricing and hedging of options on commodity futures contracts, as well as path-dependent and basket options. We also present examples of finite-difference schemes to solve some nonlinear PDEs.

Then we proceed to the review of the most successful approach to option hedging with transaction costs, the utility-based approach pioneered by Hodges and Neuberger in 1989. Judging against the best possible tradeoff between the risk and the costs of a hedging strategy, this approach seems to achieve excellent empirical performance. The asymptotic analysis of the option pricing and hedging in this approach reveals that the solution is also given by a nonlinear PDE. However, this approach has one major drawback that prevents the broad application of this approach in practice, namely, the lack of a closed-form solution. The numerical computations are cumbersome to implement and the calculations of the optimal hedging strategy are time consuming.

Using the results of asymptotic analysis we suggest a simplified parameterized functional form of the optimal hedging strategy for either a single option or a portfolio of options and a method for finding the optimal parameters.

1. Introduction

The celebrated Black and Scholes [5] model for option pricing assumes perfect capital market where it is possible to replicate the payoff of an option by constructing a self-financing dynamic trading strategy. As a consequence, the absence of arbitrage dictates that the option price be equal to the cost of setting up the replicating portfolio. In the presence of transaction costs in capital markets the continuous replication policy is not feasible. Hence, perfect hedging is impossible and the absence of arbitrage argument is no longer valid.

In this chapter we consider some approaches to option pricing and hedging in the presence of transaction costs. The distinguishing feature of all these approaches is that the solution for the option price and hedging strategy is given by a nonlinear partial differential equation (PDE). We start with a review of the Leland [17] approach which yields a nonlinear parabolic PDE for the option price, one of the first such in finance. This PDE is valid for the price of a portfolio of options as well. What is interesting here is the fact that, since the pricing PDE is nonlinear, the value of portfolio of options that is the sum of individual options is not the same as sum of values of individual options. Using the Leland's method one hedges an option with a delta calculated similarly as the Black-Scholes delta, but with an "adjusted" (modified) hedging volatility. The Leland's volatility adjustment depends on the sign of the second derivative of the option price with respect to the price of the underlying.

Since the Leland's approach to option pricing and hedging has been criticized on different grounds, we present a justification of this approach. In particular, we show that the Leland's approach minimizes, to leading order, the local risk of the hedging strategy. It turns out that the Leland's hedging with a modified hedging volatility "equalizes" the replication error across different stock paths which also reduces the total risk of a hedging strategy. Because the knowledge of the Leland's hedging strategy is very important for optimal hedging of options and option portfolios in the presence of transaction costs, we extend the Leland's method to cover the cases of pricing and hedging an option portfolio on a commodity futures contract, a portfolio of strongly path-dependent options on a stock, and options on several assets. In all the cases as a result we obtain a nonlinear PDE for the option (portfolio) price. We show that with a single underlying asset a non-linear PDE in some cases may be reduced to a linear PDE. Moreover, the hedging strategy in this case can still be intuitively interpreted as hedging with "adjusted" volatility. In contrast, in the case with several underlyings the resulting PDE is fully non-linear and the option hedging strategy can no longer be interpreted as hedging with "adjusted" volatility. Since in some cases the Leland's hedging can be suboptimal, we show how the performance of the Leland's hedging strategy can be improved. Finally, we present examples of finite-difference schemes to solve some nonlinear PDEs.

Then we proceed to the review of the most successful approach to option hedging with transaction costs, the utility-based approach pioneered by Hodges and Neuberger [14]. Judging against the best possible tradeoff between the risk and the costs of a hedging strategy, this approach seems to achieve excellent empirical performance. The asymptotic analysis of the option pricing and hedging in this approach reveals that the solution is also given by a nonlinear PDE[1]. The utility-based hedging strategy can also be interpreted as hedg-

[1]Note the immediate consequence of nonlinearity: in the utility-based approach the value of portfolio of options is not the same as sum of values of individual options.

ing with adjusted volatility and this volatility adjustment is, to large extent, similar to the Leland's volatility adjustment. However, the utility-based approach has one major drawback that prevents the broad application of this approach in practice, namely, the lack of a closed-form solution. The numerical computations are cumbersome to implement and the calculations of the optimal hedging strategy are time consuming.

Since there are no explicit solutions for the utility-based hedging with transaction costs and the numerical methods are computationally hard, for practical applications it is of major importance to use other alternatives. One of such alternatives is to hedge according to an asymptotic solution. However, Zakamouline [25] compared the performance of the asymptotic strategies against the exact strategy and found that under realistic model parameters an asymptotic strategy performs noticeably worse than that obtained from the exact numerical solution. He suggested to use an alternative to the asymptotic analysis, namely, the approximation method. This method consists in specifying a flexible functional form of the optimal hedging policy and then choosing the parameters that provide the best fit to the exact numerical solution. Zakamouline performed the approximation of the utility-based hedging strategy for a short European call option and showed empirically that his approximation strategy outperforms all the other alternative hedging strategies. However, the method suggested in Zakamouline [25] retains the major drawback of the utility-based approach since to obtain the approximation strategy for a specific option one needs to implement first the numerical computations of the utility-based hedging strategy.

In practice one often faces the problem of hedging a large portfolio of options on the same underlying asset. Consequently, the knowledge of the optimal hedging strategy for a portfolio of options is of great practical significance. First, we present a detailed description of the nature of the utility-based hedging strategy for a portfolio of options. We illustrate that the optimal hedging strategy for a portfolio of option is, in fact, rather simple and has three essential features: (i) existence of the no transaction region, (ii) optimal form of the no transaction region, and (iii) the volatility adjustment. Second, using the results of asymptotic analysis, we suggest a simplified parameterized description of the utility-based hedging strategy for a portfolio of options and a method of finding the optimal parameters of a hedging strategy. In the essence our idea is to "mimic" the essential properties of the utility-based hedging strategy by specifying a simple but flexible functional form of the optimal hedging policy. Then, we optimize the form of the hedging strategy by changing a few parameters in order to find the best possible risk-return tradeoff of the hedging strategy (basically, we need to find the optimal form of the no-transaction region and the optimal volatility adjustment). The optimization method is based on simulations of a hedging strategy and is very easy in practical implementations, though computationally time consuming. Despite the simplicity of the functional form, the optimized strategy showed to be very effective with the performance close to that of the exact utility-based strategy. We provide an empirical testing of our optimized hedging strategy against the alternative strategies and show that our strategy outperforms all the others.

The rest of the chapter is organized as follows. In Section 2 we review briefly the structure of realistic transaction costs in capital markets and how these costs are modelled. In Section 3 we present the Leland's method, its justification, extensions, and numerical methods to solve some nonlinear PDEs. In Section 4 we review the utility-based option pricing and hedging approach as well as the results of various asymptotic and approximation methods. Also in the same section we provide a detailed description of the nature of the optimal hedging strategy for a portfolio of options, introduce the simplified hedging strategy, describe the optimization methodology, and provide an empirical testing of the optimized strategies. Section 5 concludes this chapter.

2. Modelling the Transaction Costs

In this section we review briefly the structure of realistic transaction costs in capital markets. Then we present the usual treatment of transaction costs in theoretical models.

In most theoretical models a capital market is assumed to be friction-free and a share of a stock is assumed to be infinitely divisible. The latter means that an investor can either buy or sell any real number Δ of shares of a stock. In actual markets, however, trades are made in discrete increments, for example, at least one share increments, and often in 100-share (round lot) increments or blocks (of at least 10,000 shares). Transaction costs in capital markets consist of the following three main components: half-size of the quoted bid-ask spread, brokerage fees (commissions), and market impact costs. In addition there are various taxes applicable in some equity markets, delay costs, opportunity costs, etc. (see, for example, Freyre-Sanders, Guobuzaite, and Byrne [12]. If investors sell securities they do not own (short sale), then they also incur short borrowing costs. All investors face the same bid-ask spreads and market-impact costs for a trade or short borrowing of any given size and security at any given moment. In contrast, the commissions (on purchase, sale, and short borrowing) are negotiated and depend on the annual volume of trading, as well as on the investor's other trading practices. In order to model realistic transaction costs one usually distinguishes between two classes of investors: large and small (see, for example, Dermody and Prisman [11]).

Large investors are defined as those who frequently make large trades in blocks via the block trading desks or brokerage houses. Large investors usually face transaction costs schedule with no minimum fee specified. Large traders typically make ongoing agreements with the trading desks or brokerage houses to execute their trades for a flat institutional commission rate that applies to any size trade. Thus, commissions paid by large investors for trading a given stock are proportional to the number of shares traded. But marginal market impact costs for a given stock rise in the number of shares traded. In contrast, small investors are defined as those who use retail brokerage firms and often trade in 100-share round lots. For small investors there is a minimum fee on any trade. They face retail commissions rates for a given stock that decrease in the transaction size. Their total transaction costs therefore, exhibit decreasing rates for any given trade up to some particular size.

Market impact costs are closely related to liquidity: a relatively big order exerts pressure on price. Market impact costs become a problem if an investor places an order to buy or sell a quantity of shares that is large relative to a market average daily share volume. Market impact costs are less significant with liquid stocks. Liquidity refers to the ease with which a stock can be bought or sold without disturbing its price. Market impact costs are further considered to be the sum of two components: temporary and permanent price effects of trades.

Even such a short review of the structure of transaction costs in capital markets reveals that it is not so easy to model realistic transaction costs. To simplify the treatment of transaction costs, one usually assumes that transaction costs are proportional to the volume of trading. In a strict sense this assumption is valid for only large investors who trade in liquid stocks. In this situation the quoted bid-ask spread is the main component of transaction costs and the market impact costs are negligible. To realize this, consider the following. Large investors usually pay very low commissions of about 0.1% of the volume of trade. In contrast, Bessembinder [4] reports that the average quoted bid-ask half-spread amounts to 0.486% for stocks on the NYSE and 0.739% for stocks on the NASDAQ.

The formal treatment of the proportional transaction costs is the following. We denote the bid price of the stock at time t by $S_{\text{bid}}(t)$ and the ask price by $S_{\text{ask}}(t)$, such that $S_{\text{bid}}(t) < S_{\text{ask}}(t)$. We suppose that $S(t)$ is the midpoint of the bid-ask prices, and we denote by λ half-size of the ratio of the quoted bid-ask spread to the bid-ask price midpoint. Consequently, this allows us to interpret λ as proportional transaction costs such that

$$S_{\text{bid}}(t) = (1-\lambda)S(t) \text{ and } S_{\text{ask}}(t) = (1+\lambda)S(t).$$

Observe that the commissions that are proportional to the volume of trade can easily be incorporated in λ.

3. The Leland's Approach to Option Pricing and Hedging

3.1. The Leland's Method and the Derivation of the Pricing Equation

In a complete market with no transaction cost the option price equals the cost of setting up the replicating portfolio. This follows from the absence of arbitrage argument. In a market with transaction costs the option hedging is costly. The Leland's idea was to include the expected transaction costs in the cost of a replicating portfolio. That is, according to Leland, the price of an option must equal the expected costs of the replicating portfolio including the transaction costs. As a result, the market maker, who writes, for example, a European call option and constructs the replicating portfolio, should sell it with a premium (as compared to the Black-Scholes price) which offsets the expected transaction costs. By contrast, the market maker, who buys a European call option and constructs the replicating portfolio, should buy the option with a discount (again, as compared to the Black-Scholes price) to offset the expected transaction costs. In addition, Leland assumed that the revision of the replicating portfolio must occur at fixed regular intervals of length δt.

In this subsection we present briefly the model of Leland [17] using the interpretation of Hoggard, Whalley, and Wilmott [15]. We consider a continuous time economy with a risky security, for example, a stock, and a risk-free money market account which provides a constant interest rate of r. The price of the stock, S, evolves according to the following stochastic differential equation

$$dS(t) = \mu S(t)dt + \sigma S(t)dW(t),$$

where μ and σ are constant, and $W(t)$ is a standard Brownian motion. Over a sufficiently small time interval δt the change in the stock price is given by

$$\delta S(t) = \mu S(t)\delta t + \sigma S(t)\varepsilon\sqrt{\delta t} + O(\delta t^{\frac{3}{2}}),$$

where ε is drawn from the standard normal distribution. We assume that a purchase or sale of α shares of the stock incurs transaction costs $\lambda|\alpha|S(t)$ proportional to the transaction ($\lambda \geq 0$).

Leland proceeds as in the Black-Scholes model and considers a hedging portfolio, Π, at time t consisting of one option of value $V(t)$ and $-\Delta(t)$ shares of the stock

$$\Pi(t) = V(t) - \Delta(t)S(t).$$

From now on, in order to simplify the exposition, we will suppress the dependence of Π, V, Δ, and S on t. Over δt, the change in the value of the hedging portfolio will be

$$\delta\Pi = \delta V - \Delta\delta S - \lambda|\delta\Delta|(S + \delta S), \tag{1}$$

where the last term on the right hand side of (1) are the transaction costs payed at $t + \delta t$ and

$$\delta\Delta = \Delta(t + \delta t, S + \delta S) - \Delta(t, S).$$

Using a Taylor series expansion for V and disregarding the terms with $\delta t^{\frac{3}{2}}$ and higher powers of δt, we obtain

$$\delta\Pi = V_t\delta t + (V_S - \Delta)\delta S + \frac{1}{2}V_{SS}\delta S^2 - \lambda|\Delta_S\delta S|S, \tag{2}$$

where $V_t = \frac{\partial V}{\partial t}$, $V_S = \frac{\partial V}{\partial S}$, and $V_{SS} = \frac{\partial^2 V}{\partial S^2}$. Note that in addition we have used a Taylor series expansion for $\delta\Delta$

$$\delta\Delta = \Delta_t\delta t + \Delta_S\delta S + \frac{1}{2}\Delta_{SS}\delta S^2,$$

where $\Delta_t = \frac{\partial \Delta}{\partial t}$, $\Delta_S = \frac{\partial \Delta}{\partial S}$, and $\Delta_{SS} = \frac{\partial^2 \Delta}{\partial S^2}$. Here observe that the dominated term in $\delta\Delta$ is $\Delta_S\delta S$ with $\delta t^{\frac{1}{2}}$. Similarly as in the Black-Scholes model, Leland chooses to hedge with

$$\Delta = V_S,$$

which gives

$$\Delta_S = V_{SS}.$$

Finally, Leland assumes that the expected return of the hedging portfolio must equal the risk-free rate of return

$$E[\delta\Pi] = \Pi r\delta t.$$

It can be shown (see Leland [17] or Hoggard, Whalley, and Wilmott [15]) that the expected transaction costs over δt amount to

$$E[\lambda|V_{SS}\delta S|S] = \sqrt{\frac{2}{\pi}}\lambda\sigma S^2|V_{SS}|\sqrt{\delta t}, \tag{3}$$

which is based on the result $E[|\varepsilon|] = \sqrt{2/\pi}$. All this put together yields a nonlinear partial differential equation (PDE) for the value of the option

$$V_t + rSV_S + \frac{1}{2}\sigma^2 S^2\left[1 - A\,\mathrm{sign}(V_{SS})\right]V_{SS} - rV = 0, \tag{4}$$

where

$$A = \frac{\lambda}{\sigma}\sqrt{\frac{8}{\pi\delta t}},$$

and where sign$(\cdot)$ is the sign function. Observe that PDE (4) is a nonlinear parabolic partial differential equation for the option price, one of the first such in finance. It is valid for the price of a portfolio of options as well. What is interesting here is the fact that, since (4) is nonlinear, *the value of portfolio of options that is the sum of individual options is not the same as sum of values of individual options.*

The comparison of the original Black-Scholes PDE and PDE (4) allows us to introduce a new parameter

$$\sigma_m^2 = \sigma^2\left[1 - A\,\mathrm{sign}(V_{SS})\right], \tag{5}$$

and interpret it as the modified volatility. This volatility adjustment depends on the sign of the second derivative of the option price with respect to the underlying stock price (that is, the option gamma). For a general portfolio of options, the curvature of V with respect to S will not stay of one sign. However, there are some cases where we can drop the modulus sign. For example, the gamma of a short European call option is always negative which gives an increased modified volatility

$$\sigma_m^2 = \sigma^2\left[1 + \frac{\lambda}{\sigma}\sqrt{\frac{8}{\pi\delta t}}\right], \tag{6}$$

and an increased option price. By contrast, the gamma of a long European call option is always positive which gives a decreased modified volatility

$$\sigma_m^2 = \sigma^2\left[1 - \frac{\lambda}{\sigma}\sqrt{\frac{8}{\pi\delta t}}\right], \tag{7}$$

and a decreased option price. The Leland's price of a European call option is computed using the same Black-Scholes option price formula

$$V = SN(d_1) - Ke^{-r(T-t)}N(d_2), \tag{8}$$

where

$$d_1 = \frac{\ln\left(\frac{S}{K}\right) + (r + \frac{1}{2}\sigma_m^2)(T-t)}{\sigma_m\sqrt{T-t}}, \quad d_2 = d_1 - \sigma_m\sqrt{T-t},$$

K is the strike, T is the maturity of the option, and $N(\cdot)$ is the cumulative probability distribution function of a normal variable with mean 0 and variance 1. We would like to emphasize that in using the Leland's method one hedges an option with the delta of the option price calculated with modified volatility.

Of course, hedging an option according to the Leland's strategy is risky. The main argument of Leland that justifies his hedging strategy was the proof of the theorem (see Leland [17, p. 1290] which says that as $\delta t \to 0$, his strategy yields the terminal option payoff $V(T, S(T))$ almost surely inclusive of transaction costs.

3.2. Justification of the Leland's Approach

The Leland's method has been criticized on different grounds:

- Hodges and Neuberger [14] and then Davis, Panas, and Zariphopoulou [10] noted that the Leland's method is a purely "financial engineering" approach where one is exclusively concerned with the problem of replicating the payoff of an option, but not with the optimal policy. Indeed, according to Leland, the replicating portfolio is rebalanced after every δt whether or not it is optimal in any sense.

- Avellaneda and Paras [2] noted that, when the revision interval tends to zero, the Leland's approach for hedging a long European call is not well-posed as the modified volatility becomes negative (see equation (7)).

- Kabanov and Safarian [16] discovered an error in the Leland's proof of the theorem on the limiting hedging error. It turns out that the limiting hedging error is not zero as the revision interval δt tends to zero. Nevertheless, according to Kabanov and Safarian the Leland's hedging error is small and decreases as the level of transaction costs decreases. However, it is easy to observe that, in the case of a short European call option when the revision interval tends to zero, the Leland's hedge degenerates to the super-replication policy when one share of the stock is held to hedge an option. This follows from the fact that the Leland's modified volatility becomes unbounded (see equation (6)). Therefore, the limiting hedging error cannot be small irrespective of the level of transaction costs.

Despite the critique, the Leland's method does work in many cases. That is, if we compare the risk-return tradeoff provided by the Leland's strategy with the risk-return tradeoff of the Black-Scholes strategy, we discover that often the Leland's strategy outperforms the Black-Scholes strategy even when a hedger starts with the same initial value of the replicating portfolio.

The question of major theoretical and practical importance is how to compare alternative strategies and find out which strategy is better. The widely accepted method of comparison is the following (see, for example, Mohamed [19], Clewlow and Hodges [8], Martellini and Priaulet [18] and Zakamouline [24], [25]): first of all we need to decide upon a unified risk-return framework for comparison. Then we perform a simulation analysis of every hedging strategy. Take, as an example, the discrete hedging strategy of Black and Scholes. We fix the value of δt, perform simulations of the hedging strategy, and compute the risk and return of the hedging strategy for this particular δt. Then, we vary the parameter δt and span all the possible combinations of risk and return. The set of all possible combination of risk and return of a particular hedging strategy comprises the *efficient frontier* of the hedging strategy. Similarly, we can perform simulations of the Leland's strategy and find its efficient frontier in the same risk-return space. Each hedging strategy produces a different efficient frontier, that is, each hedging strategy offers a different risk-return tradeoff. Observe that a rational hedger will always prefer a strategy that minimizes the risk for a given level of return. This is in a clear analogy with the modern portfolio theory where the investor seeks to minimize the volatility of the portfolio (i.e., the risk) for a given level of returns. Consequently, some hedging strategy is better than the others if for a given level of returns it offers the lowest risk. Alternatively, some hedging strategy is better than the others if for a given level of risk it offers the highest returns. A hedging strategy may be superior with respect to the others if for any level of returns it offers the lowest risk.

The expected replication error (also known as the expected profit and loss, P&L, of a hedging strategy) seems to be the only sensible candidate for the return measure. As for the risk measure, there are many metrics of risk. In the context of option hedging, the variance of the replication error seems to be the most popular risk metrics. We provide an empirical testing of the Leland's strategy against the Black-Scholes strategy. The two hedging strategies were simulated and the results are presented in Figure 1. The option was a 1 year European call with the initial stock price $S(0) = 100$, the strike $K = 100$, the volatility $\sigma = 0.25$, and the drift $\mu = r = 0.05$. The proportional transaction costs were $\lambda = 0.01$. The simulations proceed as follows: At the beginning, the writer/buyer of a European call option receives/pays the Black-Scholes value of the option (the same for each strategy) and sets up a replicating portfolio. The stock path is simulated according to

$$S(t + \delta t) = S(t) \exp\left((\mu - 0.5\sigma^2)\delta t + \sigma\sqrt{\delta t}\, \varepsilon_t\right), \tag{9}$$

where ε_t is a normally distributed variable with mean 0 and variance 1. The rebalancing trade is performed at every δt, which takes values in $[\frac{1}{200}, \frac{1}{10}]$, and transaction costs are drawn from the money market account. Finally, at expiration, we compute the replication error, that is, the value of the replicating portfolio minus the due exercise payment. For each value of δt we generate 200,000 paths and compute the mean and the variance of the replication error.

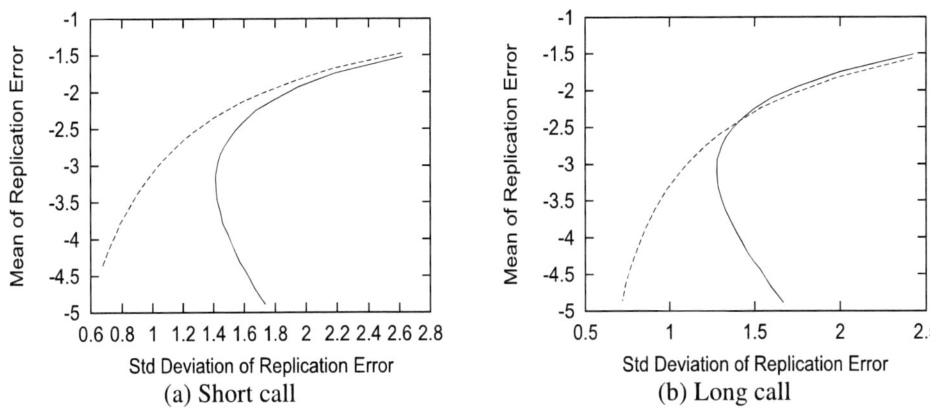

Figure 1. Comparison of the risk-return tradeoffs of the Leland's strategy (dashed line) against the Black-Scholes strategy (solid line) for different values of δt.

By studying Figure 1, we can conclude that the Leland's strategy is superior to the Black-Scholes strategy in hedging a short European call option. In addition, we can note that the Leland's strategy is better than the Black-Scholes strategy in hedging a long European call option when the rehedging interval becomes sufficiently small.

Observe that an apparent weakness of the Leland's approach is not quite justified choice of Δ. To hedge with the same Δ as in the absence of transaction costs seems to be not optimal. A standard and feasible criterion of optimality of a hedging strategy is to minimize the risk of the hedging strategy, as measured by the variance of the replication error. Now we are going to show that if we choose to minimize the local variance of the replication error, that is, if we solve the problem

$$\min_{\Delta} \mathrm{Var}[\delta\Pi - \delta V], \tag{10}$$

then the Leland's choice does to leading order minimize the local risk of the hedging strategy. Indeed, the stochastic part of δV is given by $V_S \delta S$. The stochastic part of $\delta\Pi$ is (to leading order) given by

$$\Delta\delta S + \lambda|\Delta_S \delta S|S.$$

We rewrite the quantity above as

$$\Delta\delta S \left(1 + \lambda S \frac{|\Delta_S \delta S|}{\Delta\delta S}\right).$$

Observe that the problem (10) amounts now to

$$\min_{\Delta} \mathrm{Var}\left[\Delta\delta S \left(1 + \lambda S \frac{|\Delta_S \delta S|}{\Delta\delta S}\right) - V_S \delta S\right].$$

Note that the Leland's choice $\Delta = V_S$ will be close to the optimal one if

$$\left|\lambda S \frac{V_{SS}}{V_S}\right| \ll 1.$$

Apparently, the lesser the transaction costs rate λ, the closer the Leland's hedge to the optimal one. But generally we need to examine how big/small is the value of $\left|\lambda S \frac{V_{SS}}{V_S}\right|$ with respect to 1. One possibility to evaluate the optimality of the Leland's hedge is to examine the absolute values of $\lambda S \frac{V_{SS}}{V_S}$ for some particular options. Figure 2 presents the absolute values of $\lambda S \frac{V_{SS}}{V_S}$ for a plain vanilla call and a digital cash-or-nothing call for different times to maturity and stock prices. From the figure we can conclude that the closer the maturity time, the farther the Leland's hedge could be from the optimal one. Observe that for a plain vanilla call which is not deep out of the money, we expect that the underlying price remains in the region where the absolute value of $\lambda S \frac{V_{SS}}{V_S}$ is very low (with respect to 1) irrespective of the time to maturity. This implies that for a plain vanilla call the Leland's hedge is close to the optimal one. However, for a digital call the Leland's hedge remains optimal for any time to maturity only when the underlying price is close to the strike. When the underlying price drifts away from the strike and the time is close to maturity, the Leland's hedge can be far away from the optimal one.

Zakamouline [27] considers the Leland's hedging of European call options and explains in details how the Leland's modified hedging volatility reduces the risk of a hedging strategy and might, at the same time, decrease the expected replication error of a hedging strategy. Now we present briefly the explanation. First note that the (total) replication error of a hedging strategy can be subdivided into a hedging error and transaction costs. To understand how the Leland's modified works, observe that both the hedging error and the amount of transaction costs of a discretely adjusted delta-neutral replicating strategy are

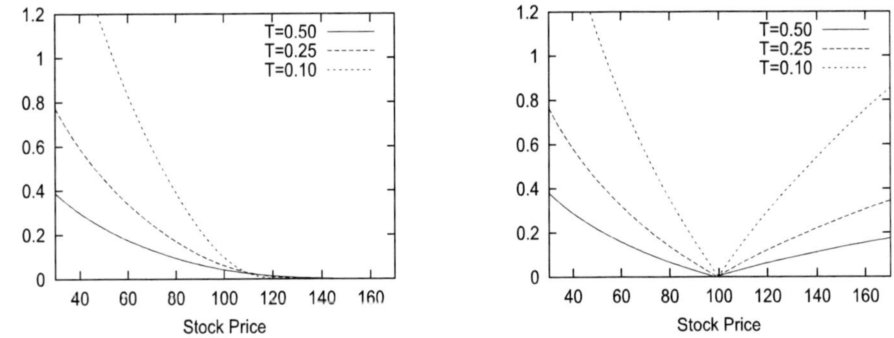

(a) Vanilla call (b) Digital call

Figure 2. The absolute values of $\lambda S \frac{V_{SS}}{V_S}$ for a plain vanilla call and a digital cash-or-nothing call for different times to maturity and stock prices. Both call options have the same strike $K = 100$, stock volatility $\sigma = 25\%$, and the level of transaction costs $\lambda = 1\%$.

path-dependent. In particular, the amount of transaction costs depends on the absolute value of the option gamma (for a motivation, see equation (3)). The hedging error depends on the values and the signs of the option theta and gamma (see Boyle and Emanuel [7]). In short, the Leland's modified hedging volatility makes the hedging error be "negatively" correlated with transaction costs: the hedging error becomes positive when transaction costs are high, and the hedging error becomes negative when transaction costs are low. Thus, the Leland's modified volatility "equalizes" the replication error across different stock paths. This reduces the risk of the hedging strategy as measured by the variance of the replication error. To illustrate the aforesaid, Figure 3 presents the simulation results of hedging a plain vanilla European call option according to the Leland's strategy as compared to hedging according to the Black-Scholes strategy.

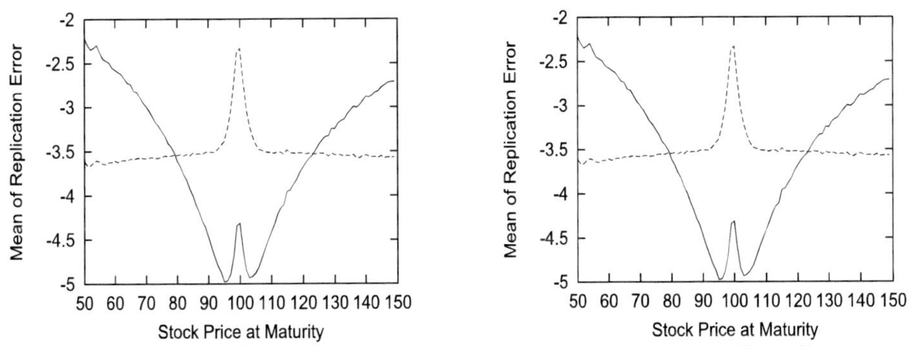

(a) Short vanilla call (b) Long vanilla call

Figure 3. Comparison of the expected replication errors of the Leland's strategy (dashed line) against the Black-Scholes strategy (solid line) across different stock prices at maturity. The option is a vanilla call with initial stock price $S = 100$, strike $K = 100$, stock volatility $\sigma = 25\%$, the level of transaction costs $\lambda = 1\%$, maturity $T = 1$ year, which is rehedged every $\delta t = 1/100$.

Note that the Leland's volatility adjustment reduces the risk of a replicating strategy. However, the reduction of risk can happen at the expense of reducing the returns of a replicating strategy. When a hedger is short gamma, the option is hedged with an increased hedging volatility. An increased hedging volatility decreases the absolute value of the option gamma in the region where the option gamma is high. This reduces the amount of hedging transaction costs (see Zakamouline [27]). By contrast, when a hedger is long gamma, the option is hedged with a decreased hedging volatility. A decreased hedging volatility increases the absolute value of the option gamma in the region where the option gamma is high. This increases the amount of hedging transaction costs. The illustration is given in Table 1. Observe that in hedging a long European call option a hedger is long gamma which increases the amount of hedging transaction costs and results in the deterioration of the risk-return tradeoff of the Leland's strategy as compared with the Black-Scholes hedging when δt is large (see also Figure 1 (b)).

Table 1. Comparison of the risk-return tradeoffs of the Leland's strategy against the Black-Scholes strategy. The option is a vanilla call with the same parameters as those in Figure 3.

Strategy	Black-Scholes	Leland
Mean of replication error	-3.632	-3.276
Std. deviation of replication error	1.443	0.938

(a) Short vanilla call

Strategy	Black-Scholes	Leland
Mean of replication error	-3.632	-4.361
Std. deviation of replication error	1.329	0.778

(b) Long vanilla call

3.3. Extensions of the Leland's Approach

As it has been shown in the previous subsection, the knowledge of the Leland's hedging strategy is important for optimal hedging of options and option portfolios in the presence of transaction costs. In this subsection we extend the Leland's option pricing and hedging method to cover the cases of pricing and hedging of a portfolio of options on a futures contract, a portfolio of strongly path-dependent options on a stock, and options on several assets. In all the cases as a result we obtain a nonlinear partial differential equation for the option portfolio price. Generally, these nonlinear partial differential equations must be solved numerically and we present later some examples of finite difference schemes to solve them. The ultimate practical goal, however, is not the solution for the option portfolio price, but the solution for the values of the Leland's hedging delta. The content of this subsection is largely based on the paper by Zakamouline [27].

3.3.1. Commodity Futures Contracts

Here we extend the method of Leland for pricing and hedging of a portfolio of options on a commodity futures contract. The futures price of a commodity is given by $F(t, T)$, where t

is the current time and T is the delivery time. The transaction costs are proportional to the volume of trading with rate λ. The dynamics of the futures price is assumed to be

$$dF(t,T) = \mu(t,T)F(t,T)dt + \sigma(t,T)F(t,T)dW(t), \tag{11}$$

where $\mu(t,T)$ and $\sigma(t,T)$ are some functions, and $W(t)$ is a standard Brownian motion. Depending on the type of commodity, there are many possible specifications for $\mu(t,T)$ and $\sigma(t,T)$. The choice of the specification of $\sigma(t,T)$ is especially important and should fit the empirically observable volatility term structure. Later on, in order to simplify the exposition, we will suppress the dependence of F, μ, and σ on T.

A crucial feature of a futures contract is daily *marking to market*. This means that at the end of a trading day the contract is settled and replaced by a new one with a new futures price. Settlement is carried out by exchanging the *undiscounted* difference between the new and old futures prices. The settlement price is usually the average of the bid and ask prices at the end of a trading day. Consequently, the settlement of a futures contract at the end of a trading day does not involve payment of transaction costs. Typically, when one enters a futures contract one needs to post a collateral on a margin account with the broker. The margin account is a usual money market account that provides a constant interest rate of r. At the end of a trading day the gained profit on a futures contract is added to this amount or the loss is written off.

Besides commodity futures contracts, we suppose that there is a trade in option contracts on futures. As usual, there are exchange tradable standard European options (calls and puts). In addition, in the over-the-counter (bilateral) market there is a trade in different exotic European-type options. In this section we consider pricing and hedging of a portfolio of European non-path-dependent options on a futures contract with maturity T. We denote the value of the option portfolio at time t by $V(t,F(t))$. The terminal payoff of the option portfolio one wishes to hedge is given by $V(T,F(T))$.

Using the Leland's method, to hedge the option position one constructs a self-financing replicating portfolio, $\Pi(t,F(t))$, consisting of the amount $\beta(t,F(t))$ in the money market account and $\Delta(t,F(t))$ futures contracts. Since entering a futures contract does not imply immediate outlay of money, we get the following relationship

$$\Pi(t,F) = \beta(t,F) = V(t,F). \tag{12}$$

Note that from now on, in order to shorten all the subsequent formulas, we also suppress the obvious dependence of F on time t. Over a sufficiently small interval δt the change in the futures price is given by

$$\delta F = \mu(t)F\delta t + \sigma(t)F\varepsilon\sqrt{\delta t} + O(\delta t^{\frac{3}{2}}),$$

where ε is drawn from the standard normal distribution. Over the same δt, the change in the value of the replicating portfolio will be

$$\delta\Pi(t,F) = \delta\beta(t,F) + \Delta(t,F)\delta F + \lambda|\delta\Delta(t,F)|(F+\delta F), \tag{13}$$

where the first term on the right hand side of (13) is the change on the money market account, the second term is a gain or loss on Δ futures contracts, and the last term is due to the inclusion of the transaction costs where

$$\delta\Delta(t,F) = \Delta(t+\delta t, F+\delta F) - \Delta(t,F).$$

We require that over δt the expected change in the value of the replicating portfolio should equal the expected change in the value of the option position

$$E[\delta\Pi(t,F)] = E[\delta V(t,F)]. \tag{14}$$

In other words, the expected hedging error over δt, including of transaction costs, should equal zero

$$E[\delta\Pi(t,F) - \delta V(t,F)] = 0.$$

Using a Taylor series expansion for $V(t,F)$ and disregarding the terms with $\delta t^{\frac{3}{2}}$ and higher powers of δt, we obtain

$$\delta V(t,F) = V_t\delta t + V_F\delta F + \frac{1}{2}\sigma^2(t)F^2 V_{FF}\delta t, \tag{15}$$

where $V_t = \frac{\partial V}{\partial t}$, $V_F = \frac{\partial V}{\partial F}$, and $V_{FF} = \frac{\partial^2 V}{\partial F^2}$. Similarly, using a Taylor series expansion for $\delta\Delta(t,F)$ we obtain

$$\delta\Delta(t,F) = \Delta_t\delta t + \Delta_F\delta F + \frac{1}{2}\sigma^2(t)F^2\Delta_{FF}\delta t,$$

where $\Delta_t = \frac{\partial\Delta}{\partial t}$, $\Delta_F = \frac{\partial\Delta}{\partial F}$, and $\Delta_{FF} = \frac{\partial^2\Delta}{\partial F^2}$. Here observe that the dominated term in $\delta\Delta(t,F)$ is $\Delta_F\delta F$ with $\delta t^{\frac{1}{2}}$. Consequently,

$$\delta\Delta(t,F) = \Delta_F\delta F. \tag{16}$$

The change in the money market account is given by

$$\delta\beta(t,F) = r\beta(t,F)\delta t = rV(t,F)\delta t, \tag{17}$$

where the last equation follows from equation (12). Putting this all together into equation (14) and keeping only the leading terms gives us

$$rV(t,F)\delta t + \Delta(t,F)E[\delta F] + \lambda E\left[|\Delta_F\delta F|\right]F = V_t\delta t + V_F E[\delta F] + \frac{1}{2}\sigma^2(t)F^2 V_{FF}\delta t. \tag{18}$$

A very crucial assumption in the Leland's method is to hedge as in the Black(-Scholes) model with

$$\Delta(t,F) = V_F,$$

which gives

$$\Delta_F = V_{FF}.$$

Consequently, the expected transaction costs are proportional to

$$E\left[|V_{FF}\delta F|\right] = |V_{FF}|E[|\delta F|] = |V_{FF}|\sigma(t)F\sqrt{\frac{2}{\pi}}\delta t^{\frac{1}{2}},$$

since $E[|\varepsilon|] = \sqrt{2/\pi}$. Equation (18) can be rewritten now as

$$V_t + \frac{1}{2}\sigma^2(t)F^2\left[1 - A(t)\,\mathrm{sign}(V_{FF})\right]V_{FF} - rV(t,F) = 0, \tag{19}$$

where

$$A(t) = \frac{\lambda}{\sigma(t)}\sqrt{\frac{8}{\pi \delta t}},$$

and where sign$(\cdot)$ is the sign function. Observe that equation (19) is a nonlinear partial differential equation for the value of the option portfolio. Observe in addition that by introducing a new parameter

$$\sigma_m^2(t) = \sigma^2(t)\left[1 - A(t)\,\text{sign}(V_{FF})\right],$$

$\sigma_m(t)$ can be interpreted as the modified hedging volatility. This volatility adjustment depends on the sign of the option portfolio gamma. With this notation the PDE (19) is completely equivalent to the original PDE of Black [6] which gives the solutions for the price and the hedging strategy for an option portfolio on a commodity futures contract in the absence of transaction costs.

Generally, to obtain the solution of the nonlinear PDE (19) one needs to solve this PDE numerically. In some cases when the gamma of an option portfolio does not change sign one, can obtain a closed-from solution. For example, the price of a single call option on a futures contract with strike K is given by

$$V(t, F) = e^{-r(T-t)}[FN(d_1) - KN(d_2)], \tag{20}$$

where

$$d_1 = \frac{\ln\left(\frac{F}{K}\right) + \frac{1}{2}\int_t^T \sigma_m^2(s)ds}{\sqrt{\int_t^T \sigma_m^2(s)ds}}, \quad d_2 = d_1 - \sqrt{\int_t^T \sigma_m^2(s)ds},$$

and $N(\cdot)$ is the cumulative probability distribution function of a normal variable with mean 0 and variance 1.

3.3.2. Path-Dependent Options

Here we present the extension of the Leland's method to pricing and hedging of a portfolio of strongly path-dependent European options on a stock. We denote the option portfolio price at time t by $V(t, S(t), Y(t))$, where $Y(t)$ is some path-dependent quantity. The terminal payoff of the option portfolio is given by $V(T, S(T), Y(T))$. The pricing and hedging of an option portfolio depends on whether $Y(T)$ is sampled continuously or discretely. In this subsection the derivations of the nonlinear PDEs follow along the lines as in the previous subsections. Consequently, to avoid a lot of repetitions, we will present only the main stages of the derivations. As before, to simplify the exposition we will often suppress the obvious dependence of S and Y on t. For the derivations of the corresponding linear PDEs in the absence of transaction costs, the interested reader can consult Wilmott [22, Chapters 17–19].

Continuous Sampling

With continuous sampling the path-dependent quantity $Y(T)$ is represented by an integral

$$Y(T) = \int_t^T f(s, S(s))ds,$$

where $f(\cdot,\cdot)$ is some function. For example, if $Y(T)$ is the arithmetic average, then

$$Y(T)=\frac{1}{T-t}\int_t^T S(s)ds.$$

To replicate the payoff of an option portfolio we construct a hedging portfolio consisting of $\Delta(t,S,Y)$ shares of the stock and the amount $\beta(t,S,Y)$ in the money market account. Therefore, we get the following equation

$$\Pi(t,S,Y)=\Delta(t,S,Y)S+\beta(t,S,Y)=V(t,S,Y). \tag{21}$$

Following Leland, we need to assure that over a small δt

$$E[\delta\Pi(t,S,Y)]) = E[\delta V(t,S,Y)]. \tag{22}$$

Using a Taylor series expansion for $V(t,S,Y)$ and disregarding the terms with high powers of δt, we obtain

$$\delta V(t,S,Y)=V_t\delta t+V_S\delta S+\frac{1}{2}\sigma^2S^2V_{SS}\delta t+f(t,S)V_Y\delta t, \tag{23}$$

where $V_Y=\frac{\partial V}{\partial Y}$ and $\delta Y=f(t,S)\delta t$. The change in the value of the replicating portfolio is given by

$$\delta\Pi(t,S,Y)=\delta\beta(t,S,Y)+\Delta(t,S,Y)\delta S+\lambda|\delta\Delta(t,S,Y)|(S+\delta S), \tag{24}$$

where

$$\delta\Delta(t,S,Y)=\Delta(t+\delta t,S+\delta S,Y+\delta Y)-\Delta(t,S,Y).$$

It is easy to show that the dominated term in $\delta\Delta(t,S,Y)$ is $\Delta_S\delta S$ with $\delta t^{\frac{1}{2}}$. Consequently,

$$\delta\Delta(t,S,Y)=\Delta_S\delta S. \tag{25}$$

The equation (21) gives us

$$\delta\beta(t,S,Y)=r\beta(t,S,Y)\delta t=r[V(t,S,Y)-\Delta(t,S,Y)S]\delta t. \tag{26}$$

Putting this all together into equation (22) and keeping only the leading terms we obtain

$$\begin{aligned} r[V(t,S,Y)-\Delta(t,S,Y)S]\delta t+\Delta(t,S,Y)E[\delta S]+\lambda E\left[|\Delta_S\delta S|\right]S \\ =V_t\delta t+V_SE[\delta S]+\frac{1}{2}\sigma^2S^2V_{SS}\delta t+f(t,S)V_Y\delta t. \end{aligned} \tag{27}$$

By following the choice of Leland

$$\Delta(t,S,Y)=V_S,$$

the equation (27) can be rewritten as

$$V_t+rSV_S+\frac{1}{2}\sigma^2S^2\left[1-A\,\mathrm{sign}(V_{SS})\right]V_{SS}+f(t,S)V_Y-rV(t,S,Y)=0, \tag{28}$$

where

$$A = \frac{\lambda}{\sigma}\sqrt{\frac{8}{\pi \delta t}}.$$

Again

$$\sigma_m^2 = \sigma^2 \left[1 - A \operatorname{sign}(V_{SS})\right]$$

can be interpreted as the modified hedging volatility. Generally, to obtain the solution of the nonlinear PDE (28), one needs to solve it numerically. In some simple cases where the option portfolio gamma does not change sign, using the similarity reduction (see Wilmott [22, Chapters 19 and 20]) one can obtain closed-form solutions. For example, it is very straightforward to modify the pricing formulas (just replacing the original σ by σ_m) for geometric average rate calls and puts, floating strike lookback calls and puts, fixed strike lookback calls and puts, and some other options.

Discrete Sampling

In this case the path-dependent quantity $Y(T)$ is measured on the sampling dates t_k using an updating rule

$$Y(t_k) = f(t_k, S(t_k), Y(t_{k-1})),$$

where $f(\cdot,\cdot,\cdot)$ is some function. For example, for an Asian option with a discretely sampled arithmetic average the updating rule is given by

$$Y(t_k) = \frac{1}{k} S(t_k) + \frac{k-1}{k} Y(t_{k-1}).$$

For a lookback option with a discretely sampled maximum the updating rule is given by

$$Y(t_k) = \max(S(t_k), Y(t_{k-1})).$$

The derivation of the nonlinear PDE for the option portfolio price in the discrete sampling case follows along the lines as in the continuous sampling case with two important distinctions. In contrast to the continuous sampling case, during the open time interval (t_k, t_{k+1}) the option portfolio price is an explicit function of (t, S) only since Y is constant here, hence $\delta Y = 0$. Consequently, during (t_k, t_{k+1}) the option portfolio pricing equation becomes

$$V_t + rSV_S + \frac{1}{2}\sigma^2 S^2 \left[1 - A \operatorname{sign}(V_{SS})\right] V_{SS} - rV(t, S, Y) = 0, \tag{29}$$

which is the same as equation (4). Across a sampling date t_k the option portfolio price must be continuous. This means that

$$V(t_k^+, S(t_k^+), Y(t_k)) = V(t_k^-, S(t_k^-), Y(t_{k-1})), \tag{30}$$

where the superscripts $\pm$ denote the times just before and just after t_k. The equation (30) is commonly denoted as the *jump condition.*

To illustrate the advantages of the Leland's hedging strategy over the Black-Scholes discrete hedging strategy, we consider the hedging of a portfolio consisting of two fixed strike lookback call options with discrete sampling: a short call with strike K_1 and a long call with strike K_2. Consequently, the option portfolio payoff is given by

$$V(T, S(T), S_{\max}(T)) = \max(S_{\max}(T) - K_2, 0) - \max(S_{\max}(T) - K_1, 0),$$

where $S_{\max}(T)$ is the maximum stock price over the life of the options. The model parameters we use are the following: the initial stock price $S(0) = 100$, the stock volatility $\sigma = 25\%$, the expected stock returns $\mu = 10\%$, and the risk-free interest rate $r = 5\%$. The time to maturity for both the options is $T = 0.5$ years. The strike prices are $K_1 = 90$ and $K_2 = 110$. The proportional transaction costs rate $\lambda = 1\%$. The sampling period for computing the maximum is one week.

The simulations of the two hedging strategies proceed in the similar manner as that described in Section 3.2.: At the beginning, regardless of the hedging strategy, the hedger of the option portfolio receives the Black-Scholes value of the option position (that is, the value in the absence of transaction costs) and sets up a replicating portfolio. The stock path is simulated according to equation (9). At each δt the hedger rebalances the replicating portfolio to have the correct $\Delta(t, S, S_{\max})$, and transaction costs are drawn from the money market account. To obtain the value of $\Delta(t, S, S_{\max})$ for the Leland's strategy, we solve the nonlinear PDE (29) with a jump condition given by (30). By knowing the values of $V(t, S, S_{\max})$ we calculate $\Delta(t, S, S_{\max})$ using

$$\Delta(t, S, S_{\max}) = \frac{V(t, S + \delta S, S_{\max}) - V(t, S - \delta S, S_{\max})}{2\delta S}. \tag{31}$$

The Black-Scholes values of $\Delta(t, S, S_{\max})$ are obtained by solving the same nonlinear PDE (29) where we set $A = 0$. Then $\Delta(t, S, S_{\max})$ is also computed in accordance with (31).

We provide the simulation analysis where we vary $\delta t \in [\frac{1}{200}, \frac{1}{5}]$ and the results are presented in Figure 4. By studying the figure we come to the obvious conclusion that the Leland's strategy is superior to the Black-Scholes strategy. As compared with the Black-Scholes strategy and irrespective of the value of δt, the Leland's strategy has a lower variance of the replication error and, at the same time (at least for this option portfolio), has greater expected returns. In particular, in hedging such an option portfolio the Leland's strategy might be up to 50% less risky as compared with the Black-Scholes strategy at the same level of the replication error.

3.3.3. Basket Options

In this subsection we extend the Leland's method to cover the case of pricing and hedging of options on many underlyings in the presence of transaction costs. These options are called basket options or rainbow options. We consider the case where each underlying is a stock. The total numbers of stocks is n. The price of each stock follows a diffusion process given by

$$dS_i(t) = \mu_i S_i(t)dt + \sigma_i S_i(t)dW_i(t), \quad i = 1, 2, \ldots, n,$$

where μ_i and σ_i are, respectively, the mean and volatility of the ith stock returns per unit of time, and $W_i(t)$ is a standard Brownian motion. The instantaneous correlation between stock i and stock j is ρ_{ij} which implies $dW_i(t)dW_j(t) = \rho_{ij}dt$. As before, we assume

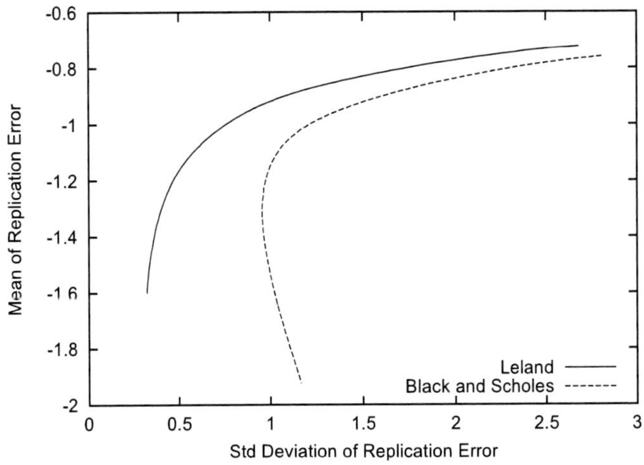

Figure 4. Efficient frontiers for the Leland's and Black-Scholes hedging strategies in hedging a portfolio consisting of two fixed strike lookback call options with discrete sampling.

that transaction costs are proportional to the volume of trading with rate λ_i for stock i. We denote the basket option price at time t by $V(t, S_1, \ldots, S_n)$.

Now to replicate the payoff of the basket option we set up a replicating portfolio consisting of $\Delta_i(t, S_1, \ldots, S_n)$ shares of each stock i and the amount $\beta(t, S_1, \ldots, S_n)$ in the money market account. Therefore, we get the following equation

$$\Pi(t, S_1, \ldots, S_n) = \beta(t, S_1, \ldots, S_n) + \sum_{i=1}^{n} \Delta_i(t, S_1, \ldots, S_n) S_i = V(t, S_1, \ldots, S_n). \quad (32)$$

Later on, in order to simplify the exposition, we will often suppress the obvious dependence of Π, V, Δ_i, and β on t and $S_1, \ldots, S_n$.

As before, using a Taylor series expansion for $V(t, S_1, \ldots, S_n)$ and disregarding the terms with high powers of δt, we obtain

$$\delta V(t, S_1, \ldots, S_n) = V_t \delta t + \sum_{i=1}^{n} V_{S_i} \delta S_i + \frac{1}{2} \sum_{i=1}^{n} \sum_{j=1}^{n} \sigma_i \sigma_j \rho_{ij} S_i S_j V_{S_i S_j} \delta t, \quad (33)$$

where $V_{S_i} = \frac{\partial V}{\partial S_i}$ and $V_{S_i S_j} = \frac{\partial^2 V}{\partial S_i \partial S_j}$. The change in the value of the replicating portfolio is given by

$$\delta \Pi(t, S_1, \ldots, S_n) = \delta \beta + \sum_{i=1}^{n} \Delta_i \delta S + \sum_{i=1}^{n} \lambda_i |\delta \Delta_i| (S_i + \delta S_i), \quad (34)$$

where

$$\delta \Delta_i(t, S_1, \ldots, S_n) = \Delta_i(t + \delta t, S_1 + \delta S_1, \ldots, S_n + \delta S_n) - \Delta_i(t, S_1, \ldots, S_n).$$

Recall that we are looking for the dominated terms in $\delta\Delta_i$ with $\delta t^{\frac{1}{2}}$. This gives us

$$\delta\Delta_i(t, S_1, \ldots, S_n) = \sum_{j=1}^{n} \Delta_{iS_j}\delta S_j, \tag{35}$$

where $\Delta_{iS_j} = \frac{\partial \Delta_i}{\partial S_j}$. Now using the Leland's condition $E[\delta\Pi] = E[\delta V]$ and keeping only the leading terms we arrive to the following equation

$$\begin{aligned} r\left[V - \sum_{i=1}^{n}\Delta_i S_i\right]\delta t + \sum_{i=1}^{n}\Delta_i E[\delta S_i] + \sum_{i=1}^{n}\lambda_i E\left[\left|\sum_{j=1}^{n}\Delta_{iS_j}\delta S_j\right| S_i\right] \\ = V_t\delta t + \sum_{i=1}^{n} V_{S_i}E[\delta S_i] + \frac{1}{2}\sum_{i=1}^{n}\sum_{j=1}^{n}\sigma_i\sigma_j\rho_{ij}S_iS_jV_{S_iS_j}\delta t. \end{aligned} \tag{36}$$

The Leland's choice to hedge with $\Delta_i = V_{S_i}$ gives us

$$\Delta_{iS_j} = V_{S_iS_j}.$$

The expected transaction costs over δt (disregarding the higher powers of δt) are given by

$$\sum_{i=1}^{n}\lambda_i E\left[\left|\sum_{j=1}^{n} V_{S_iS_j}\sigma_j S_j\varepsilon_j\right|\right] S_i\delta t^{\frac{1}{2}},$$

where ε_j, $j = 1, \ldots, n$, are standard normal variables which are mutually correlated. Note that

$$\sum_{j=1}^{n} V_{S_iS_j}\sigma_j S_j\varepsilon_j$$

is a normal variable which has zero mean and standard deviation given by

$$\Theta_i = \sqrt{\left|\sum_{j=1}^{n}\sum_{k=1}^{n} V_{S_iS_j}V_{S_iS_k}\sigma_j\sigma_k\rho_{jk}S_jS_k\right|}. \tag{37}$$

Consequently, the expected transaction costs over δt amount to

$$\sum_{i=1}^{n}\lambda_i\Theta_i S_i\sqrt{\frac{2}{\pi}}\delta t^{\frac{1}{2}}.$$

Now the equation (36) can be rewritten as

$$V_t + r\sum_{i=1}^{n} S_iV_{S_i} + \frac{1}{2}\sum_{i=1}^{n}\sum_{j=1}^{n}\sigma_i\sigma_j\rho_{ij}S_iS_jV_{S_iS_j} - \sqrt{\frac{2}{\pi\delta t}}\sum_{i=1}^{n}\lambda_i\Theta_iS_i - rV = 0. \tag{38}$$

Observe that equation (38) contains a non-linear term with Θ_i. Note in addition that the non-linear PDEs, which we derived in the previous subsections, reduce to linear PDEs in cases where the sign of the second derivative with respect to the price of the underlying remains constant. That is, with single underlying the non-linearity feature was caused by the term with the absolute value of the second derivative. Here, in the multi-asset case, the non-linear PDE can never be reduced to a linear PDE. Also observe that, unlike the case with a single underlying, with several underlying assets the option hedging strategy is difficult to interpret as hedging with "adjusted" volatility.

3.4. A Simple Improvement of the Leland's Hedging Strategy

As we have illustrated in Section 3.2., the Leland's modification of the hedging volatility reduces the variance of the replication error and in the majority of cases improves the risk-return tradeoff of the hedging strategy as compared to the outright Black-Scholes hedging (at least, when δt is small). However, in some cases the Leland's choice to hedge with $\Delta = V_S$ is suboptimal. In this section we present a simple method that improves the risk-return tradeoff of the Leland's hedging strategy. The idea is to find the optimal value of the modified volatility which is determined by the Leland's number A, see equation (5). In particular, using Monte-Carlo simulations as the underlying numerical method, we want to solve the following problem

$$\min_{A>0} \text{Var}[RE(\sigma_m(A))], \tag{39}$$

where $RE(\sigma_m(A))$ is the replication error which depends on the magnitude of the volatility adjustment, σ_m, which is a function of parameter A. In words, our optimization problem is to find the value of A for which the risk of the hedging strategy will be the lowest possible. The outright implementation of the numerical algorithm to solve this optimization problem is as follows: start with $A = 0$, simulate the hedging strategy n times and estimate the risk of the strategy. Then change the value of A by some step ΔA, simulate the new strategy and estimate the risk again. Proceed in such a manner and find the optimal $A^* = i\Delta A$, $i = 0, 1, 2, \ldots$, which gives the strategy with the lowest possible risk. Note that in order to find the delta of the hedging strategy we generally need to solve numerically a particular nonlinear PDE. Various numerical schemes can be used to solve these PDEs using some finite-difference method. The simplest explicit finite-difference scheme is presented in the subsequent section. Hoggard, Whalley, and Wilmott [15] also describe a numerical method that uses an implicit finite-difference scheme.

Our experiments show that it is always possible to improve the risk-return tradeoff of the original Leland's strategy by finding the optimal modification of volatility. In some simple cases as, for example, hedging a plain vanilla call or put, the improvement is often insignificant. However, in hedging a complex option position one can sometimes improve the Leland strategy considerably. As an example, in Figure 5 we present the results of our improvement of the Leland's strategy for hedging a digital cash-or-nothing call option. Hedging such an option is a well-known difficult problem for which the standard Black-Scholes delta hedging yields poor results, see Avellaneda and Paras [2]. As it follows from Figure 5 (b), hedging a short digital call according to the original Leland's strategy yields generally even more poor results than the Black-Scholes hedging. However, it is possible to find the optimal hedging volatility which substantially improves the Leland's hedging strategy in this case.

The simulations of the Black-Scholes and the original Leland's strategies were conducted similarly as in Section 3.2.. In all the cases the initial value of the hedging portfolio was given the option value in the absence of transaction costs

$$V(t, S) = XN(d_1 - \sigma\sqrt{T-t}),$$

where X is the option payoff if the stock price ends up above the strike price. Both the simulations and the improvement algorithm were implemented on C++. It is worth to mention that both the algorithms are very easy to implement. We used the number of simulations $n = 200000$ and the step size $\Delta A = 0.025$. Usually it takes no more than 10 min on a computer with a 2.4 GHz processor to find the optimal hedging volatility for a given value of δt. The higher the value of δt the shorter the time needed to find the optimal volatility adjustment.

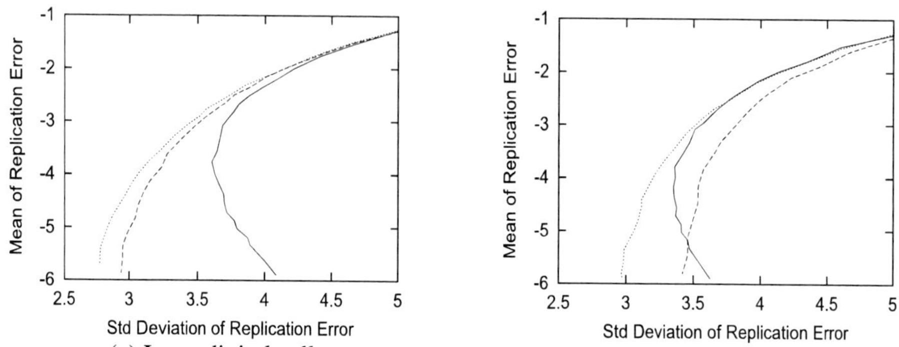

(a) Long digital call
(b) Short digital call

Figure 5. Comparison of the risk-return tradeoffs of the improved Leland's strategy (dotted line) against the original Leland's strategy (dashed line) and the Black-Scholes strategy (solid line) for different values of δt. The option is a short/long digital cash-or-nothing call which pays a fixed amount of $X = 25$ and the rest of the model parameters as in Figure 1 (a).

3.5. Numerical Solutions of the Nonlinear Partial Differential Equations

In this subsection we present the methods to obtain the solutions of some nonlinear PDEs derived in the previous subsections using finite difference schemes. To solve the nonlinear PDEs (4) and (19), we employ an explicit finite difference scheme which is extremely simple in a practical implementation. It is widely known that an explicit finite difference scheme can be unstable. To satisfy the stability condition one needs to choose a rather small time step. Consequently, the disadvantage of this method is a reduced time speed and a higher memory requirement. However, since the problem is two-dimensional and modern computers are quite powerful nowadays, the computation of the solution takes almost no time.

To implement the solution of either PDE (28) or PDE (29) we have to work in a three dimensional space. In this case the implementation of an explicit finite difference scheme is cumbersome as the memory requirement and, hence, the computational time are enormous. Consequently, one needs to implement an implicit finite difference scheme and our choice here is the Crank-Nicolson method. Since due to some practical and legal reasons path-dependent quantities are rarely measured continuously, we describe only the implementation of the solution of the nonlinear PDE (29).

Finally we make some comments on the implementation of the numerical solution of PDE (38).

3.5.1. Numerical Solution of the Nonlinear PDE for a Portfolio of Options on a Stock

To solve the nonlinear PDE (4), we employ a simple explicit finite difference scheme. We define a grid $(i\delta S, j\delta t)$ on the (S,t)-space and denote $V_i^j = V(i\delta S, j\delta t)$. Using the explicit finite difference scheme, we approximate the derivatives as follows:

$$\frac{\partial V}{\partial t} \approx \frac{V_i^j - V_i^{j-1}}{\delta t}, \quad \frac{\partial V}{\partial S} \approx \frac{V_{i+1}^j - V_{i-1}^j}{2\delta S}, \quad \frac{\partial^2 V}{\partial S^2} \approx \frac{V_{i+1}^j - 2V_i^j + V_{i-1}^j}{\delta S^2}.$$

We denote $S_i = i\delta S$. Moreover, we denote and calculate the sign function according to

$$Z_i^j = \text{sign}\left(\frac{\partial^2 V}{\partial S^2}\right) = \text{sign}\left(V_{i+1}^j - 2V_i^j + V_{i-1}^j\right)$$

Consequently, after putting into (4) the approximations for the derivatives we obtain

$$\frac{V_i^j - V_i^{j-1}}{\delta t} + \frac{\sigma^2 S_i^2}{2}\left[1 - AZ_i^j\right]\frac{V_{i+1}^j - 2V_i^j + V_{i-1}^j}{\delta S^2} + rS_i\frac{V_{i+1}^j - V_{i-1}^j}{2\delta S} - rV_i^j = 0. \quad (40)$$

Now we multiply (40) by δt, move V_i^{j-1} to the right, and collect the terms with V_{i-1}^j, V_i^j, and V_{i+1}^j. Having done this, we get

$$\begin{aligned} V_i^{j-1} = {} & V_{i+1}^j\left(S_i^2\left[1 - AZ_i^j\right]\frac{\sigma^2}{2}\frac{\delta t}{\delta S^2} + S_i\frac{r\delta t}{2\delta S}\right) \\ & + V_i^j\left(1 - 2S_i^2\left[1 - AZ_i^j\right]\frac{\sigma^2}{2}\frac{\delta t}{\delta S^2} - r\delta t\right) \\ & + V_{i-1}^j\left(S_i^2\left[1 - AZ_i^j\right]\frac{\sigma^2}{2}\frac{\delta t}{\delta S^2} - S_i\frac{r\delta t}{2\delta S}\right). \end{aligned} \quad (41)$$

As usually in an explicit finite difference scheme, we interpret (41) as calculating an unknown value of V_i^{j-1} using three known values of V_{i-1}^j, V_i^j, and V_{i+1}^j. Finally, we define zero boundary condition as

$$V_0^{j-1} = V_0^j(1 - r\delta t),$$

and the far boundary condition as

$$V_n^j = 2V_{n-1}^j - V_{n-2}^j.$$

At last, recall the stability condition for an explicit finite difference scheme

$$\delta t \le \frac{\delta S^2}{\sigma^2 S_n^2}.$$

3.5.2. Numerical Solution of the Nonlinear PDE for a Portfolio of Options on a Futures Contract

To solve the nonlinear PDE (19), we employ the same simple explicit finite difference scheme as in the previous subsection. We define a grid $(j\delta t, i\delta F)$ on the (t,F)-space and denote $V_i^j = V(j\delta t, i\delta F)$, $F_i = i\delta F$, $A_j = A(j\delta t)$, and $\sigma_j = \sigma(j\delta t)$. Using the explicit finite difference scheme, we approximate the derivatives as follows:

$$\frac{\partial V}{\partial t} \approx \frac{V_i^j - V_i^{j-1}}{\delta t}, \quad \frac{\partial^2 V}{\partial F^2} \approx \frac{V_{i+1}^j - 2V_i^j + V_{i-1}^j}{\delta F^2}.$$

Moreover, we denote and calculate the sign function according to

$$Z_i^j = \text{sign}\left(\frac{\partial^2 V}{\partial F^2}\right) = \text{sign}\left(V_{i+1}^j - 2V_i^j + V_{i-1}^j\right).$$

Consequently, after putting into (19) the approximations for the derivatives, we obtain

$$\frac{V_i^j - V_i^{j-1}}{\delta t} + \frac{\sigma_j^2 F_i^2}{2}\left[1 - A_j Z_i^j\right]\frac{V_{i+1}^j - 2V_i^j + V_{i-1}^j}{\delta F^2} - rV_i^j = 0. \tag{42}$$

Now we multiply (42) by δt, move V_i^{j-1} to the right, and collect the terms with V_{i-1}^j, V_i^j, and V_{i+1}^j. Having done this, we get

$$\begin{aligned} V_i^{j-1} = {} & V_{i+1}^j\left(F_i^2\left[1 - A_j Z_i^j\right]\frac{\sigma_j^2}{2}\frac{\delta t}{\delta F^2}\right) \\ & + V_i^j\left(1 - 2F_i^2\left[1 - A_j Z_i^j\right]\frac{\sigma_j^2}{2}\frac{\delta t}{\delta F^2} - r\delta t\right) \\ & + V_{i-1}^j\left(F_i^2\left[1 - A_j Z_i^j\right]\frac{\sigma_j^2}{2}\frac{\delta t}{\delta F^2}\right). \end{aligned} \tag{43}$$

As usually in an explicit finite difference scheme, we interpret (43) as calculating an unknown value of V_i^{j-1} using three known values of V_{i-1}^j, V_i^j, and V_{i+1}^j. Finally, we define zero boundary condition as

$$V_0^{j-1} = V_0^j(1 - r\delta t),$$

and the far boundary condition as

$$V_n^j = 2V_{n-1}^j - V_{n-2}^j,$$

where F_n is the upper bound for F.

3.5.3. Numerical Solution of the Nonlinear PDE for a Portfolio of Strongly Path-Dependent Options on a Stock with Discrete Sampling

To solve the nonlinear PDE (29) with a jump condition given by (30), we employ the Crank-Nicolson finite difference scheme. We define a grid $(j\delta t, i\delta S, m\delta Y)$ on the (t, S, Y)-space and denote $V_i^j = V(j\delta t, i\delta F)$ and $S_i = i\delta S$. Recall that between the sampling dates (t_k, t_{k+1}) for each value of $m\delta Y$ we need to solve the nonlinear PDE (29). Using the Crank-Nicolson finite difference scheme, we approximate the derivatives as follows:

$$\frac{\partial V}{\partial t} \approx \frac{V_i^{j+1} - V_i^j}{\delta t},$$

$$\frac{\partial V}{\partial S} \approx \frac{V_{i+1}^{j+1} - V_{i-1}^{j+1}}{2\delta S} + \frac{V_{i+1}^{j} - V_{i-1}^{j}}{2\delta S},$$

$$\frac{\partial^2 V}{\partial S^2} \approx \frac{V_{i+1}^{j+1} - 2V_{i}^{j+1} + V_{i-1}^{j+1}}{\delta S^2} + \frac{V_{i+1}^{j} - 2V_{i}^{j} + V_{i-1}^{j}}{\delta S^2}.$$

Moreover, we denote and calculate the sign function according to

$$Z_i^{j+1} = \text{sign}\left(\frac{\partial^2 V}{\partial S^2}\right) = \text{sign}\left(V_{i+1}^{j+1} - 2V_i^{j+1} + V_{i-1}^{j+1}\right).$$

Consequently, after putting into (29) the approximations for the derivatives, we obtain

$$\begin{aligned} &\frac{V_i^{j+1} - V_i^j}{\delta t} + \frac{1}{2}a_i\left(\frac{V_{i+1}^{j+1} - 2V_{i}^{j+1} + V_{i-1}^{j+1}}{\delta S^2} + \frac{V_{i+1}^{j} - 2V_{i}^{j} + V_{i-1}^{j}}{\delta S^2}\right) \\ &+\frac{1}{2}b_i\left(\frac{V_{i+1}^{j+1} - V_{i-1}^{j+1}}{2\delta S} + \frac{V_{i+1}^{j} - V_{i-1}^{j}}{2\delta S}\right) - \frac{1}{2}r(V_i^{j+1} + V_i^j) = 0, \end{aligned} \tag{44}$$

where

$$a_i = \frac{\sigma^2}{2}S_i^2\left[1 - AZ_i^{j+1}\right] \quad \text{and} \quad b_i = rS_i.$$

Collecting the terms we find

$$D_i V_{i-1}^j - (1 + B_i)V_i^j + C_i V_{i+1}^j = -D_i V_{i-1}^{j+1} - (1 - B_i)V_i^{j+1} - C_i V_{i+1}^{j+1}, \tag{45}$$

where

$$\begin{aligned} D_i &= \frac{1}{2}a_i\frac{\delta t}{\delta S^2} - \frac{1}{2}b_i\frac{\delta t}{2\delta S}, \\ B_i &= a_i\frac{\delta t}{\delta S^2} + \frac{1}{2}r\delta t, \\ C_i &= \frac{1}{2}a_i\frac{\delta t}{\delta S^2} + \frac{1}{2}b_i\frac{\delta t}{2\delta S}. \end{aligned}$$

As before, we define zero boundary condition as

$$V_0^j = V_0^{j+1}(1 - r\delta t),$$

and the far boundary condition as

$$V_n^j = 2V_{n-1}^j - V_{n-2}^j.$$

Using these boundary conditions, the Crank-Nicolson method can be written in matrix form

$$\begin{bmatrix} -(1+B_1) & C_1 & 0 & 0 & \cdots \\ D_2 & -(1+B_2) & C_2 & 0 & \cdots \\ \vdots & \vdots & \vdots & \vdots & \ddots \\ \cdots & 0 & 0 & (D_{n-1} - C_{n-1}) & (2C_{n-1} - (1 + B_{n-1})) \end{bmatrix} \begin{bmatrix} V_1^j \\ V_2^j \\ \vdots \\ V_{n-1}^j \end{bmatrix}$$

$$
= \begin{bmatrix} -D_1(2 - r\delta t)V_0^{j+1} - (1 - B_1)V_1^{j+1} - C_1 V_2^{j+1} \\ -D_2 V_1^{j+1} - (1 - B_2)V_2^{j+1} - C_2 V_3^{j+1} \\ \vdots \\ -D_{n-1} V_{n-2}^{j+1} - (1 - B_{n-1})V_{n-1}^{j+1} - C_{n-1} V_n^{j+1} \end{bmatrix} \tag{46}
$$

Consequently, the Crank-Nicolson finite difference scheme is interpreted as calculating the unknown values of V at time step j using the known values of V at time step $j+1$. Note that finding the unknown values of V involves the solution of a system of linear equations (46). For the methods of solutions, the interested reader can consult, for example, Wilmott [22, Chapter 63]. Finally, across a sampling date t_k we need to implement the jump condition (30). That is, to advance the solution of the PDE we have to apply the jump condition, where we must compute the right side of (30) from the left side. Here we need to use some sort of interpolation (for example, linear) to enforce (30) since the values of V on the left side have to be evaluated in the points which are not, in general, grid points in Y direction.

3.5.4. Numerical Solution of the Nonlinear PDE for an Option on Several Assets

With n underlying assets we need to solve numerically an $n+1$ dimensional problem. Due to the "curse of dimensionality", in practice one uses finite difference methods when the problem is either two or three-dimensional and hardly ever in higher dimensions. In the absence of transaction costs one can give a probabilistic interpretation for the PDE for an option price. This motivates to use the Monte-Carlo approach to the solution of an option pricing problem in higher dimensions. However, in the presence of transaction costs a probabilistic interpretation is not possible and a finite difference method seems to be the only solution.

Consequently, the numerical solution of the non-linear PDE (38) is feasible with no more than three underlying assets. The easiest case is an option on two underlying assets, for example, option to exchange one asset for another, or exchange option which pays off at maturity

$$
V(T, S_1(T), S_2(T)) = \max(S_1(T) - S_2(T), 0).
$$

The simplest and most straightforward method is an explicit finance difference scheme when one uses central differences for all the derivatives in (38). The advantage of this method is that it is easy to program. The main disadvantage comes in instability and lack of speed. The use of implicit finite difference schemes (as, for example, Alternating Direction Implicit method) seems not to be possible as the implicit method requires solutions of a set of equations with many unknowns. When a PDE is linear the equations are linear and have a nice form. By contrast, when a PDE is non-linear, one needs to solve a set of non-linear equations which is very complicated. We believe that the most efficient method to solve this PDE is the Hopscotch method which does not require the solution of any simultaneous equations.

4. Utility-Based Option Pricing and Hedging

4.1. The Utility-Based Approach

There has been proposed a variety of alternative approaches to option pricing and hedging with transaction costs. A great deal of them are concerned with the "financial engineering" problem of either replicating or super-replicating the option payoff. These approaches are mainly preference-free models where rehedging occurs at some discrete time intervals whether or not it is optimal in any sense. However, common sense tells us that an "optimal" hedging policy should achieve the best possible tradeoff between the risk and the costs of replication. Recognizing the fact that risk preferences differ among the individuals, the following conclusion becomes obvious: in pricing and hedging options one must consider the hedger's attitude toward risk.

In modern finance it is customary to describe risk preferences by a utility function. The expected utility theory maintains that individuals behave as if they were maximizing the expectation of some utility function of the possible outcomes. Hodges and Neuberger [14] pioneered the option pricing and hedging approach based on this theory. The key idea behind the utility-based approach is the indifference argument: The writing price of an option is defined as the amount of money that makes the hedger indifferent, in terms of expected utility, between trading in the market with and without writing the option. In many respects such an option price is determined in a similar manner to a certainty equivalent within the expected utility framework, which is a well grounded pricing principle in economics. The difference in the two trading strategies, with and without the option, is interpreted as "hedging" the option.

Initially, the utility-based approach was applied to the pricing and hedging of single options. However, this method can be easily applied for pricing and hedging a portfolio of options. The starting point for the utility-based option pricing and hedging approach is to consider the optimal portfolio selection problem of the hedger who faces transaction costs and maximizes expected utility of his terminal wealth. The hedger has a finite horizon $[t, T]$. For the simplicity of the exposition we assume that there are no transaction costs at terminal time T. The hedger has the amount x_t in the bank account and y_t shares of the stock at time t. We define the value function of the hedger with no options as

$$J_0(t, x_t, y_t, S_t) = \max E_t[U(x_T + y_T S_T)], \tag{47}$$

where $U(z)$ is the hedger's utility function. Similarly, the value function of the hedger who, for example, sells the option portfolio is defined by

$$J_1(t, x_t, y_t, S_t) = \max E_t[U(x_T + y_T S_T - V(T, S_T))]. \tag{48}$$

Finally, the price of the option portfolio is defined as the compensation p such that

$$J_1(t, x_t + p, y_t, S_t) = J_0(t, x_t, y_t, S_t). \tag{49}$$

The solutions to problems (47), (48), and (49) provide the unique price and, above all, the optimal hedging strategy. The existence and uniqueness of the solutions were rigorously proved by Davis, Panas, and Zariphopoulou [10].

Unfortunately, there are no closed-form solutions to all these problems. As a result, the solutions have to be obtained by numerical methods. The numerical algorithm is cumbersome to implement and the computation of the optimal hedging strategy is time consuming. In case of a general utility function one needs to solve numerically an optimal stochastic control problem in four dimensions $(t \times x \times y \times S)$. To reduce the dimensionality of the problem one usually uses the negative exponential utility function

$$U(z) = -\exp(-\gamma z); \quad \gamma > 0,$$

where γ is a measure of the hedger's (constant) absolute risk aversion. This particular choice of utility function might seem restrictive. However, as it was conjectured by Davis, Panas, and Zariphopoulou [10] and showed in Andersen and Damgaard [1], an option price is approximately invariant to the specific form of the hedger's utility function, and mainly only the level of absolute risk aversion plays an important role. The great advantage of this utility function is that the hedger's strategy does not depend on his holdings in the bank account, x, and, therefore, the dimensionality of the problem is reduced by one. So far the problem was solved[2] mainly in three[3] dimensions, that is, assuming a negative exponential utility of the hedger. Below we present the description of the nature of the optimal hedging strategy in this case.

The numerical calculations show that when the hedger's risk aversion is rather low, the hedger implements mainly a so-called "static" hedge, which consists in buying Δ shares of the stock at time t and holding them until the option maturity T. When the hedger's risk aversion increases, he starts to rebalance the hedging portfolio in between (t, T). Recall that in the framework of the utility-based hedging approach the hedging strategy is defined as the difference, $\Delta(\tau) = y_1(\tau) - y_0(\tau)$, $\tau \in [t, T]$, between the hedger's optimal trading strategies with and without options. When the hedger's risk aversion is moderate, it is impossible to give a concise description of the optimal hedging strategy. However, when the hedger's risk aversion is rather high, then we can assume that $y_0(\tau) \equiv 0$ (that is, the hedger does not invest in the underlying stock in the absence of option contracts) and the optimal hedging strategy can be conveniently described as $\Delta(\tau) = y_1(\tau)$. In the latter case the numerical calculations show that the optimal hedge ratio Δ is constrained to evolve between two boundaries, Δ_l and Δ_u, such that $\Delta_l < \Delta_u$. As long as the hedge lies within these two boundaries, $\Delta_l \leq \Delta \leq \Delta_u$, no rebalancing of the hedging portfolio takes place. That is why the region between the two boundaries is commonly denoted as the "no transaction region". As soon as the hedge ratio goes out of the no transaction region, a rebalancing occurs in order to bring the hedge to the nearest boundary of the no transaction region. In other words, if Δ moves below Δ_l, one should immediately transact to bring it back to Δ_l. Similarly, if Δ moves above Δ_u, a rebalancing trade occurs to bring it back to Δ_u. Figures 6 and 7 illustrate the optimal hedging strategy.

[2]For implementations of numerical algorithms, the interested reader can consult Davis and Panas [9] and Clewlow and Hodges [8].

[3]Andersen and Damgaard [1] solved the problem assuming a hedger with a power utility function, that is, in four dimensions. However, they managed to solve the problem with only nine periods of trading.

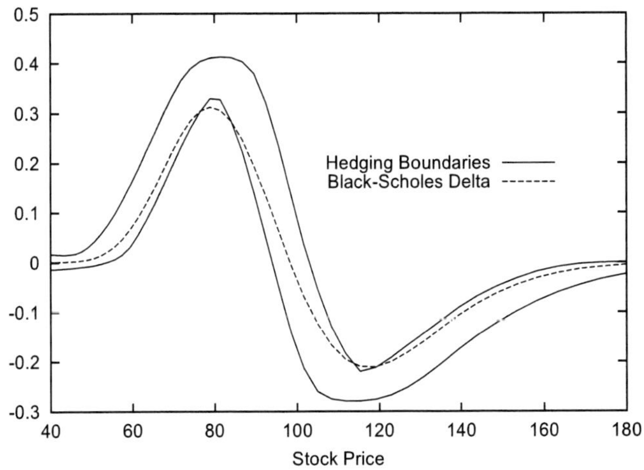

Figure 6. Optimal hedging strategy versus the Black-Scholes delta for a short Butterfly Spread and the following model parameters: $\gamma = 2.0$, $\lambda = 0.01$, $S_t = 100$, $\sigma = 0.25$, $\mu = r = 0.05$, $T - t = 0.5$, and exercise prices 80, 100, and 120.

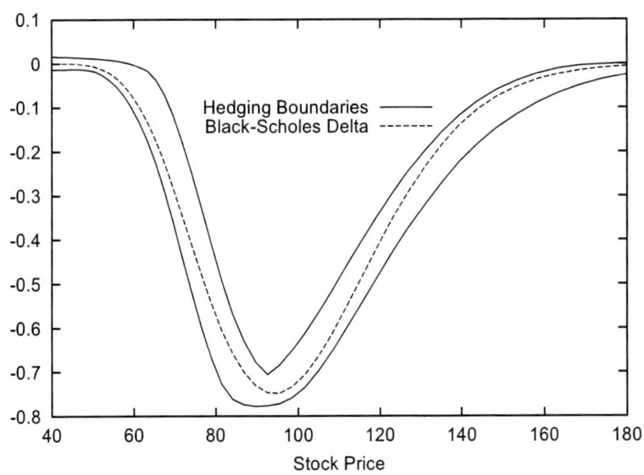

Figure 7. Optimal hedging strategy versus the Black-Scholes delta for a long Bull Spread with exercise prices 80 and 120, and the rest of the model parameters as in Figure 6.

4.2. Asymptotic Analysis of the Utility-Based Approach

4.2.1. The Asymptotic Analysis of Whalley and Wilmott

Since there are no explicit solutions for the utility-based hedging strategy with transaction costs and the numerical methods are computationally hard, for practical applications it is of major importance to use other alternatives. One of such alternatives is to obtain an asymptotic solution. In asymptotic analysis one studies the solution to a problem when some parameters in the problem assume large or small values.

Whalley and Wilmott [21] were the first to provide an asymptotic analysis of the model of Hodges and Neuberger [14] assuming that transaction costs are small. Using formal matched asymptotics, they showed that the boundaries of the no transaction region are given by

$$\Delta = \frac{\partial V}{\partial S} \pm H_{ww} = \frac{\partial V}{\partial S} \pm \left(\frac{3}{2} \frac{e^{-r(T-t)} \lambda S \Gamma^2}{\gamma} \right)^{\frac{1}{3}}, \tag{50}$$

where, $\frac{\partial V}{\partial S}$ is the Black-Scholes hedge, and $\Gamma = \frac{\partial^2 V}{\partial S^2}$ is the option gamma. It is important to note that the asymptotic method of Whalley and Wilmott [21] is of general applicability. Even though the authors considered the pricing and hedging of a short European call option, the final expressions for the option price and hedging strategy are valid for any option, including an option portfolio as a special case of a complex option.

It is easy to see that the optimal hedging bandwidth is not constant, but depends in a natural way on a number of parameters. As $\lambda \to 0$, the optimal hedge approaches the Black-Scholes hedge. As γ increases, the hedging bandwidth decreases in order to decrease the risk of the hedging portfolio. The dependence of the hedging bandwidth on the option gamma is also natural, as we expect to rehedge more often in regions with high gamma. Moreover, it agrees quite well with the results of exact numerical computations, see Figures 8 and 9. Note, however, that the option portfolio gamma approaches zero as the option position goes farther either out-of-the-money ($S \to 0$) or in-the-money ($S \to \infty$). This means, in particular, that the size of the no transaction region in the asymptotic strategy also approaches zero. On the contrary, the exact numerics show (see Figures 8 and 9) that, when the option portfolio gamma goes to zero, the size of the no transaction region times the stock price approaches a constant value.

4.2.2. The Asymptotic Analysis of Barles and Soner

Barles and Soner [3] performed an alternative asymptotic analysis of the same model assuming that both the transaction costs and the hedger's risk tolerance are small. They found that to find the option value one needs to solve the following nonlinear PDE

$$\frac{\partial V}{\partial t} + rS\frac{\partial V}{\partial S} + \frac{1}{2}\sigma^2 S^2 \left[1 - f\left(e^{r(T-t)}\lambda^2 \gamma S^2 \Gamma\right)\right] \frac{\partial^2 V}{\partial S^2} - rV = 0. \tag{51}$$

Formally the option value is equal to the Black-Scholes value with an "adjusted" variable volatility given by

$$\sigma_m^2 = \sigma^2[1 - A_{bs}] = \sigma^2 \left[1 - f\left(e^{r(T-t)}\lambda^2 \gamma S^2 \Gamma\right)\right], \tag{52}$$

where $f(\cdot)$ is some function. The optimal hedging strategy consists in keeping the hedge ratio inside the no transaction region given by

$$\Delta = \frac{\partial V(\sigma_m)}{\partial S} \pm H_{bs} = \frac{\partial V(\sigma_m)}{\partial S} \pm \frac{1}{\lambda \gamma S} g\left(\lambda^2 \gamma S^2 \Gamma\right), \tag{53}$$

where $g(\cdot)$ is some function and $\frac{\partial V(\sigma_m)}{\partial S}$ is the Black-Scholes hedge with an adjusted volatility.

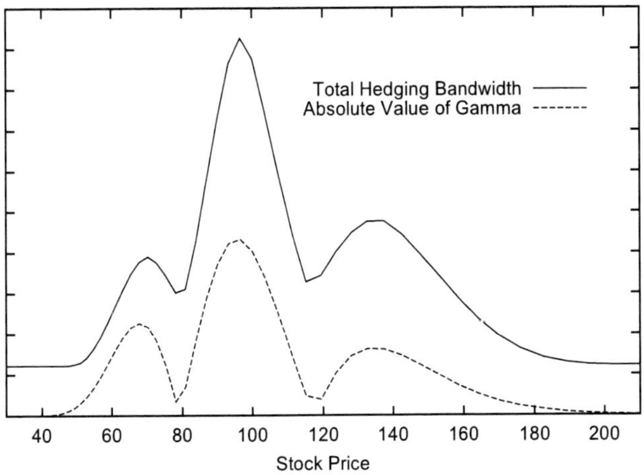

Figure 8. The form of the optimal hedging bandwidth $(\Delta_u - \Delta_l)S$, obtained using the exact numerics, versus the absolute value of the option portfolio gamma times the stock price, $|\Gamma|S$, for a short Butterfly Spread and the following model parameters: $\gamma = 0.1$, $\lambda = 0.01$, $S_t = 100$, $\sigma = 0.25$, $\mu = r = 0.05$, $T - t = 0.5$, and exercise prices 80, 100, and 120.

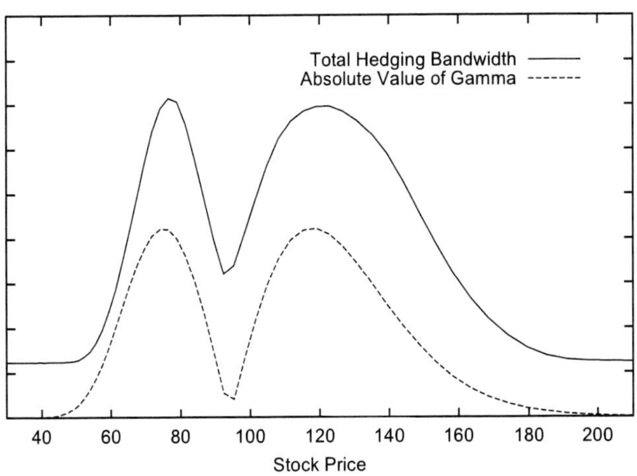

Figure 9. The form of the optimal hedging bandwidth $(\Delta_u - \Delta_l)S$, obtained using the exact numerics, versus the absolute value of the option portfolio gamma times the stock price, $|\Gamma|S$, for a long Bull Spread with exercise prices 80 and 120, and the rest of the model parameters as in Figure 8.

According to the results of the Barles and Soner asymptotic analysis, the optimal hedging strategy has two important features: a particular form of the hedging bandwidth and the volatility adjustment. The comparative statics for the Barles and Soner hedging bandwidth is similar to that of the Whalley and Wilmott one: the hedging bandwidth increases when

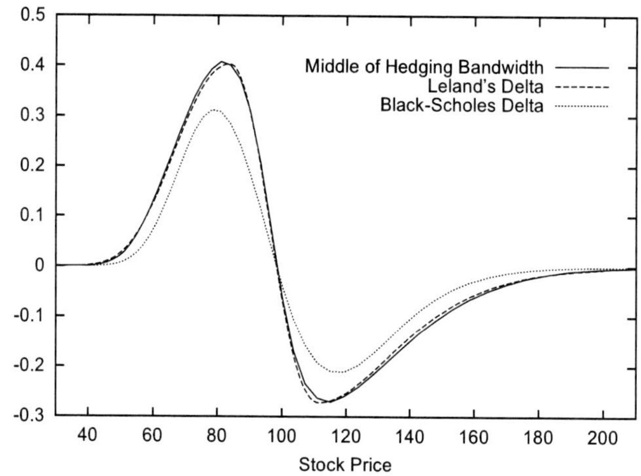

Figure 10. The middle of the optimal hedging bandwidth, obtained using the exact numerics, versus the Black-Scholes delta and the Leland's delta with $A = 0.55$ for a short Butterfly Spread and the following model parameters: $\gamma = 3.0$, $\lambda = 0.01$, $S_t = 100$, $\sigma = 0.25$, $\mu = r = 0.05$, $T - t = 0.5$, and exercise prices 80, 100, and 120.

either the level of the transaction costs, the hedger's risk tolerance, or the option gamma increases. The volatility adjustment means, in particular, that the middle of the hedging bandwidth does not coincide with the Black-Scholes hedge. This essential feature of the utility-based hedging strategy was already observed by Hodges and Neuberger [14] and emphasized by Clewlow and Hodges [8] and further by Zakamouline [24] using the results of Monte-Carlo simulations of hedging strategies with and without the optimal volatility adjustment.

Unfortunately, the functions $f(\cdot)$ and $g(\cdot)$ depend on the option payoff and Barles and Soner gave their forms only for the case of a plain vanilla call option. Consequently, one cannot use[4] the asymptotic strategy of Barles and Soner for hedging any option or a portfolio of options. However, Clewlow and Hodges [8] and then Barles and Soner [3] noted that the volatility adjustment in the utility-based model is similar to the Leland [17] volatility adjustment (This is also clearly seen from the comparison of equations (51) and (4)). This agrees quite well with the results of the exact numerical computations. Figures 10 and 11 illustrate that the numerically computed middle of the hedging bandwidth in the utility-based hedging is very close to the Leland's hedging delta when we appropriately choose the value of parameter A in the PDE (4). Note that A determines the "degree" of volatility adjustment.

4.2.3. The Approximation Method of Zakamouline

Recall that in asymptotic analysis one studies the limiting behavior of the optimal hedging policy as one or several parameters of the problem approach zero. Even though asymptotic

[4] Another reason for not using an asymptotic solution is the fact that under realistic model parameters an asymptotic solution deviates noticeable from the exact solution, see [25].

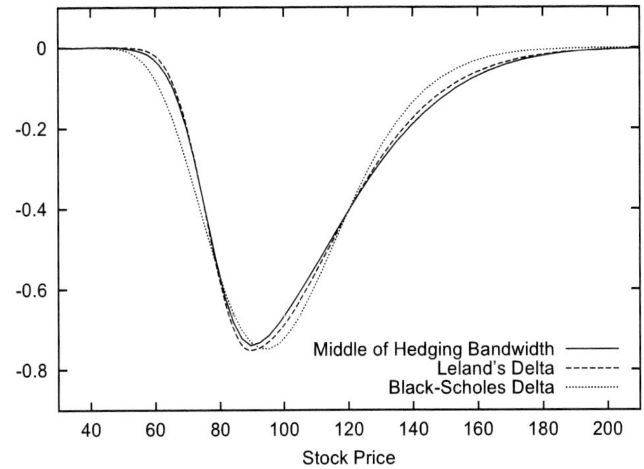

Figure 11. The middle of the optimal hedging bandwidth, obtained using the exact numerics, versus the Black-Scholes delta and the Leland's delta with $A = 0.55$ for a long Bull Spread with exercise prices 80 and 120, and the rest of the model parameters as in Figure 10.

analysis can reveal the underlying structure of the solution, under realistic parameters this method provide not quite accurate results. Zakamouline [25] compared the performance of asymptotic strategies against the exact strategy and found that under realistic model parameters an asymptotic strategy performs noticeably worse than that obtained from the exact numerical solution. The explanation lies in the fact that, as compared to the exact numerical solution, under realistic parameters the size of the hedging bandwidth and the volatility adjustment obtained from asymptotic analysis are overvalued. What is more important, an asymptotic solution showed to be unable to sustain the correct interrelationship between the size of the hedging bandwidth and the degree of volatility adjustment. The significance of the correct interrelationship could hardly be overemphasized: the empirical testing of the hedging strategies revealed that either undervaluation or overvaluation of the volatility adjustment (with respect to the size of the hedging bandwidth) results in a drastic deterioration of the performance of a hedging strategy.

Zakamouline [24], [25] summarized the stylized facts about the nature of the utility-based hedging strategy and suggested a general specification of the optimal hedging policy for a single plain-vanilla European option. The careful visual inspection of the numerically calculated optimal hedging policy together with the insights from asymptotic analysis advocate for the following general specification of the optimal hedging strategy

$$\Delta = \frac{\partial V(\sigma_m)}{\partial S} \pm (H_1 + H_0),$$

where σ_m is the adjusted volatility given by

$$\sigma_m^2 = \sigma^2 (1 - A_\sigma).$$

The term H_1 is closely related to H_{ww} in the Whalley and Wilmott hedging strategy (see equation (50)) and to H_{bs} in the Barles and Soner hedging strategy (see equation (53)).

The main feature of H_1 is that this term depends on the option gamma. Recall that when the option gamma goes to zero, the size of the no transaction region times the stock price approaches a constant value. It turns out that this constant value is actually the size of the no transaction region in the optimal portfolio selection problem without options. To reflect this feature of the optimal hedging policy, Zakamouline introduced the term H_0. Finally, A_σ is closely related to A_{bs} in the Barles and Soner volatility adjustment (see equation (52)) and to A in the Leland's volatility adjustment (see equation (5)).

Zakamouline [25] suggested to use an alternative to the asymptotic analysis, namely, the approximation method. Under approximation it is meant the following: one has a rather slow and cumbersome way to compute the optimal hedging policy with transaction costs and wants to replace it with simple and efficient approximating function(s). To do this, one first specifies a flexible functional form of the optimal hedging policy. Then, given a functional form, the parameters are chosen to provide the best fit to the exact numerical solution. This second stage is known as "model calibration". In short, Zakamouline assumed the following functional form of the approximating function for H_0, H_1, and A_σ:

$$H_0 = \alpha \sigma^{\beta_1} \lambda^{\beta_2} \gamma^{\beta_3} S^{\beta_4} (T-t)^{\beta_5},$$

$$H_1 = A_\sigma = \alpha \sigma^{\beta_1} \lambda^{\beta_2} \gamma^{\beta_3} S^{\beta_4} \Gamma^{\beta_5} e^{-\beta_6 r(T-t)} (T-t)^{\beta_7},$$

where $\alpha, \beta_1, \ldots, \beta_k$ are parameters to be chosen in order to achieve the best fit. The results of his estimations of the best-fit parameters for hedging a short European call option, after some rounding off the values of parameters α and β_i, give the following approximating functions

$$H_0 = \frac{\lambda}{\gamma S \sigma^2 (T-t)}, \tag{54}$$

$$H_1 = 1.12\, \lambda^{0.31} (T-t)^{0.05} \left(\frac{e^{-r(T-t)}}{\sigma} \right)^{0.25} \left(\frac{|\Gamma|}{\gamma} \right)^{0.5}, \tag{55}$$

$$A_\sigma = -4.76 \frac{\lambda^{0.78}}{(T-t)^{0.02}} \left(\frac{e^{-r(T-t)}}{\sigma} \right)^{0.25} \left(\gamma S^2 |\Gamma| \right)^{0.15}.$$

Then, Zakamouline tested empirically his approximation strategy against both the asymptotic strategies and some other well-known strategies. He evaluated the performance of alternative hedging strategies within three different unified risk-returns frameworks[5] and found that his approximation strategy outperforms all the others.

4.3. Optimal Hedging of Options and Option Portfolios Using a Simplified Utility-Based Hedging Strategy

The approximation methodology presented in Zakamouline [25] produces a clearly superior result as applied to the problem of finding a closed-form expressions for the optimal hedging strategy for a specific option or an option portfolio. However, this methodology

[5]These frameworks are: (1) mean replication error - variance of replication error; (2) mean replication error - Value at Risk of replication error; (3) mean replication error - Expected Shortfall of replication error.

does retain the main disadvantage one would like to get rid off: to approximate the optimal strategy for a specific option one needs to start with the numerical calculations of the exact hedging strategy for a large set of the model parameters. Since the numerical algorithm is cumbersome to implement and the calculation of the optimal hedging strategy is time consuming, the approximation methodology is unlikely to be commonly used by the practitioners.

In the preceding subsections we presented the description of the nature of the optimal hedging strategy for a portfolio of options. This presentation suggests that the optimal hedging strategy is *rather simple* and has three essential features:

- The presence of the no transaction region such that if the hedge ratio lies outside of the no transaction region a rebalancing occurs in order to bring the hedge to the nearest boundary of the no transaction region (see Figures 6 and 7).

- The form of the no transaction region is mainly derived from the form of the absolute value of the option portfolio gamma. In addition we know that when the option portfolio gamma tends to zero, the size of the no transaction region tends to some constant (see Figures 8 and 9)).

- The middle of the no transaction region does not coincide with the Black-Scholes delta. This feature of the optimal hedging strategy can be conveniently described as a modified hedging volatility (see Figures 10 and 11).

To reflect all three essential features of the utility-based hedging strategy, we propose the following simplified description of the utility-based hedging strategy for any option portfolio

$$\Delta = \frac{\partial V(\sigma_m)}{\partial S} \pm \left(h_1 |\Gamma| + \frac{h_0}{S(T-t)} \right), \tag{56}$$

where the delta of the option portfolio is obtained by solving the following PDE with a proper boundary condition

$$\frac{\partial V}{\partial t} + rS\frac{\partial V}{\partial S} + \frac{1}{2}\sigma^2 S^2 \left[1 - a_\sigma \text{sign}\left(\frac{\partial^2 V}{\partial S^2} \right) \right] \frac{\partial^2 V}{\partial S^2} - rV = 0. \tag{57}$$

Observe that the simplified utility-based hedging strategy is specified by a triple of (constant) parameters (h_1, h_0, a_σ). Now we would like to present a method of finding the optimal triple of parameters of the simplified utility-based hedging strategy.

To motivate for our method, recall how we compare alternative hedging strategies and find out which strategy is better (for a detailed description, see Section 3.2.): first of all we need to decide upon a unified risk-return framework for comparison. Then we perform a simulation analysis of every hedging strategy and the set of all possible combinations of risk and return of a particular hedging strategy comprises the efficient frontier of the hedging strategy. Each hedging strategy produces different efficient frontier, that is, each hedging strategy offers a different risk-return tradeoff. Observe that a rational hedger will always prefer a strategy that minimizes the risk for a given level of return. Consequently, if we choose a suitable risk-return framework, there are some optimal combinations of the triple of parameters (h_1, h_0, a_σ) of the simplified utility-based hedging strategy that constitute the efficient frontier of the strategy. The essential feature of a point belonging to the efficient frontier is that the risk-return tradeoff provided by this point cannot be improved. That is, any other combinations of (h_1, h_0, a_σ) which produce a lower risk have a lower return, or, conversely, any other combinations of (h_1, h_0, a_σ) which produce a higher return have a higher risk. Thus, the main idea behind our method of finding the optimal parameters of the simplified hedging strategy is to use the simulation analysis to find the efficient frontier of the hedging strategy in some risk-return space.

Before proceeding to the formal presentation of our method, we would like to discuss briefly some suitable risk-return frameworks. The expected replication error is the only sensible candidate for the return measure. As to the risk measure, there are many metrics of risk. In the context of option hedging, the most popular risk metrics are: the variance of replication error; the Value at Risk of replication error; and the Expected Shortfall of replication error. Zakamouline [25] showed that the utility-based hedging strategy exhibits the best risk-return tradeoff with any (of the three) risk metrics. Thus, there is a good reason to believe that the choice of risk metrics might be done arbitrary, that is, using a particular hedger's preferences for risk measurement. Moreover, our study shows that the distribution of the hedging error of the utility-based hedging strategy is, to a large extent, symmetrical. This is another reason to believe that any risk metrics would do.

Now we turn to the formal presentation of our optimization method in some chosen risk-return space where we search for the combinations of the triple of parameters (h_1, h_0, a_σ) which belong to the efficient frontier. We choose some triple of parameters of the hedging strategy, (h_1, h_0, a_σ), perform path simulations, and estimate the return, $\eta = \eta(h_1, h_0, a_\sigma)$, and the risk, $\rho = \rho(h_1, h_0, a_\sigma)$, associated with this particular hedging strategy. Observe that for any risk averse hedger the hedging strategy with parameters (h_1', h_0', a_σ') is considered to be better than the strategy with parameters (h_1, h_0, a_σ) if

$$\eta(h_1', h_0', a_\sigma') \geq \eta(h_1, h_0, a_\sigma) \text{ and } \rho(h_1', h_0', a_\sigma') \leq \rho(h_1, h_0, a_\sigma).$$

That is, if the strategy with (h_1', h_0', a_σ') provides either higher return with less or equal risk, or less risk with higher or equal return than the strategy with (h_1, h_0, a_σ).

The optimization method we propose is based on a sequential improvement of the risk-return tradeoff of a hedging strategy. That is, starting with some (h_1, h_0, a_σ) we search for a new $(h_1', h_0', a_\sigma') = (h_1 + \Delta h_1, h_0 + \Delta h_0, a_\sigma + \Delta a_\sigma)$ such that the risk-return tradeoff of the strategy with (h_1', h_0', a_σ') is better than the strategy with (h_1, h_0, a_σ). The most general procedure for finding the risk-return improvement step could be interpreted as finding a unit vector $\mathbf{u} = \langle a, b, c \rangle$ such that

$$D_{\mathbf{u}}\eta(h_1, h_0, a_\sigma) \geq 0, \tag{58}$$

and

$$D_{\mathbf{u}}\rho(h_1, h_0, a_\sigma) \leq 0, \tag{59}$$

where $D_{\mathbf{u}}\eta(h_1, h_0, a_\sigma)$ and $D_{\mathbf{u}}\rho(h_1, h_0, a_\sigma)$ are the directional derivatives of $\eta(h_1, h_0, a_\sigma)$ and $\rho(h_1, h_0, a_\sigma)$, respectively, in the direction of the unit vector **u**. The reader is reminded that the directional derivative of some function $f(x, y, z)$ in the direction of the unit vector

$\mathbf{u} = \langle a, b, c \rangle$ is defined as

$$D_{\mathbf{u}} f(x,y,z) = \lim_{h \to 0} \frac{f(x+ha, y+hb, z+hc) - f(x,y,z)}{h}$$
$$= f_x(x,y,z)a + f_y(x,y,z)b + f_z(x,y,z)c = \nabla f(x,y,z)\mathbf{u}.$$

Consequently, to implement an improvement step from the point (h_1, h_0, a_σ) we need first to find the partial derivatives of η and ρ at this point and then to find a vector $\mathbf{u}$ such that both (58) and (59) are satisfied. Note that the vector $\mathbf{u}$ is not unique. However, since our goal is to find the points on the efficient frontier, the actual path from some (h_1, h_0, a_σ) to (h_1', h_0', a_σ') belonging to the efficient frontier is not important.

A simple practical realization of the optimization method is based on the implementation of an improvement step with respect to (h_1, h_0) and a consequent improvement step with respect to (h_1, a_σ). Note that the starting point for our optimization method is the risk-return tradeoff given by the benchmark strategy $(h_1, 0, 0)$. The introduction of $h_0 > 0$ and $a_\sigma > 0$ helps to improve the risk-return tradeoff of the benchmark strategy. Suppose that some (h_1, h_0, a_σ) is an improvement of the benchmark strategy and $\eta(h_1, h_0, a_\sigma)$ and $\rho(h_1, h_0, a_\sigma)$ are the corresponding return and the risk of this improvement. An improvement step with respect to (h_1, h_0) is carried out in the following manner. We define the step sizes Δh_1 and Δh_0, implement the simulations and find the risk-returns tradeoffs in the points with $(h_1 + \Delta h_1, h_0, a_\sigma)$ and $(h_1, h_0 + \Delta h_0, a_\sigma)$. Then we estimate the derivatives of η with respect to h_1 and h_0

$$\frac{\partial \eta}{\partial h_1} = \frac{\eta(h_1 + \Delta h_1, h_0, a_\sigma) - \eta(h_1, h_0, a_\sigma)}{\Delta h_1},$$

$$\frac{\partial \eta}{\partial h_0} = \frac{\eta(h_1, h_0 + \Delta h_0, a_\sigma) - \eta(h_1, h_0, a_\sigma)}{\Delta h_0}.$$

Similarly, we estimate the derivatives of ρ with respect to h_1 and h_0. Now given a step size δh_0 we want to find a step size δh_1 such that

$$\delta \eta = \eta(h_1 + \delta h_1, h_0 + \delta h_0, a_\sigma) - \eta(h_1, h_0, a_\sigma) > 0,$$

$$\delta \rho = \rho(h_1 + \delta h_1, h_0 + \delta h_0, a_\sigma) - \rho(h_1, h_0, a_\sigma) < 0.$$

The step size δh_1 is not unique. That is, there is some degree of freedom in determining δh_1. Therefore, we need to impose an additional condition in order to determine the step size δh_1. For example, this condition could be

$$\delta \eta = -\delta \rho.$$

It is easy to check that this condition gives us

$$\delta h_1 = -\frac{\frac{\partial \eta}{\partial h_0} + \frac{\partial \rho}{\partial h_0}}{\frac{\partial \eta}{\partial h_1} + \frac{\partial \rho}{\partial h_1}} \delta h_0.$$

In addition we need to check that

$$\delta \rho = \frac{\partial \rho}{\partial h_1} \delta h_1 + \frac{\partial \rho}{\partial h_0} \delta h_0 < 0. \tag{60}$$

Alternatively, we can check that $\delta\eta > 0$. If we are able to find the step size δh_1 such that the condition (60) is satisfied, we conclude that an improvement step with respect to (h_1, h_0) is feasible and the risk-return tradeoff of the point $(h_1 + \delta h_1, h_0 + \delta h_0, a_\sigma)$ becomes a new improvement of the benchmark strategy.

Similarly we can implement an improvement step with respect to (h_1, a_σ). In this case given the step size δa_σ we also search for δh_1 such that

$$\delta\eta = \eta(h_1 + \delta h_1, h_0, a_\sigma + \delta a_\sigma) - \eta(h_1, h_0, a_\sigma) > 0,$$

$$\delta\rho = \rho(h_1 + \delta h_1, h_0, a_\sigma + \delta a_\sigma) - \rho(h_1, h_0, a_\sigma) < 0.$$

To summarize, the optimization algorithm we propose is, in principle, extremely simple. It is largely based on simulations of a hedging strategy that is determined by some triple of parameters and, at the end, computations of the risk and return of a hedging strategy. In addition, one needs to solve numerically the PDE (57) (which is the same as (4)) using some finite difference method.

It should be noted, however, that the computational time for finding the optimal parameters (h_1, h_0, a_σ) of the risk-return optimized hedging strategy by starting from some benchmark $(h_1, 0, 0)$ is substantially longer than the computational time for finding numerically the utility-based hedging strategy for some value of the hedger's risk aversion γ. The reason is that in order to estimate the risk and return of a hedging strategy with sufficiently high precision one needs to simulate a great number of stock paths. For the purpose of comparison, the numerical computation of the utility-based hedging strategy for a portfolio of options in a 250-periods model takes 5-10 minutes[6] on a computer with a 3.0 GHz processor, given that the program is implemented on C++. We implemented the above described algorithm and some results are presented below. To estimate the risk and return of a hedging strategy with acceptable precision (determined by the step sizes Δh_1, Δh_0 and Δa_σ) we simulated 200,000 paths, where each path consists of 250 equally spaced trading dates over the life of an option portfolio. In this case given that the algorithm is also implemented on C++, the computational time for finding the optimal parameters (h_1, h_0, a_σ) amounts, on average, to an hour. However, the risk-return optimization algorithm can be easily extended to cover the cases of path-depended option portfolios and jump-diffusion or stochastic volatility processes for the price of the underlying stock. In all these cases the numerical computations of the utility-based hedging strategy become extremely time consuming (if ever feasible): With path-dependent options the dimensionality of the problem increases, and with either jump-diffusion or stochastic volatility process for the stock price it can be no longer approximated by a simple binomial recombining tree.

Finally in this subsection we provide an empirical testing of our optimized strategy against: the Black and Scholes strategy at fixed regular intervals; the Leland's strategy; the Whalley and Wilmott asymptotic strategy; and three other well-known strategies. These three strategies are:

- The hedging to a fixed bandwidth around delta strategy. This is a commonly used simplification of the utility-based hedging strategy (see, for example, Martellini and

[6]Note that the computational time depends on the specified precision of computation.

Priaulet [18]). The strategy prescribes to rehedge when the hedge ratio moves outside of the prescribed tolerance from the corresponding Black-Scholes delta. More formally, the boundaries of the no transaction region are defined by

$$\Delta = \frac{\partial V}{\partial S} \pm H,$$

where $\frac{\partial V}{\partial S}$ is the middle of the hedging bandwidth that equals the Black-Scholes hedge, and H is some constant which gives a constant size of the hedging bandwidth.

- The delta tolerance strategy[7] which prescribes rehedging to the Black-Scholes delta when the hedge ratio moves outside of the prescribed tolerance from the perfect hedge position. More formally, the series of stopping times is recursively given by

$$\tau_1 = t, \quad \tau_{i+1} = \inf\left\{\tau_i < \tau < T : \left|\Delta - \frac{\partial V}{\partial S}\right| > h\right\}, \quad i = 1, 2, \ldots,$$

where $\frac{\partial V}{\partial S}$ is the Black-Scholes hedge, and h is a given constant tolerance;

- The asset tolerance strategy[8] which prescribes rehedging to the Black-Scholes delta when the percentage change in the value of the underlying asset exceeds the prescribed tolerance. More formally, the series of stopping times is recursively given by

$$\tau_1 = t, \quad \tau_{i+1} = \inf\left\{\tau_i < \tau < T : \left|\frac{S(\tau) - S(\tau_i)}{S(\tau_i)}\right| > h\right\}, \quad i = 1, 2, \ldots,$$

where h is a given constant percentage.

For a more detailed description of these strategies, the interested reader can consult [23].

We provide the simulation results for a long Bull Spread (one long call with strike K_1 and one short call with strike $K_2 > K_1$, both call options with asset settlements) and a short Butterfly Spread (one long call with strike K_1, one long call with strike K_3, and 2 short calls with strike K_2 such that $K_1 < K_2 < K_3$, all call options with asset settlements). The model parameters are: $T = 1$ year to maturity, $S_0 = 100$, the volatility $\sigma = 0.25$, and the drift $\mu = r = 0.05$. The proportional transaction costs were $\lambda = 0.01$. The exercise prices for the long Bull Spread are $K_1 = 80$ and $K_2 = 120$. The exercise prices for the short Butterfly Spread are $K_1 = 80$, $K_2 = 100$, and $K_3 = 120$.

The simulations proceed in the same manner at that in Section 3.2.: at the beginning, the writer of an option portfolio receives the Black-Scholes value of the portfolio and sets up a replicating portfolio. The underlying path of the stock is simulated according to equation (9). At each δt a check is made to see if the option needs to be rehedged. If so, the rebalancing trade is performed and transaction costs are drawn from the money market account. Finally, at expiration, we compute the replication error, that is, the cash value of the replicating portfolio minus the due exercise payment. As a risk measure we use the variance of the replication error.

[7] This strategy seems to be first suggested by Whalley and Wilmott [20].

[8] This strategy was first suggested by Henrotte [13].

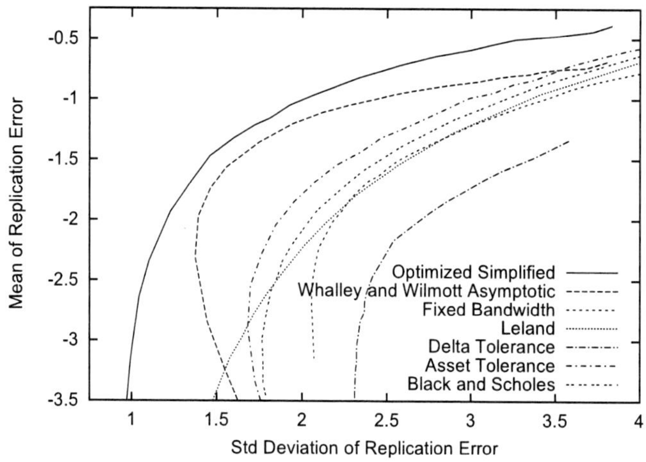

Figure 12. Comparison of hedging strategies for a short Butterfly Spread.

For the Black and Scholes and Leland's strategies we vary the parameter $\delta t \in [\frac{1}{200}, \frac{1}{5}]$ (rehedging interval). For the other strategies, each path consists of 250 equally spaced trading dates over the life of the option. In the delta tolerance, asset tolerance, and hedging to a fixed bandwidth strategies, H and h take values in $[0.01, 0.35]$. For each value of the parameter of these hedging strategies we generate 100,000 paths and compute the mean and variance of the replication error. By varying the value of the parameter of a hedging strategy, we span all the possible combinations of risk and return. For the simplified utility-based strategy we varied the initial value of the parameter h_1 in $[0.025, 2.5]$.

Figures 12 and 13 summarize the results of simulations. For both the option portfolios our optimized strategies outperform all the others. In addition, we have carried on similar simulations for different values of the model parameters r, μ, σ, T, and λ, different types of option portfolios, and different risk metrics (such as Value at Risk and Expected Shortfall). Due to the space limitations, the results of these simulations are not presented. Generally, the relative performance of different hedging strategies depends on the composition of an option portfolio. Some typical examples are presented in Zakamouline [23]. However, in all the cases our optimized simplified utility-based strategy outperforms all the other alternative strategies.

5. Conclusion

In this chapter we considered some approaches to option pricing and hedging in the presence of transaction costs. The distinguishing feature of all these approaches is that the solution for the option price and hedging strategy is given by a nonlinear PDE. When it comes to the pricing of options and option portfolios with transaction costs, the big difference between linear and nonlinear models is the following: whereas in the linear models the value of portfolio of options that is the sum of individual options is the same as sum of values of individual options, in the nonlinear models it is not. When it comes to the differences

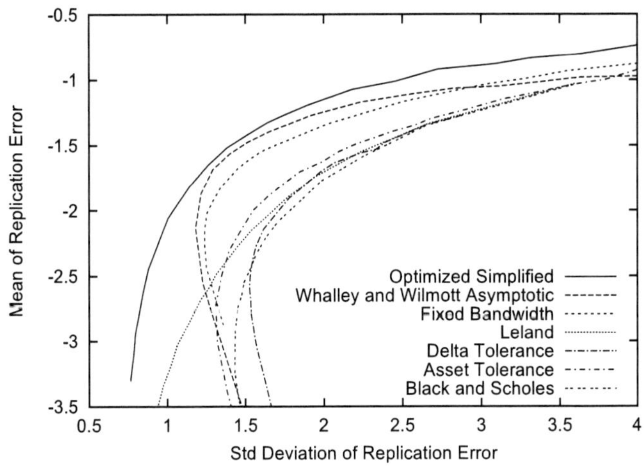

Figure 13. Comparison of hedging strategies for a long Bull Spread.

in option hedging between nonlinear and linear models, in most cases the option hedging strategy in a nonlinear model can be interpreted as hedging with adjusted volatility. The appropriately chosen adjusted hedging volatility in a nonlinear model can greatly improve the risk-return tradeoff of a hedging strategy as compared with that in a linear model.

Acknowledgements

A part of this chapter is based on the results that were originally published by Zakamouline in [25], [26] and [27]. The author would like to thank Saita Francesco, Mete Soner, Paul Wilmott, and several anonymous referees for their helpful comments. The material presented in this chapter has also benefited from comments by participants of the Stochastic Finance 2004 Conference, the Financial Management Association International 2005 European Conference, the Quantitative Methods in Finance 2005 Conference, and the International Conference of Numerical Analysis and Applied Mathematics 2006.

References

[1] E.D. Andersen and A. Damgaard, Utility-based Option Pricing with Proportional Transaction Costs and Diversification Problems: an Interior-Point Optimization Approach, *Appl. Numer. Math.* **29** (1999), 395–422.

[2] M. Avellaneda and A. Paras, Optimal Hedging Portfolios for Derivative Securities in the Presence of Large Transaction Costs, *Appl. Math. Finance* **1** (1994), 165–193.

[3] G. Barles and H.M. Soner, Option Pricing with Transaction Costs and a Nonlinear Black-Scholes Equation, *Finance Stochastics* **2** (1998), 369–397.

[4] H. Bessembinder, Issues in Assessing Trades Execution Costs, *J. Financial Markets* **6** (2003), 233–257.

[5] F. Black and M. Scholes, The Pricing of Options and Corporate Liabilities, *J. Polit. Econ.* **81** (1973), 637–659.

[6] F. Black, The Pricing of Commodity Contracts, *J. Financial Economics* **3** (1976), 167–179.

[7] P.P. Boyle and D. Emanuel, Discretely Adjusted Option Hedges, *J. Financial Economics* **8** (1980), 259–282.

[8] L. Clewlow and S.D. Hodges, Optimal Delta-Hedging under Transaction Costs, *J. Economic Dynamics Control* **21** (1997), 1353–1376.

[9] M.H.A. Davis and V.G. Panas, The Writing Price of a European Contingent Claim under Proportional Transaction Costs, *Comput. Appl. Math.* **13** (1994), 115–137.

[10] M.H.A. Davis, V.G. Panas, and T. Zariphopoulou, European Option Pricing with Transaction Costs, *SIAM J. Control Optim.* **31** (1993), 470–493.

[11] J.C. Dermody and E.Z. Prisman, No Arbitrage and Valuation in Markets with Realistic Transaction Costs, *J. Financial and Quantitative Analysis* **28** (1993), 65–80.

[12] A. Freyre-Sanders, R. Guobuzaite, and K. Byrne, A Review of Trading Cost Model: Reducing Transaction Costs, *J. Investing* **13** (2004), 93–116.

[13] P. Henrotte, Transaction Costs and Duplication Strategies, *Working Paper*, Stanford University and HEC, 1993.

[14] S.D. Hodges and A. Neuberger, Optimal Replication of Contingent Claims under Transaction Costs, *Rev. Futures Markets* **8** (1989), 222–239.

[15] T. Hoggard, A.E. Whalley, and P. Wilmott, Hedging Option Portfolios in the Presence of Transaction Costs, *Adv. Futures and Options Research* **7** (1994), 21–35.

[16] Y.M. Kabanov and M.M. Safarian, On Leland's Strategy of Option Pricing with Transaction Costs, *Finance Stochastics* **1** (1997), 239–250.

[17] H. Leland, Option Pricing and Replication with Transaction Costs, *J. Finance* **5** (1985), 1283–1301.

[18] L. Martellini and P. Priaulet, Competing Methods for Option Hedging in the Presence of Transaction Costs, *J. Derivatives* **9** (2002), 26–38.

[19] B. Mohamed, Simulations of Transaction Costs and Optimal Rehedging, *Appl. Math. Finance* **1** (1994), 49–63.

[20] A.E. Whalley and P. Wilmott, Counting the Costs, *Risk* **6** (1993), 59–66.

[21] A.E. Whalley and P. Wilmott, An Asymptotic Analysis of an Optimal Hedging Model for Option Pricing with Transaction Costs, *Math. Finance* **7** (1997), 307–324.

[22] P. Wilmott, *Paul Wilmott on Quatitative Finance*, John Wiley, England, 2000.

[23] V. Zakamouline, Dynamic Hedging of Complex Option Positions with Transaction Costs, *Working paper*, Bodø Graduate School of Business, 2005.

[24] V. Zakamouline, European Option Pricing and Hedging with both Fixed and Proportional Transaction Costs, *J. Economic Dynamics and Control* **30** (2006), 1–25.

[25] V. Zakamouline, Efficient Analytic Approximation of the Optimal Hedging Strategy for a European Call Option with Transaction Costs, *Quantitative Finance* **6** (2006), 435–445.

[26] V. Zakamouline, Yet Another Note on the Leland's Option Hedging Strategy with Transaction Costs, *WILMOTT Magazine,* March 2006, 70–78.

[27] V. Zakamouline, Hedging of Option Portfolios and Options on Several Assets with Transaction Costs and Nonlinear Partial Differential Equations, *Int. J. Contemporary Mathematical Sciences* **3** (2008), 159–180.

In: Nonlinear Models in Mathematical Finance...
Editor: Matthias Ehrhardt, pp. 67-100
ISBN 978-1-60456-931-5

Chapter 2

UTILITY INDIFFERENCE PRICING WITH MARKET INCOMPLETENESS

Michael Monoyios
Mathematical Institute, University of Oxford,
24–29 St Giles', Oxford OX1 3LB, UK

Abstract

Utility indifference pricing and hedging theory is presented, showing how it leads to linear or to non-linear pricing rules for contingent claims. Convex duality is first used to derive probabilistic representations for exponential utility-based prices, in a general setting with locally bounded semi-martingale price processes. The indifference price for a finite number of claims gives a non-linear pricing rule, which reduces to a linear pricing rule as the number of claims tends to zero, resulting in the so-called marginal utility-based price of the claim. Applications to basis risk models with log-normal price processes, under full and partial information scenarios are then worked out in detail. In the full information case, a claim on a non-traded asset is priced and hedged using a correlated traded asset. The resulting hedge requires knowledge of the drift parameters of the asset price processes, which are very difficult to estimate with any precision. This leads naturally to a further application, a partial information problem, with the drift parameters assumed to be random variables whose values are revealed to the hedger in a Bayesian fashion via a filtering algorithm. The indifference price is given by the solution to a non-linear PDE, reducing to a linear PDE for the marginal price when the number of claims becomes infinitesimally small.

1. Introduction

This chapter presents theory and examples of utility indifference pricing and hedging, a method for managing risk from trading contingent claims in incomplete markets, that has become a classical tool in the valuation of non-hedgeable claims. For a position in n claims, it results in a non-linear pricing rule, which reduces to a linear pricing rule in the limit of a small position in the claim. In this section we qualitatively describe the main idea and highlight some of the literature on the technique. This review is not intended to be exhaustive, but the reader will find ample material and further references in the citations.

Suppose we have a European contingent claim C, a contract which pays the random amount $C \geq 0$ almost surely (a.s.) at some future time $T > 0$. The typical example is where C is dependent on the trajectory of a vector $S = (S_t)_{0 \leq t \leq T}$, which is the price process of a set of $d+1$ traded securities: $S = (S^0, S^1, \ldots, S^d)$. The zeroth asset S^0 is assumed riskless. For ease of exposition, we shall assume zero interest rates throughout this chapter, so we normalise the riskless asset price to be $S_t^0 = 1$ for all $t \in [0, T]$.

As is well-known, in a complete financial market every contingent claim C can be perfectly replicated by a controlled portfolio of the traded securities: a portfolio with wealth process $X = (X_t)_{0 \leq t \leq T}$ exists satisfying $X_T = C$ a.s. In this case, no-arbitrage arguments imply the option price at time 0 is the initial value X_0 of the replicating portfolio, given by $X_0 = E^Q C$, where E^Q denotes expectation under the unique martingale measure Q of the complete market. The hedging strategy for a short position in the claim is simply to hold the replicating portfolio. The classical example of such a replication procedure is the Black-Scholes (BS) [3] option pricing model, where S is the price process of a single stock following a geometric Brownian motion.

By definition, an incomplete financial market is one in which not all claims C can be replicated. In this case writing a claim entails genuine risk, and the pricing and hedging of the claim can only be carried out by specifying the agent's preferences towards such risk. Classically, economists have done this by specifying the agent's utility function U. This was the inspiration for Hodges and Neuberger [12] to introduce the concept of utility indifference pricing, in the context of option pricing under transaction costs in the BS model. In this methodology, the agent seeks to solve for a claim price which, when incorporated into the initial wealth, results in a maximal expected utility when trading claims that is the same as the maximum utility in the absence of claims. The associated hedging strategy is the difference in the agent's optimal stock strategies with and without the random endowment of the claim payoff at the terminal time.

Utility indifference prices usually result in price bounds which mark the bid and ask prices at which a utility maximising investor would be prepared to buy or sell claims. They are nonlinear pricing rules (as we will see in later sections) in that the price for the claim nC ($n \in \mathbb{R}$) is not n times the price for 1 claim. The utility-based price bounds are invariably within the no-arbitrage price bounds, the latter usually being quite wide and therefore of limited practical use.

A marginal version of utility-based pricing, based on classical economic ideas of incremental pricing, was developed by Davis [4]. This gives the utility indifference price for an infinitesimal position in claims, the so-called marginal utility-based price (MUBP), which is (under fairly mild conditions) a unique price within the no-arbitrage interval, and within the bid and ask utility-based prices for a finite position in claims. The marginal price is given by a linear pricing rule, as we shall see. This linear pricing rule will emerge naturally from considering the dual problem to the agent's primal utility maximisation problem.

The utility indifference pricing technique has received much attention in the academic literature. The earliest applications were to transaction cost models. Following Hodges and Neuberger [12], Davis, Panas and Zariphopoulou [6] further developed the application to the BS model with proportional transaction costs. An asymptotic analysis of the Davis, Panas and Zariphopoulou [6] model, valid for small transaction costs, was carried out by Whalley and Wilmott [33]. Monoyios [22, 23] computed the MUBP and associated hedge in a BS model with proportional transaction costs. The marginal price depended on the agent's initial stock holding, and lay within bounds which marked the agent's indifference price for a single claim.

Karatzas and Kou [15] analysed the MUBP in continuous markets with Itô processes for stock prices, and with portfolio constraints. They showed how it nearly always lies within the no-arbitrage bounds. Subsequent developments were to more specific incomplete market scenarios. In particular, a number of papers studied so-called "basis risk" models, in which a claim on a non-traded asset is optimally hedged with a correlated traded asset. These models were studied by [5, 11, 24, 28] among others. In particular, Monoyios [24] showed how exponential utility-based hedging could outperform a "naive" hedge which took the traded asset as a perfect proxy for the non-traded asset. We shall describe basis risk models in some detail in Section 3..

Other applications of utility-based hedging have been to stochastic volatility models [32] and to the pricing of volatility derivatives [10]. The general theory of utility-based pricing, with a particular emphasis on relations with the dual to the primal utility maximisation problem, has been studied by Delbaen et al [7], by Becherer [1], and by Kramkov and co-authors [13, 20, 21].

The rest of the chapter is as follows. In the next section we present the theory of utility-based pricing and hedging in a general semi-martingale setting, and use results from the dual approach to optimal investment to derive probabilistic representations for utility-based prices. In Section 3. we apply the methodology to a basis risk model with full information, in which a claim on a non-traded asset is hedged with a correlated traded asset. The resulting optimal hedge requires knowledge of the asset price drifts, which are virtually impossible to estimate accurately. This motivates Section 4., in which drift parameter uncertainty is acknowledged by modelling the drifts as random variables with a prior distribution, which is updated using a Kalman filter. The resulting model becomes a full information model with random drifts, and we derive representations for utility-based prices and hedges in this case. The resulting hedge is approximated using the MUBP as a pricing rule, and shown to out-perform the naive approach of taking the traded asset as a perfect proxy for the non-traded one. We end with some concluding remarks.

2. Utility-Based Pricing and Hedging: The General Set-up

Here we present the theory of utility-based pricing in a fairly general semi-martingale setting, with frictionless markets, meaning no transaction costs on trading of basic securities such as stocks.

We start with a locally bounded vector semi-martingale $S = (S_t)_{0\le t\le T}$ on a filtered probability space $(\Omega, \mathcal{F}, \mathbb{F}, P)$. The filtration $\mathbb{F} := (\mathcal{F}_t)_{0\le t\le T}$ satisfies the usual conditions, and we take $\mathcal{F}_T = \mathcal{F}$.

There are $d+1$ assets: one savings account and d stocks, so we write $S = (S^0, \ldots, S^d)$, with each S^i a locally bounded semi-martingale, $S^i = (S^i_t)_{0\le t\le T}$. The zeroth asset is riskless, with $S^0_t = 1$ for all $t \in [0, T]$, so we are in addition assuming the interest rate is zero, as we shall do throughout this chapter.

A probability measure Q is called an equivalent martingale measure if it is equivalent to P and if S is a local martingale under Q. We denote by $\mathcal{M}$ the family of all such measures and assume that

$$\mathcal{M} \neq \emptyset.$$

This condition is essentially equivalent to the absence of arbitrage opportunities in the market; see Delbaen and Schachermayer [8] for precise statements and more details.

A self-financing portfolio is defined by a pair (x, H) where $x \in \mathbb{R}$ defines the initial capital and $H = (H^i)_{i=1}^d$ is a predictable and S-integrable process specifying the number of shares of each of the stocks held in the portfolio. Hence, for each $i = 1, \ldots, d$, $H^i = (H^i_t)_{0 \leq t \leq T}$. The value process of a self-financing portfolio evolves in time as the stochastic integral of the process H with respect to the stock price:

$$X_t := x + (H \cdot S)_t = x + \int_0^t H_s dS_s, \quad t \in [0, T].$$

(Here, $\int_0^t H_s dS_s \equiv \sum_{i=1}^d \int_0^t H^i_s dS^i_s$.) The market is, in general, incomplete, so not every contingent claim C can be replicated by a self-financing portfolio. Then there is no unique martingale measure, and the possible no-arbitrage prices span an interval given by

$$\left(\inf_{Q \in \mathcal{M}} E^Q C, \, \sup_{Q \in \mathcal{M}} E^Q C \right). \tag{1}$$

This was shown by Kramkov [18]. When the market is complete, there exists a unique self-financing portfolio satisfying $X_T = C$. In this case we say the strategy H replicates C, and H is the unique hedging strategy for the claim. In this case, there is a unique martingale measure Q, the interval (1) reduces to a single point, and the no-arbitrage price of the claim at time 0 is $p_0^{\mathrm{NA}} := E^Q C$.

In an incomplete market, one is faced with choosing one of the martingale measures $Q \in \mathcal{M}$ as a pricing measure. At first sight, this choice appears to have little to do with optimal investment. But the incompleteness means that selling a claim entails opening oneself up to non-zero terminal risk, as represented by the difference $X_T - C$, where X_T is the terminal wealth of any self-financing portfolio. The question arises as to how one should deal with the residual risk $X_T - C$. This can only be answered by specifying the risk preferences of the financial agent selling the claim. This will be done via a utility function. We shall see that a possible pricing measure then emerges naturally, via the dual to a primal utility maximisation problem.

We consider an economic agent whose preferences over terminal wealth are represented by a utility function $U : \mathbb{R} \to \mathbb{R}$, which is assumed to be strictly increasing, strictly concave, continuously differentiable, and is assumed to satisfy the Inada conditions:

$$U'(-\infty) = \lim_{x \to -\infty} U'(x) = \infty, \quad U'(\infty) = \lim_{x \to \infty} U'(x) = 0.$$

The utility function that we shall employ in this chapter is the exponential utility function:

$$U(x) = -\exp(-\alpha x), \quad \alpha > 0, \tag{2}$$

with constant risk aversion parameter α.

The convex conjugate of the agent's utility function is defined to be the Legendre transform of the convex function $-U(-\cdot)$. That is,

$$V(\eta) := \sup_{x \in \mathbb{R}}[U(x) - x\eta], \quad \eta > 0. \tag{3}$$

The conjugate function is then a continuously differentiable, strictly decreasing and strictly convex function satisfying $-V'(-\infty) = \infty$, $V'(\infty) = 0$.

The supremum in (3) is achieved by $x = x^*$ satisfying

$$U'(x^*) = \eta \Leftrightarrow x^* = I(\eta), \tag{4}$$

where I is the inverse of U'. Then $V(\eta)$ may be written as

$$V(\eta) = U[I(\eta)] - \eta I(\eta). \tag{5}$$

Further, from (3) we have the inequality

$$V(\eta) \geq U(x) - x\eta, \text{ with equality iff } \quad x = x^* \quad \text{such that } U'(x^*) = \eta. \tag{6}$$

Also, differentiating (5) gives

$$V'(\eta) = -I(\eta), \tag{7}$$

so that the agent's marginal utility is the inverse of minus the gradient of the convex conjugate:

$$U'(-V'(x)) = x.$$

We note that the defining duality relation (3) is equivalent to the bidual relation

$$U(x) = \inf_{\eta > 0}[V(\eta) + x\eta], \quad x \in \mathbb{R}, \tag{8}$$

since this gives that the value of η achieving the above infimum is η^* satisfying

$$V'(\eta^*) = -x,$$

or, by (7),

$$I(\eta^*) = x \Leftrightarrow U'(x) = \eta^*,$$

which is (4). Note also that the bidual relation (8) implies the inequality (6).

For the exponential utility function, the convex conjugate is given by

$$V(\eta) = \frac{\eta}{\alpha}\left(\log\left(\frac{\eta}{\alpha}\right) - 1\right). \tag{9}$$

For a martingale measure $Q \in \mathcal{M}$, define the relative entropy $\mathcal{H}(Q, P)$ between Q and P by

$$\mathcal{H}(Q, P) := E\left[\frac{dQ}{dP}\log\frac{dQ}{dP}\right].$$

This quantity will play a role in the dual to a primal utility maximisation problem under exponential utility. Denote the set of martingale measures with finite relative entropy by $\mathcal{M}_f$.

For an agent with exponential utility function, we we follow Becherer [1] and define the set $\mathcal{A}$ of admissible strategies H by

$$\mathcal{A} := \{H : (H \cdot S) \text{ is a } Q\text{-martingale for all } Q \in \mathcal{M}_f\}. \tag{10}$$

Other choices for $\mathcal{A}$ are possible. For instance, one may follow Schachermayer [31] and consider strategies with wealth bounded from below, and then maximise over the $L^1(P)$-closure of the set of all random variables $U(\Gamma)$ such that Γ can be super-hedged by some trading strategy. The good news is that these choices lead to the same solution for the dual to the primal utility maximisation problem [7, 14, 31].

For an initial capital x, the agent's primal value function is $u(x)$ defined by

$$u(x) := \sup_{H\in\mathcal{A}} EU(X_T) = \sup_{H\in\mathcal{A}} EU\left(x + \int_0^T H_t dS_t\right). \tag{11}$$

Define, for any $Q \in \mathcal{M}$, the change of measure martingale $Z = (Z_t)_{0\le t\le T}$ by

$$Z_t := \left.\frac{dQ}{dP}\right|_{\mathcal{F}_t}.$$

Define the dual value function $v(\eta)$ by

$$v(\eta) := \inf_{Q\in\mathcal{M}} EV(\eta Z_T). \tag{12}$$

We shall assume that this function is finite valued. A celebrated body of work has shown that the primal value function may be computed by solving the dual problem. This work culminated in the paper by Kramkov and Schachermayer [19], who considered semi-martingale processes S, following earlier work by several authors based on Itô price processes, notably Karatzas et al [16]. The monograph by Karatzas and Shreve [17] contains an authoritative account of this theory for continuous models based on Itô price processes. For an application of duality theory in two-factor incomplete markets, in which explicit representations for the dual minimiser are obtained, see Monoyios [25].

We assume that there is a unique dual optimiser attaining the infimum in (12). This will be the case in the models we tackle in subsequent sections. Denote the dual optimiser by Q^*, the associated change of measure martingale by Z^*, the optimal terminal wealth attaining the supremum in (11) by X_T^*, and the optimal trading strategy by H^*.

The main results of the dual approach to optimal investment are summarised in Theorem 1 below.

Theorem 1 *1. The primal and dual value functions $u(x)$ and $v(\eta)$ in (11) and (12) are conjugate:*

$$v(\eta) = \sup_{x\in\mathbb{R}}[u(x) - x\eta], \qquad u(x) = \inf_{\eta>0}[v(\eta) + x\eta],$$

so that $u'(x) = \eta$ (equivalently, $v'(\eta) = -x$);

2. *The optimal terminal wealth X_T^* and optimal dual minimiser Q^* are unique and related by*

$$U'(X_T^*) = \eta \frac{dQ^*}{dP}, \quad equivalently, \quad X_T^* = I\left(\eta \frac{dQ^*}{dP}\right);$$

3. *The following properties for $u'(x)$ and $v'(\eta)$ hold true:*

$$u'(x) = EU'(X_T^*), \qquad v'(\eta) = E^{Q^*} V'\left(\eta \frac{dQ^*}{dP}\right). \tag{13}$$

For a proof of this theorem, see [19].

To define utility-based prices, we introduce a utility maximisation problem with a random terminal endowment involving a claim C. Define the primal value function with random endowment, $u^{(n)}$, as the maximum expected utility when receiving a terminal payoff of n claims:

$$u^{(n)}(x) := \sup_{H \in \mathcal{A}} EU(X_T + nC) = \sup_{H \in \mathcal{A}} EU\left(x + \int_0^T H_t dS_t + nC\right). \tag{14}$$

Clearly, the value function $u(x)$ of (11) coincides with $u^{(0)}(x)$.

Define the dual problem to (14), with value function $v^{(n)}$, by

$$v^{(n)}(\eta) := \inf_{Q \in \mathcal{M}} E\left[V(\eta Z_T) + \eta Z_T nC)\right]. \tag{15}$$

Clearly, the value function $v(\eta)$ of (12) coincides with $v^{(0)}(\eta)$.

Denote the dual optimiser in (15) by $Q^{*,n}$, the associated change of measure martingale by $Z^{*,n}$, the optimal terminal wealth attaining the supremum in (14) by $X_T^{*,n}$, and the optimal trading strategy by $H^{*,n}$. Then the corresponding optimal quantities of the problems (11,12) without claims are given by $Q^* \equiv Q^{*,0}$, and similarly for Z^*, X_T^*, H^*.

For the value functions $u^{(n)}, v^{(n)}$ involving a random terminal endowment, and with exponential utility, a similar duality result to Theorem 1 has been obtained by Delbaen et al [7].

Theorem 2 *Given an exponential utility function $U(x) = -\exp(-\alpha x)$, we have:*

1. *The value functions $u^{(n)}(x)$ and $v^{(n)}(\eta)$ of (14) and (15) are conjugate:*

$$v^{(n)}(\eta) = \sup_{x \in \mathbb{R}}[u^{(n)}(x) - x\eta], \qquad u^{(n)}(x) = \inf_{\eta > 0}[v^{(n)}(\eta) + x\eta],$$

so that $u_x^{(n)}(x) = \eta$ (equivalently, $v_\eta^{(n)}(\eta) = -x$);

2. *The optimal terminal wealth $X_T^{*,n}$ and optimal dual minimiser $Q^{*,n}$ are unique and related by*

$$U'(X_T^{*,n} + nC) = \eta \frac{dQ^{*,n}}{dP}, \quad equivalently, \quad X_T^{*,n} + nC = I\left(\eta \frac{dQ^{*,n}}{dP}\right); \tag{16}$$

3. *The following properties for $u_x^{(n)}(x)$ and $v_\eta^{(n)}(\eta)$ hold true:*

$$u_x^{(n)}(x) = EU'(X_T^{*,n} + nC), \qquad v_\eta^{(n)}(\eta) = E^{Q^{*,n}}\left[V'\left(\eta\frac{dQ^*}{dP}\right) + nC\right]. \quad (17)$$

We will use these theorems to derive properties of the indifference prices, the major objects of interest in this chapter, which we now define.

Definition 1 (Utility indifference price) *The* utility indifference price *per claim, $p^{(n)}$, for a random endowment of n claims, is defined by*

$$u^{(n)}(x - np^{(n)}) = u^{(0)}(x). \quad (18)$$

In other words, when issuing or purchasing claims, the agent ensures that the price per claim results in no loss of utility compared with the alternative strategy of not writing or buying any claims.

The optimal hedging strategy is defined as the difference of the optimal trading strategies with and without the random endowment of n claims.

Definition 2 (Optimal hedging strategy) *The optimal hedging strategy for n units of the claim is $H^{(\mathrm{H})} := (H_t^{(\mathrm{H})})_{0\le t\le T}$ given by*

$$H_t^{(\mathrm{H})} := H_t^{*,n} - H_t^{*,0}, \quad 0 \le t \le T.$$

The *marginal utility-based price* (MUBP), $\hat{p}$, of the claim, may be defined in a number of ways. One is that it corresponds to a price which solves (18) as $n \to 0$:

Definition 3 (Marginal price) *The marginal utility-based price of the claim at time 0 is $\hat{p}$ defined by*

$$\hat{p} := \lim_{n\to 0} p^{(n)}. \quad (19)$$

It is well known that with exponential utility the marginal price is also equivalent to the limit of the indifference price as risk aversion goes to zero. Under appropriate conditions (satisfied in this model) it is given by the expectation of the payoff under the optimal measure of the dual to the problem without the claim. We shall derive these results shortly.

Definition 3 is equivalent to the definition below, which was the original definition due to Davis [4]. Suppose one diverts an amount c of the initial capital into the purchase or sale of claims. If the unit price per claim is p then the number of claims traded is c/p. We consider the value function $u^{(c/p)}(x)$ and make the following definition.

Definition 4 (Marginal price [4]) *The* marginal price *$\hat{p}$ of the claim is the price which solves*

$$\left.\frac{\partial u^{(c/p)}}{\partial c}(x - c)\right|_{c=0} = 0. \quad (20)$$

In [4] it is shown that under suitable smoothness conditions $\hat{p}$ is given by the following theorem.

Theorem 3 *The marginal price is given by*

$$\hat{p} = \frac{E\left[U'\left(X_T^*\right)C\right]}{u'(x)}. \tag{21}$$

In the next subsection we shall show the equivalence of (21) and (19), by showing that both representations result in

$$\hat{p} = E^{Q^*}C = E^{Q^E}C,$$

where Q^E is the minimal entropy measure, which is the optimal dual measure Q^* of the dual problem (12) for exponential utility, and is defined by

$$Q^E := \arg\min_{Q \in \mathcal{M}} \mathcal{H}(Q, P).$$

2.1. Dual Representations for Exponential Utility-Based Prices

Using the fundamental duality results in Theorem 2 we can obtain the following representation for the primal value function $u^{(n)}$, from which dual representations for the indifference prices follow.

Theorem 4 *The value function $u^{(n)}(x)$ in (14) has the representation*

$$u^{(n)}(x) = -\exp\left[-\alpha x - \inf_{Q \in \mathcal{M}}\left(\mathcal{H}(Q, P) + \alpha n E^Q C\right)\right].$$

Proof Using the definition (15) of the dual value function $v^{(n)}$, and the formula (9) for the conjugate function corresponding to exponential utility, we have

$$v^{(n)}(\eta) = V(\eta) + \frac{\eta}{\alpha}\inf_{Q \in \mathcal{M}}\left(\mathcal{H}(Q, P) + \alpha n E^Q C\right). \tag{22}$$

The duality between $u^{(n)}$ and $v^{(n)}$ in Theorem 2 implies $v^{(n)}(\eta) = -x$, so (22) gives that $\eta > 0$ is given by

$$\eta = \alpha \exp\left[-\alpha x - \inf_{Q \in \mathcal{M}}\left(\mathcal{H}(Q, P) + \alpha n E^Q C\right)\right].$$

Using this representation for η in (16) we find that the optimal terminal wealth satisfies

$$\exp(-\alpha X_T^{*,n}) = Z_T^{*,n} \exp\left[-\alpha x - \inf_{Q \in \mathcal{M}}\left(\mathcal{H}(Q, P) + \alpha n E^Q C\right)\right].$$

The result then follows from $u^{(n)}(x) = EU(X_T^{*,n})$.

□

In particular, we have that the primal value function in the absence of claims is given by

$$u^{(0)}(x) = -\exp\left(-\alpha x - \mathcal{H}(Q^E, P)\right),$$

where Q^E is the minimal entropy measure, the optimal dual measure of the problem without claims.

We then have the following immediate corollary, from the definitions of the indifference price.

Corollary 1 *The indifference price $p^{(n)}$ is given by*

$$p^{(n)} = \frac{1}{\alpha n}\left[\inf_{Q\in\mathcal{M}}\left(\mathcal{H}(Q,P) + \alpha n E^Q C\right) - \mathcal{H}(Q^E, P)\right].$$

In particular, for $n > 0$, we have that the utility indifference bid price per claim for n claims is given by

$$p_{\text{bid}}^{(n)} := p^{(n)} = \inf_{Q\in\mathcal{M}}\left[E^Q C + \frac{1}{\alpha n}\left(\mathcal{H}(Q,P) - \mathcal{H}(Q^E,P)\right)\right], \quad n > 0. \tag{23}$$

From this, we see that, in the limit $\alpha \to \infty$, we obtain

$$\lim_{\alpha\to\infty} p_{\text{bid}}^{(n)} = \inf_{Q\in\mathcal{M}} E^Q C,$$

which is the lower bound of the no-arbitrage interval (1).

Similarly, for $n < 0$, say $n = -m$, with $m > 0$, we obtain that the utility indifference ask price per claim for m claims is given by

$$p_{\text{ask}}^{(m)} := p^{(-m)} = \sup_{Q\in\mathcal{M}}\left[E^Q C - \frac{1}{\alpha m}\left(\mathcal{H}(Q,P) - \mathcal{H}(Q^E,P)\right)\right], \quad m > 0. \tag{24}$$

In the limit $\alpha \to \infty$, we obtain

$$\lim_{\alpha\to\infty} p_{\text{ask}}^{(m)} = \sup_{Q\in\mathcal{M}} E^Q C,$$

which is the upper bound of the no-arbitrage interval (1). We conclude that the possible indifference prices for exponential utility span the entire no-arbitrage interval, as we vary the risk aversion parameter.

For the marginal price, we have the following equivalences.

Theorem 5 *The marginal price has the equivalent representations*

$$\hat{p} := \lim_{n\to 0} p^{(n)} = \lim_{\alpha\to 0} p^{(n)} = E^{Q^E} C, \tag{25}$$

and these coincide with the representation (21).

Proof First, (25) follows by letting $\alpha \to 0$ or $n \downarrow 0$ in (23) and letting $\alpha \to 0$ or $m \downarrow 0$ in (24). The equivalence with (21) follows from Theorem 1, which implies that

$$\frac{U'(X_T^*)}{u'(x)} = \frac{dQ^*}{dP},$$

and the fact that for exponential utility, $Q^* = Q^E$, the minimal entropy measure.

□

Theorem 5 confirms that the marginal indifference price is governed by a linear pricing rule, as given by an expectation of the payoff under the optimal dual measure.

3. Basis Risk Model

Here is a quintessential example of an incomplete market in which the ideas of utility-based pricing given in the previous section can be illustrated with great clarity and explicit solutions. A number of papers ([5,11,24,26,28] to name but a few) have studied such basis risk models.

The setting is a filtered probability space $(\Omega, \mathcal{F}, \mathbb{F} := (\mathcal{F}_t)_{0 \le t \le T}, P)$, where the filtration $\mathbb{F}$ is the P-augmentation of that generated by a two-dimensional Brownian motion $(B, B^{\perp})$. A traded stock price $S := (S_t)_{0 \le t \le T}$ follows a log-Brownian process given by

$$dS_t = \sigma S_t(\lambda dt + dB_t) =: \sigma S_t d\xi_t, \tag{26}$$

where $\sigma > 0$ and λ are known constants. For simplicity, the interest rate is taken to be 0. The process ξ in (26) defined by $d\xi_t := \lambda dt + dB_t$ will play a role as one component of an "observation process" in a partial information model in the next section, when λ will be treated as a random variable rather than as a known constant.

A non-traded asset price $Y := (Y_t)_{0 \le t \le T}$ follows the correlated log-Brownian motion

$$dY_t = \beta Y_t(\theta dt + dW_t) =: \beta Y_t d\zeta_t, \tag{27}$$

with $\beta > 0$ and θ known constants. The Brownian motion W is correlated with B according to

$$d[B, W]_t = \rho dt, \quad W = \rho B + \sqrt{1 - \rho^2} B^{\perp}, \quad \rho \in [-1, 1],$$

and the process ζ, given by $d\zeta_t := \theta dt + dW_t$, will act as the second component of an observation process in a partial information model in the next section, when θ will be considered a random variable. We shall refer to the Sharpe ratios λ (respectively, θ) as the drift of S (respectively, Y), for brevity.

A European contingent claim pays the non-negative random variable $h(Y_T)$ at time T, where h is a bounded continuous function. If $|\rho| = 1$, the model is complete and a BS-style perfect hedge is possible (as we shall show). But for $|\rho| \neq 1$ the market is incomplete.

Examples of underlying assets that are either not traded (or are difficult to trade) include weather indices or baskets of many stocks. There is no tradeable asset which can be used to perfectly replicate the claim payoff. Traders may resort to using a correlated traded asset to hedge the claim, where the correlation is presumed to be close to 1, in effect taking the traded asset as a perfect proxy for the non-traded one. A typical case is the hedging of a basket option using a futures contract on a stock index, where the composition of the basket and the index are not identical.

The set $\mathcal{M}$ of local martingale measures Q is defined via the density process $Z = (Z_t)_{0\le t\le T}$ given by

$$Z_t := \mathcal{E}(-\lambda\cdot B - \psi\cdot B^{\perp})_t, \quad 0\le t\le T,$$

where $\mathcal{E}$ denotes the stochastic exponential, and $\psi = (\psi_t)_{0\le t\le T}$ is a process satisfying $\int_0^T \psi_t^2 dt < \infty$ a.s. If, in addition, Z is a martingale, then we may define probability measures Q equivalent to P by

$$\left.\frac{dQ}{dP}\right|_{\mathcal{F}_t} = Z_t.$$

The set $\mathcal{M}$ of martingale measures is then in one-to-one correspondence with the set of integrands ψ.

The minimal martingale measure Q^M, which will feature in many of our formulae, corresponds to $\psi = 0$, so has density process with respect to P given by

$$\left.\frac{dQ^M}{dP}\right|_{\mathcal{F}_t} = \mathcal{E}\left(-\lambda\cdot B\right)_t, \quad 0\le t\le T.$$

Under Q^M, (S, Y) follow

$$\begin{aligned} dS_t &= \sigma S_t dB_t^{Q^M}, \\ dY_t &= \beta Y_t\left[(\theta-\rho\lambda)\,dt + dW_t^{Q^M}\right], \end{aligned} \tag{28}$$

where B^{Q^M}, W^{Q^M} are correlated Brownian motions under Q^M. The stock price S is a local Q^M-martingale but this is not the case for the non-traded asset.

3.1. Perfect Correlation Case

In the perfect correlation case, $\rho = 1$, Y is effectively a traded asset, so no-arbitrage requires the Q^M-drift of Y to be 0. Therefore, given σ, β, in the $\rho = 1$ case the Sharpe ratios of the assets are related by

$$\theta = \lambda. \tag{29}$$

In fact, with $\rho = 1$, $W = B$, so we have

$$Y_t = Y_0\left(\frac{S_t}{S_0}\right)^{\beta/\sigma}\exp\left[\frac{1}{2}\beta(\sigma-\beta)\right].$$

We show that in this case the market becomes complete, and perfect hedging is possible. Let the claim price process be $P(t, Y_t), 0\le t\le T$, where $P : [0, T]\times\mathbb{R}^+ \to \mathbb{R}^+$ is smooth enough to apply the Itô formula, so that

$$dP(t, Y_t) = \left[P_t(t, Y_t) + \mathcal{A}_Y P(t, Y_t)\right]dt + \beta Y_t P_y(t, Y_t) dB_t,$$

where $\mathcal{A}_Y$ is the generator of the process Y in (27), and we have used the fact that $W = B$ if $\rho = 1$.

A self-financing portfolio with initial capital x and $(H_t)_{0\le t\le T}$ shares of stock at time $t \in [0,T]$ has wealth process $X = (X_t)_{0\le t\le T}$ given by

$$X_t = x + \int_0^t H_s dS_s.$$

To hedge a purchase of n claims, one replicates a position in $-n$ claims, so we require $X_T = -nh(Y_T)$ a.s., and in particular

$$X_t = -nP(t, Y_t), \quad 0 \le t \le T, \quad dX_t = -ndP(t, Y_t).$$

Then the perfect hedge is $H_t^{(\mathrm{P})}$ (the superscript P denoting "perfect") units of S at $t \in [0,T]$, where

$$H_t^{(\mathrm{P})} = -n\frac{\beta}{\sigma}\frac{Y_t}{S_t}\frac{\partial}{\partial y}P(t, Y_t), \tag{30}$$

and the claim pricing function $P(t,y)$ then satisfies

$$P_t(t,y) + \frac{1}{2}\beta^2 y^2 P_{yy}(t,y) = 0, \quad P(T,y) = h(y),$$

where we have used the no-arbitrage condition (29). This is the BS partial differential equation (PDE), so

$$P(t, Y_t) = \mathrm{BS}(t, Y_t; \beta),$$

where $\mathrm{BS}(t,y;\beta)$ denotes the BS formula at time t for underlying asset price y and volatility β. An important feature of (30) is that the perfect hedge does not require knowledge of the values of the drifts λ, θ.

3.2. Utility-Indifference Valuation and Hedging

Now suppose the correlation is not perfect, so that the market is incomplete. We embed the problem in a utility maximisation framework. Let the agent have risk preferences expressed via the exponential utility function (2). The agent maximises expected utility of terminal wealth at time T, with a random endowment of n units of claim payoff. Define $\pi = (\pi_t)_{0\le t\le T}$ as the process of wealth in the stock, so that $\pi_t := H_t S_t$. Given a starting time $t \in [0,T]$ the objective to be maximised is

$$J^{(n)}(t,x,y;\pi) = E[U(X_T + nh(Y_T))|X_t = x, Y_t = y].$$

The value function is $u^{(n)}(t,x,y)$, defined by

$$u^{(n)}(t,x,y) := \sup_{\pi\in\mathcal{A}} J^{(n)}(t,x,y;\pi), \tag{31}$$

$$u^{(n)}(T,x,y) = U(x + nh(y)). \tag{32}$$

Denote the optimal trading strategy that achieves the supremum in (31) by $\pi^{*,n}$, and denote the optimal wealth process by $X^{*,n}$.

We assume the random endowment $nh(Y_T)$ is bounded below. This ensures the maximum utility in (31) is well-defined.

The utility-based price at $t \in [0,T]$ is given by the analogue of Definition 1 for a starting time $t \in [0,T]$. Given $X_t = x, Y_t = y$, the price per claim is $p^{(n)}(t,x,y)$, defined by

$$u^{(n)}(t, x - np^{(n)}(t,x,y), y) = u^{(0)}(t,x,y).$$

We allow for possible dependence on t,x,y of $p^{(n)}$ in the above definition, but with exponential preferences it turns out that there is no dependence on x, as we shall see.

The optimal hedging strategy is given by Definition 2. In terms of the variable $\pi := HS$, we have that the optimal hedging strategy for n units of the claim is $\pi^{(\mathrm{H})} := (\pi_t^{(\mathrm{H})})_{0\le t\le T}$ given by

$$\pi_t^{(\mathrm{H})} := \pi_t^{*,n} - \pi_t^{*,0}, \quad 0 \le t \le T. \tag{33}$$

The solution to the optimisation problem (31) is well-known, using a so-called distortion transformation (see Zariphopoulou [34]) to linearise the Hamilton-Jacobi-Bellman (HJB) equation for $u^{(n)}$. See [24] for more details of the computation in this model.

The HJB equation for the value function $u^{(n)}$ is

$$u_t^{(n)} + \mathcal{A}_Y u^{(n)} - \frac{\left(\lambda u_x^{(n)} + \rho\beta y u_{xy}^{(n)}\right)^2}{2u_{xx}^{(n)}} = 0. \tag{34}$$

The optimal trading strategy $\pi^{*,n}$ is given by $\pi_t^{*,n} = \Pi^{*,n}(t, X_t^{*,n}, Y_t)$, where the function $\Pi^{*,n} : [0,T] \times \mathbb{R} \times \mathbb{R}^+$ is given by

$$\Pi^{*,n}(t,x,y) := -\left(\frac{\lambda u_x^{(n)} + \rho\beta y u_{xy}^{(n)}}{\sigma u_{xx}^{(n)}}\right). \tag{35}$$

We have the following representation for the value function and indifference price.

Proposition 1 *[24] The value function $u \equiv u^{(n)}$ and indifference price $p \equiv p^{(n)}$, given $X_t = x, Y_t = y$ for $t \in [0,T]$, are given by*

$$\begin{aligned} u^{(n)}(t,x,y) &= -e^{-\alpha x - \frac{1}{2}\lambda^2(T-t)}\left[F(t,Y)\right]^{1/(1-\rho^2)}, \\ F(t,y) &= E^{Q^M}\left[\exp\left(-\alpha(1-\rho^2)nh(Y_T)\right)\middle|\, Y_t = y\right], \\ p^{(n)}(t,y) &= -\frac{1}{\alpha(1-\rho^2)n}\log F(t,y). \end{aligned} \tag{36}$$

The function $F(t,y)$ satisfies a linear PDE by virtue of the stochastic representation (36) and the Feynman-Kac theorem. It is easy to verify that the value function as given in the proposition then satisfies the HJB equation (34). The indifference price formula then follows from its definition.

The indifference pricing function $p^{(n)}(t,y)$ satisfies

$$p_t^{(n)} + \beta(\theta - \rho\lambda)yp_y^{(n)} + \frac{1}{2}\beta^2 y^2 p_{yy}^{(n)} - \frac{1}{2}\beta^2 y^2 n\alpha(1-\rho^2)(p_y^{(n)})^2 = 0,$$

with $p^{(n)}(T,y) = h(y)$. This is a semi-linear PDE, and in this sense the indifference pricing methodology constitutes a non-linear pricing rule.

For $n = 0$, the indifference pricing PDE becomes linear, and by the Feynman-Kac Theorem we obtain the following representation for the marginal price $\hat{p}(t,y) := \lim_{n\to 0} p^{(n)}(t,y)$:

$$\hat{p}(t,y) = E^{Q^M}[h(Y_T)|Y_t = y].$$

This is a special case of the general representation in Theorem 5, since in the basis risk model the minimal entropy measure Q^E coincides with the minimal martingale measure Q^M. This is because the relative entropy between a martingale measure $Q \in \mathcal{M}$ and P is given by

$$\mathcal{H}(Q,P) = E^Q\left[\frac{1}{2}\left(\lambda^2 T + \int_0^T \psi_t^2 dt\right)\right],$$

and this is clearly minimised by $\psi = 0$.

Given the form of the value function, it is easy to show that the expression (35) for the optimal control loses its dependence on x. Then, applying Definition 2 gives the optimal hedging strategy for a position in n claims (see [24] for further details of this derivation).

Proposition 2 *The optimal hedging strategy for a position in n claims is to hold $H_t^{(\mathrm{H})}$ shares at $t \in [0,T]$, given by*

$$H_t^{(\mathrm{H})} = -n\rho \frac{\beta}{\sigma} \frac{Y_t}{S_t} \frac{\partial p^{(n)}}{\partial y}(t, Y_t). \tag{37}$$

We note that for $\rho = 1$ we recover the perfect delta hedge (30), and the claim price then satisfies the BS PDE.

3.3. Residual Risk Process

Suppose the agent trades n claims at time 0 for price $p^{(n)}(0,Y_0)$ per claim. The agent hedges this position over $[0,T]$ using the strategy $(H_t^{(\mathrm{H})})_{0\le t\le T}$. Her overall position has value process $R := (R_t)_{0\le t\le T}$ given by $R_t = X_t^{(\mathrm{H})} + np^{(n)}(t,Y_t)$, so that

$$dR_t = dX_t^{(\mathrm{H})} + ndp^{(n)}(t,Y_t), \tag{38}$$

where $X^{(\mathrm{H})} = (X_t^{(\mathrm{H})})_{0\le t\le T}$ is the value of the hedging portfolio in S, satisfying

$$\begin{aligned} dX_t^{(\mathrm{H})} &= H_t^{(\mathrm{H})} dS_t, \\ X_0^{(\mathrm{H})} &= -np^{(n)}(0,Y_0). \end{aligned}$$

Using this in (38) along with the Itô formula and the PDE satisfied by $p^{(n)}(t,y)$, we obtain

$$dR_t = \frac{1}{2}\beta^2 n^2 \alpha(1-\rho^2)Y_t^2\left(p_y^{(n)}\right)^2(t,Y_t)dt + \beta n\sqrt{1-\rho^2}Y_t p_y^{(n)}(t,Y_t)dB_t^{\perp}, \tag{39}$$

with $R_0 = 0$. We call R the *residual risk* (or hedging error) process. The term in $dB_t^{\perp}$, orthogonal to the Brownian increments driving the stock price, is interpreted as the unhedgeable component of risk. If $\rho = 1$ we see that the process R becomes riskless (recall that the interest rate is zero), reflecting the fact that the market incompleteness disappears in this case.

3.4. Power Series Expansions for the Indifference Price and Hedge

We are interested in analysing the distribution of the terminal hedging error R_T. This is not possible in closed form, so our approach is to use the SDE (39) to simulate R over many asset price histories, and compute the distribution of terminal hedging error R_T. This programme was carried out in [24] and [26].

To simulate R via (39) efficiently, one may use analytic approximations for $p^{(n)}(t,y)$ and $p_y^{(n)}(t,y)$, in the form of power series expansions in powers of $a := -\alpha(1-\rho^2)n$. These arise from a Taylor expansion of the indifference pricing function

$$p^{(n)}(t,y) = \frac{1}{a}\log E^{Q^M}\left[\exp\left(ah(Y_T)\right)\middle| Y_t = y\right]. \tag{40}$$

For a random variable X, recall that its cumulant generating function (CGF) is $\Psi_X(a) := \log E\exp(aX)$. Using linearity of the expectation operator, it is not hard to see that the CGF has a Taylor expansion of the form

$$\Psi_X(a) = \sum_{j=1}^{\infty}\frac{1}{j!}k_j(X)a^j,$$

where $k_j(X) \equiv k_j$ is the jth *cumulant* of X. The cumulants are related to the central moments of X. For instance, writing

$$m_j(X) := E(X^j), \quad \mu_j(X) := E[(X-m_1)^j], \quad j \in \mathbb{N},$$

for the jth raw and central moments, it is not hard to show that the first three cumulants are the mean, variance and skewness:

$$\begin{aligned} k_1(X) &= m_1(X), \\ k_2(X) &= \mu_2(X), \\ k_3(X) &= \mu_3(X). \end{aligned}$$

The first two cumulants being equal to the mean and variance implies that the low order approximation to the optimal hedge is a mean-variance hedging strategy, as pointed out by Kramkov and Sirbu [21].

Since the pricing function (40) is proportional to the cumulant generating function of the payoff under the minimal measure, it is easy to generate an analytic formula for the indifference pricing function. In [26], Monoyios gives the following representation.

Proposition 3 *The indifference pricing function $p^{(n)}(t,y)$ has the power series expansion*

$$p^{(n)}(t,y) = \sum_{j=1}^{5}\frac{1}{j!}k_j\left(h(Y_T)\right)a^{j-1} + O(a^5), \tag{41}$$

where $a = -\alpha(1-\rho^2)n$ and k_j is the jth cumulant of the payoff under Q^M, conditional on $Y_t = y$. The expansion is valid for model parameters satisfying

$$E^{Q^M}\left[\exp(ah(Y_T))\middle| Y_t = y\right] \le 2. \tag{42}$$

This means one can produce an accurate perturbation series for $p^{(n)}(t,y)$, as a series of BS-type formulae, which can be differentiated term by term to give an analytic approximation for $p_y^{(n)}(t,y)$. In particular, the leading order term in the price expansion is Davis' [4] marginal price. Once again, this shows how the non-linear pricing rule given by the indifference price reduces to the linear pricing rule of the marginal price as the number of claims tends to zero.

The terms in the expansion depend ultimately on the moments $m_j := E^{Q^M}[h^j(Y_T)|Y_t = y]$, $j \in \mathbb{N}$ and (in the case of $p_y^{(n)}(t,y)$) on their partial derivatives $\partial m_j := \partial m_j/\partial y$, $j \in \mathbb{N}$. These are easy to compute (in the case of a put option we give some results shortly) since, under Q^M, and conditional upon $Y_t = y$, $\log Y_T$ is normally distributed: with $\mathrm{N}(m, \Sigma^2)$ denoting the normal probability law with mean m and variance Σ^2, we have

$$\begin{aligned} \log Y_T &\sim \mathrm{N}\left(\log y + b - \tfrac{1}{2}\Sigma^2, \Sigma^2\right), \\ b &= \beta(\theta - \rho\lambda)(T-t), \\ \Sigma^2 &= \beta^2(T-t). \end{aligned} \tag{43}$$

For the optimal hedging strategy, the explicit results are obtained by differentiating (41) with respect to y. If we denote by $\partial\kappa_j$ the partial derivative of κ_j with respect to y:

$$\partial\kappa_j \equiv \frac{\partial\kappa_j}{\partial y},$$

where κ_j denotes any of m_j, μ_j, k_j, then we have:

Corollary 2 *The partial derivative of the indifference price $p^{(n)}(t,y)$ with respect to y has the power series expansion*

$$\frac{\partial p^{(n)}}{\partial y}(t,y) = \sum_{j=1}^{5} \frac{1}{j!}(\partial k_j)a^{j-1} + O(a^5).$$

The partial derivatives of the cumulants are related to $\mu_j, \partial\mu_j$. For instance, up to $j = 3$ we have

$$\begin{aligned} \partial k_1 &= \partial m_1, \\ \partial k_2 &= \partial\mu_2, \\ \partial k_3 &= \partial\mu_3. \end{aligned}$$

(See [26] for full details and more more formulae.)

The significance of the expansions is that they give easily computed closed form approximations for the indifference price and optimal hedge. In the specific case of a put option, we have the following formulae for the raw moments of the payoff under the minimal measure Q^M.

Lemma 1 *For a put option, $h(y) = (K - y)^+$, where $K > 0$ is the strike price, the jth moment $m_j := E^{Q^M}[h^j(Y_T)|Y_t = y]$, $j \in \mathbb{N}$, is given by*

$$m_j = \sum_{\ell=0}^{j} \binom{j}{\ell} (-y)^\ell K^{(j-\ell)} \exp\left[\ell\left(b + \frac{1}{2}(\ell-1)\Sigma^2\right)\right] \Phi(-d_1 - (\ell-1)\Sigma),$$

where $\Phi(\cdot)$ *denotes the standard cumulative normal distribution function, and where*

$$\begin{aligned} d_1 &= \frac{1}{\Sigma}\left[\log\left(\frac{y}{K}\right) + b + \frac{1}{2}\Sigma^2\right] \\ b &= \beta(\theta - \rho\lambda)(T-t), \\ \Sigma^2 &= \beta^2(T-t). \end{aligned}$$

Proof For the put payoff, we have, for $j \in \mathbb{N}$,

$$\begin{aligned} (h(Y_T))^j &= ((K-Y_T)^+)^j \\ &= (K-Y_T)^j I_{\{Y_T \le K\}} \\ &= \sum_{\ell=0}^{j} \binom{j}{\ell} (-1)^\ell K^{(j-\ell)} Y_T^\ell I_{\{Y_T \le K\}}, \end{aligned}$$

where $I_{\{Y_T \le K\}}$ denotes the indicator function of the event $\{Y_T \le K\}$. Given the lognormal distribution (43) of Y_T, it is easy to show that

$$E^{Q^M}\left[Y_T^\ell I_{\{Y_T \le K\}} \middle| Y_t = y\right] = y^\ell \exp\left(\ell\left(b + \frac{1}{2}(\ell-1)\Sigma^2\right)\right)\Phi(-d_1 - (\ell-1)\Sigma),$$

from which the result follows.

□

Lemma 2 *Let* $j \in \mathbb{N}$. *For a put option payoff,* $h(y) = (K-y)^+$, ∂m_j *is given by*

$$\partial m_j = -\sum_{\ell=1}^{j} \binom{j}{\ell} (-y)^{(\ell-1)} K^{(j-\ell)} \exp\left(\ell\left(b + \frac{1}{2}(\ell-1)\Sigma^2\right)\right) \ell N(-d_1 - (\ell-1)\Sigma).$$

Proof This is a straightforward exercise in differentiation.

□

This power series expansions for $p^{(n)}(t,y)$ and $p_y^{(n)}(t,y)$ give a closed form and extremely accurate (see [24]) computation of the optimal price and hedging strategy, with the leading order term in the expansion for the price being the marginal price, $\hat{p}(t,y) = E^{Q^M}[h(Y_T)|Y_t = y]$, of the claim.

3.5. Optimal versus Naive Hedging

In [24, 26], a comparison was made between hedging a claim with the optimal strategy versus with the BS-style "naive" strategy (30) which takes S as a good proxy for Y.

In the BS-style hedge, let us repeat the calculation leading to the residual risk SDE (39), but with the claim traded at the BS price $P(0, Y_0) = \mathrm{BS}(0, Y_0)$ per claim and hedged using the $\rho \to 1$ limit of hedging formula (even though true value of ρ is *not* equal to 1). We then obtain the "naive" hedging error process R^N, following

$$dR_t^N = n\beta Y_t(\theta - \lambda)P_y(t, Y_t)dt + n\beta Y_t P_y(t, Y_t)[(\rho - 1)dB_t + \sqrt{1-\rho^2}dB_t^\perp].$$

Once again, we note that this is not riskless, but becomes so if the true value of ρ is indeed 1. The "naive" trader hopes this proves a good approximation.

For the case when the agent sells a put option ($n = -1$) on the non-traded asset, in [26] Monoyios generated optimal and naive hedging error distributions using 10,000 asset price histories. These showed that the optimal hedge error distribution has a higher mean, lower standard deviation, and a higher median, than the naive hedge error distribution. The increased median, in particular, showed how the relative frequency of profits over losses is increased when hedging optimally. We shall see some examples of this type of simulation in the next section, in the context of a partial information model.

Thus, the hedging strategy in (37) is, at first sight, superior to the BS-style hedge (30) But from (28) we see that the exponential hedge requires knowledge of λ, θ, which are impossible to estimate with any degree of accuracy (see Rogers [30] or Monoyios [26]). This can ruin the effectiveness of indifference hedging, as shown in [26].

It is therefore difficult to draw any meaningful conclusions on the effectiveness of utility-based hedging in this model without relaxing the assumption that the agent knows the true values of the drifts. This is the subject of the next section.

4. Partial Information Basis Risk Model

In the basis risk model of the previous section, we now assume the hedger does not know the values of the return parameters λ, θ, so these are considered to be random variables. Equivalently, the agent cannot observe the Brownian motions B, W driving the asset prices, so is required to use strategies adapted to the observation filtration $\hat{\mathbb{F}}$ generated by asset returns. This is a *partial information* scenario.

Partial information problems under various scenarios have been studied by a number of authors, usually in the context of optimal investment. Examples include Rogers [30], Björk, Davis and Landén [2], Nagai and Peng [29], and Monoyios [27], who treats the case of exponential hedging of basis risk, the subject of this section.

4.1. Choice of Prior

We shall take the the two-dimensional random variable

$$U := \begin{pmatrix} \lambda \\ \theta \end{pmatrix}$$

to have a Gaussian distribution which will be updated as the agent attempts to filter the values of the drifts from asset observations during the hedging interval $[0, T]$.

The choice of Gaussian prior is motivated by the idea that the agent has some past observations of S, Y before time 0, uses these to obtain classical point estimates of the drifts, and the joint distribution of the estimators is used as the prior in a Bayesian framework. Ultimately, in order to obtain explicit solutions, we shall assume that the agent uses observations before time 0 of equal length for both assets. Throughout, we make the approximation that the asset price observations are continuous, so that σ, β, ρ are known from the quadratic variation and co-variation of S, Y. This is because our goal here is to focus on the severest problem of drift parameter uncertainty.

Consider an observer with data for S over a time interval of length t_S, and for Y over a window of length t_Y, who considers λ and θ as *constants*, and records the returns dS_t/S_t and dY_t/Y_t in order to estimate the values of the drifts. The best estimator of, say, λ is $\bar{\lambda}(t_S)$ given by

$$\begin{aligned}\bar{\lambda}(t_S) &= \frac{1}{t_S}\int_{t_0}^{t_0+t_S}\frac{dS_u}{S_u} \\ &= \lambda + \frac{B_{t_0+t_S}}{t_S} \\ &\sim \mathrm{N}\left(\lambda, \frac{1}{t_S}\right),\end{aligned}$$

The estimator of λ is normally distributed, with a similar computation for the estimator of θ. The estimator, $(\bar{\lambda}, \bar{\theta})$, of the (supposed constant) vector (λ, θ) is bivariate normal. Defining $\mathrm{v}_0 := 1/t_S$ and $\mathrm{w}_0 := 1/t_Y$ it is easily checked that

$$\begin{pmatrix}\bar{\lambda} \\ \bar{\theta}\end{pmatrix} \sim \mathrm{N}(M, C),$$

where the mean vector M and covariance matrix C are given by

$$M = \begin{pmatrix}\lambda \\ \theta\end{pmatrix}, \quad C = \begin{pmatrix}\mathrm{v}_0 & \rho\min(\mathrm{v}_0, \mathrm{w}_0) \\ \rho\min(\mathrm{v}_0, \mathrm{w}_0) & \mathrm{w}_0\end{pmatrix}. \tag{44}$$

With this in mind, we shall suppose that (λ, θ), now considered as a *random variable*, is bivariate normal according to

$$\lambda \sim \mathrm{N}(\lambda_0, \mathrm{v}_0), \quad \theta \sim \mathrm{N}(\theta_0, \mathrm{w}_0), \quad \mathrm{cov}(\lambda, \theta) = \mathrm{c}_0 := \rho\min(\mathrm{v}_0, \mathrm{w}_0).$$

This distribution will be updated via subsequent observations of

$$\xi_t := \frac{1}{\sigma}\int_0^t \frac{dS_u}{S_u} = \lambda t + B_t, \qquad \zeta_t := \frac{1}{\beta}\int_0^t \frac{dY_u}{Y_u} = \theta t + W_t$$

over the hedging interval $[0, T]$.

4.2. Kalman-Bucy Filter

We are firmly within the realm of a two-dimensional Kalman filtering problem, which we treat as follows. Define the observation filtration by

$$\hat{\mathbb{F}} := (\hat{\mathcal{F}}_t)_{0\le t\le T}, \qquad \hat{\mathcal{F}}_t = \sigma(\xi_s, \zeta_s; 0 \le s \le t).$$

The *observation process*, O, and unobservable *signal process*, U, are defined by

$$O := \begin{pmatrix}\xi_t \\ \zeta_t\end{pmatrix}_{0\le t\le T}, \qquad U := \begin{pmatrix}\lambda \\ \theta\end{pmatrix},$$

satisfying the stochastic differential equations

$$dO_t = Udt + Dd\mathbf{B}_t, \qquad dU = \begin{pmatrix} 0 \\ 0 \end{pmatrix},$$

where

$$D = \begin{pmatrix} 1 & 0 \\ \rho & \sqrt{1-\rho^2} \end{pmatrix}, \qquad \mathbf{B}_t = \begin{pmatrix} B_t \\ B_t^{\perp} \end{pmatrix}.$$

The optimal filter is $\hat{U}_t := E[U|\hat{\mathcal{F}}_t], 0 \le t \le T$, a two-dimensional process defining the best estimates of λ and θ given observations up to time $t \in [0, T]$:

$$\hat{U}_t \equiv \begin{pmatrix} \hat{\lambda}_t \\ \hat{\theta}_t \end{pmatrix} := \begin{pmatrix} E[\lambda|\hat{\mathcal{F}}_t] \\ E[\theta|\hat{\mathcal{F}}_t] \end{pmatrix}, \qquad \begin{pmatrix} \hat{\lambda}_0 \\ \hat{\theta}_0 \end{pmatrix} = \begin{pmatrix} \lambda_0 \\ \theta_0 \end{pmatrix}. \tag{45}$$

The solution to this filtering problem converts the partial information model to a full information model with random drifts, given in the following proposition. To avoid a proliferation of different symbols, we abuse notation and write, for example, $\hat{\lambda}_t \equiv \hat{\lambda}(t, S_t)$ when a process $\hat{\lambda}$ is a function of time and current stock price.

Proposition 4 *The partial information model is equivalent to a full information model in which the asset price dynamics in the observation filtration $\hat{\mathbb{F}}$ are*

$$dS_t = \sigma S_t(\hat{\lambda}_t dt + d\hat{B}_t), \tag{46}$$
$$dY_t = \beta Y_t(\hat{\theta}_t dt + d\hat{W}_t), \tag{47}$$

where $\hat{B}, \hat{W}$ are $\hat{\mathbb{F}}$-Brownian motions with correlation ρ, and the random drifts $\hat{\lambda}, \hat{\theta}$ are $\hat{\mathbb{F}}$-adpated processes.

If λ and θ have common initial variance v_0, then $\hat{\lambda}, \hat{\theta}$ are given by

$$\begin{pmatrix} \hat{\lambda}_t \\ \hat{\theta}_t \end{pmatrix} = \begin{pmatrix} \lambda_0 \\ \theta_0 \end{pmatrix} + \int_0^t \mathrm{v}_u \begin{pmatrix} d\hat{B}_s \\ d\hat{W}_u \end{pmatrix}, \quad 0 \le t \le T, \tag{48}$$

where $(\mathrm{v}_t)_{0 \le t \le T}$ is the deterministic function

$$\mathrm{v}_t = \frac{\mathrm{v}_0}{1 + \mathrm{v}_0 t}.$$

Equivalently, $\hat{\lambda}, \hat{\theta}$ are given as functions of time and current asset price by

$$\hat{\lambda}_t = \hat{\lambda}(t, S_t) = \frac{\lambda_0 + \mathrm{v}_0 \xi_t}{1 + \mathrm{v}_0 t}, \qquad \hat{\theta}_t = \hat{\theta}(t, Y_t) = \frac{\theta_0 + \mathrm{v}_0 \zeta_t}{1 + \mathrm{v}_0 t}, \tag{49}$$

with

$$\xi_t = \frac{1}{\sigma} \log\left(\frac{S_t}{S_0}\right) + \frac{1}{2}\sigma t, \qquad \zeta_t = \frac{1}{\beta} \log\left(\frac{Y_t}{Y_0}\right) + \frac{1}{2}\beta t. \tag{50}$$

Proof By the Kalman-Bucy filter, for example Theorem V.9.2 in Fleming and Rishel [9], $\hat{U}$ satisfies the stochastic differential equation

$$d\hat{U}_t = V_t \left(DD^T\right)^{-1} (dO_t - \hat{U}_t dt) =: V_t \left(DD^T\right)^{-1} dN_t, \tag{51}$$

where $(N_t)_{0\le t\le T}$ is the innovations process, defined by

$$\begin{aligned} N_t &:= O_t - \int_0^t \hat{U}_s ds \\ &= \begin{pmatrix} \xi_t - \int_0^t \hat{\lambda}_s ds \\ \zeta_t - \int_0^t \hat{\theta}_s ds \end{pmatrix} \\ &=: \begin{pmatrix} \hat{B}_t \\ \hat{W}_t \end{pmatrix}, \end{aligned} \tag{52}$$

and classical filtering theory implies that $\hat{B}, \hat{W}$ are $\hat{\mathbb{F}}$-Brownian motions with correlation ρ. The deterministic matrix function V is the conditional variance-covariance matrix defined by

$$V_t := E\left[(U - \hat{U}_t)(U - \hat{U}_t)^{\mathrm{T}} \middle| \hat{\mathcal{F}}_t\right] = E\left[(U - \hat{U}_t)(U - \hat{U}_t)^{\mathrm{T}}\right],$$

(T denoting transpose) where the last equality follows because the error $U - \hat{U}_t$ is independent of $\hat{\mathcal{F}}_t$ (Theorem V.9.2 in [9] again).

Using (52), and writing dS_t in terms of $d\xi_t$, as in (26), gives the dynamics (46) of S in the observation filtration; (47) is established similarly.

The matrix $V = (V_t)_{0\le t\le T}$ satisfies the Riccati equation

$$\frac{dV_t}{dt} = -V_t \left(DD^T\right)^{-1} V_t, \qquad V_0 = C,$$

where C is the covariance matrix in (44). Then $R_t := V_t^{-1}$ satisfies the Lyapunov equation

$$\frac{dR_t}{dt} = \left(DD^T\right)^{-1}.$$

Define the elements of the conditional covariance matrix by

$$V_t =: \begin{pmatrix} \mathrm{v}_t & \mathrm{c}_t \\ \mathrm{c}_t & \mathrm{w}_t \end{pmatrix}.$$

Then the filtering equation (51) is a pair of coupled stochastic differential equations:

$$\begin{aligned} \begin{pmatrix} d\hat{\lambda}_t \\ d\hat{\theta}_t \end{pmatrix} &= \frac{1}{1-\rho^2} \begin{pmatrix} \mathrm{v}_t - \rho\mathrm{c}_t & \mathrm{c}_t - \rho\mathrm{v}_t \\ \mathrm{c}_t - \rho\mathrm{w}_t & \mathrm{w}_t - \rho\mathrm{c}_t \end{pmatrix} \begin{pmatrix} d\xi_t - \hat{\lambda}_t dt \\ d\zeta_t - \hat{\theta}_t dt \end{pmatrix} \\ &= \frac{1}{1-\rho^2} \begin{pmatrix} \mathrm{v}_t - \rho\mathrm{c}_t & \mathrm{c}_t - \rho\mathrm{v}_t \\ \mathrm{c}_t - \rho\mathrm{w}_t & \mathrm{w}_t - \rho\mathrm{c}_t \end{pmatrix} \begin{pmatrix} d\hat{B}_t \\ d\hat{W}_t \end{pmatrix}. \end{aligned}$$

Solving the Lyapunov equation yields 3 equations for $\mathrm{v}_t, \mathrm{w}_t, \mathrm{c}_t$:

$$\begin{aligned} \frac{\mathrm{v}_t}{\mathrm{v}_t\mathrm{w}_t - \mathrm{c}_t^2} - \frac{\mathrm{v}_0}{\mathrm{v}_0\mathrm{w}_0 - \mathrm{c}_0^2} &= \frac{t}{1-\rho^2}, \\ \frac{\mathrm{w}_t}{\mathrm{v}_t\mathrm{w}_t - \mathrm{c}_t^2} - \frac{\mathrm{w}_0}{\mathrm{v}_0\mathrm{w}_0 - \mathrm{c}_0^2} &= \frac{t}{1-\rho^2}, \\ \frac{\mathrm{c}_t}{\mathrm{v}_t\mathrm{w}_t - \mathrm{c}_t^2} - \frac{\mathrm{c}_0}{\mathrm{v}_0\mathrm{w}_0 - \mathrm{c}_0^2} &= \frac{\rho t}{1-\rho^2}, \end{aligned} \tag{53}$$

where we have written $\mathrm{c}_0 \equiv \rho \min(\mathrm{v}_0, \mathrm{w}_0)$ for brevity.

Now make the simplification $\mathrm{w}_0 = \mathrm{v}_0$. From the discussion on Section 4.1., we see that this corresponds to using past observations over the same length of time, $t_S = t_Y$, for both S and Y in fixing the prior. Then $\mathrm{c}_0 = \rho\mathrm{v}_0$, and the solution to the system of equations (53) gives the entries of the matrix V_t as

$$\mathrm{v}_t = \frac{\mathrm{v}_0}{1+\mathrm{v}_0 t}, \quad \mathrm{w}_t = \mathrm{v}_t, \quad \mathrm{c}_t = \rho\mathrm{v}_t.$$

With this simplification, the equation for the optimal filter simplifies to

$$\begin{pmatrix} d\hat{\lambda}_t \\ d\hat{\theta}_t \end{pmatrix} = \mathrm{v}_t \begin{pmatrix} d\xi_t - \hat{\lambda}_t dt \\ d\zeta_t - \hat{\theta}_t dt \end{pmatrix} = \mathrm{v}_t \begin{pmatrix} d\hat{B}_t \\ d\hat{W}_t \end{pmatrix}, \tag{54}$$

which, along with the initial condition in (45), yields (48) and (49).

Finally, the expression in (50) for ξ_t follows easily from the solution of (26) for S:

$$\log\left(\frac{S_t}{S_0}\right) = \sigma\left(\xi_t - \frac{1}{2}\sigma^2 t\right),$$

and a similar calculation gives the formula for ζ_t.

□

With Proposition 4 we may now treat the model as a full information model with random drift parameters $(\hat{\lambda}_t, \hat{\theta}_t)$, and this is done in the next section.

4.3. Indifference Hedging with Random Drifts

On the stochastic basis $(\Omega, \hat{\mathcal{F}}, \hat{\mathbb{F}}, P)$, the wealth process associated with trading strategy $\pi := (\pi_t)_{0\le t\le T}$, an $\hat{\mathbb{F}}$-adapted process satisfying $\int_0^T \pi_t^2 dt < \infty$ a.s., is $X^\pi \equiv X := (X_t)_{0\le t\le T}$, satisfying

$$dX_t = \sigma\pi_t(\hat{\lambda}_t dt + d\hat{B}_t). \tag{55}$$

We use an exponential utility function, $U(x) = -\exp(-\alpha x)$, $x \in \mathbb{R}$, $\alpha > 0$. The primal value function $u^{(n)}$ is defined once again as the maximum expected utility of wealth at T from trading S and receiving n units of the claim on Y, when starting at time $t \in [0,T]$:

$$u^{(n)}(t,x,s,y) := \sup_{\pi\in\mathcal{A}} E[U(X_T + nh(Y_T))|X_t = x, S_t = s, Y_t = y], \tag{56}$$

where $\mathcal{A}$ denotes the set of admissible trading strategies. The dynamics of the state variables X, S, Y are given by (55) and (46,47). The set of admissible strategies is defined as in (10). Once again denote the optimal trading strategy by $\pi^* \equiv \pi^{*,n}$, and the optimal wealth process by $X^* \equiv X^{*,n}$.

The utility indifference price and hedge for a position in n claims are defined in the classical manner as earlier. In this case, we have that the indifference price per claim at $t \in [0,T]$, given $X_t = x, S_t = s, Y_t = y$, is $p^{(n)}$ given by

$$u^{(n)}(t, x - np^{(n)}(t,x,s,y), s, y) = u^{(0)}(t,x,s).$$

The optimal hedging strategy, $\pi^{\mathrm{H}} := \left(\pi_t^{\mathrm{H}}\right)_{0 \le t \le T}$, is defined by

$$\pi_t^{\mathrm{H}} := \pi_t^{*,n} - \pi_t^{*,0}, \quad 0 \le t \le T.$$

As before, with exponential utility the indifference price will not in fact depend on the initial cash wealth x, so we shall write $p^{(n)}(t,x,s,y) \equiv p^{(n)}(t,s,y)$ from now on.

For small positions in the claim, the marginal utility-based price of the claim at $t \in [0,T]$ is $\hat{p}(t,s,y)$ defined by

$$\hat{p}(t,s,y) := \lim_{n \to 0} p^{(n)}(t,s,y).$$

We have seen that with exponential utility the marginal price is also equivalent to the limit of the indifference price as risk aversion goes to zero. Under appropriate conditions (satisfied in this model) it is given by the expectation of the payoff under the optimal measure of the dual to the problem without the claim. As we shall see in the next section, in our case this measure will be the minimal martingale measure Q^M, and we shall obtain the representation $\hat{p}(t,s,y) = E^{Q^M}[h(Y_T)|S_t = s, Y_t = y]$.

4.4. The Dual Problem

We shall attack the primal utility maximisation problem (56) via its dual problem.

The class $\mathcal{M}$ of local martingale measures for this model are measures Q with density processes defined by

$$Z_t := \left.\frac{dQ}{dP}\right|_{\mathcal{F}_t} = \mathcal{E}(-\hat{\lambda} \cdot \hat{B} - \psi \cdot \hat{B}^{\perp})_t, \quad 0 \le t \le T,$$

for integrands ψ satisfying $\int_0^T \psi_t^2 dt < \infty$ a.s. ($\int_0^T \hat{\lambda}_t^2 dt < \infty$ is not hard to show). For $\psi = 0$ we obtain the minimal martingale measure Q^M, with density process $Z_t^{Q^M} = \mathcal{E}(-\hat{\lambda} \cdot \hat{B})_t$, for $t \in [0,T]$.

The change of measure density $Z_t^{Q^M}$ satisfies $dZ_t^{Q^M} = -\hat{\lambda}_t Z_t^{Q^M} d\hat{B}_t$ and, since $\hat{\lambda}_t = \hat{\lambda}(t, S_t)$ is a function of S_t, then so is $Z_t^{Q^M}$. The relevance of this is that the dual value function will be a function of the current asset prices at any initial time, as we shall see.

Under $Q \in \mathcal{M}$, $(\hat{B}^Q, \hat{B}^{\perp,Q})$ is two-dimensional Brownian motion, where

$$d\hat{B}_t^Q := d\hat{B}_t^Q + \hat{\lambda}_t dt, \qquad d\hat{B}_t^{\perp,Q} := d\hat{B}_t^{\perp} + \psi_t dt.$$

Further, under $Q \in \mathcal{M}$, the asset prices and random drifts satisfy (with $\bar{\rho} := \sqrt{1-\rho^2}$)

$$\begin{aligned} dS_t &= \sigma S_t d\hat{B}_t^Q, \\ dY_t &= \beta Y_t[(\hat{\theta}_t - \rho\hat{\lambda}_t - \bar{\rho}\psi_t)dt + d\hat{W}_t^Q], \\ d\hat{\lambda}_t &= \mathrm{v}_t[-\hat{\lambda}_t dt + d\hat{B}_t^Q], \\ d\hat{\theta}_t &= \mathrm{v}_t[-(\rho\hat{\lambda}_t + \bar{\rho}\psi_t)dt + d\hat{W}_t^Q], \end{aligned}$$

where $\hat{W}^Q = \rho\hat{B}^Q + \bar{\rho}\hat{B}^{\perp,Q}$.

The *dual value function* $v \equiv v^{(n)}$ is defined by

$$v^{(n)}(t,\eta,s,y) := \inf_{Q\in\mathcal{M}} E\left[V\left(\eta\frac{Z_T}{Z_t}\right) + \eta\frac{Z_T}{Z_t}nh(Y_T)\middle| S_t = s, Y_t = y\right],$$

where V is the convex conjugate of the utility function U.

The primal value function $u^{(n)}(t,x,s,y)$ is then recovered from the bidual relation

$$u^{(n)}(t,x,s,y) = \inf_{\eta>0}[v^{(n)}(t,\eta,s,y) + x\eta].$$

Expressing the density process $(Z_t)_{0\le t\le T}$ in terms of Q-Brownian motions and using the form of the convex conjugate function V it is easy to obtain the dual value function in the form

$$v^{(n)}(t,\eta,s,y) = V(\eta) + \frac{\eta}{\alpha}\inf_{Q\in\mathcal{M}} E^Q\left[\log\left(\frac{Z_T}{Z_t}\right) + \alpha nh(Y_T)\middle| S_t = s, Y_t = y\right], \quad (57)$$

with

$$\log\left(\frac{Z_T}{Z_t}\right) = -\int_t^T \hat{\lambda}_u d\hat{B}_u^Q - \int_t^T \psi_u d\hat{B}_u^{\perp,Q} + \frac{1}{2}\int_t^T \left(\hat{\lambda}_u^2 + \psi_u^2\right)du.$$

It is not difficult to see that $E^Q\int_0^t \hat{\lambda}_u^2 du < \infty$ for all $t \in [0,T]$. If, in addition,

$$E^Q\int_0^t \psi_u^2 du < \infty, \quad t \in [0,T], \quad (58)$$

then the stochastic integrals in (57) will have zero expectation. Denoting by $\mathcal{M}'$ the subset of $\mathcal{M}$ for which (58) holds, we clearly have

$$\begin{aligned} &\inf_{Q\in\mathcal{M}} E^Q\left[\log\left(\frac{Z_T}{Z_t}\right) + \alpha nh(Y_T)\middle| S_t = s, Y_t = y\right] \\ &= \inf_{Q\in\mathcal{M}'} E^Q\left[\log\left(\frac{Z_T}{Z_t}\right) + \alpha nh(Y_T)\middle| S_t = s, Y_t = y\right]. \end{aligned}$$

Then the dual value function decomposes as

$$v^{(n)}(t,\eta,s,y) = V(\eta) + \frac{\eta}{\alpha}\left(H(t,s) + G(t,s,y)\right),$$

where

$$H(t,s) := E^Q\left[\frac{1}{2}\int_t^T \hat{\lambda}_u^2 du\middle| S_t = s, Y_t = y\right],$$

and where $G(t,s,y)$ is the value function of a stochastic control problem:

$$G(t,s,y) := \inf_{\psi} E^Q \left[\frac{1}{2}\int_t^T \psi_u^2 du + \alpha n h(Y_T) \middle| S_t = s, Y_t = y \right]. \quad (59)$$

Note also that for $n = 0$, then $G = 0$, and we also have the dual value function for $n \neq 0$ given directly in terms of its counterpart for no claim on Y:

$$v^{(n)}(t,\eta,s,y) = v^{(0)}(t,\eta,s,y) + \frac{\eta}{\alpha}G(t,s,y),$$

which will be useful in obtaining a representation for the indifference price. The primal value function is then recovered from the bidual relation as

$$u^{(n)}(t,x,s,y) = u^{(0)}(t,x,s)\exp[-G(t,s,y)], \quad (60)$$

where

$$u^{(0)}(t,x,s) = U(x)\exp[-H(t,s)]. \quad (61)$$

Applying the definition of the indifference price then gives the following representation.

Theorem 6 *The indifference price at* $t \in [0,T]$ *is given by*

$$p^{(n)}(t,s,y) = \frac{1}{\alpha n}G(t,s,y),$$

where the value function G is defined in (59).

Remark 1 Theorem 6 is a special case of the general dual representation for the indifference price, Corollary 1, since the relative entropy $\mathcal{H}(Q,P)$ between a measure $Q \in \mathcal{M}$ and P is given by

$$\mathcal{H}(Q,P) = E^Q \left[\frac{1}{2}\int_0^T \left(\hat{\lambda}_t^2 + \psi_t^2\right) dt \right].$$

The Hamilton-Jacobi-Bellman (HJB) equation for $G(t,s,y)$ is

$$G_t + \mathcal{A}_{S,Y}^{Q^M} G + \min_{\psi}\left[\frac{1}{2}\psi^2 - \beta\bar{\rho}\psi y G_y\right] = 0 \quad G(T,s,y) = \alpha n h(y),$$

where $\mathcal{A}_{S,Y}^{Q^M}$ is the generator of (S,Y) under the minimal measure:

$$\mathcal{A}_{S,Y}^{Q^M} G = \frac{1}{2}s^s G_{ss} + \beta(\hat{\theta} - \rho\hat{\lambda})yG_y + \frac{1}{2}\beta^2 y^2 G_{yy} + \rho\sigma\beta s y G_{sy}.$$

The optimal control is $\psi_t^* \equiv \psi^*(t,S_t,Y_t)$ where

$$\psi^*(t,s,y) = \bar{\rho}\beta y G_y(t,s,y)$$

so the HJB equation is the semi-linear PDE

$$G_t + \mathcal{A}_{S,Y}^{Q^M} G - \frac{1}{2}(1-\rho^2)\beta^2 y^2 G_y^2 = 0.$$

Hence, the indifference price $p^{(n)}$ satisfies

$$p_t^{(n)} + \mathcal{A}_{S,Y}^{QM} p^{(n)} - \frac{1}{2}\alpha n(1-\rho^2)\beta^2 y^2 \left(p_y^{(n)}\right)^2 = 0. \quad p^{(n)}(T,s,y) = h(y)$$

The optimal hedging strategy is obtained easily, since the HJB equation for primal value function gives the optimal trading strategy $\pi^{*,n}$ in terms of derivatives of $G(t,s,y)$, and hence in terms of derivatives of the indifference price, and we have the following theorem.

Theorem 7 *The optimal hedging strategy is to hold* $(\Delta_t^{\mathrm{H}})_{0\le t\le T}$ *shares of* S *at time* $t \in [0,T]$, *given by*

$$\Delta_t^{\mathrm{H}} = -n\left(p_s^{(n)}(t,S_t,Y_t) + \rho\frac{\beta}{\sigma}\frac{Y_t}{S_t}p_y^{(n)}(t,S_t,Y_t)\right).$$

Note that, compared with the full information case, the optimal hedging strategy contains an additional term involving a derivative with respect to the stock price variable. This in turn has resulted from the extra dimension of the associated control problems. The drift parameter uncertainty has led to additional risk and hence to an extra hedging term to counteract the added risk.

Proof of Theorem 7 The HJB equation for the primal value function is

$$u_t^{(n)} + \max_{\pi} \mathcal{A}_{X,S,Y} u^{(n)} = 0,$$

where $\mathcal{A}_{X,S,Y}$ is the generator of (X,S,Y) under P. Performing the maximisation over π yields the optimal Markov control as $\pi_t^{*,n} = \pi^{*,n}(t, X_t^{*,n}, S_t, Y_t)$, where

$$\pi^{*,n}(t,x,s,y) = -\left(\frac{\hat{\lambda}u_x^{(n)} + \sigma s u_{xs}^{(n)} + \rho\beta y u_{xy}^{(n)}}{\sigma u_{xx}^{(n)}}\right),$$

and where the arguments of the functions on the right-hand-side are omitted for brevity. For the case $n = 0$ there is no dependence on y in the value function $u^{(0)}$, and we have $\pi_t^{*,0} = \pi^{*,0}(t, X_t^{*,0}, S_t)$, where

$$\pi^{*,0}(t,x,s) = -\left(\frac{\hat{\lambda}u_x^{(0)} + \sigma s u_{xs}^{(0)}}{\sigma u_{xx}^{(0)}}\right).$$

Applying the definition of the optimal hedging strategy along with the representations (60) and (61) for the value functions, gives the result.

□

4.4.1. Linear Approximation for the Indifference Price

To obtain analytic results and hence conduct a simulation study of the effectiveness of the optimal hedging strategy, we may approximate the indifference price by the marginal price. For $n = 0$ the indifference price PDE becomes linear, and the Feynman-Kac theorem gives the marginal price as follows.

Corollary 3 *The marginal price at $t \in [0, T]$ is given by*

$$\hat{p}(t, s, y) = E^{Q^M}[h(Y_T)|S_t = s, Y_t = y].$$

This is perfectly consistent with the general result of Theorem 5 since the minimal entropy measure Q^E coincides with Q^M, as can be seen from Remark 1.

The marginal price (and hence the optimal trading strategy) can be computed in analytic form since, under Q^M, $\log Y_T$ is Gaussian. We have the following result.

Proposition 5 *Under Q^M, conditional on $S_t = s, Y_t = y$,*

$$\begin{aligned}
\log Y_T &\sim N(m, \Sigma^2) \\
m &= \log y + \beta\left(\hat{\theta}(t, y) - \rho\hat{\lambda}(t, s) - \frac{1}{2}\beta\right)(T - t) \\
\Sigma^2 &= \left[\rho^2 + (1 - \rho^2)\mathrm{v}_t(T - t)\right]\beta^2(T - t)
\end{aligned}$$

Proof This is established by computing the SDEs for Y and for $\hat{\theta}_t - \rho\hat{\lambda}_t$ under Q^M. Indeed, applying the Itô formula to $\log Y_t$ under Q^M, we obtain, for $t < T$,

$$\log Y_T = \log Y_t + \beta\int_t^T \left(\hat{\theta}_u - \rho\hat{\lambda}_u\right) du - \frac{1}{2}\beta^2(T - t) + \beta\int_t^T d\hat{W}_u^{Q^M}, \quad (62)$$

where $\hat{W}^{Q^M}$ is a Brownian motion under Q^M. The dynamics of $\hat{\theta}_t - \rho\hat{\lambda}_t$ under Q^M are

$$d(\hat{\theta}_t - \rho\hat{\lambda}_t) = \bar{\rho}\mathrm{v}_t d\hat{B}_t^{\perp, Q^M},$$

where $\hat{B}^{\perp, Q^M}$ is a Q^M-Brownian motion perpendicular to that driving the stock, and related to $\hat{W}^{Q^M}$ by $\hat{W}^{Q^M} = \rho\hat{B}^{Q^M} + \bar{\rho}\hat{B}^{\perp, Q^M}$, where $\hat{B}^{Q^M}$ is the Brownian motion driving S. Hence, for $u > t$, after changing the order of integration in a double integral, we obtain

$$\int_t^T \left(\hat{\theta}_u - \rho\hat{\lambda}_u\right) du = \left(\hat{\theta}_t - \rho\hat{\lambda}_t\right)(T - t) + \bar{\rho}\int_t^T \mathrm{v}_u(T - u) d\hat{B}_u^{\perp, Q^M}.$$

This can be inserted into (62) to yield the desired result.

□

We are thus able to obtain BS-style formulae for the price and hedge. For a put option of strike K we obtain the following explicit formulae for the marginal price and the associated optimal hedging strategy, where Φ denotes the standard cumulative normal distribution function.

Corollary 4 *With m and Σ as in Proposition 5, define $b \equiv b(t,s,y)$ by*

$$m = \log y + b - \frac{1}{2}\Sigma^2.$$

Then the marginal price at time $t \in [0,T]$ of a put option with payoff $(K - Y_T)^+$ is $\hat{p}(t, S_t, Y_t)$, where

$$\begin{aligned} \hat{p}(t,s,y) &= K\Phi(-d_1 + \Sigma) - ye^b\Phi(-d_1), \\ d_1 &= \frac{1}{\Sigma}\left[\log\left(\frac{y}{K}\right) + b + \frac{1}{2}\Sigma^2\right]. \end{aligned}$$

The optimal hedging strategy given by Theorem 7 with $\hat{p}$ as an approximation to the indifference price is $\hat{\Delta}_t \equiv \hat{\Delta}(t, S_t, Y_t)$, where

$$\hat{\Delta}(t,s,y) = n\rho\frac{\beta}{\sigma}\frac{y}{s}e^b\Phi(-d_1).$$

4.5. Performance of the Optimal Strategy

To assess the performance of the optimal strategy based on the marginal price, we conduct the following simulation experiment.

Using chosen values for the "true" drifts λ, θ, we generate asset price paths S, Y over a time frame t_0 (1 year) and use this "data" to estimate the asset drifts, and so set a prior distribution at time 0. We then set initial asset prices S_0, Y_0 and generate a price history over $[0,T]$.

We suppose a put option of strike K is sold at time 0 for $\hat{p}(0, S_0, Y_0)$ and optimally hedged over $[0,T]$ (T = 1year), incorporating updating from filtering, and using daily rebalancing. In this way, we generate a terminal hedging error. We repeat this procedure of setting a prior using past "data" and then hedging over $[0,T]$, over many paths (with the same values of S_0, Y_0 in each simulation) to generate a terminal hedging error distribution. Note that we used a new set of "past data" to set the prior on each simulation. The idea is to allow for occasions where the mean of prior distribution is of variable quality in relation to the true values of the drifts.

We repeat the hedging error computation over the same asset price histories using the BS-style hedge (30), and also with a hedge in the absence of filtering, where we used the initial estimates of the asset drifts to compute the hedge throughout the hedging time frame. This uses the hedge in (37) and the approximation formulae from Monoyios [26].

The parameters were set as below. The initial BS price and delta are denoted by BS_0 and Δ_0^{BS}. The notation $\bar{p}_0$ denotes the average marginal price that the put was sold for (recall that each simulation run results in a different initial prior). The average mean for the prior distribution of λ is denoted $\bar{\lambda}_0$ (and similarly for θ); "NF" denotes "no filtering".

$$\begin{array}{lll} \lambda = 0.5, & \sigma = 0.22, & \theta = 0.45, \quad \beta = 0.18, \quad \rho = 0.85 \\ S_0 = 100, & Y_0 = 100, & K = 100 \\ \mathrm{BS}_0 = 7.17, & \Delta_0^{\mathrm{BS}} = -0.46 & \\ \bar{p}_0 = 7.20, & \bar{\Delta}_0 = -0.32 & \\ \bar{p}_0^{\mathrm{NF}} = 7.33, & \bar{\Delta}_0^{\mathrm{NF}} = -0.47 & \\ \bar{\lambda}_0 = 0.11, & \bar{\theta}_0 = 0.09. & \end{array}$$

Table 1. Hedging error statistics for Figure 1 (as percentage of premium)

	Mean	SD	Median
Optimal Hedge	2.09	62.8	12.3
BS Hedge	-17.6	80.5	-11.6
Unfiltered Hedge	-15.7	79.4	-10.1

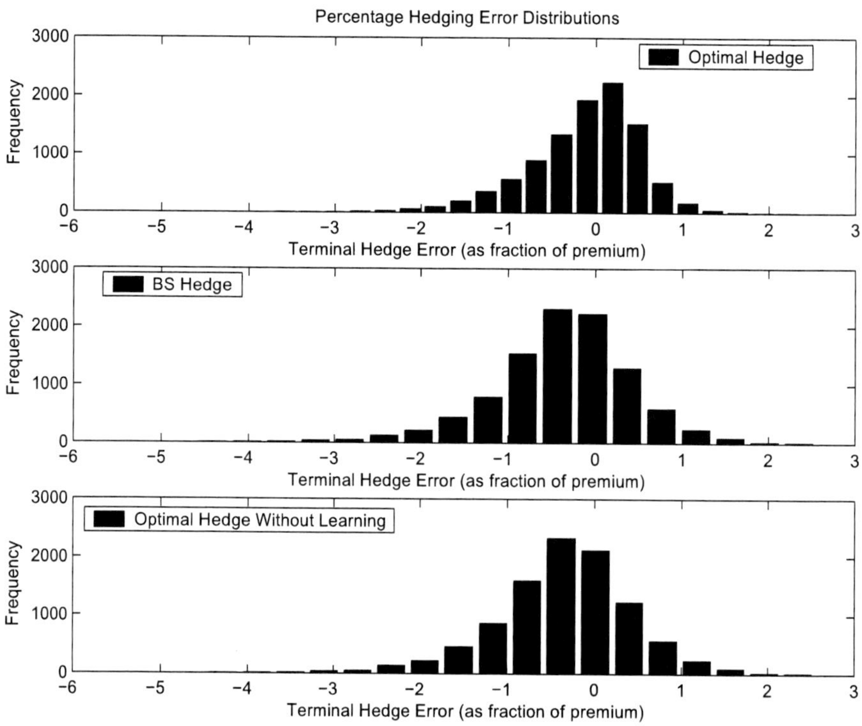

Figure 1. Percentage hedging error distribution

The results are shown in Figure 1 and Table 1. They clearly show that the optimal hedge with learning produces a hedging error distribution with higher mean, lower standard deviation and, significantly, a higher median (all as percentages of the initial option premium), than either the BS hedge or the hedge without learning. Thus, the frequency of profits over losses is increased by the optimal hedging program incorporating learning.

Varying some parameters slightly gave the results in Figure 2 and Table 2. The results are still favourable, even if one sells the option for a lower value than the BS price, showing that the improvement in hedging performance is not due to starting with a higher wealth in the initial hedging portfolio.

The conclusion is that optimal hedging combined with a filtering algorithm to deal with drift parameter uncertainty can indeed give improved hedging performance over methods which take S as a perfect proxy for Y.

Table 2. Hedging error statistics for Figure 2 (as percentage of premium). The parameters are the same as those for Table 1 except for $\lambda = 0.4$, $\beta = 0.23$, $\rho = 0.9$. In this case the BS price was 9.16, the average optimal price was 8.99 and the average price without filtering was 9.12.

	Mean	SD	Median
Optimal Hedge	7.09	51.8	13.1
BS Hedge	5.57	52.8	10.0
Unfiltered Hedge	5.16	53.1	9.60

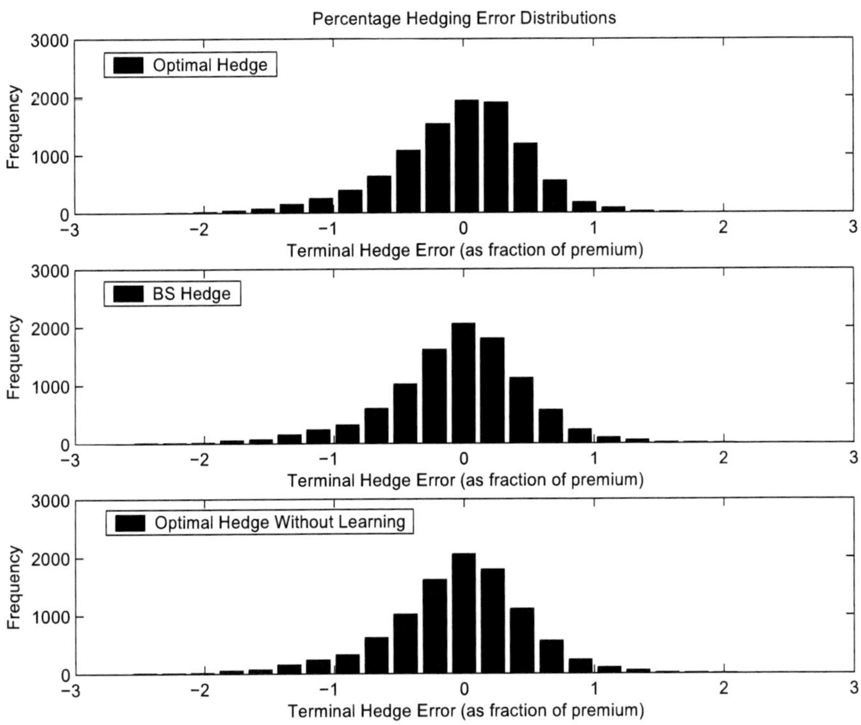

Figure 2. Percentage hedging error distribution

Conclusion

In this chapter we analysed utility-based pricing and hedging in incomplete security markets. We have derived probabilistic representations for indifference prices. In general these are non-linear pricing rules reducing to linear pricing rules as the number of claims becomes small. We computed prices and optimal hedging strategies in basis risk models, under full and partial information scenarios. these showed how optimal hedging can indeed outperform hedging methods based on complete market approximations. This is therefore of relevance to practitioners as well as to academics. The take-up of utility-based pricing by practitioners has been somewhat limited, given the need to specify a utility function. It is possible that the results of this chapter can show the potential benefits of applying such techniques in practice.

Acknowledgements

Thanks to Matthias Ehrhardt for the invitation to contribute to this volume.

References

[1] D. Becherer Utility-indifference hedging and valuation via reaction-diffusion systems, *Proc. R. Soc. Lond. A* **460** (2004), 27–51.

[2] T. Björk, M.H.A. Davis and C. Landén, *Optimal investment under partial information*, Preprint, 2008.

[3] F. Black and M. Scholes, The pricing of options and corporate liabilities, *J. Political Economy* **81** (1973), 637–659.

[4] M.H.A. Davis *Option pricing in incomplete markets,* in "Mathematics of Derivative Securities" (M.A.H. Dempster and S.R. Pliska, eds.), Cambridge University Press, 1997.

[5] M.H.A. Davis, *Optimal hedging with basis risk,* in "From Stochastic Calculus to Mathematical Finance : The Shiryaev Festschrift" (Y. Kabanov R. Lipster and J. Stoyanov, eds.), Springer, 2006.

[6] M.H.A. Davis, V.G. Panas and T. Zariphopoulou, European option pricing with transaction costs, *SIAM J. Control Optim.* **31** (1993), 470–493.

[7] F. Delbaen, P. Grandits, T. Rheinländer, D. Samperi, M. Schweizer and C. Stricker, Exponential hedging and entropic penalties, *Math. Finance* **12** (2002), 99–123.

[8] F. Delbaen and W. Schachermayer, A general version of the fundamental theorem of asset pricing, *Mathematische Annalen* **300** (1994), 463–520.

[9] W.H. Fleming and R.W. Rishel, *Deterministic and Stochastic Optimal Control*, Springer-Verlag, 1975.

[10] M.R. Grasselli and T.R. Hurd, Indifference pricing and hedging of volatility derivatives, *Appl. Math. Finance* **14** (2007), 303–317.

[11] V. Henderson, Valuation of claims on nontraded assets using utility maximization, *Math. Finance* **12** (2002) 351–373.

[12] S.D. Hodges and A. Neuberger, Optimal replication of contingent claims under transaction costs, *Rev. Futures Markets* **8** (1989), 222–239.

[13] J. Hugonnier, D. Kramkov and W. Schachermayer, On utility-based pricing of contingent claims in incomplete markets, *Math. Finance* **15** (2005), 203–212.

[14] Y. Kabanov and C. Stricker On the optimal portfolio for the exponential utility maximization: remarks to the six-author paper, *Math. Finance* **12** (2002), 125–134.

[15] I. Karatzas and S.G. Kou, Pricing contingent claims with constrained portfolios, *Annals of Applied Probability* **6** (1996), 321–369.

[16] I. Karatzas, J.P. Lehoczky, S.E. Shreve and G-L. Xu, Martingale and duality methods for utility maximization in an incomplete market, *SIAM J. Control Optim.* **29** (1991), 702–730.

[17] I. Karatzas and S.E. Shreve, *Methods of Mathematical Finance*, Springer, 1998.

[18] D. Kramkov, Optional decomposition of supermartingales and hedging contingent claims in incomplete security markets, *Probability Theory and Related Fields* **105** (1996), 459–479.

[19] D. Kramkov and W. Schachermayer, The asymptotic elasticity of utility functions and optimal investment in incomplete markets, *Annals of Applied Probability* **9** (1999), 904–950.

[20] D. Kramkov and M. Sirbu, Sensitivity analysis of utility-based prices and risk-tolerance wealth processes, *Annals of Applied Probability* **16** (2006), 2140–2194.

[21] D. Kramkov and M. Sirbu, Asymptotic analysis of utility-based hedging strategies for small number of contingent claims, *Stochastic Processes and their Applications* **117** (2007), 1606–1620.

[22] M. Monoyios, Efficient option pricing with transaction costs, *J. Comput. Finance* **7** (2003), 107–128.

[23] M. Monoyios, Option pricing with transaction costs using a Markov chain approximation, *J. Economic Dynamics and Control* **28** (2004), 889–913.

[24] M. Monoyios, Performance of utility-based strategies for hedging basis risk, *Quantitative Finance* **4** (2004), 245–255.

[25] M. Monoyios, Characterisation of optimal dual measures via distortion, *Decisions in Economics and Finance* **29** (2006), 95–119.

[26] M. Monoyios, Optimal hedging and parameter uncertainty, *IMA J. Management Mathematics* **18** (2007), 331–351.

[27] M. Monoyios, *Marginal utility-based valuation and hedging of claims on non-traded assets with partial information*, Preprint, 2008.

[28] M. Musiela and T. Zariphopoulou, An example of indifference prices under exponential preferences, *Finance & Stochastics* **8** (2004), 229–239.

[29] H. Nagai and S. Peng, Risk-sensitive dynamic portfolio optimization with partial information on infinite time horizon, *Annals of Applied Probability* **12** (2002), 173–195.

[30] L.C.G. Rogers, The relaxed investor and parameter uncertainty, *Finance & Stochastics* **5** (2001), 131–154.

[31] W. Schachermayer, Optimal investment in incomplete markets when wealth may become negative, *Annals of Applied Probability* **11** (2001), 694–734.

[32] R. Sircar and T. Zariphopoulou, Bounds and asymptotic approximations for utility prices when volatility is random, *SIAM J. Control Optim.* **43** (2005), 1328–1353.

[33] A.E. Whalley, and P. Wilmott, An asymptotic analysis of an optimal hedging model for option pricing with transaction costs, *Math. Finance* **7** (1997), 307–324.

[34] T. Zariphopoulou, A solution approach to valuation with unhedgeable risks, *Finance & Stochastics* **5** (2001), 61–82.

PART II.
ANALYTIC SOLUTIONS

In: Nonlinear Models in Mathematical Finance...
Editor: Matthias Ehrhardt, pp. 103-129
ISBN 978-1-60456-931-5

Chapter 3

PRICING OPTIONS IN ILLIQUID MARKETS: SYMMETRY REDUCTIONS AND EXACT SOLUTIONS

***Ljudmila A. Bordag*[1] *and Rüdiger Frey*[2]**
[1]IDE, Halmstad University, Box 823, 301 18 Halmstad, Sweden
[2]Department of Mathematics, Universität Leipzig,
Postfach 10 09 20, 04009 Leipzig, Germany

Abstract

This chapter is concerned with nonlinear Black-Scholes equations arising in certain option pricing models with a large trader and/or transaction costs. In the first part we give an overview of existing option pricing models with frictions. While the financial setup differs between models, it turns out that in many of these models derivative prices can be characterized by fully nonlinear versions of the standard parabolic Black-Scholes equation. In the second part of this chapter we study a typical nonlinear Black-Scholes equation using methods from Lie group analysis. The equation possesses a rich symmetry group. By introducing invariant variables, invariant solutions can therefore be characterized in terms of solutions to ordinary differential equations. Finally, we discuss properties and applications of these solutions.

1. Introduction

Standard derivative pricing theory is based on the assumption of frictionless markets. In particular, it is assumed that there are no transaction costs and that all investors are *small* relative to the market so that they can buy arbitrarily large quantities of the underlying assets without affecting its price (perfectly liquid or elastic markets). Given the scale of hedging activities on many financial markets this is clearly unrealistic. Hence in recent years a number of models for studying the pricing and the hedging of derivative securities in illiquid markets or in the presence of transaction costs have been developed. [1] The financial framework that is being used differs substantially between model classes. However, as shown below, in many of these models derivative prices can be characterized by fully nonlinear versions of the standard parabolic Black-Scholes equation; moreover, these nonlinear

[1]We discuss the relevant literature in the body of this chapter.

Black-Scholes equations have a very similar structure. This makes these equations a useful reference point for studying derivative asset analysis in illiquid markets.

The nonlinear Black-Scholes equations derived in the literature contain a parameter representing the size of the transaction costs or of the trader's impact on the market. Usually it is expected that the equations reduce to the linear Black-Scholes PDE if this parameter becomes small; put differently, the nonlinear equations are considered as *perturbations* of the linear Black-Scholes equation. From an analytical point of view these equations can be divided roughly into two classes: equations with regular perturbations and equations with singular perturbations. Equations of the latter type are fully nonlinear, and there are so far relatively few papers where equations with singular perturbations are studied by analytical methods. On the other hand, an analytical study of these nonlinear equations makes it possible to determine the scope of applications of different models represented by fully nonlinear partial differential equations (PDEs).

In this chapter we therefore apply Lie group analysis to a typical nonlinear Black-Scholes equation. It turns out that the equation possesses a rich Lie symmetry group which allows us to introduce invariant variables and to reduce the corresponding PDEs to ordinary differential equations (ODEs). It is even possible to find exact invariant solutions to the ensuing ODEs. We show that most of the exact solutions for a given nonlinear equation have no counterpart in the linear Black-Scholes case; they intrinsically reflect a nonlinearity of the equation. The last part of this chapter is devoted to applications. We study properties of solutions and discuss the sensitivity with respect to model parameters; in particular, we show that some solutions approximate typical financial derivatives relatively closely.

This chapter is organized as follows: the overview of existing option pricing models with limited market liquidity can be found in Section 2.; the analytic properties of nonlinear Black-Scholes equations are studied in Section 3; applications are discussed in Section 4.

2. Illiquid Markets and Nonlinear Black-Scholes Equations

In order to motivate the subsequent analysis we present a brief synopsis of three different frameworks for modeling illiquid markets. We group them under the labels *quadratic transaction-cost models*; *reduced-form SDE-models*; *reaction-function* or *equilibrium models*. In particular, we show that the value function of a certain type of self-financing strategies (so called *Markovian* strategies) must be a solution of a fully nonlinear version of the standard Black-Scholes equation. In all models there will be two assets, a risk-free money-market account B which is perfectly liquid and a risky and illiquid asset S (the stock), modelled on some filtered probability space $(\Omega, \mathcal{F}, \{\mathcal{F}_t\}, P)$. Without loss of generality, we use the money market account as numeraire; hence $B_t \equiv 1$, and interest rates can be taken equal to zero.

2.1. (Quadratic) Transaction-Cost Models

The predominant model in this class has been put forward by Cetin, Jarrow and Protter [7], [8]. In this model there is a *fundamental* stock price process S^0 following geometric Brownian motion,

$$dS_t^0 = \mu S_t^0 dt + \sigma S_t^0 dW_t \tag{1}$$

for constants $\mu \in \mathbb{R}$, $\sigma > 0$ and a standard Brownian motion W. The *transaction price* to be paid at time t for trading α shares is

$$\bar{S}_t(\alpha) = e^{\rho\alpha} S_t^0, \quad \alpha \in \mathbb{R}, \ \rho > 0, \tag{2}$$

where ρ is a liquidity parameter.

Intuitively, in the model (2) a trader has to pay a spread whose size relative to the fundamental price equals $S_t^0(e^{\rho\alpha} - 1)$, so that the spread depends on the amount α to be traded. As shown in [7], this leads to transaction costs which are proportional to the *quadratic variation* of the stock trading strategy. In order to explain this statement in more detail, we consider a self-financing trading strategy $(\Phi_t, \eta_t)_{t\geq 0}$ giving the number of stocks and the position in the money market for predictable stochastic processes Φ and η. The *value* of this strategy at time t equals $V_t = \Phi_t S_t^0 + \eta_t$. Note that V_t is the so-called *paper value* of the position; under (2) the liquidation value of the strategy (the amount of money the large trader receives if he actually liquidates his stock position) is typically lower than V_t. For a detailed discussion of this point we refer to Bank and Baum [1].

In order to motivate the form of the dynamics of the paper value we consider a simple predictable strategy of the form $\Phi_t(\omega) = \sum_i \phi_i(\omega) 1_{(t_i, t_{i+1}]}(t)$ for deterministic time points $0 = t_0 < t_1 < \dots$. Then the self-financing condition in t_i reads

$$\begin{aligned} \eta_{t_i} - \eta_{t_{i-1}} &= -(\Phi_{t_i} - \Phi_{t_{i-1}})\, \bar{S}_{t_i}(\Phi_{t_i} - \Phi_{t_{i-1}}) \\ &= -\left(\Phi_{t_i} - \Phi_{t_{i-1}}\right)\left(S_{t_i}^0 + \rho S_{t_i}^0(\Phi_{t_i} - \Phi_{t_{i-1}})\right) + o\left((\Phi_{t_i} - \Phi_{t_{i-1}})^2\right) . \end{aligned}$$

Note that in the last line we have used that $\frac{d}{d\alpha}_{|\alpha=0} \bar{S}_t(\alpha) = S_t^0$. Hence the change in the value of the portfolio equals

$$\begin{aligned} V_{t_i} - V_{t_{i-1}} &= \Phi_{t_i} S_{t_{i+1}}^0 - \Phi_{t_{i-1}} S_{t_i}^0 + \eta_{t_i} - \eta_{t_{i-1}} \\ &= \Phi_{t_i}(S_{t_{i+1}}^0 - S_{t_i}^0) - \rho S_{t_i}^0 \left(\Phi_{t_i} - \Phi_{t_{i-1}}\right)^2 + o\left((\Phi_{t_i} - \Phi_{t_{i-1}})^2\right) . \end{aligned}$$

This suggests that for a continuous semi-martingale Φ with quadratic variation[2] $[\Phi]_t$ the wealth dynamics of a self-financing strategy becomes

$$dV_t = \Phi_t dS_t^0 - \rho S_t^0 d[\Phi]_t. \tag{3}$$

This is in fact true as is shown in Theorem A3 of [7]; see also [1]. Note that the last term in (3) represents the extra transaction cost due to the limited liquidity of the market.

Consider now a *Markovian strategy*, that is a trading strategy of the form $\Phi_t = \phi(t, S_t^0)$ for a smooth function ϕ. In this case Φ is a semi-martingale with quadratic variation given by

$$[\Phi]_t = \int_0^t \left(\phi_S(s, S_s^0)\sigma S_s^0\right)^2 ds\,;$$

see for instance Chapter 2 of [29]. Combining this with (3) yields the following dynamics of V_t

$$dV_t = \phi(t, S_t^0) dS_t^0 - \rho S_t^0 \left(\phi_S(t, S_t^0)\sigma S_t^0\right)^2 dt. \tag{4}$$

[2] See for instance Chapter 2 of [29] for a definition of quadratic variation.

Suppose now that u and ϕ are smooth functions such that $u(t, S_t^0)$ gives the value of a self-financing trading strategy with stock position $\phi(t, S_t^0)$. According to the Itô formula, the process $(u(t, S_t^0))_{t\geq 0}$ has dynamics

$$du(t, S_t^0) = u_S(t, S_t^0)dS_t^0 + \left(u_t(t, S_t^0) + \frac{1}{2}\sigma^2 (S_t^0)^2 u_{SS}(t, S_t^0)\right) dt\,. \qquad (5)$$

By the uniqueness of semi-martingale decompositions it is immediate that the dt-terms in (4) and (5) have to coincide, so that u must satisfy the equation $u_t + \frac{1}{2}\sigma^2 S^2 u_{SS} + \rho S^3 \sigma^2 \phi_S = 0$. Moreover, we must have that $\phi \equiv u_S$. The last relation gives $\phi_S = u_{SS}$, so that we obtain the following nonlinear PDE for u

$$u_t + \frac{1}{2}\sigma^2 S^2 u_{SS}\left(1 + 2\rho S u_{SS}\right) = 0\,. \qquad (6)$$

When pricing a derivative security with maturity date T and payoff $h(S_T)$ for some function $h : [0,\infty) \to \mathbb{R}$ we have to add the terminal condition $u(T, S) = h(S)$, $S \geq 0$. For instance, in case of a European call option with strike price $K > 0$ we have $h(S) = \max\{S - K, 0\}$.

Note that the original paper [7] goes further in the analysis of quadratic transaction cost models. To begin with, a general framework is proposed that contains (2) as special (but typical) case. Moreover, conditions for absence of arbitrage for a general class of trading strategies - containing Markovian trading strategies as special case - are given, and a notion of approximative market completeness is studied.

2.2. Reduced-Form SDE Models

Under this modeling approach it is assumed that investors are *large traders* in the sense that their trading activity affects equilibrium stock prices. More precisely, given a liquidity parameter $\rho \geq 0$ and a semi-martingale Φ representing the stock trading strategy of a given trader, it is assumed that the stock price satisfies the stochastic differential equation (SDE)

$$dS_t = \sigma S_t dW_t + \rho S_t d\Phi_t\,. \qquad (7)$$

The intuitive interpretation is as follows: given that the investor buys (sells) stock ($\Delta\Phi_t > 0$) the stock price is pushed (downward) upward by $\rho S_{t-}\Delta\Phi_t$; the strength of this price impact depends on the parameter ρ. Note that for $\rho = 0$ the asset price simply follows a Black-Scholes model with reference volatility σ. The model (7) and slight variants thereof are studied among others in [11], [12], [18], [23], [28] or [14].

In the sequel we denote the asset price process which results if a large trader uses a particular trading strategy Φ by S^Φ. Suppose as before that the trading strategy is Markovian, i.e. of the form $\Phi_t = \phi(t, S_t)$ for a smooth function ϕ and that ϕ satisfies the constraint

$$1 - \rho S\phi_S(t, S) > 0 \text{ for all } (t, S);$$

given a liquidity parameter ρ, this last condition basically limits the permissible variations in the stock trading strategy of the large trader. Applying the Itô formula to (7) shows that S^ϕ is an Itô process with dynamics

$$dS_t^\phi = v^\phi(t, S_t^\phi)S_t^\phi dW_t + b^\phi(t, S_t^\phi)S_t^\phi dt$$

with *adjusted volatility* given by

$$v^{\phi}(t,S) = \frac{\sigma}{1-\rho S\phi_S(t,S)}; \tag{8}$$

see for instance [11] for a detailed derivation. Note that this adjusted volatility is increased (decreased) relative to the reference volatility σ if $\phi_S > 0$ ($\phi_S < 0$).

In the model (7), a portfolio with stock trading strategy Φ and value V is termed *self-financing*, if satisfies the equation $dV_t = \Phi_t dS_t^{\Phi}$. Note that the form of the strategy Φ affects the dynamics of S; this *feedback effect* will give rise to nonlinearities in the wealth dynamics as we now show. Suppose that $V_t = u(t, S_t)$ and $\Phi_t = \phi(t, S_t)$ for smooth functions u and ϕ. As before, applying the Itô formula to the process $(u(t, S_t^{\phi}))_{t\geq 0}$ yields $\phi \equiv u_S$. Moreover, u must satisfy the relation $u_t + \frac{1}{2}(v^{\phi})^2(t,S)S^2 u_{SS} = 0$. Using (8) and the relation $\phi_S = u_{SS}$ we thus obtain the following fully nonlinear PDE for $u(t,S)$

$$u_t + \frac{1}{2}\frac{\sigma^2}{(1-\rho S u_{SS})^2} S^2 u_{SS} = 0\,. \tag{9}$$

Again, for pricing derivative securities a terminal condition corresponding to the particular payoff at hand needs to be added.

2.3. Equilibrium or Reaction-Function Models

Here the model primitive is a smooth *reaction function* ψ that gives the equilibrium stock price S_t at time t as function of some fundamental value F_t and the stock position of a large trader. A reaction function can be seen as reduced-form representation of an economic equilibrium model, such as the models proposed in [13], [27] or [30]. In these models there are two types of traders in the market: ordinary investors and a large investor. The overall supply of the stock is normalized to one. The normalized stock demand of the ordinary investors at time t is modelled as a function $D(F_t, S_t)$ where S_t is the proposed price of the stock. The normalized stock demand of the large investor is written the form $\rho\Phi_t$; $\rho \geq 0$ is a parameter that measures the size of the trader's position relative to the total supply of the stock. The equilibrium price S_t is then determined by the market clearing condition

$$D(F_t, S_t) + \rho\Phi_t = 1\,. \tag{10}$$

Under suitable assumptions on D, equation (10) admits a unique solution. Hence S_t can be expressed as a function ψ of F_t and $\rho\Phi_t$, so that $S_t = \psi(F_t, \rho\Phi_t)$. For instance we have in [27] that $\psi(f,\alpha) = f\exp(\alpha)$; the model used in [13] and [32] leads to the reaction function $\psi(f,\alpha) = f/(1-\alpha)$. The reaction-function approach is also used in [19] and in [10].

Now we turn to the characterization of self-financing hedging strategies in reaction-function models. Throughout we assume that the fundamental-value process F follows a geometric Brownian motion with volatility σ as in (1). Moreover, we assume that the reaction function is of the form $\psi(f,\alpha) = fg(\alpha)$ for some increasing function g. This holds for the specific examples introduced above and, more generally, for any model where $D(f,s) = U(f/s)$ for a strictly increasing function $U\colon (0,\infty) \to \mathbb{R}$ with suitable range.

Assuming as before that the normalized trading strategy of the large trader is of the form $\rho\phi(t,S)$ for a smooth function ϕ, we get from Itô's formula that

$$dS_t = g(\rho\phi(t,S_t))\,dF_t + \rho F_t g_\alpha(\rho\phi(t,S_t))\phi_S(t,S_t)\,dS_t + b(t,S_t)\,dt \tag{11}$$

(since $S_t = g(\rho\phi(t,S_t))F_t$); the precise form of $b(t,S_t)$ is irrelevant for our purposes. Assume now that

$$\left(1 - \rho F_t g_\alpha(\rho\phi(t,S_t))\phi_S(t,S_t)\right) > 0 \quad \text{a.s.}; \tag{12}$$

as before this can be viewed as an upper bound on the permissible variations of the large trader's strategy. Rearranging and integrating $\left(1-\rho F_t g_\alpha(\rho\phi(t,S_t))\phi_S(t,S_t)\right)^{-1}$ over both sides of equation (11) gives the following dynamics of S:

$$dS_t = \frac{1}{1-\rho\frac{g_\alpha(\rho\phi(t,S_t))}{g(\rho\phi(t,S_t))}S_t\phi_S(t,S_t)}\sigma S_t dW_t + \tilde{b}(t,S_t)dt\,; \tag{13}$$

again the precise form of $\tilde{b}$ is irrelevant.

A similar reasoning as in the case of the reduced-form SDE models now gives the following PDE for the value function $u(t,S)$ of a self-financing strategy

$$u_t + \frac{1}{2}\frac{\sigma^2}{\left(1-\rho\frac{g_\alpha(\rho u_S)}{g(\rho u_S)}Su_{SS}\right)^2}S^2 u_{SS} = 0\,. \tag{14}$$

In particular, for $g(\alpha) = \exp(\alpha)$ we have $g = g_\alpha$ and (14) reduces to equation (9); for $g(\alpha) = 1/(1-\alpha)$ as in [13], [32], we get the PDE

$$u_t + \frac{1}{2}\frac{\sigma^2(1-\rho u_S)^2}{(1-\rho u_S - \rho S u_{SS})^2}S^2 u_{SS} = 0\,. \tag{15}$$

A thorough analysis of the dynamics of self-financing strategies in generalized reaction-function models via the *Itô-Wentzell formula* can be found in [1].

2.4. Nonlinear Black-Scholes Equations

The nonlinear PDEs (6), (9), (14) and (15) are all of the form

$$u_t + \frac{1}{2}\sigma^2 S^2 v(\rho u_S, \rho S u_{SS})u_{SS} = 0\,, \tag{16}$$

where $v(0,0) = 1$. Since ρ is often considered to be small, it is of interest to replace v with its first order Taylor approximation around $\rho = 0$. It is immediately seen that for the equations (9) and (15) this linearization is given by $v(\rho u_S, \rho S u_{SS}) \approx 1+\rho S u_{SS}$; replacing $v(\rho u_S, \rho S u_{SS})$ with this first order Taylor approximation in (9) and (15) thus immediately leads to the PDE (6).

Note that (16) is a fully nonlinear equation in the sense that the coefficient of the highest derivative is a nonlinear function of this derivative. A similar feature can be observed for the limiting price in certain transaction cost models under a proper re-scaling of transaction cost and trading frequency; see for instance [2] or [16]. Nonlinear PDEs for incomplete markets obtained via exponential utility indifference hedging such as [3] or [24] on the other hand are quasi-linear equations in the sense that the highest derivative u_{SS} enters the equation in a linear way, similar to the well-known reaction-diffusion equations arising in physics or chemistry. From an analytical point of view the nonlinearities arising in both cases are quite different. In the latter quasi-linear case we have to do with a regular perturbation of the classical Black-Scholes (BS) equation but in the case (14) we have to do with a singular perturbation, meaning, in particularly, that the highest derivative is included in the perturbation [20]. In case of a regular perturbation it is typical to look for a representation of a solutions to a quasi-linear equation in the form

$$u\,(S,t) = u_{BS}(S,t) + \sum_{n=1}^{N} \rho u_1(S,t) + O(\rho^{N+1}). \tag{17}$$

For many forms of nonlinearities the uniform convergence as $\rho \to 0$ of the expansions of type (17) can be established. If the perturbation is of the singular type, the asymptotic expansion of the form (17) typically breaks down for some S, t and some $N \geq 0$. It can as well happens that the definition domain for the solutions of type (17) include just one point or that it is empty. It is very important to obtain explicit solutions for such models because in these case we can not hope to get good approximations for solutions by expansions of the type (17) in the whole region. This issue is taken up in the remainder of the chapter.

3. Invariant Solutions for a Nonlinear Black-Scholes Equation

We have seen in the previous section that the form of the equation (16) is typical for nonlinear PDEs arising in pricing equations for derivatives in illiquid markets. One of the methods to study properties of solutions to such type of fully nonlinear PDEs is the Lie group analysis of these equations. This approach can be applied to all equations listed before, i.e., to (6), (9) and (15). The first results were achieved for the equation (9). In [6] a special family of invariant solutions to the equation (9) was studied; in particular, the explicit solutions were used as test case for various numerical methods. Later on the slightly modified equation (32) was studied in [5] where the complete family of invariant solutions was described. The symmetry groups of the equations (6) and (15) were found in [4] and [31] correspondingly.

Our goal in this section is to investigate the nonlinear Black-Scholes equation (9) using analytical methods. Using the symmetry group and its invariants the PDE (9) can be reduced in special cases to ODEs. In the present chapter we study *all* invariant solutions to equation (9).

In the next paragraph 3.1. we provide a description of the Lie group method in application to the differential equations. This short introduction can by no means replace any study of classical books devoted to the method but it will allowed us to introduce necessary notations.

3.1. Lie Group Analysis of Differential Equations

In this paragraph we formulate all definitions for the case of one PDE of two independent and one dependent variables. It makes ideas transparent and easier accessible for somebody who does not used this method before. On the other side these formulations are general enough to study the equations of type (16). As a practical and easy introduction to this theory one can take as well the book [33]. The most general formulations can be found in books cited below.

Let us introduce a two-dimensional space X of independent variables $(S, t) \in X$ and a one-dimensional space of dependent variables $u \in U$. In our case X is isomorphic to a two-dimensional Euclidean space $\mathbb{R}^2$ and U to $\mathbb{R}$. We consider the space $U_{(1)}$ endowed with differential coordinates (u_S, u_t) which represent all the first derivatives of the variable u on S and t. Analogously, we introduce the space $U_{(2)}$ endowed with differential coordinates (u_{SS}, u_{St}, u_{tt}) which represent all the second order derivatives of the dependent variable u on S and t. We can continue with this procedure and introduce spaces of the type $U_{(n)}$, $n > 2$. As long we study an equation of the order two it will be sufficient to introduce $U_{(n)}$ up to the order $n = 2$.

We denote by M an open subset of the Cartesian product $M \subset X \times U$. The space M is an underlying space with elements denoted by (x, u) with $x = (S, t) \in X, \ \ u \in U$. We denote an element of a vector field on M by

$$V = \xi(S,t,u)\frac{\partial}{\partial S} + \tau(S,t,u)\frac{\partial}{\partial t} + \phi(S,t,u)\frac{\partial}{\partial u}, \tag{18}$$

where $\xi(S,t,u)$,$\tau(S,t,u)$ and $\phi(S,t,u)$ are smooth functions of their arguments, $V \in \mathcal{D}iff(M)$. The operators (18) are called as well *infinitesimal generators*.

The differential equation (16) is of the second order and to represent this equation as an algebraic relation on an appropriate space we introduce a second order jet bundle $M^{(2)}$ of the space M, i.e.,

$$M^{(2)} = M \times U_{(1)} \times U_{(2)} \tag{19}$$

with a natural contact structure [15], [25]. We label the coordinates in the jet bundle $M^{(2)}$ by $w = (S, t, u, u_S, u_t, u_{SS}, u_{St}, u_{tt}) \in M^{(2)}$.

The corresponding vector fields on $M^{(2)}$ have the form

$$\begin{aligned} pr^{(2)}V = {} & \xi(S,t,u)\frac{\partial}{\partial S} + \tau(S,t,u)\frac{\partial}{\partial t} + \phi(S,t,u)\frac{\partial}{\partial u} \\ & + \phi^S(S,t,u)\frac{\partial}{\partial u_S} + \phi^t(S,t,u)\frac{\partial}{\partial u_t} \\ & + \phi^{SS}(S,t,u)\frac{\partial}{\partial u_{SS}} + \phi^{St}(S,t,u)\frac{\partial}{\partial u_{St}} + \phi^{tt}(S,t,u)\frac{\partial}{\partial u_{tt}}. \end{aligned} \tag{20}$$

The vector fields $pr^{(2)}V$ are called the second prolongation of the vector fields V. Here the smooth functions $\phi^S(S,t,u)$, $\phi^t(S,t,u)$, $\phi^{SS}(S,t,u)$, $\phi^{St}(S,t,u)$ and $\phi^{tt}(S,t,u)$ are uniquely defined by the functions $\xi(S,t,u), \tau(S,t,u)$ and $\phi(S,t,u)$ (18) using the prolongation procedure (see [26], [25], [33], [15], [17]).

Remark 3.1. In the very simplified way we can present the idea behind this prolongation procedure as follows. We introduce some equivalence relation on the space of smooth functions defined on some open subset of X. Let a point x belong to this open subset. We call the functions equivalent in the point x if all their derivatives up to the order n coincide in the point $x \in X$. We call this equivalence class attached to the point x $n-$jet. It means at the same time as we introduced jets we introduced on the space a natural contact structure and prolonged our space by $U_{(n)},\ n \geq 1$.

Because of the very special form of the equation (9) we will use in our further calculations the exact form of the coefficients $\phi^t(S,t,u)$ and $\phi^{SS}(S,t,u)$ only. The coefficient $\phi^t(S,t,u)$ can be defined by the formula

$$\phi^t(S,t,u) = \phi_t + u_t\phi_u - u_S\xi_t - u_Su_t\xi_u - u_t\tau_t - (u_t)^2\tau_u, \tag{21}$$

and the coefficient $\phi^{SS}(S,t,u)$ by the expression

$$\begin{aligned} \phi^{SS}(S,t,u) = {} & \phi_{SS} + 2u_S\phi_{Su} + u_{SS}\phi_u \\ & + (u_S)^2\phi_{uu} - 2u_{SS}\xi_S - u_S\xi_{SS} - 2(u_S)^2\xi_{Su} \\ & - 3u_Su_{SS}\xi_u - (u_S)^3\xi_{uu} - 2u_{St}\tau_S - u_t\tau_{SS} \\ & - 2u_Su_t\tau_{Su} - (u_tu_{SS} + 2u_Su_{St})\tau_u - (u_S)^2u_t\tau_{uu}, \end{aligned} \tag{22}$$

where the subscripts by ξ, τ, ϕ denote corresponding partial derivatives. If we would study equation of the type (16) then we would need as well the form of the coefficient $\phi^S(S,t,u)$ which has a similar structure to the listed above.

Remark 3.2. The jet bundle $M^{(2)}$ is an example of a locally trivial smooth vector bundle $(M^{(2)}, \pi, X)$ where π is a smooth map $\pi \to X$ and X is the base space (see [21]).

In the jet bundle $M^{(2)}$ equation (16) is equivalent to the relation

$$\Delta(w) = 0, \quad w \in M^{(2)}, \tag{23}$$

where we denote by Δ the following function

$$\Delta(S,t,u,u_S,u_t,u_{SS},u_{St},u_{tt}) = u_t + \frac{1}{2}\sigma^2S^2v(\rho u_S, \rho S u_{SS})u_{SS}. \tag{24}$$

We identify the algebraic relation (23) with its solution manifold L_Δ defined by

$$L_\Delta = \{w \in M^{(2)} | \Delta(w) = 0\} \subset M^{(2)}. \tag{25}$$

Let us consider an action of a Lie-point group on our differential equation and its solutions. We define a symmetry group G_Δ of equation (23) by

$$G_\Delta = \{g \in \mathrm{Diff}(\mathrm{M}^{(2)}) |\ \mathrm{g}:\ \ \mathrm{L}_\Delta \to \mathrm{L}_\Delta\}, \tag{26}$$

consequently we are interested in a subgroup of $\mathrm{Diff}(\mathrm{M}^{(2)})$ which is compatible with the structure of L_Δ.

We follow as usual the idea of Sophus Lie [22]: instead to determine directly a complicated structure of the Lie-point symmetry group G_Δ we first determine the corresponding symmetry Lie algebra $\mathcal{D}iff_\Delta(M^{(2)}) \subset \mathcal{D}iff(M^{(2)})$ and then use the main Lie theorem to obtain G_Δ and its invariants.

The symmetry algebra $\mathcal{D}iff_{\Delta}(M^{(2)})$ of the second order differential equation (23) can be found as a solution to the determining equations

$$pr^{(2)}V(\Delta) = 0 \; (mod(\Delta = 0)), \tag{27}$$

the expression $mod(\Delta = 0)$ means that the equation (27) should be satisfied on the solution manifold L_{Δ} only.

If the infinitesimal generators $pr^{(2)}V \in \mathcal{D}iff_{\Delta}(M^{(2)})$ exist then they have the structure of type (20) and form a Lie algebra $\mathcal{D}iff_{\Delta}(M^{(2)})$. All of these infinitesimal generators are uniquely defined as prolongations of operators of the type (18). It is well known [22], [26], [25] that the prolongation procedure preserves the Lie algebraic structure and we can take into account for further studies the algebra $\mathcal{D}iff_{\Delta}(M)$ only. The algebra $\mathcal{D}iff_{\Delta}(M^{(2)})$ has the same algebraic structure as $\mathcal{D}iff_{\Delta}(M)$.

The symmetry algebra $\mathcal{D}iff_{\Delta}(M)$ defines by the Lie equations the corresponding symmetry group G_{Δ} of the equation (23)(see [22], [26], [25]). The Lie group of transformations $G_{\Delta}(M)$ acting on the space M induce in unique manner the transformations on $M^{(2)}$ which form a group denoted by G_{Δ}.

To find the global form of point transformations for the solutions to equation (16) corresponding to this symmetry group $G_{\Delta}(M)$ we just integrate the system of ODEs. The Lie equations in our case take a form

$$\frac{d\tilde{S}}{d\epsilon} = \xi(\tilde{S}, \tilde{t}, u), \tag{28}$$

$$\frac{d\tilde{t}}{d\epsilon} = \tau(\tilde{S}, \tilde{t}, \tilde{u}), \tag{29}$$

$$\frac{d\tilde{u}}{d\epsilon} = \phi(\tilde{S}, \tilde{t}, \tilde{u}), \tag{30}$$

with initial conditions

$$\tilde{S}|_{\epsilon=0} = S, \quad \tilde{t}|_{\epsilon=0} = t, \quad \tilde{u}|_{\epsilon=0} = u. \tag{31}$$

Here the variables $\tilde{S}, \tilde{t}$ and $\tilde{u}$ denote values S, t, u after a symmetry transformation. The parameter ϵ describes a motion along an orbit of the group. Usually it is possible to solve the autonomous system of equations (28) in closed analytical form.

If we have global transformations which form the symmetry group $G_{\Delta}(M)$ it is rather easy to obtain analytical expressions which are invariant under the action of this Lie group, i.e. invariants of the group. One can obtain invariants as well on a direct way by solving a system of equations which follows from the Lie equations. Both methods are equivalent and provide the same set of invariants. The form of invariants is not unique because any function of invariants is as well an invariant. Always we obtain one and the same number of functionally independent invariants. In case of ODEs invariants are the first integrals and if we have a rich set of invariants with some additional properties then we can obtain solutions to the ODEs without any additional integration procedure. In case of PDEs the situation is much more complicated and we cannot hope to obtain in some sense general solutions to a PDE. Using the invariants we can introduce invariant variables and reduce the partial differential equation to some set of ODEs. Each of these ODEs gives rise to a family of group-invariant solutions to the partial differential equation. The famous Black-Scholes formula for the Call option is one of the examples for the invariant solutions of these type.

3.2. Previous Results

In this paragraph we give a short overview of the results presented in [5]. These results will be used in the next paragraph to provide the complete set of invariant solutions to the equation (9).

In the paper [5] the slightly more general equation

$$u_t + \frac{\sigma^2 S^2}{2} \frac{u_{SS}}{(1 - \rho\lambda(S) S u_{SS})^2} = 0, \tag{32}$$

with a continuous function $\lambda : (0, \infty) \to (0, \infty)$ is studied (note that for $\lambda \equiv 1$ this equation reduces to (9)) and the following theorems were proved.

Theorem 3.1. *The differential equation (32) with an arbitrary function $\lambda(S)$ possesses a trivial three dimensional Lie algebra $Diff_\Delta(M)$ spanned by infinitesimal generators*

$$V_1 = \frac{\partial}{\partial t}, \quad V_2 = S\frac{\partial}{\partial u}, \quad V_3 = \frac{\partial}{\partial u}.$$

Only for the special form of the function $\lambda(S) \equiv \omega S^k$, where $\omega, k \in \mathbb{R}$ equation (32) *admits a non-trivial four dimensional Lie algebra spanned by generators*

$$V_1 = \frac{\partial}{\partial t}, \quad V_2 = S\frac{\partial}{\partial u}, \quad V_3 = \frac{\partial}{\partial u}, \quad V_4 = S\frac{\partial}{\partial S} + (1-k)u\frac{\partial}{\partial u}. \tag{33}$$

The proof of this theorem is rather easy. We solve the system of the determining equations (27) where Δ is defined by (32). The calculations are tedious but straightforward. Finally, we obtain the admissible form of the coefficients ξ, τ, ϕ in (18). We notice that the four dimensional Lie algebra (33) has a three dimensional Abelian subalgebra and in case $k = 0, 1$ a two dimensional center. In these two cases it is possible to find invariant solutions in closed analytical form.

On the next step we use the Lie algebra (33) to obtain the symmetry group of the studied equation. The Lie equations (28) in case of the Lie algebra (33) are simple. The results of these calculations are presented in the form of the following theorem.

Theorem 3.2. *The action of the symmetry group G_Δ of (32) with an arbitrary function $\lambda(S)$ is given by*

$$\tilde{S} = S, \tag{34}$$

$$\tilde{t} = t + \epsilon_1, \tag{35}$$

$$\tilde{u} = u + S\epsilon_2 + \epsilon_3, \qquad \epsilon_1, \epsilon_2, \epsilon_3 \in \mathbb{R}. \tag{36}$$

If the function $\lambda(S)$ has the special form $\lambda(S) = \omega S^k$, $k, \omega \in \mathbb{R}$, equation (32) takes the special form

$$u_t + \frac{\sigma^2 S^2}{2} \frac{u_{SS}}{(1 - bS^{k+1} u_{SS})^2} = 0, \quad b = \omega\rho. \tag{37}$$

In that case the symmetry group G_Δ has the richer structure

$$\begin{aligned} \tilde{S} &= Se^{\epsilon_1}, && (38)\\ \tilde{t} &= t + \epsilon_2, \\ \tilde{u} &= ue^{(1-k)\epsilon_1} + \frac{\epsilon_3}{k} Se^{\epsilon_1}(1 - e^{-k\epsilon_1}) \\ &+ \frac{\epsilon_4}{(1-k)}(e^{(1-k)\epsilon_1} - 1), \quad k \neq 0, k \neq 1, && (39)\\ \tilde{u} &= ue^{\epsilon_1} + S\epsilon_3 e^{\epsilon_1} + \epsilon_4(e^{\epsilon_1} - 1), \quad k = 0, \\ \tilde{u} &= u + \epsilon_3 S(e^{\epsilon_1} - 1) + \epsilon_4, \quad k = 1. && (40)\\ &\epsilon_1, \epsilon_2, \epsilon_3, \epsilon_4 \in \mathbb{R}. \end{aligned}$$

As we will see later, the solutions found in [5] for the case $\lambda(S) = S$, i.e. for the case $k = 1$ can be used to obtain the complete set of invariant solutions to equation (9), i.e. for the case $k = 0$.

For $\lambda(S) = \omega S^k$ with $k = 0, 1$ we can use (38) to obtain a new independent invariant variable z and (40) to obtain a new dependent variable v; these variables are given by

$$\begin{aligned} z &= \log S + at, \quad a \neq 0, \\ v &= u\, S^{(k-1)}, \quad k = 0, 1. \end{aligned} \qquad (41)$$

Using these variables we can reduce the PDE (37) to the ODE

$$v_z + q\frac{v_{zz} + \xi v_z}{(1 - b(v_{zz} + \xi v_z))^2} = 0, \qquad (42)$$

where $q = \frac{\sigma^2}{2a}$, $a \neq 0$, $b = \omega\rho \neq 0$, $\xi = (-1)^k$, $k = 0, 1$. It is a straightforward consequence of Theorem 3.2 that these are the only nontrivial invariant variables. In the sequel we will determine explicit solutions for (42) for the case $k = 0$ which corresponds to the original equation (9).

Remark 3.3. Equation (42) with $k = 0$ and $k = 1$ are related to each other. Note however, that the relation between the corresponding solutions is not so straightforward, because (42) is nonlinear and we need real valued solutions. Hence the results from [5] — where (42) was studied for the case $k = 1$ — do not carry over directly to the present case $k = 0$, so that a detailed analysis of the case $k = 0$ is necessary.

3.3. Symmetry Reductions of the Equation (9)

We use the previous results and continue with an analytical treatment of the equation (42) in case $k = 0$. To find families of invariant solution to (9) we introduce a new dependent variable

$$y(z) = v_z(z). \qquad (43)$$

If we assume that the denominator of the equation (42) is different from zero, we can multiply both terms of equation (42) by the denominator of the second term and obtain

$$yy_z^2 + 2\xi\left(y^2 - \xi\frac{1}{b}y + \xi\frac{q}{2b^2}\right)y_z + \left(y^2 - \xi\frac{2}{b}y + \left(\frac{1+\xi q}{b^2}\right)\right)y = 0, \quad b \neq 0. \qquad (44)$$

We denote the left hand side of this equation by $F(y, y_z)$. The equation (44) can possess exceptional solutions which are the solutions of a system

$$\frac{\partial F(y, y_z)}{\partial y_z} = 0, \qquad F(y, y_z) = 0. \tag{45}$$

The first equation in this system defines a discriminant curve which has the form

$$y(z) = \frac{q}{4b}. \tag{46}$$

If this curve is also a solution of the original equation (44) then we obtain an exceptional solution to equation (44). We obtain an exceptional solution if $q = -4\xi$, i.e. $a = \xi\sigma^2/8$. It has the form

$$y(z) = -\frac{\xi}{b}, \quad \xi = (-1)^k, \quad k = 0, 1. \tag{47}$$

This solution belongs to the family of solutions (49) by the specified value of the parameter q. In all other cases the equation (44) does not possess any exceptional solutions.

Hence the set of solutions of equation (44) is the union of the solution sets of the following equations

$$y = 0, \tag{48}$$

$$y = \frac{\xi \pm \sqrt{q}}{b}, \tag{49}$$

$$y_z = \left(-\xi y^2 + \frac{1}{b}y - \frac{q}{2b^2} - \sqrt{\frac{q}{b^3}\left(\frac{q}{4b} - y\right)}\right)\frac{1}{y}, \quad y \neq 0, \tag{50}$$

$$y_z = \left(-\xi y^2 + \frac{1}{b}y - \frac{q}{2b^2} + \sqrt{\frac{q}{b^3}\left(\frac{q}{4b} - y\right)}\right)\frac{1}{y}, \quad y \neq 0, \tag{51}$$

where one of the solutions (49) is an exceptional solution (47) by $q = -4\xi$ for $k = 1$. In the case $k = 0$ we have $\xi = 1$ so that solutions of (49) are complex valued functions. We denote the right hand side of equations (50), (51) by $f(y)$. The Lipschitz condition for equations of the type $y_z = f(y)$ is satisfied in all points where the derivative $\frac{\partial f}{\partial y}$ exists and is bounded. It is easy to see that this condition will not be satisfied by

$$y = 0, \ \ y = \frac{q}{4b}, \ \ y = \infty. \tag{52}$$

Hence on the lines (52) the uniqueness of solutions of equations (50), (51) can be lost. We will study in detail the behavior of solutions in the neighborhood of the lines (52). For this purpose we look at the equation (44) from another point of view. If we assume now that z, y, y_z are complex variables and denote

$$y(z) = \zeta, \ \ y_z(z) = w, \ \ \zeta, w \in \mathbb{C}, \tag{53}$$

then the equation (44) takes the form

$$F(\zeta, w) = \zeta w^2 + 2\xi\left(\zeta^2 - \xi\frac{1}{b}\zeta + \xi\frac{q}{2b^2}\right)w + \left(\zeta^2 - \xi\frac{2}{b}\zeta + \frac{1 + \xi q}{b^2}\right)\zeta = 0, \tag{54}$$

where $b \neq 0$. The equation (54) is an algebraic relation in $\mathbb{C}^2$ and defines a plane curve in this space. The polynomial $F(\zeta, w)$ is an irreducible polynomial if at all roots $w_r(z)$ of $F(\zeta, w_r)$ either the partial derivative $F_\zeta(\zeta, w_r)$ or $F_w(\zeta, w_r)$ are non zero. It is easy to prove that the polynomial (54) is irreducible.

We can treat equation (54) as an algebraic relation which defines a Riemannian surface $\Gamma\,:\,F(\zeta, w) = 0$ of $w = w(\zeta)$ as a compact manifold over the ζ-sphere. The function $w(\zeta)$ is uniquely analytically extended over the Riemann surface Γ of two sheets over the $\zeta-$sphere. We find all singular or branch points of $w(\zeta)$ if we study the roots of the first coefficient of the polynomial $F(\zeta, w)$, the common roots of equations

$$F(\zeta, w) = 0, \quad F_w(\zeta, w) = 0, \qquad \zeta, w \in \mathbb{C} \cup \infty. \tag{55}$$

and the point $\zeta = \infty$. The set of singular or branch points consists of the points

$$\zeta_1 = 0, \;\; \zeta_2 = \frac{q}{4b}, \;\; \zeta_3 = \infty. \tag{56}$$

As expected we got the same set of points as in the real case (52) by the study of the Lipschitz condition but now the behavior of solutions at the points is more visible.

The points ζ_2, ζ_3 are the branch points at which two sheets of Γ are glued on. We remark that

$$w(\zeta_2) = -\frac{1}{b}\left(1 - \xi\frac{q}{4}\right) - t\frac{4\xi}{\sqrt{-bq}} + \cdots, \qquad t^2 = \zeta - \frac{q}{4b}, \tag{57}$$

where t is a local parameter in the neighborhood of ζ_2. For the special value of $q = 4\xi$, i.e. $k = 0$, the value $w(\zeta_2)$ is equal to zero.

At the point $\zeta_3 = \infty$ we have

$$w(\zeta) = -\frac{\xi}{t^2} + \frac{1}{b} - \xi t\sqrt{\frac{-q}{4b^3}}, \quad t^2 = \frac{1}{\zeta}, \quad \zeta \to \infty,$$

where t is a local parameter in the neighborhood of ζ_3. At the point $\zeta_1 = 0$ the function $w(\zeta)$ has the following behavior

$$w(\zeta) \;\sim\; -\frac{q}{b^2}\frac{1}{\zeta}, \quad \zeta \to \zeta_1 = 0, \text{ on the principal sheet,} \tag{58}$$

$$w(\zeta) \;\sim\; -\left(\xi + \frac{1}{q}\right)\zeta, \quad \zeta \to \zeta_1 = 0,\; q \neq 1, \text{ on the second sheet,} \tag{59}$$

$$w(\zeta) \;\sim\; -2b^2\zeta^2, \quad \zeta \to \zeta_1 = 0,\; q = 1, \text{ on the second sheet.} \tag{60}$$

Any solution $w(\zeta)$ of an irreducible algebraic equation (54) is meromorphic on this compact Riemann surface Γ of the genus 0 and has a pole of the order one correspondingly (58) over the point $\zeta_1 = 0$ and the pole of the second order over $\zeta_3 = \infty$. It means also that the meromorphic function $w(\zeta)$ cannot be defined on a manifold of less than 2 sheets over the ζ sphere.

To solve the differential equations (50) and (51) from this point of view it is equivalent to integrate on Γ a differential of the type $\frac{\mathrm{d}\zeta}{w(\zeta)}$ and then to solve an Abel's inverse problem of degenerated type

$$\int \frac{\mathrm{d}\zeta}{w(\zeta)} = z + \text{const.} \tag{61}$$

The integration can be done very easily because we can introduce a uniformizing parameter on the Riemann surface Γ and represent the integral (61) in terms of rational functions merged possibly with logarithmic terms.

To realize this program we introduce a new variable (our uniformizing parameter p) by

$$\zeta = \frac{q(1-p^2)}{4b}, \tag{62}$$

$$w = \frac{\xi(p-1)(q(1+p)^2+4\xi)}{4b(p+1)}. \tag{63}$$

Then the equations (50) and (51) will take the form

$$2q\xi \int \frac{p(p+1)\mathrm{d}p}{(p-1)(q(p+1)^2+4\xi)} = z + \mathrm{const}, \tag{64}$$

$$2q\xi \int \frac{p(p-1)\mathrm{d}p}{(p+1)(q(p-1)^2+4\xi)} = z + \mathrm{const}. \tag{65}$$

The integration procedure of equations (64), (65) gives rise to relations defining a complete set of first order differential equations. In order to see that these are first order ODEs recall that from the substitutions (53) and (43) we have

$$p = \sqrt{1 - \frac{4b}{q} v_z}. \tag{66}$$

We summarize all these results in the following theorem.

Theorem 3.3. *The equation (42) for arbitrary values of the parameters $q, b \neq 0$ can be reduced to the set of first order differential equations which consists of the equations*

$$v_z = 0, \quad v_z = (1 \pm 2)/b, \tag{67}$$

and equations (50), (51) where y is defined by the substitution (43). The complete set of solutions of the equation (42) coincides with the union of solutions of these equations.

To solve equations (64), (65) exactly we have first to integrate and then invert these formulas in order to obtain an exact representation of p as a function of z. If an exact formula for the function $p = p(z)$ is found we can use the substitution (66) to obtain an explicit ODE of the type $v_z(z) = f(z)$ or another suitable type; in that case it is possible to solve the problem completely. However, for an arbitrary value of the parameter q inversion is impossible, and we have just an implicit representation for the solutions of the equation (42) as solutions of the implicit first order differential equations.

3.4. Exact Invariant Solutions for a Fixed Relation between S and t

For a special value of the parameter q, namely for $q = -4$, we can integrate and invert the equations (64) and (65). For $q = -4$ the relation between the variables S and t is fixed in the form

$$z = \log S - \frac{\sigma^2}{8} t, \tag{68}$$

equation (64) takes the form

$$(p-1)^2(p+2) = 2\,c\,\exp\left(-3z/2\right), \tag{69}$$

and correspondingly the equation (65) becomes

$$(p+1)^2(p-2) = 2\,c\,\exp\left(-3z/2\right), \tag{70}$$

where c an arbitrary constant. It is easy to see that the equations (69) and (70) are connected by a transformation

$$p \to -p, \quad c \to -c. \tag{71}$$

This symmetry arises from the symmetry of the underlying Riemann surface Γ (54) and corresponds to a change of the sheets on Γ. Using these symmetry properties we can prove the following theorem.

Theorem 3.4. *The second order differential equation*

$$v_z - 4\frac{v_{zz}+v_z}{(1-b(v_{zz}+v_z))^2} = 0, \tag{72}$$

is exactly integrable for any value of the parameter b. The complete set of solutions for $b \neq 0$ is given by the union of solutions (76), (78)-(81) and solutions

$$v(z) = d, \quad v(z) = \frac{3}{b}z + d, \quad v(z) = -\frac{1}{b}z + d, \tag{73}$$

where d is an arbitrary constant. The last solution in (73) corresponds to the exceptional solution of equation (44).

For $b = 0$ equation (72) is linear and its solutions are given by $v(z) = d_1 + d_2\exp-(3z/4)$, where d_1, d_2 are arbitrary constants.

Proof. Because of the symmetry (71) it is sufficient to study either the equations (69) or (70) for $c \in \mathbb{R}$ or both these equations for $c > 0$. The value $c = 0$ can be excluded because it complies with the constant value of $p(z)$ and correspondingly constant value of $v_z(z)$, but all such cases are studied before and the solutions are given by (73).

We will study equation (70) in case $c \in \mathbb{R}\setminus\{0\}$ and obtain on this way the complete class of exact solutions for equations (69)-(70) and on this way for the equation (72).

Equation (70) for $c > 0$ has one real root only. It leads to an ODE of the form

$$\begin{aligned} v_z(z) = \frac{1}{b}\Bigg(1 &+ \left(1 + c e^{-\frac{3z}{2}} + \sqrt{2ce^{-\frac{3z}{2}} + c^2e^{-3z}}\right)^{-\frac{2}{3}} \\ &+ \left(1 + c e^{-\frac{3z}{2}} + \sqrt{2ce^{-\frac{3z}{2}} + c^2e^{-3z}}\right)^{\frac{2}{3}}\Bigg), \quad c > 0. \end{aligned} \tag{74}$$

Equation (74) can be exactly integrated if we use an Euler substitution and introduce a new independent variable

$$\tau = 2\left(1 + c e^{\frac{-3z}{2}} + \sqrt{2ce^{\frac{-3z}{2}} + c^2e^{-3z}}\right). \tag{75}$$

The corresponding solution is given by

$$\begin{aligned} v_r(z) &= -\frac{1}{b}\left(1+c\,e^{-\frac{3\,z}{2}}+\sqrt{2\,c\,e^{-\frac{3\,z}{2}}+c^2\,e^{-3\,z}}\right)^{-\frac{2}{3}} \\ &-\frac{1}{b}\left(1+c\,e^{-\frac{3\,z}{2}}+\sqrt{2\,c\,e^{-\frac{3\,z}{2}}+c^2\,e^{-3\,z}}\right)^{\frac{2}{3}} \\ &-\frac{2}{b}\log\left(\left(1+c\,e^{-\frac{3\,z}{2}}+\sqrt{2\,c\,e^{-\frac{3\,z}{2}}+c^2\,e^{-3\,z}}\right)^{-\frac{1}{3}}\right. \\ &\left.+\left(1+c\,e^{-\frac{3\,z}{2}}+\sqrt{2\,c\,e^{-\frac{3\,z}{2}}+c^2\,e^{-3\,z}}\right)^{\frac{1}{3}}-2\right)+d\,, \end{aligned} \tag{76}$$

where $d \in \mathbb{R}$ is an arbitrary constant. If in the right hand side of equation (70) the parameter c satisfies the inequality $c < 0$ and the variable z chosen in the region

$$z \in \left(-\frac{2}{3}\ln\frac{2}{|c|}, \infty\right) \tag{77}$$

then the equation on p possesses maximal three real roots. These three roots of the cubic equation (70) give rise to three differential equations of the type $v_z = -(1-p^2(z))/b$. These equations can be exactly solved and we find correspondingly three solutions $v_i(z),\ i = 1, 2, 3$.

The first solution is given by the expression

$$\begin{aligned} v_1(z) &= -\frac{z}{b}-\frac{2}{b}\cos\left(\frac{2}{3}\arccos\left(1-|c|\,e^{\frac{-3\,z}{2}}\right)\right) \\ &-\frac{4}{3b}\log\left(1+2\cos\left(\frac{1}{3}\arccos\left(1-|c|\,e^{\frac{-3\,z}{2}}\right)\right)\right) \\ &-\frac{16}{3b}\log\left(\sin\left(\frac{1}{6}\arccos\left(1-|c|\,e^{\frac{-3\,z}{2}}\right)\right)\right)+d, \end{aligned} \tag{78}$$

where $d \in \mathbb{R}$ is an arbitrary constant. The second solution is given by the formula

$$\begin{aligned} v_2(z) &= -\frac{z}{b}-\frac{2}{b}\cos\left(\frac{2}{3}\pi-\frac{2}{3}\arccos\left(1-|c|\,e^{\frac{-3\,z}{2}}\right)\right) \\ &-\frac{4}{3b}\log\left(-1+2\cos\left(\frac{1}{3}\pi-\frac{1}{3}\arccos\left(1-|c|\,e^{\frac{-3\,z}{2}}\right)\right)\right) \\ &-\frac{16}{3b}\log\left(\sin\left(\frac{1}{6}\pi-\frac{1}{6}\arccos\left(1-|c|\,e^{\frac{-3\,z}{2}}\right)\right)\right)+d, \end{aligned} \tag{79}$$

where $d \in \mathbb{R}$ is an arbitrary constant. The first and second solutions are defined up to the point $z = -\frac{2}{3}\ln\frac{2}{|c|}$ where they coincide (see Fig. 1).

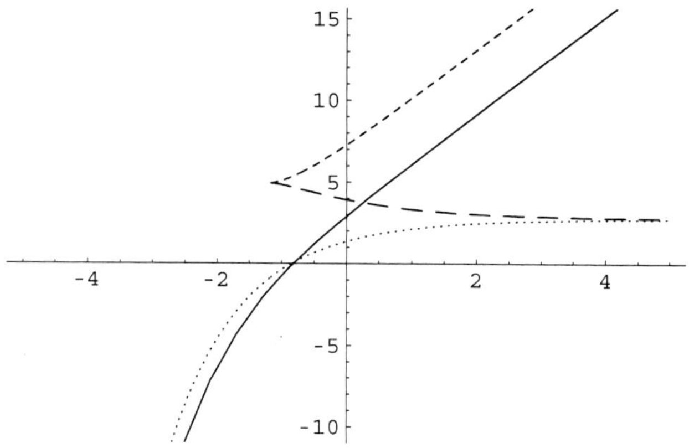

Figure 1. Plot of the solution $v(z)$ given in (76) (thick solid line); $v_1(z)$ from (78) (short dashed line); $v_2(z)$ from (79) (long dashed line) and the third solution $v_{3,1}(z), v_{3,2}(z)$ from (80), (81) (thin solid line). The parameters takes the values $|c| = 0.35$, $q = -4$, $d = 0$, $b = 1$ and the variable $z \in (-5, 4.5)$.

The third solution for $z > -\frac{2}{3} \ln \frac{2}{|c|}$ is given by the formula

$$\begin{aligned} v_{3,2}(z) = & -\frac{z}{b} - \frac{2}{b} \cos\left(\frac{2}{3}\pi + \frac{2}{3} \arccos\left(1 - |c|\, e^{\frac{-3z}{2}}\right)\right) \\ & - \frac{4}{3b} \log\left(-1 + 2\cos\left(\frac{1}{3}\pi + \frac{1}{3} \arccos\left(-1 + |c|\, e^{\frac{-3z}{2}}\right)\right)\right) \\ & - \frac{16}{3b} \log\left(\cos\left(\frac{1}{6}\pi + \frac{1}{6} \arccos\left(-1 + |c|\, e^{\frac{-3z}{2}}\right)\right)\right) + d, \end{aligned} \tag{80}$$

where $d \in \mathbb{R}$ is an arbitrary constant. In the case $z < \frac{2}{3} \ln \frac{2}{|c|}$ the polynomial (70) has one real root and the corresponding solution can be represented by the formula

$$\begin{aligned} v_{3,1}(z) = & -\frac{z}{b} - \frac{2}{b} \cosh\left(\frac{2}{3} \operatorname{arccosh}\left(-1 + |c|\, e^{\frac{-3z}{2}}\right)\right) \\ & - \frac{16}{3\,b} \log\left(\cosh\left(\frac{1}{6} \operatorname{arccosh}\left(-1 + |c|\, e^{\frac{-3z}{2}}\right)\right)\right) \\ & - \frac{4}{3\,b} \log\left(-1 + 2\cosh\left(\frac{1}{3} \operatorname{arccosh}\left(-1 + |c|\, e^{\frac{-3z}{2}}\right)\right)\right) + d. \end{aligned} \tag{81}$$

The third solution is represented by formulas $v_{3,2}(z)$ and $v_{3,1}(z)$ for different values of the variable z.

One of the sets of solutions (76), (78) -(81) for fixed parameters b, c, d is represented in Fig. 1. The first solution (74) and the third solution given by both (80) and (81) are defined for any values of z. The solutions $v_1(z)$ and $v_2(z)$ cannot be continued to the left after the point $z = -\frac{2}{3} \ln \frac{2}{|c|}$ where they coincide.

If we keep in mind that $z = \log S - \frac{\sigma^2}{8} t$ and $u(S, t) = S\, v(z)$ we can represent the complete set of exact invariant solution of equation (9). The solution (76) gives rise to an

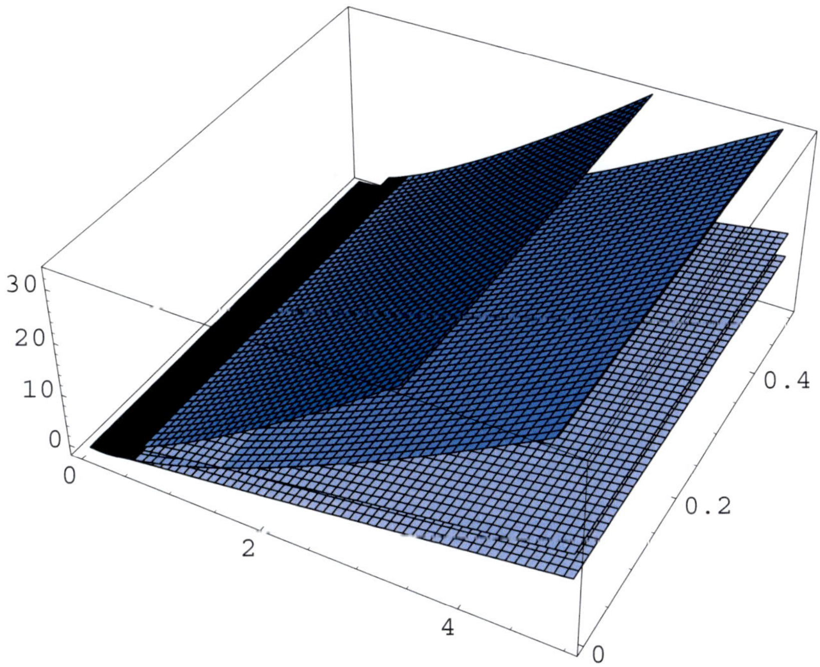

Figure 2. Plot of solutions $u_r(S,t), u_1(S,t), u_2(S,t)$ and $u_{3,1}(S,t), u_{3,2}(S,t)$ for the parameters $\sigma = 0.4,\ |c| = 0.5, q = -4, b = 1.0, d = 0$. The variables S, t lie in intervals $S \in (0., 5.)$ and $t \in [0, 0.5]$. All invariant solutions change slowly in t-direction.

invariant solution $u_r(S,t)$ in the form

$$\begin{aligned} u_r(S,t) &= -\frac{1}{\omega\rho} S \left(1 + c\, S^{-\frac{3}{2}} e^{\frac{3\sigma^2}{16} t} + \sqrt{2\, c\, S^{-\frac{3}{2}} e^{\frac{3\sigma^2}{16} t} + c^2\, S^{-3} e^{\frac{3\sigma^2}{8} t}}\right)^{-\frac{2}{3}} \\ &- \frac{1}{\omega\rho} S \left(1 + c\, S^{-\frac{3}{2}} e^{\frac{3\sigma^2}{16} t} + \sqrt{2\, c\, S^{-\frac{3}{2}} e^{\frac{3\sigma^2}{16} t} + c^2\, S^{-3} e^{\frac{3\sigma^2}{8} t}}\right)^{\frac{2}{3}} \\ &- \frac{2}{\omega\rho} S \log \left(\left(1 + c\, S^{-\frac{3}{2}} e^{\frac{3\sigma^2}{16} t} + \sqrt{2\, c\, S^{-\frac{3}{2}} e^{\frac{3\sigma^2}{16} t} + c^2\, S^{-3} e^{\frac{3\sigma^2}{8} t}}\right)^{-\frac{1}{3}} \right. \\ &+ \left. \left(1 + c\, S^{-\frac{3}{2}} e^{\frac{3\sigma^2}{16} t} + \sqrt{2\, c\, S^{-\frac{3}{2}} e^{\frac{3\sigma^2}{16} t} + c^2\, S^{-3} e^{\frac{3\sigma^2}{8} t}}\right)^{\frac{1}{3}} - 2 \right) + d\, S + d_2 \end{aligned} \tag{82}$$

where $d,\ d_2 \in \mathbb{R}$, $c > 0$. This solution was obtained and studied in [6]. We describe now other invariant solutions from the complete set of invariant solutions.

In the case $c < 0$ we can obtain correspondingly three real solutions if

$$S \geq \left(\frac{|c|}{2}\right)^{2/3} \exp\left(\frac{\sigma^2}{8} t\right). \tag{83}$$

The first solution is represented by

$$
\begin{aligned}
u_1(S,t) = &-\frac{1}{\omega\rho}S\left(\log S-\frac{\sigma^2}{8}t\right)-\frac{2}{\omega\rho}S\cos\left(\frac{2}{3}\arccos\left(1-|c|\,S^{-\frac{3}{2}}e^{\frac{3\sigma^2}{16}t}\right)\right)\\
&-\frac{4}{3\omega\rho}S\log\left(1+2\cos\left(\frac{1}{3}\arccos\left(1-|c|\,S^{-\frac{3}{2}}e^{\frac{3\sigma^2}{16}t}\right)\right)\right)\\
&-\frac{16}{3\omega\rho}S\log\left(\sin\left(\frac{1}{6}\arccos\left(1-|c|\,S^{-\frac{3}{2}}e^{\frac{3\sigma^2}{16}t}\right)\right)\right)+d\,S+d_2,
\end{aligned}
\tag{84}
$$

where $d,\ d_2\in\mathbb{R}$, $c<0$. The second solution is given by the formula

$$
\begin{aligned}
u_2(S,t) &= -\frac{1}{\omega\rho}S\left(\log S-\frac{\sigma^2}{8}t\right)-\frac{2}{\omega\rho}S\cos\left(\frac{2}{3}\pi+\frac{2}{3}\arccos\left(-1+|c|\,S^{-\frac{3}{2}}e^{\frac{3\sigma^2}{16}t}\right)\right)\\
&- \frac{4}{3\omega\rho}S\log\left(1+2\cos\left(\frac{1}{3}\pi+\frac{1}{3}\arccos\left(-1+|c|\,S^{-\frac{3}{2}}e^{\frac{3\sigma^2}{16}t}\right)\right)\right)\\
&- \frac{16}{3\omega\rho}S\log\left(\sin\left(\frac{1}{6}\pi+\frac{1}{6}\arccos\left(-1+|c|\,S^{-\frac{3}{2}}e^{\frac{3\sigma^2}{16}t}\right)\right)\right)+d\,S+d_2.
\end{aligned}
\tag{85}
$$

where $d\ ,\ d_2\in\mathbb{R}$, $c<0$. The first and second solutions are defined for the variables under conditions (83). They coincide along the curve

$$
S=\left(\frac{|c|}{2}\right)^{2/3}\exp\left(\frac{\sigma^2}{8}t\right)
$$

and cannot be continued further. The third solution is defined by

$$
\begin{aligned}
u_{3,2}(S,t) &= -\frac{1}{\omega\rho}S\left(\log S-\frac{\sigma^2}{8}t\right)\\
&- \frac{2}{\omega\rho}S\cos\left(\frac{2}{3}\arccos\left(-1+|c|\,S^{-\frac{3}{2}}e^{\frac{3\sigma^2}{16}t}\right)\right)\\
&- \frac{4}{3\omega\rho}S\log\left(-1+2\cos\left(\frac{1}{3}\arccos\left(-1+|c|\,S^{-\frac{3}{2}}e^{\frac{3\sigma^2}{16}t}\right)\right)\right)\\
&- \frac{16}{3\omega\rho}S\log\left(\cos\left(\frac{1}{6}\arccos\left(-1+|c|\,S^{-\frac{3}{2}}e^{\frac{3\sigma^2}{16}t}\right)\right)\right)+d\,S+d_2,
\end{aligned}
\tag{86}
$$

where $d,\, d_2\in\mathbb{R}$ and S,t satisfy the condition (83).

In case $0<S<\left(\frac{|c|}{2}\right)^{\frac{2}{3}}\exp\left(\frac{\sigma^2}{8}t\right)$ the third solution can be represented by the formula

$$
\begin{aligned}
u_{3,1}(S,t) &= -\frac{1}{\omega\rho}S\left(\log S-\frac{\sigma^2}{8}t\right)\\
&- \frac{2}{\omega\rho}S\cosh\left(\frac{2}{3}\operatorname{arccosh}\left(-1+|c|\,S^{-\frac{3}{2}}e^{\frac{3\sigma^2}{16}t}\right)\right)\\
&- \frac{16}{3\,\omega\rho}S\log\left(\cosh\left(\frac{1}{6}\operatorname{arccosh}\left(-1+|c|\,S^{-\frac{3}{2}}e^{\frac{3\sigma^2}{16}t}\right)\right)\right)\\
&- \frac{4}{3\,\omega\rho}S\log\left(-1+2\cosh\left(\frac{1}{3}\operatorname{arccosh}\left(-1+|c|\,S^{-\frac{3}{2}}e^{\frac{3\sigma^2}{16}t}\right)\right)\right)\\
&+ d\,S+d_2,
\end{aligned}
\tag{87}
$$

where $d,\ d_2 \in \mathbb{R}$. The solution $u(S,t)$ (82) and the third solution given by $u_{3,1}$, $u_{3,2}$ (86), (87) are defined for all values of variables t and $S > 0$. They have a common intersection curve of the type $S = \text{const.}\ \exp(\frac{\sigma^2}{8}t)$. The typical behavior of all these invariant solutions is represented in Fig. 2.

We summarize the previous results in the following theorem which describes the set of invariant solutions of equation (9).

Theorem 3.5. *The invariant solutions of equation (9) can be defined by the set of first order ordinary differential equations listed in Theorem 3.3.*

If moreover the parameter $q = -4$, or equivalent in the substitution (41) we chose $a = -\sigma^2/8$ then the complete set of invariant solutions of (9) can be found exactly. This set of invariant solutions is given by formulas (82)–(87) and by solutions

$$u(S,t) = d\ S, \quad u(S,t) = \frac{3}{b}\ S\ (\log S - \sigma^2 \frac{t}{8}),$$
$$u(S,t) = -\frac{1}{b}\ S\ (\log S - \sigma^2 \frac{t}{8}),$$

where d denotes an arbitrary constant. This set of invariant solutions is unique up to the transformations of the symmetry group G_Δ given by Theorem 3.2.

The solutions $u_r(S,t)$, (82), $u_1(S,t)$, (84), $u_2(S,t)$, (85), $u_{3,1}(S,t)$, (86), $u_{3,2}(S,t)$ (87), have no one counterpart in a linear case. If in the equation (9) the parameter $\rho = 0$ the equation becomes linear. If the parameter $\rho \to 0$, then equation (9) and correspondingly equation (37) will be reduced to the linear Black-Scholes equation but the solutions (82)-(87) which we obtained here will be completely blown up by $\rho \to 0$ because of the factor $1/b = 1/(\omega\rho)$ in the formulas (82)-(87). This phenomena was described as well in [6], [9], where the solution $u_r(S,t)$ was studied and for the complete set of invariant solutions of equation (37) with $k = 1$ in [5].

4. Properties of Solutions and Parameter-Sensitivity

We study the properties of solutions, keeping in mind that because of the symmetry properties (see Theorem 3.2) of the equation (9) we can add to each solution a linear function of the variable S.

4.1. Dependence on the Constant c and Terminal Payoff

First we study the dependence of the solution u_2 on the arbitrary constant c. The constant c is the first constant of integration of the ODEs (50), (51). This dependence is illustrated in Figure 3. We see on Fig. 3 that for different values of the constant c the domains of the solutions are different. The dependence of the solutions $u_{3,1}, u_{3,2}(S,t)$ on the constant c is exemplified in Figure 4.

We obtain a typical terminal payoff function for the solutions (82)-(87) if we just fix $t = T$. By changing the value of the constant c and by adding a linear function of S we are able to modify terminal payoff function for the solutions. Hence we can approximate typical payoff profiles of financial derivatives quite well.

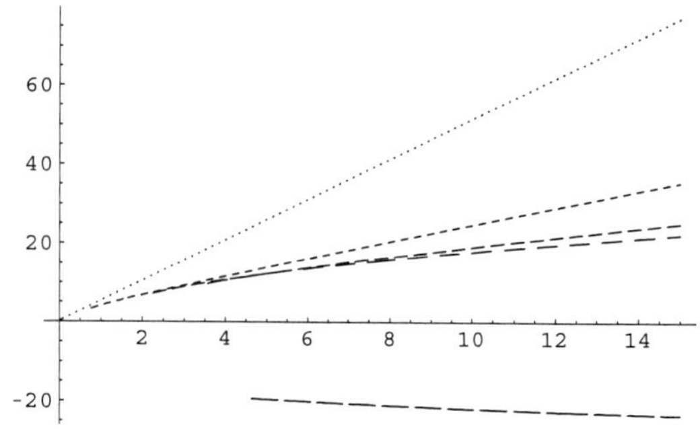

Figure 3. Plot of solutions $u_2(S,t)$ where $|c| = 0.01, 1., 5., 10., 20., \sigma = 0.4, q = -4$, $b = 1.0, d = 0$. The variables S lie in intervals $S \in (0., 15.)$ and $t = 1.$.
The curves going from up to down with the growing value of the parameter $|c|$.

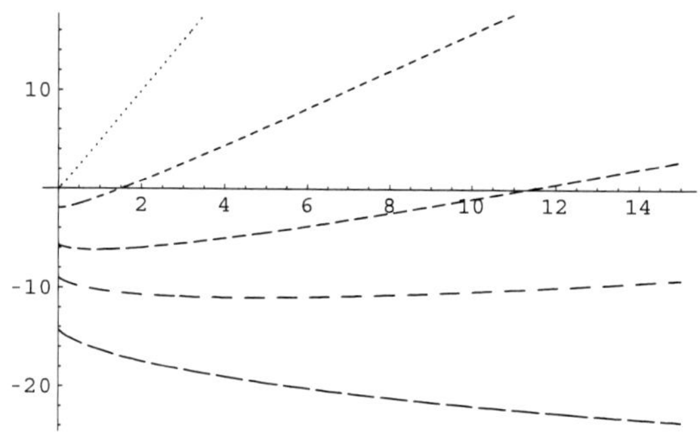

Figure 4. Plot of solutions $u_3(S,t)$ where $|c| = 0.01, 1., 5., 10., 20., \sigma = 0.4, q = -4$, $b = 1.0, d = 0$. The variables S lie in intervals $S \in (0., 15.)$ and $t = 1.$.
The curves going from up to down with the growing value of the parameter $|c|$.

4.2. Dependence on Time

All solutions depend weakly on time because of the substitution (68) all invariant solutions depends on the combination $\sigma^2 t$. As long as we take the volatility σ to be small we obtain a dampened dependence of the solutions on time.

In Fig. 5 we can see this dependence for examples of the solution $u_{3,1}(S,t), u_{3,2}(S,t)$.

4.3. Dependence on the Parameter ρ

All solutions found in this work have the form

$$u(S,t) = w(S,t)/\rho, \tag{88}$$

where $w(S,t)$ is a smooth function of S, t. Hence the function $w(S,t)$ solves the equation (9) with $\rho = 1$,

$$w_t + \frac{\sigma^2 S^2}{2} \frac{w_{SS}}{(1 - S w_{SS})^2} = 0. \tag{89}$$

Because of this relation any ρ-dependence of invariant solutions of (9) is trivial. In particular, if the terminal conditions are fixed, $u(S,T) = h(S)$, then the value $u(S,t)$ will increase

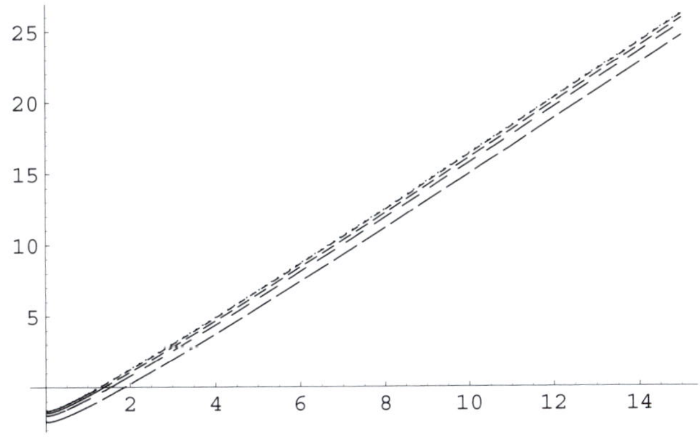

Figure 5. Plot of solutions $u_{3,1}(S,t), u_{3,2}(S,t)$ for the parameters $\sigma = 0.4$, $|c| = 10., q = -4, b = 1.0, d = 0$. The variables S lie in intervals $S \in (0., 15.)$ and $t = 0.01, 1., 5., 10., 20.$. The highest level corresponds to $t = 0.01$ and the lowest to $t = 20.$.

if the value of the parameter ρ increases. This dependence of hedge costs on the position of the large trader on the market is very natural.

4.4. Dependence on the Asset Price S

In practice one use often delta-hedging to reduce the sensitivity of a portfolio to the movements of an underlying asset. Hence it is important to know the value Δ defined by $\Delta = \frac{\partial u}{\partial S}$, where u denotes the value of the derivative product or of a portfolio. Using the exact formulas for the invariant solutions we can easily calculate Δ as a function of S and t. The Δ corresponding to the solution u_2 is presented in Fig. 6

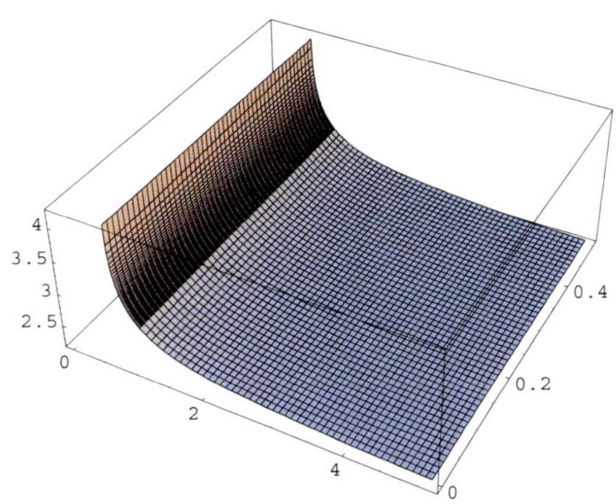

Figure 6. Plot of Δ for $u_2(S,t)$ and $\sigma = 0.4$, $|c| = 1., q = -4, b = 1.0, d = 0$. The variables lie in intervals $S \in (0., 5.)$ and $t = [0., 0.5]$.

The Δ of the solution $u_{3,1}, u_{3,2}$ is represented on the Fig. 7.

We see in both cases the strong dependence on S for small value of S. If $S \to \infty$, in both cases the Δ tends to a constant which is independent of time and the constant c.

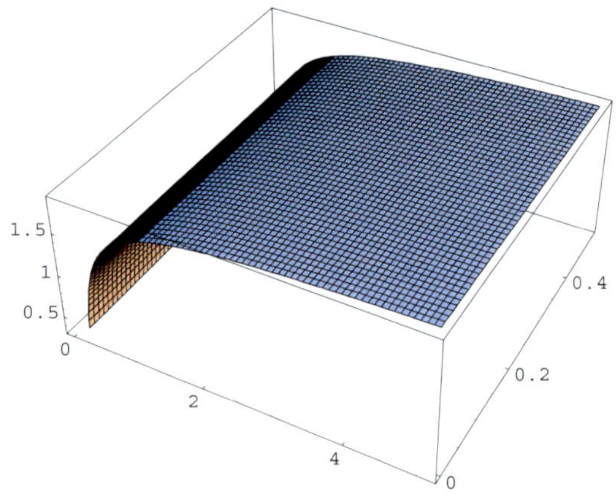

Figure 7. Plot of the Δ for $u_{3,1}(S,t), u_{3,2}(S,t)$ and $\sigma = 0.4$, $|c| = 1., q = -4, b = 1.0, d = 0$. The variables lie in intervals $S \in (0., 5.)$ and $t \in [0.01, 0.5]$.

4.5. Asymptotic Behavior of Invariant Solutions

If S is large enough we have four well defined solutions $u_r(S,t)$, $u_1(S,t)$, $u_2(S,t)$, $u_{3,2}(S,t)$. The asymptotic behavior solutions $u_r(S,t)$ from (82) and u_1 from (84) coincides in the main terms as $S \to \infty$ and is given by the formula

$$u_r(S,t),\ u_1(S,t) \sim \frac{1}{b}\left(3S\ln S + \text{const.}S + \mathcal{O}(S^{-1/2})\right). \tag{90}$$

The exact formula for the asymptotic behavior of the function $u_1(S,t)$ for $S \to \infty$ is given by

$$\begin{aligned} u_1(S,t) \quad &\sim \quad \frac{3}{b}\,S\log(S) + \frac{1}{b}S\left(4\,\log(3) - 2 - \frac{8}{3}\,\log\left(\frac{2}{|c|}\right) - \frac{3}{8}\sigma^2\,t\right) \\ &+ \quad \frac{2^4}{3^3 b}\,|c|\,e^{\frac{3\,\sigma^2\,t}{16}}\,S^{-\frac{1}{2}} + \mathcal{O}\left(S^{-\frac{5}{4}}\right). \end{aligned} \tag{91}$$

We see that for $S \to \infty$, the main term does not depend on time or on the value of the constant c. Moreover, this term cannot be changed by adding a linear function of S to the solution.

The main terms of the solutions $u_2(S,t)$ from (85) and $u_{3,2}$ from (87) behave similar to each other as $S \to \infty$; this behavior is given by the formulas

$$\begin{aligned} u_2(S,t) \quad &\sim \quad \frac{1}{b}\left(1 + \frac{2}{3}\,\log\left(\frac{2^7}{3^3|c|}\right)\right)\,S \qquad (92) \\ &+ \quad \frac{2^3}{3b}\sqrt{\frac{2|c|}{3}}\,e^{\frac{3\,\sigma^2\,t}{32}}\,S^{\frac{1}{4}} - \frac{2^3}{3^3 b}\,|c|\,e^{\frac{3\,\sigma^2\,t}{16}}\,S^{-\frac{1}{2}} + \mathcal{O}\left(S^{-\frac{5}{4}}\right) \end{aligned}$$

and

$$\begin{aligned} u_{3,2}(S,t) \quad &\sim \quad \frac{1}{b}\left(1 + \frac{2}{3}\,\log\left(\frac{2^7}{3^3|c|}\right)\right)\,S \qquad (93) \\ &- \quad \frac{2^3}{3b}\sqrt{\frac{2|c|}{3}}\,e^{\frac{3\,\sigma^2\,t}{32}}\,S^{\frac{1}{4}} - \frac{2^3}{3^3 b}\,|c|\,e^{\frac{3\,\sigma^2\,t}{16}}\,S^{-\frac{1}{2}} + \mathcal{O}\left(S^{-\frac{5}{4}}\right). \qquad (94) \end{aligned}$$

Note that the main term in formulas (92) and (93) depends on S linearly and has a coefficient which depends on the constant c only. Hence we can change the asymptotic behavior of the solutions $u_2(S,t)$ and $u_{3,1}(S,t)$, $u_{3,2}(S,t)$ by simply adding a linear function of S to the solutions.

For S in a neighborhood of $S \to 0$ there exist just two non trivial real invariant solutions of equation (9), i.e. the known solution $u_r(S,1)$ and the new solution $u_{3,1}(S,t)$. Using the exact formulas for the last solution we retain the first term and obtain as $S \to 0$ for the solution $u_{3,1}(S,t)$

$$u_{3,1}(S,t) \sim \frac{-14}{b} S \ln(S) + \mathcal{O}(S). \tag{95}$$

The main term in this formula do not depend on t or constant c and can be changed by adding a linear function of S to the solution.

Conclusion

In this chapter we obtained explicit solutions for the equation (9) and studied their analytical properties. These solutions are useful for a number of reasons. To begin with, while the payoffs of these similarity solutions cannot be chosen arbitrary, the payoffs can be modified using embedded constants to tailor a given portfolio reasonably well. For some values of the parameters c, d we obtain a payoff typical for futures, for other values c, d the payoff is very similar to the form of calls (see Figure 3 and Figure 4). Moreover, the explicit solutions can be used as benchmark for different numerical methods.

An important difference between the case of the linear Black-Scholes equation and these nonlinear cases can be noticed if we consider the asymptotic for $S \to \infty$. In the linear case the price of a Call option satisfies $u(S,t) \to \text{const} \cdot S$. In the nonlinear case the families of similarity solutions u_2 and u_3 which approximate the payoff of a Call option well on a finite interval $[0, \bar{S}]$, grow faster than linear as $S \to \infty$; see the formulas (92) and (93) for details. This reflects the fact that in illiquid markets option hedging is more expensive than in the standard case of perfectly liquid markets.

Invariant solutions can be used as benchmarks for different numerical methods or as a starting point for stability investigations of numerical schemes. We refer to [6] for further information on this issue.

Acknowledgements

The authors are grateful to Albert N. Shiryaev for interesting and fruitful discussions.

References

[1] P. Bank and D. Baum, Hedging and portfolio optimization in financial markets with a large trader, *Math. Finance* **14** (2004), 1–18.

[2] G. Barles and M. Soner, Option pricing with transaction costs and a nonlinear Black-Scholes equation, *Finance Stoch.* **2** (1998), 369–397.

[3] D. Becherer, Utility indifference hedging and valuation via reaction-diffusion systems, *R. Soc. Lond. Proc. Ser. A Math. Phys. Eng. Sci.* **460** (2004), 27–51.

[4] M.V. Bobrov, *The fair price valuation in illiquid markets,* Master's Thesis in Financial Mathematics, Technical report IDE0738, Halmstad University, Sweden, 2007.

[5] L. A. Bordag, *On option-valuation in illiquid markets: invariant solutions to a nonlinear model,* in A. Sarychev A. Shiryaev, M. Guerra and M. R. Grossinho (eds): Mathematical Control Theory and Finance, Springer, (2008), 71–94.

[6] L.A. Bordag and A.Y. Chmakova, Explicit solutions for a nonlinear model of financial derivatives, *Int. J. Theoret. Appl. Finance* **10** (2007), 1–21.

[7] U. Cetin, R. Jarrow and P. Protter, Liquidity risk and arbitrage pricing theory, *Finance Stoch.* **8** (2004), 311–341.

[8] U. Cetin, R. Jarrow and P. Protter, Pricing options in an extended Black-Scholes economy with illiquidity, *Rev. Financial Studies* **19** (2006), 493–529.

[9] A.Y. Chmakova, *Symmetriereduktionen und explizite Lösungen für ein nichtlineares Modell eines Preisbildungsprozesses in illiquiden Märkten,* Ph.D. thesis, BTU Cottbus, Germany, 2005 (German).

[10] R. Frey, Perfect option replication for a large trader, *Finance Stoch.* **2** (1998), 115–148.

[11] R. Frey, *Market illiquidity as a source of model risk in dynamic hedging,* in "Model Risk" (R. Gibson, ed.), 125–136, Risk Publications, London, 2000.

[12] R. Frey and P. Patie, *Risk management for derivatives in illiquid markets: a simulation study,* in "Advances in Finance and Stochastics" (K. Sandmann and P. Schönbucher, eds), Springer, Berlin, 2002.

[13] R. Frey and A. Stremme, Market volatility and feedback effects from dynamic hedging, *Math. Finance* **7** (1997), 351–374.

[14] K. Glover, P. Duck and D. Newton, *On nonlinear models of markets with finite liquidity: some cautionary notes*, Preprint, University of Manchester (2007).

[15] G. Gaeta, Nonlinear Symmetries and Nonlinear Equations, *Mathematics and its Applications* **299**, Kluwer Academic Publishers, Dordrecht, 1994.

[16] S. Howison,Matched asymptotic expansions in financial engineering, *J. Engineering Math.* **53** (2005), 385–406.

[17] N.H. Ibragimov, *Elementary Lie Group Analysis and Ordinary Differential Equations,* John Wiley & Sons, Chichester, USA, 1999.

[18] M. Jandačka and D. Ševčovič, On the risk-adjusted pricing-methodology-based valuation of vanilla options and explanation of the volatility smile, *J. Appl. Math.* **3** (2005), 253–258.

[19] R.A. Jarrow, Derivative securities markets, market manipulation and option pricing theory, *J. Finan. Quant. Anal.* **29** (1994), 241–261.

[20] R.S. Johnson, *Singular perturbation theory: Mathematical and analytical techniques with applications to engineering*, Springer, New York, 2004.

[21] I.S. Krasil'shchik, A. M. Vinogradov, et al., Symmetries and conservation laws for differential equations of Mathematical Physics, *Amer. Math. Soc.,* Providence, RI, 1999.

[22] S. Lie, *Vorlesungen über Differentialgleichungen mit bekannten infinitesimalen Transformationen,* Teubner, Leipzig, 1912 (German).

[23] H. Liu and J. Yong, Option pricing with an illiquid underlying asset market, *J. Economic Dynamics Control,* **29** (2005), 2125–2156.

[24] M. Musiela and T. Zariphopoulou, An example of indifference prices under exponential preferences, *Finance Stoch.* **8** (2004), 229–239.

[25] P.J. Olver, *Application of Lie groups to differential equations*, Springer-Verlag, New York, USA, 1986.

[26] L.V. Ovsiannikov, *Group Analysis of Differential Equations*, Academic Press, New York, USA, 1982.

[27] E. Platen and M. Schweizer, On feedback effects from hedging derivatives, *Math. Finance* **8** (1998), 67–84.

[28] U. Polte, *On Hedging and Pricing of Derivatives in Incomplete Markets – a PDE Approach*, PhD-thesis, Universität Leipzig, 2007.

[29] P. Protter, Stochastic Integration and Differential Equations: A New Approach, *Applications of Mathematics,* Springer, Berlin, 2nd ed., 2004.

[30] P. Schönbucher and P. Wilmott, *The feedback-effect of hedging in illiquid markets, SIAM J. Appl. Math.* **61** (2000), 232–272.

[31] N. Sergeeva, *The feedback effects in illiquid markets, hedging strategies of large traders,* Master's Thesis in Financial Mathematics, Technical report IDE0741, Halmstad University, Sweden, 2007.

[32] R. Sircar and G. Papanicolaou, General Black-Scholes models accounting for increased market volatility from hedging strategies, *Appl. Math. Finance* **5** (1998), 45–82.

[33] H. Stephani, *Differentialgleichungen: Symmetrien und Lösungsmethoden*, Spektrum Akademischer Verlag, Heidelberg, Germany, 1994.

In: Nonlinear Models in Mathematical Finance...
Editor: Matthias Ehrhardt, pp. 131-146
ISBN 978-1-60456-931-5

Chapter 4

DISTRIBUTIONAL SOLUTIONS TO AN INTEGRO-DIFFERENTIAL PARABOLIC PROBLEM ARISING IN FINANCIAL MATHEMATICS

Maria Christina Mariani and Michael Eydenberg
Department of Mathematical Sciences,
New Mexico State University, USA

Abstract

In this chapter, we study solutions to a parabolic integro-differential operator based on the Black-Scholes model of option pricing in financial mathematics. We obtain solutions under suitable conditions using the method of upper and lower solutions in conjunction with standard methods for solving nonlinear parabolic differential equations. We also consider solutions for a constant-coefficient divergence-form operator when the integration term is in the form of a convolution. This result is applied to obtain distribution-valued solutions to the problem in a simple setting.

1. Introduction

In recent years there has been an increasing interest on problems arising in financial mathematics and in particular on option pricing. The standard approach to this problem leads to the study of equations of parabolic type. In financial mathematics, usually the Black-Scholes model ([6], [8], [9], [10], [13], [14], [15], [22]) is used for pricing derivatives by means of a reversed-time parabolic partial differential equation. In this model, an important quantity is the *volatility*, which is a measure of the fluctuation (risk) in the asset prices; it corresponds to the diffusion coefficient in the Black-Scholes equation. In the standard Black-Scholes model, a basic assumption is that the volatility is constant. Several models that have been proposed in recent years, however, allow the volatility to be non-constant or a stochastic variable. For instance, in [12] a model with stochastic volatility is proposed. In this model the underlying security S follows, as in the Black-Scholes model, a stochastic process

$$dS_t = \mu S_t dt + \sigma_t S_t Z_t,$$

where Z is a standard Brownian motion. Unlike the classical model, the variance $v(t) = \sigma^2(t)$ also follows a stochastic process given by

$$dv_t = k(\theta - v(t))dt + \gamma\sqrt{v_t}dW_t,$$

where W is another standard Brownian motion. The correlation coefficient between W and Z is denoted by ρ:

$$E(dZ_t, dW_t) = \rho dt.$$

This leads to a generalized Black-Scholes equation:

$$\frac{1}{2}vS^2U_{SS} + \rho\gamma vSU_{vS} + \frac{1}{2}v\gamma^2U_{vv} + rSU_S + [k(\theta - v) - \lambda v]U_v - rU + U_t = 0.$$

A similar model has been considered in [4], for which the stationary equation has been studied in [1].

More general models with stochastic volatility have been considered for example in [5], where the following problem is derived from the Feynman-Kac relation:

$$u_t = \frac{1}{2}Tr\left(M(x,\tau)D^2u\right) + q(x,\tau) \cdot Du,$$
$$u(x,0) = u_0(x)$$

for some diffusion matrix M and a payoff function u_0.

Recently, several authors have studied the development and analysis of mathematical models to enhance understanding of extreme events in financial markets. This will be undertaken through several specific problems in the mathematics of risk management:

1. The analysis of generalized Black-Scholes models, like the ones including transaction costs, see [2] and the references therein.

2. The development and application of tools from mathematical physics to analyze market dynamics leading to a "crash", and the corresponding matching with tools from mathematical finance, see [16], [17], [18] [23], [24] [25] and the references therein.

3. The analysis of asset-price dynamics in models that capture the possibility of sudden, large changes in prices — i.e., "jumps".

Both the importance and difficulties of jumps in asset-price dynamics has been recognized at least since the classical work of Nobel laureate Robert Merton in his 1976 paper "Option Pricing When Underlying Returns are Discontinuous", published in the Journal of Financial Economics (pp. 125–144). From the perspective of both mathematical and financial theory, it has been more convenient to work with models in which asset prices are described by pure diffusion processes evolving continuously in time. Researchers and practitioners have generally favored these models more for their simplicity than their realism. In particular, the pricing of derivative securities in this framework can be reduced to the solution of partial differential equations, allowing the application of an extensive body of mathematical knowledge from that field.

More recently, there has been a resurgence of interest in asset-price models that capture jumps as well as diffusions. In this setting, the prices of derivative securities are described not by partial differential equations but rather by partial integro-differential equations, the integral terms arising from jumps.

The Black-Scholes models with jumps arise from the fact that the Brownian random-walk doesn't fit the financial data representing large fluctuations. The necessity of taking into account large market movements and a sudden change of information (i.e. a jump) has led to the study of partial integro-differential equations (PIDE) in which an integral term models the jump.

In [3] and [22], the following PIDE on the variables t and S is obtained:

$$\frac{1}{2}\sigma^2 S^2 F_{SS} + (r - \lambda k)\, S F_S - F_t - rF + \lambda \epsilon \left(F\left(SY, t\right) - F(S,t)\right) = 0. \qquad (1)$$

Here, r denotes the riskless rate, λ the jump intensity, and $k = \epsilon\,(Y - 1)$, where ϵ is the expectation operator and the random variable $Y - 1$ measures the percentage change in the stock price if the jump occurs. The jump itself is modeled by a Poisson process.

The following PIDE is a generalization of (1) for N assets with prices $S_1, \ldots, S_N$:

$$\sum_{i=1}^{N} \frac{1}{2}\sigma_i^2 S_i^2 F_{S_i S_i} + \sum_{i \neq j} \frac{1}{2}\rho_{ij}\sigma_i\sigma_j S_i S_j F_{S_i S_j} + \sum_{i=1}^{N} (r - \lambda k_i)\, S_i F_{S_i} - F_\tau - rF +$$

$$\lambda \int \left[F(S_1 Y_1, \ldots, S_d Y_d, t) - F\left(S_1, \ldots, S_d, t\right)\right] g\left(Y_1, \ldots, Y_d\right) dY_1 \ldots dY_d = 0$$

with

$$\rho_{ij} dt = \epsilon\left(dz_i, dz_j\right)$$

the correlation coefficients. We note that the case in which F is increasing and all jumps are negative corresponds to the evolution of a call option near a crash.

When the volatility is stochastic, we can consider the process

$$\begin{aligned} dS &= S\sigma dZ + S\mu dt \\ d\sigma &= \beta\sigma dW + \alpha\sigma dt \end{aligned}$$

where Z and W are two standard Brownian motions both with correlation coefficient ρ. If $F(S, \sigma, t)$ is the price of an option depending on the price of the asset S, then by Ito's lemma [14], we have

$$dF(S, \sigma, t) = F_S dS + F_\sigma d\sigma + \mathcal{L}F dt$$

where $\mathcal{L}$ is given by

$$\mathcal{L} = \partial_t + \frac{1}{2}\sigma^2 S^2 D_{SS} + \frac{1}{2}\beta^2\sigma^2 D_{\sigma\sigma} + \rho\sigma^2 S\beta D_{S\sigma}.$$

Under an appropriate choice of the portfolio the stochastic term of this equation vanishes (cf. [4]). If we include the jump, we obtain the following PIDE, which is a generalization of (1) for stochastic volatility:

$$\frac{\partial F}{\partial t} = \frac{1}{2}\sigma^2 S^2 \frac{\partial^2 F}{\partial S^2} + \frac{1}{2}\sigma^2\beta^2 \frac{\partial^2 F}{\partial \sigma^2} + \rho\sigma^2\beta S \frac{\partial^2 F}{\partial S \partial \sigma} - \frac{1}{2}\rho\sigma^2\beta \frac{\partial F}{\partial \sigma} - rF + (r - \lambda k) S \frac{\partial F}{\partial S}$$

$$+ \lambda \int_{\mathbb{R}^2} [F(SY_1, \sigma Y_2, t) - F(S, \sigma, t)] g(Y_1, Y_2) dY_1 dY_2 \quad (2)$$

In some of these models the high frequency data have been described by a Levy-like stochastic process, and again the partial integro-differential equations have been widely used for modeling the presence of jumps and critical stochastic processes.

Integro-differential models in exponential Levy models, where the market price of an asset is represented as the exponential of a Levy stochastic process, have also been discussed by several authors (cf. [7] and [11]). The integro-differential equation for a European option is

$$\frac{\partial C}{\partial \tau} + L^S C = rC$$

with final condition

$$C(T, S) = \widetilde{H}(S),$$

where L^S is the integro-differential operator defined by

$$L^S f = rxf_x + \frac{\sigma^2 x^2}{2} f_{xx} + \int \left[f\left(xe^y\right) - f(x) - x\left(e^y - 1\right) f_x\right] v\left(dy\right)$$

with $v\left(dy\right)$ a positive Radon measure.

If we introduce the change of variables given by $x = \ln(S)$, $t = T - \tau$, $u(x,t) = C(S, \tau)$, we obtain the problem

$$\frac{\sigma^2}{2} u_{xx} + \left(r - \frac{\sigma^2}{2}\right) u_t - ru - u_t + I(u) = L(u) - u_t + I(u) = 0$$

with initial condition

$$u(x, 0) = u_0(x),$$

and the term $I(u)$ modeling the jumps is

$$I(u) = \int \left[u(e^{x+y}) - u(e^x) - \left(e^y - 1\right) u_x\right] v\left(dy\right).$$

We recall that in this case, the evolution of a call option near a crash is given by a spectrally negative Levy process.

In a more general context, the previous discussion motivates us to consider general integro-differential parabolic problems. This work is devoted to the study of the solutions to the following general partial integro-differential equation in a bounded convex domain $\Omega \subset \mathbb{R}^{n+1}$:

$$\begin{aligned} Lu &= G\left(x, t, u\right) \text{ in } \Omega \\ u &= u_0 \text{ on } \mathcal{P}\Omega, \end{aligned} \tag{3}$$

where $\mathcal{P}\Omega$ is the parabolic boundary of Ω. Here, L is a parabolic differential operator of the form

$$Lu = \sum_{ij} a_{ij} D_i D_j u + \sum_i b_i D_i u + cu - D_t u, \tag{4}$$

where $1 \leq i, j \leq n$ and D_i indicates a derivative with respect to the space variable x_i. The coefficients $a_{ij}(x,t)$, $b_i(x,t)$, and $c(x,t)$ belong to the Hölder space $C_{\delta,\delta/2}(\Omega)$ for some $\delta \in (0,1)$ and satisfy the conditions

$$\begin{aligned} \lambda |\xi|^2 \leq \sum_{ij} a_{ij}(x,t)\, \xi_i \xi_j \leq \Lambda |\xi|^2 \quad & \text{for some } 0 < \lambda \leq \Lambda \text{ and any } \xi \in \mathbb{R}^n, \\ ||b_i||_\infty < C, & \\ c \leq 0. & \end{aligned} \tag{5}$$

Observe that we may extend the coefficients to define elements of $C_{\delta,\delta/2}\left(\overline{\Omega}\right)$ since the domain Ω is convex [20].

The integral term $G(x,t,u)$ has the form

$$G(x,t,u) = \int_\Omega g(x,t,y,s,u(y,s))dyds \tag{6}$$

where $g(x,t,y,s,r) \in C(\Omega \times \Omega \times \mathbb{R})$ is decreasing with respect to r. Using the method of upper and lower solutions, we prove the existence of solutions to (3) in the case that g satisfies a specific Hölder regularity condition.

In a previous paper [19], we proved an existence theorem for solutions in $\mathcal{D}'(\Omega)$ to the parabolics problem

$$\widetilde{L}u = F \text{ in } \Omega$$

where $\Omega \subset \mathbb{R}^{n+1}$ is a bounded, smooth, convex domain and $F = D^\alpha f$ in the sense of distributions for some $f \in C\left(\overline{\Omega}\right)$.

The operator $\widetilde{L}$ is in divergence-form and written as

$$\widetilde{L}u = \sum_{ij} D_i\left(a_{ij} D_j u\right) + \sum_i b_i D_i u + cu - D_t u, \tag{7}$$

where, in addition to satisfying (5), the coefficients are constant and the matrix a_{ij} is symmetric. Continuing in this vein, we are interested in studying whether integro-differential problems of the form (3) can also be solved in the context of distributions. This presents a new difficulty in that there is no natural way of interpreting a general integral operator of the form (6) when the kernel g is defined as a distribution. However, one setting in which this does make sense is in the context of convolution: given a domain $\Omega \subset \mathbb{R}^{n+1}$, let $\varepsilon > 0$ and define $\Omega_\varepsilon = \{x \in \Omega : d(x, \partial\Omega) > \varepsilon\}$. Then if f is a continuous function with support contained in the ball $B_\varepsilon = \{x : |x| < \varepsilon\}$ and $v \in \mathcal{D}'(\Omega)$, the convolution product $f * v$ defines an element of $\mathcal{D}'(\Omega_\varepsilon)$ with action on a test function ϕ given by

$$f * v(\phi) = \int_{B_\varepsilon} f(x,t) v_{y,s}\left(\phi(x+y, t+s)\right) dxdt,$$

where $v_{y,s}\left(\phi(x+y,t+s)\right)$ indicates the action of v on the function $\phi(x+y,t+s) \in \mathcal{D}(\Omega)$ with the variables x and t held constant. This motivates a study of a variant of (3) to find weak solutions to the parabolic problem

$$\begin{aligned} \widetilde{L}u &= f * u \text{ in } \Omega_\varepsilon \\ u &= u_0 \text{ on } \Omega \backslash \Omega_\varepsilon \end{aligned} \tag{8}$$

when $f \in C_0(B_\varepsilon)$ and $u_0 \in C(\overline{\Omega})$. We obtain solutions to (8) in the case that f is continuously differentiable, nonpositive, and satisfies a sufficiently small L^1-estimate. The solutions themselves are weak in the sense described by [21] and [19], and in the constant coefficient case they lead to a natural setting in which $\mathcal{D}'(\Omega)$-valued solutions to (8) can be described.

2. Solutions for the Integro-Differential Problem (3)

Let $\Omega \subset \mathbb{R}^{n+1}$ be a bounded domain, and let L be a parabolic differential operator of the form (4). Our focus here is to obtain solutions to (3) where $u_0 \in C_{2+\delta,1+\delta/2}(\mathcal{P}\Omega)$ and the integrand $g : \Omega \times \Omega \times \mathbb{R} \mapsto \mathbb{R}$ satisfies certain regularity conditions. We begin with the following convention for describing the Hölder spaces of interest to us:

Definition 1 *Let $I \subset \mathbb{R}$ be an interval (possibly unbounded) and choose $\delta \in (0,1)$. Define the parabolic Hölder space $C_{\delta,\delta/2,\delta}(\Omega \times I)$ as follows: given $(x,t),(x',t') \in \Omega$ and $y, y' \in I$, we define the parabolic distance $p((x,t,y),(x',t',y'))$ by*

$$p((x,t,y),(x',t',y')) = \left(|x-x'|^2 + |y-y'|^2\right)^{1/2} + |t-t'|^{1/2}$$

and $C_{\delta,\delta/2,\delta}(\Omega \times I)$ to be those functions $f : \Omega \times I \mapsto \mathbb{R}$ for which

$$||f||_\infty + \sup_{(x,t,y)\neq(x',t',y')} \frac{|f(x,t,y) - f(x',t',y')|}{[p((x,t,y),(x',t',y'))]^\delta} < \infty.$$

Observe that if $g \in C_{\delta,\delta/2,\delta}(\Omega \times I)$, then $g(\cdot,\cdot,y) \in C_{\delta',\delta'/2}(\Omega)$ for every $\delta' \in (0,\delta]$ and $y \in I$. We will need the following result concerning the continuity of the norms $||g(\cdot,\cdot,y)||_{\delta',\delta'/2}$:

Lemma 2 *Let $f \in C_{\delta,\delta/2,\delta}(\Omega \times I)$ and assume the sequence $y_k \to y_0$ in I. Then for any $0 < \delta' < \delta$, we have that $f(\cdot,\cdot,y_k) \to f(\cdot,\cdot,y_0)$ in $C_{\delta',\delta'/2}(\Omega)$.*

Proof. Choose any $\varepsilon > 0$, and let $(x_{1,k}, t_{1,k})$ and $(x_{2,k}, t_{2,k})$ be such that

$$\frac{|f(x_{1,k},t_{1,k},y_k) - f(x_{1,k},t_{1,k},y_0) - f(x_{2,k},t_{2,k},y_k) + f(x_{2,k},t_{2,k},y_0)|}{[p((x_{1,k},t_{1,k}),(x_{2,k},t_{2,k}))]^{\delta'}} > C_k - \varepsilon \quad (9)$$

where $C_k = ||f(\cdot,\cdot,y_k) - f(\cdot,\cdot,y_0)||_{\delta',\delta'/2}$. Now, there is a constant C for which

$$\frac{|f(x_{1,k},t_{1,k},y_k) - f(x_{2,k},t_{2,k},y_k)|}{[p((x_{1,k},t_{1,k}),(x_{2,k},t_{2,k}))]^{\delta'}} \leq C\,[p((x_{1,k},t_{1,k}),(x_{2,k},t_{2,k}))]^{\delta-\delta'}$$

uniformly for all k (including $k=0$). Thus, there is a γ such that if

$$p((x_{1,k},t_{1,k}),(x_{2,k},t_{2,k})) < \gamma,$$

then

$$\frac{|f(x_{1,k},t_{1,k},y_k) - f(x_{1,k},t_{1,k},y_0) - f(x_{2,k},t_{2,k},y_k) + f(x_{2,k},t_{2,k},y_0)|}{[p((x_{1,k},t_{1,k}),(x_{2,k},t_{2,k}))]^{\delta'}} < \varepsilon.$$

If, however, we have that

$$p((x_{1,k}, t_{1,k}), (x_{2,k}, t_{2,k})) \geq \gamma \tag{10}$$

then consider the term

$$\frac{|f(x_{1,k}, t_{1,k}, y_k) - f(x_{1,k}, t_{1,k}, y_0) - f(x_{2,k}, t_{2,k}, y_k) + f(x_{2,k}, t_{2,k}, y_0)|}{\gamma^{\delta'}}. \tag{11}$$

Since $f \in C_{\delta,\delta/2,\delta}(\Omega \times I)$ it follows that f is uniformly continuous on bounded subsets of $\Omega \times I$, hence $f(\cdot,\cdot,y_k) \to f(\cdot,\cdot,y_0)$ uniformly. Thus if (10) holds and k is sufficiently large, we have that (11) is bounded from above by ε. We conclude from this and (9) that $C_k < 2\varepsilon$ for sufficiently large k, and the lemma is proved. ■

We now establish the existence of solutions to (3) using the method of upper and lower solutions. We recall that a smooth function u is called an upper solution to (3) if

$$Lu \leq \int_\Omega g(x,t,y,s,u(y,s))\,dyds \quad \text{in } \Omega$$
$$u \geq u_0 \quad \text{in } \mathcal{P}\Omega.$$

Lower solutions are defined similarly. We also have the following standard existence result, cf. [20] and [21]:

Theorem 3 *Let $\Omega \subset \mathbb{R}^{n+1}$ be a bounded domain, and let $f \in C_{\delta,\delta/2}(\Omega)$ and $u_0 \in C_{2+\delta,1+\delta/2}(\mathcal{P}\Omega)$. Then there is a unique $u \in C_{2+\delta,1+\delta/2}(\Omega)$ satisfying*

$$\begin{aligned} Lu &= f \quad \text{in } \Omega \\ u &= u_0 \quad \text{on } \mathcal{P}\Omega. \end{aligned} \tag{12}$$

Furthermore, there is a constant C independent of f and u_0 for which

$$||u||_{2+\delta,1+\delta/2} \leq C\left(||f||_{\delta,\delta/2} + ||u_0||_{2+\delta,1+\delta/2}\right). \tag{13}$$

Note also that if α is a lower solution and β an upper solution to (12), then by the maximum principle we have $\alpha \leq u \leq \beta$ in Ω. Our next step is to prove that upper and lower solutions exist for (3) when $g(x,t,y,s,r)$ is decreasing with respect to the r-variable:

Lemma 4 *Let $\Omega \subset \mathbb{R}^{n+1}$ be a bounded domain. Assume $g \in C(\Omega \times \Omega \times \mathbb{R})$ satisfies the following: g is bounded on $(\Omega \times \Omega \times I)$ for every bounded interval $I \subset \mathbb{R}$, and for each $(x,t,y,s) \in \Omega \times \Omega$,*

$$g(x,t,y,s,r_1) \geq g(x,t,y,s,r_2)$$

if $r_1 \leq r_2$. Then given any $u_0 \in C_{2+\delta,1+\delta/2}(\mathcal{P}\Omega)$, there exist upper and lower solutions to (3).

Proof. Write $\tau = \inf_{(x,t)\in\Omega} t$. Set $\alpha = -M(t-\tau) - ||u_0||_\infty$, where M is to be chosen later. Then $\alpha \leq u_0$ on $\mathcal{P}\Omega$ and $g(x,t,y,s,\alpha(y,s)) \in L^1(\Omega \times \Omega)$. By a direct calculation, we have

$$L\alpha = M - c(M(t-\tau) + ||u_0||_\infty). \tag{14}$$

Furthermore, for $M > 1$ we have that

$$\int_{\Omega} g(x,t,y,s,\alpha(y,s))dyds$$

is bounded from above by the function

$$\int_{\Omega} g(x,t,y,s,-(s-\tau)-||u_0||_{\infty})dyds \tag{15}$$

which is itself bounded on Ω. Thus, by taking M sufficiently large we have (14)≥(15) pointwise, and α is a lower solution. A similar approach shows that $\beta = N(t-\tau) + ||u_0||_{\infty}$ is an upper solution for sufficiently large N, and the lemma is proved. ■

We are now ready to give the proof of existence for solutions to (3):

Theorem 5 *Let $\Omega \subset \mathbb{R}^{n+1}$ be a bounded domain and assume $g \in C(\Omega \times \Omega \times \mathbb{R})$ satisfies the hypothesis of Lemma 4 as well as the following: for each bounded interval $I \subset \mathbb{R}$ there is a $\delta_I \in (\delta, 1)$ and a function $f_I \in L^1(\Omega)$ such that*

$$||g_I(\cdot,\cdot,y,s,\cdot)||_{\delta_I,\delta_I/2,\delta_I} \leq f_I(y,s)$$

where g_I is the restriction of g to $\Omega \times \Omega \times I$. Then for any $u_0 \in C_{2+\delta,1+\delta/2}(\mathcal{P}\Omega)$, there is a solution $u \in C_{2+\delta,1+\delta/2}(\Omega)$ to (3).

Proof. Let α, β be the upper and lower solutions from Lemma 4, set $u_1 = \alpha$ and define

$$U_1 = \int_{\Omega} g(x,t,y,s,u_1(y,s))dyds.$$

Observe that $U_1 \in C_{\delta,\delta/2}(\Omega)$, so from Theorem 3 we may define $u_2 \in C_{2+\delta,1+\delta/2}(\Omega)$ to be the solution to

$$\begin{aligned} Lu &= U_1 \text{ in } \Omega \\ u &= u_0 \text{ on } \mathcal{P}\Omega. \end{aligned}$$

Since $L(u_1 - u_2) \geq 0$ while $u_1 - u_2 \leq 0$ on $\mathcal{P}\Omega$, we have from the maximum principle that $\alpha = u_1 \leq u_2$ in Ω. Similarly, we have $u_2 \leq \beta$. Hence, as u_2 is bounded it follows from the hypothesis that

$$U_2 = \int_{\Omega} g(x,t,y,s,u_2(y,s))dyds$$

defines an element of $C_{\delta,\delta/2}(\Omega)$, so we may inductively define $u_{n+1} \in C_{2+\delta,1+\delta/2}(\Omega)$ as the solution to

$$\begin{aligned} Lu &= U_n \text{ in } \Omega \\ u &= u_0 \text{ on } \mathcal{P}\Omega \end{aligned}$$

where

$$U_n = \int_{\Omega} g(x,t,y,s,u_n(y,s))dyds \in C_{\delta,\delta/2}(\Omega). \tag{16}$$

By induction, we have that $u_n \geq u_{n-1}$ and so

$$L\left(u_{n-}u_{n+1}\right) = \int_{\Omega} g(x,t,y,s,u_{n-1}(y,s)) - g(x,t,y,s,u_n(y,s))dyds \geq 0.$$

Thus $\alpha \leq u_n \leq u_{n+1}$ by the maximum principle. Similarly, we have

$$L\left(u_{n+1} - \beta\right) = \int_{\Omega} g(x,t,y,s,u_n(y,s)) - g(x,t,y,s,\beta(y,s))dyds \geq 0$$

so that $u_{n+1} \leq \beta$. Hence we may define the pointwise limit $u = \lim_n u_n$.

We claim that $u_n \to u$ in $C_{2+\delta,1+\delta/2}\left(\Omega\right)$. By estimate (13), it suffices to show that the functions (16) converge to

$$U = \int_{\Omega} g(x,t,y,s,u(y,s))dyds \tag{17}$$

in $C_{\delta,\delta/2}\left(\Omega\right)$. Let $\alpha_0 = \inf_{(x,t)\in\Omega} \alpha$ and $\beta_0 = \sup_{(x,t)\in\Omega} \beta$, and fix $I = [\alpha_0, \beta_0]$. Then we may replace g with g_I in (16) and (17). We first show that $U_n \to U$ uniformly. Since g_I is bounded, it follows that the U_n and U are uniformly bounded. Furthermore, we have for each $(x,t) \neq (x',t')$ in Ω the estimate

$$\frac{|U_n(x,t) - U_n(x',t')|}{\left[p\left((x,t),(x',t')\right)\right]^{\delta}} \leq C_I \frac{|U_n(x,t) - U_n(x',t')|}{\left[p\left((x,t),(x',t')\right)\right]^{\delta_I}} \leq C_I \left\|f_I\right\|_1.$$

Thus, by (13) it follows that the u_n are equicontinuous, and so there is a uniformly convergence subsequence u_{n_m}. Since the u_n are increasing it follows that the u_n converge to u uniformly. Now, for each $(y,s) \in \Omega$ we have

$$|g_I(x,t,y,s,u_n(y,s)) - g_I(x,t,y,s,u(y,s))| \leq f_I(y,s)\left|u_n(y,s) - u(y,s)\right|^{\delta_I/2}.$$

Thus, for sufficiently large n we have that $\left\|U_n - U\right\|_{\infty} \leq \left\|f_I\right\|_1 \varepsilon$ for any $\varepsilon > 0$, and so $U_n \to U$ uniformly as desired.

For the Hölder norms, we have the estimate

$$\frac{|g_I(x,t,y,s,u_n(y,s)) - g_I(x,t,y,s,u(y,s)) - g_I(x',t',y,s,u_n(y,s)) + g_I(x',t',y,s,u(y,s))|}{\left[p((x,t),(x',t'))\right]^{\delta}} \leq C_n\left(y,s\right),$$

where

$$C_n\left(y,s\right) = \left\|g\left(\cdot,\cdot,y,s,u_n(y,s)\right) - g\left(\cdot,\cdot,y,s,u(y,s)\right)\right\|_{\delta,\delta/2}.$$

By Lemma 2, we see that the $C_n(y,s)$ converge to 0 pointwise in Ω as $n \to \infty$. As the C_n are bounded by $2f_I\left(y,s\right) \in L^1\left(\Omega\right)$ it follows from Dominated Convergence that the U_n converge in $C_{\delta,\delta/2}\left(\Omega\right)$. Thus $u \in C_{2+\delta,1+\delta/2}\left(\Omega\right)$ is a classical solution to (3), and the theorem is proved. ■

3. Solutions for the Convolution Problem (8)

We now turn our attention to the convolution problem (8) for the divergence-form parabolic operator (7) with constant coefficients. The situation is as follows: for $\Omega \subset \mathbb{R}^{n+1}$ a bounded, smooth domain and $\varepsilon > 0$, define $B_\varepsilon = \{x : |x| < \varepsilon\}$ and $\Omega_\varepsilon = \{x : d(x, \partial\Omega) > \varepsilon\}$. Then given $f \in C_0(B_\varepsilon)$, we ask whether one can find a u for which $\widetilde{L}u = f * u$ in the weak sense on elements of $C_0^2(\Omega_\varepsilon)$. In particular, we seek a u such that for all $\phi \in C_0^2(\Omega_\varepsilon)$,

$$\widetilde{L}u(\phi) = \int_{\Omega_\varepsilon} \left[\int_{B_\varepsilon} f(y,s)u(x-y,t-s)dyds \right] \phi(x,t)\, dxdt,$$

where $\widetilde{L}u$ is the functional with action given by

$$\widetilde{L}u(\phi) = \int_{\Omega_\varepsilon} \sum_{ij} -a_{ij}(D_j u)(D_i \phi) + \sum_i b_i (D_i u)\phi + cu\phi - (D_t u)\phi\, dxdt. \tag{18}$$

The space of functions from which we seek solutions is the Sobolev space $W^{1,2}(\Omega)$. From [21], we have the following definition:

Definition 6 *Given any $f \in W^{1,\infty}(\Omega)$ and $u_0 \in W^{1,2}(\Omega)$, we say that u is a* weak $W^{1,2}$-solution *to the problem*

$$\begin{aligned} \widetilde{L}u &= f \quad \text{in } \Omega \\ u &= u_0 \quad \text{on } \partial\Omega \end{aligned} \tag{19}$$

if (18) *satisfies*

$$\widetilde{L}u(\phi) = \int_{\Omega_\varepsilon} f\phi dxdt$$

and $u - u_0 \in W_0^{1,2}(\Omega)$, the closure of $C_0^\infty(\Omega)$ with respect to the $W^{1,2}$-norm.

From our earlier work [19], and the results of [21], we have the following basic existence result:

Theorem 7 *Let $\Omega = B(R)$ be a ball of radius R. There is a $K > 0$ depending on $\widetilde{L}$ such that if $R < K$, then there is a unique weak $W^{1,2}$-solution to (19) for any $f \in W^{1,\infty}(\Omega)$ and $u_0 \in W^{1,2}(\Omega)$. Furthermore, $u \in C(\Omega)$ and there is a constant C independent of of f and u_0 for which*

$$||D^\alpha u||_\infty \leq C\left(||f||_{1,\infty} + ||u_0||_{1,\infty}\right) \tag{20}$$

where $|\alpha| = 1$.

Furthermore, given any bounded domain $\Omega \subset \mathbb{R}^{n+1}$ it is possible to find a $u \in C(\Omega)$ for which $\widetilde{L}u = f$ in the weak sense by means of the Perron process. We refer to Theorem 3.23 of [21] for the details, but there is a relevant fact from that proof that we wish to emphasize and state as a lemma:

Lemma 8 *Let $u \in C(\Omega)$ be the solution to $\widetilde{L}u = f$ obtained via the Perron process, where $f \in W^{1,\infty}(\Omega)$. Then for any $x \in \Omega$, there is a ball B_x such that the restriction $u|_{B_x} \in C\left(\overline{B_x}\right) \cap W^{1,\infty}(B_x)$. Furthermore, we have an estimate*

$$||u|_{B_x}||_{1,\infty} \leq C_x\left(||u|_{B_x}||_\infty + ||f|_{B_x}||_{1,\infty}\right), \tag{21}$$

where C_x is independent of u and f.

Our first step in extending these results to (8) is to define and establish the existence of upper and lower solutions, using an approach similar to that of Lemma 4. To do this, we consider a family $f_\varepsilon \in C_0(B_\varepsilon)$ of convolution kernels with diminishing L^1-norm, as follows:

Lemma 9 *Let $f_\varepsilon \in C_0(B_\varepsilon)$ be a family of functions for which $||f_\varepsilon||_1 \rightarrow 0$ as $\varepsilon \rightarrow 0$. Then for any $u_0 \in C\left(\overline{\Omega}\right)$, if $\varepsilon > 0$ is sufficiently small then there are functions $\alpha \leq \beta \in C^\infty\left(\mathbb{R}^{n+1}\right)$ that satisfy*

$$\begin{aligned} \alpha &\leq u_0 \text{ on } \overline{\Omega} \\ \beta &\geq u_0 \text{ on } \overline{\Omega}, \end{aligned}$$

and for all $(x,t) \in \Omega_\varepsilon$

$$\begin{aligned} \widetilde{L}\alpha(x,t) &\geq \int_{B_\varepsilon} f_\varepsilon(y,s)\alpha(x-y,t-s)\,dyds \\ \widetilde{L}\beta(x,t) &\leq \int_{B_\varepsilon} f_\varepsilon(y,s)\beta(x-y,t-s)\,dyds. \end{aligned} \tag{22}$$

Proof. For $u_0 \in C\left(\overline{\Omega}\right)$, let $T = \sup_{(x,t)\in\Omega} t$ and $\tau = \inf_{(x,t)\in\Omega} t$ and set $\alpha(x,t) = -(t-\tau) - ||u_0||_\infty$. Then $\alpha \leq u_0$ on $\overline{\Omega}$. We also have

$$\widetilde{L}\alpha = 1 - c\left(t - \tau + ||u_0||_\infty\right).$$

On the other hand, a straightforward calculation for an arbitrary $f \in C_0(B_\varepsilon)$ gives

$$\left|\int_{B_\varepsilon} f_\varepsilon(y,s)\alpha(x-y,t-s)dzds\right| \leq ||f_\varepsilon||_1\left(T - \tau + \varepsilon + ||u_0||_\infty\right).$$

Thus, for the family f_ε given in the hypothesis, it follows that α satisfies the first part of (22) for sufficiently small $\varepsilon > 0$. In this case, we set $\beta = -\alpha$, and the lemma is proved. ∎

As in Lemma 4, we refer to α as a lower solution and β as an upper solution. From their existence, we obtain the following result for solutions to (8) in small balls $\Omega = B(R)$:

Theorem 10 *Let $\Omega = B(R)$ with $R < K$, and assume the family $f_\varepsilon \in C_0^1(B_\varepsilon)$ is nonpositive and satisfies the hypothesis of Lemma 9. Then for any $u_0 \in C^2\left(\overline{\Omega}\right)$ and sufficiently small $\varepsilon > 0$ there is a weak $W^{1,2}$-solution to*

$$\begin{aligned} \widetilde{L}u &= f_\varepsilon * u \text{ in } \Omega_\varepsilon \\ u &= u_0 \text{ on } \partial\Omega_\varepsilon \end{aligned}$$

with $u = u_0$ in $\Omega \backslash \Omega_\varepsilon$.

Proof. Given such u_0, let $\varepsilon > 0$ be as in Lemma 9 and let α, β be the corresponding lower and upper solutions. From Theorem 7, we may set $u_1 = \alpha$ and define $u_2 \in W^{1,2}(\Omega_\varepsilon)$ as the weak $W^{1,2}$-solution to

$$\begin{aligned} \widetilde{L}u_2 &= f_\varepsilon * u_1 \quad \text{in } \Omega_\varepsilon \\ u_2 &= u_0 \quad \text{on } \partial\Omega_\varepsilon. \end{aligned} \tag{23}$$

Observe that this problem is well-posed as the function

$$U_1 = f_\varepsilon * u_1$$

belongs to $W^{1,\infty}(\Omega_\varepsilon)$ and the restriction of u_0 to $\overline{\Omega}_\varepsilon$ lies in $C^2\left(\overline{\Omega}_\varepsilon\right)$. Also, $\widetilde{L}\alpha - \widetilde{L}u_2 \geq 0$ and $\alpha - u_2 \leq 0$ on $\partial\Omega_\varepsilon$ in the sense of Lemma 2 of [19] (cf. also Lemma 3.15 of [21]), so by the result of that lemma it follows that $u_2 \geq \alpha$ on Ω_ε. Furthermore, as $f_\varepsilon \leq 0$ we find that

$$\begin{aligned} \widetilde{L}u_2 &= \int_{B_\varepsilon} f_\varepsilon(y,s)\alpha(x-y,t-s)dyds \\ &\geq \int_{B_\varepsilon} f_\varepsilon(y,s)\beta(x-y,t-s)dyds \\ &\geq \widetilde{L}\beta \end{aligned}$$

and $u_2 - \beta \leq 0$ on $\partial\Omega_\varepsilon$, so we in fact have $\alpha \leq u_2 \leq \beta$ on Ω_ε. Now, since $u_2 - u_0 \in W_0^{1,2}(\Omega_\varepsilon)$, it follows that we may extend u_2 to define an element $\widetilde{u}_2 \in W^{1,2}(\Omega)$ by setting $\widetilde{u}_2 = u_0$ on $\Omega \backslash \Omega_\varepsilon$. Note that the inequality $\alpha \leq \widetilde{u}_2 \leq \beta$ holds on Ω by Lemma 9, and also that

$$U_2 = f_\varepsilon * \widetilde{u}_2 \tag{24}$$

belongs to $W^{1,\infty}(\Omega_\varepsilon)$. Hence, we may define u_{n+1} for $n \geq 1$ inductively as the weak $W^{1,2}$-solution to

$$\begin{aligned} \widetilde{L}u_{n+1} &= U_n \quad \text{in } \Omega_\varepsilon \\ u_{n+1} &= u_0 \quad \text{on } \partial\Omega_\varepsilon, \end{aligned}$$

where

$$U_n = f_\varepsilon * \widetilde{u}_n \in W^{1,\infty}(\Omega_\varepsilon)$$

and $\widetilde{u}_n$ denotes the extension of u_n to $W^{1,2}(\Omega)$ by u_0. By induction, we have

$$\begin{aligned} \widetilde{L}u_{n+1} &= \int_{B_\varepsilon} f_\varepsilon(y,s)\widetilde{u}_n(x-y,t-s)\,dyds \\ &\leq \int_{B_\varepsilon} f_\varepsilon(y,s)\widetilde{u}_{n-1}(x-y,t-s)\,dyds \\ &= \widetilde{L}u_n \end{aligned}$$

and $u_n - u_{n+1} \leq 0$ on $\partial\Omega_\varepsilon$, hence $u_{n+1} \geq u_n$ on Ω_ε. Similar calculations show that $\alpha \leq u_n \leq u_{n+1} \leq \beta$ on Ω_ε and $\alpha \leq \widetilde{u}_n \leq \widetilde{u}_{n+1} \leq \beta$ on Ω.

Hence, we may define the pointwise limit $\widetilde{u}(x,t) = \lim \widetilde{u}_n(x,t)$. From (20), we have that the u_n are equicontinuous and thus there is a uniformly convergent subsequence u_{n_m}. As the u_n are monotone, it follows that the u_n converge to u uniformly, as do the $\widetilde{u}_n$ to $\widetilde{u}$.

If we now define

$$U = f_\varepsilon * \widetilde{u} \in W^{1,\infty}(\Omega_\varepsilon),$$

then the uniform convergence of the $\widetilde{u}_n$ implies that $U_i \to U$ in $W^{1,\infty}(\Omega_\varepsilon)$. This, along with (20), means that $u_i \to u$ in $W^{1,\infty}(\Omega_\varepsilon)$ and therefore also in $W^{1,2}(\Omega_\varepsilon)$. This in turn implies that $u - u_0 \in W_0^{1,2}(\Omega_\varepsilon)$, hence the extension $\widetilde{u} \in W^{1,2}(\Omega)$. Finally, a straightforward application of Dominated Convergence shows that $\widetilde{L}\widetilde{u} = f_\varepsilon * \widetilde{u}$ in the weak sense on $C_0^\infty(\Omega_\varepsilon)$, and the theorem is proved. ■

To obtain an existence result for general bounded domains Ω, we adapt the proof of Theorem 10 with the solutions to (19) implied by the Perron process:

Corollary 11 *Let Ω be a bounded domain, and assume the family $f_\varepsilon \in C_0^1(B_\varepsilon)$ is nonpositive and satisfies the hypothesis of Lemma 9. Then for any $u_0 \in C_0^2(\Omega)$ and sufficiently small $\varepsilon > 0$, there is a weak solution u to $\widetilde{L}u = f * u$ in Ω_ε.*

Proof. Let $\varepsilon > 0$ and α, β be as in Lemma 22. In analogy to Theorem 10, we define $u_2 \in C(\Omega_\varepsilon)$ as the solution to (23) obtained via the Perron process and set $\widetilde{u}_2 = 0$ on $\Omega \backslash \Omega_\varepsilon$. We again have that $\alpha \leq \widetilde{u}_2 \leq \beta$ on Ω while the regularity of f_ε implies that U_2 defined by (24) describes an element of $W^{1,\infty}(\Omega_\varepsilon)$. The inductive argument proceeds just as it did before to give an increasing sequence of functions $\widetilde{u}_n$ that are continuous in Ω_ε, equal to 0 on $\Omega \backslash \Omega_\varepsilon$, and which satisfy $\alpha \leq \widetilde{u}_n \leq \beta$ in Ω.

Let $\widetilde{u} = \lim \widetilde{u}_n$. To show that $\widetilde{u}$ defines a weak solution, observe from Lemma 8 that for each $x \in \Omega_\varepsilon$, there is a ball $B_x \subset \Omega_\varepsilon$ containing x such that $\widetilde{u}_n|_{B_x} \in C(B_x) \cap W^{1,\infty}(B_x)$. Suppressing the index x for brevity, it follows from (21) that

$$||\widetilde{u}_n|_B||_{1,\infty} \leq C\left(||\widetilde{u}_n|_B||_\infty + ||U_{n-1}|_B||_{1,\infty}\right)$$

where the U_n and U are defined as in Theorem 10. Since $||\widetilde{u}_n||_\infty$ and $||U_n||_{1,\infty}$ are uniformly bounded, we find that the the $\widetilde{u}|_B$ are uniformly bounded and equicontinuous. Hence, there is a subsequence $\widetilde{u}_{n_m}$ which converges uniformly to $\widetilde{u}$ on B, and so by monotonicty the $\widetilde{u}_n$ converge uniformly to $\widetilde{u}$ on B. Since $B = B_x$ was chosen arbitrarily, we have in fact that the $\widetilde{u}_n$ converge to $\widetilde{u}$ uniformly on Ω. It follows from (8) again that the $\widetilde{u}_n$ converge to $\widetilde{u}$ in $W^{1,2}(B)$. Hence, by Dominated Convergence, we have

$$\widetilde{L}\widetilde{u}(\phi) = f_\varepsilon * \widetilde{u}(\phi)$$

for any $\phi \in C_0^\infty(B)$, a result which can be extended to all $\phi \in C_0^\infty(\Omega_\varepsilon)$ via an appropriate partition of unity. ■

Remark 12 *Theorem 10 and Corollary 11 can be used to obtain distribution-valued solutions to (8) in the following sense: given a function $f \in C_0(B_\varepsilon)$ and a distribution $v = D^\alpha g \in \mathcal{D}'(\Omega)$ where $g \in C(\Omega)$, we may define the action of $f * v$ on a test function*

$\phi \in \mathcal{D}(\Omega_\varepsilon)$ via the integral

$$\begin{aligned} f * v(\phi) &= \int_{B_\varepsilon} f(x,t) v_{y,s}\left(\phi(x+y,t+s)\right) dxdt \\ &= (-1)^{|\alpha|} \int_{B_\varepsilon} \left[\int_{\Omega_\varepsilon - (x,t)} f(x,t) g(y,s) D^\alpha \phi(x+y,t+s) dyds \right] dxdt \\ &= (-1)^{|\alpha|} \int_{B_\varepsilon} \left[\int_{\Omega_\varepsilon} f(x,t) g(y-x,s-t) D^\alpha \phi(y,s) dyds \right] dxdt \\ &= (-1)^{|\alpha|} \int_{\Omega_\varepsilon} \left[\int_{B_\varepsilon} f(x,t) g(y-x,s-t) dxdt \right] D^\alpha \phi(y,s) dyds. \end{aligned} \quad (25)$$

Now, given a family of functions $f_\varepsilon \in C_0^1(B_\varepsilon)$ that satisfy the hypothesis of Lemma 22 and $u_0 = 0$, let u be the solution to $\widetilde{L}u = f_\varepsilon * u$ implied by Corollary 11. Then it is easily checked that the distribution $v \in \mathcal{D}'(\Omega_\varepsilon)$ defined as $v = D^\alpha u$ satisfies $\widetilde{L}v = f * v$ in the sense of distributions.

Acknowledgements

Dr. Mariani is extremely grateful to Professor Paul Glasserman for his comments and thoughtful suggestions. This work was partially supported by ADVANCE NSF, and Mini-grant College of Arts and Sciences, NMSU.

References

[1] P. Amster, C. Averbuj and M.C. Mariani, Solutions to a Stationary Nonlinear Black-Scholes Type Equation, *J. Math. Anal. Appl.* **276** (2002), 231–238.

[2] P. Amster, C.G. Averbuj, M.C. Mariani and D. Rial, A Black-Scholes Option Pricing Model with Transaction Costs, *J. Math. Anal. Appl.* **303** (2005), 689–695.

[3] L. Andersen and J. Andreasen, Jumps Diffusion Models: Volatility Smile Fitting and Numerical Methods for Pricing, *Rev. Derivatives Research*, **4** (2000), 231–262.

[4] M. Avellaneda and Y. Zhu, Risk-Neutral Stochastic Volatility Model, *Int. J. Theor. Appl. Finance* **1** (1998), 289–310.

[5] H. Berestycki, J. Busca and I. Florent, Computing the Implied Volatility in Stochastic Volatility Models, *Commun. Pure. and Appl. Math.* **57** (2004) 1352–1373.

[6] F. Black and M. Scholes, The Pricing of Options and Corporate Liabilities, *J. Political Econ.* **81** (1973), 637–659.

[7] R. Cont. and P. Tankov, *Financial Modeling with Jumps Processes*, Chapman and Hall, CRC Press (2003).

[8] D. Duffie, *Dynamic Asset Pricing Theory*, Princeton University Press (1996).

[9] B. Düring and A. Jüngel, A quasilinear parabolic equation with quadratic growth of the gradient modeling incomplete financial markets, *Nonlin. Anal.* **62** (2005), 519–544.

[10] P. Glasserman, *Monte Carlo Methods in Financial Engineering: Pure Jump Levy Processes for Asset Price Modeling*, Springer (2000).

[11] H. Geman, Pure Jump Levy Processes for Asset Price Modeling, *J. Banking and Finance* **26** (2002), 1297–1316.

[12] S.L. Heston, A Closed-Form Solution for Options with Stochastic Volatility with Applications to Bond and Currency Options, *Review of Financial Studies* **6** (1993), 327–343.

[13] J.C. Hull, *Options, Futures, and Other derivatives*, Prentice-Hall Inc. (1997).

[14] N. Ikeda and S. Watanabe, *Stochastic Differential Equations and Diffusion Processes*, North-Holland (1989).

[15] R.A. Jarrow, *Modeling Fixed-Income Securities and Interest Rate Options*, McGraw-Hill (1997).

[16] M.C. Mariani and Y. Liu, A new Analysis of Intermittence, Scale Invariance and Characteristic Scales applied to the Behavior of Financial Indices near a Crash, *Physica A* **367** (2006) 345–352.

[17] M.C. Mariani and Y. Liu, Normalized Truncated Levy walks applied to the study of Financial Indices . *Physica A* **377** (2007) 590–598.

[18] M.C. Mariani and Y. Liu, A new analysis of the effects of the Asian Crisis of 1997 on Emergent market, *Physica A* **380** (2007) 307–316.

[19] M.S. Eydenberg and M.C. Mariani, *Distribution-Valued Weak Solutions to a Parabolic Problem Arising in Financial Mathematics*, Submitted.

[20] N.V. Krylov, Lectures on Elliptic and Parabolic Equations in Hölder Spaces, *Graduate Studies in Math.* **12** AMS (1996).

[21] G.M. Lieberman, *Second Order Parabolic Differential Equations*, World Scientific (1996).

[22] R.C. Merton, *Continuous-Time Finance*, Blackwell, Cambridge (2000).

[23] R.N. Mantegna and H.E. Stanley, *An Introduction to Econophysics: Correlations and Complexity in Finance*, Cambridge University Press, Cambridge 1999.

[24] H.E. Stanley, V. Afanasev, L.A.N. Amaral, S.V. Buldyrev, A.L. Golderberger, S. Havlin, H. Leschorn, P. Maass, R.N. Mantegna, C.-K. Peng, P.A. Prince, M.A. Salinger, M.H.R. Stanley and G.M. Viswanathan, Anomalous Fluctuations in the Dynamics of Complex Systems: From DNA and Physiology to Econophysics. *Physica A* **224** (1996) 302–321.

[25] H.E. Stanley, L.A.N. Amaral, D. Canning, P. Gopikrishnan, Y. Lee and Y. Liu, Econophysics: Can Physicists Contribute to the Science of Economics?, *Physica A* **269** (1999) 156–169.

PART III.
NUMERICAL TREATMENT OF NONLINEAR BLACK-SCHOLES EQUATIONS

In: Nonlinear Models in Mathematical Finance...
Editor: Matthias Ehrhardt, pp. 149-171 ISBN 978-1-60456-931-5

Chapter 5

A Semidiscretization Method for Solving Nonlinear Black-Scholes Equations: Numerical Analysis and Computing

Lucas Jódar, Rafael Company and José Ramón Pintos
Instituto Universitario de Matemática Multidisciplinar
Universidad Politécnica de Valencia
Camino de Vera s/n, 46022 Valencia, Spain

Abstract

This chapter deals with the numerical solution of nonlinear Black-Scholes equations modeling option pricing problems in markets with transaction costs. Our approach is based on the semidiscretization technique with respect to the underlying asset variable, allowing several discretization schemes and it is applicable to distinct models.

The proposed method is purely nonlinear in the sense that it does not involve any linearization step, neither in the computation nor in the numerical analysis of the solution.

We pay special attention to two representative models. The first, is the one introduced by Hoggard, Whalley and Wilmott and developed by Parás and Avellaneda where the adjusted volatility depends on the sign of the Gamma of the option price. The second selected model is due to Barles and Soner which involves the nonlinear correction of the volatility function Ψ. This function Ψ depends on the Gamma of the option price in a more complex way apart from the time and the asset variable.

Consistency and stability of the proposed numerical solution are studied and illustrative numerical examples are included.

1. Introduction

It is well-known that the Black-Scholes (B-S) model is acceptable in idealized financial markets where one assumes that volatility is observable or transaction costs are not taken into account.

These unrealistic assumptions have been shown too restrictive in practice and this fact motivates the search of alternative models expressed by parabolic diffusion-convection equations for the option value V as a function of the underlying security S and the time

$$V_t + \frac{1}{2}\left(\sigma(S,t,V_S,V_{SS})\right)^2 S^2 V_{SS} + rSV_S - rV = 0, \quad S > 0, \quad t \in [0,T[, \qquad (1)$$

with final and boundary conditions. Here $r \geq 0$ denotes the riskless interest rate and σ means the volatility. Equations of this type arise frequently in mathematical finance, not only in option pricing, for instance in the modeling of transaction costs [4, 7], optimal portfolios in incomplete markets [15] and inverse problems [12].

For the particular case where the volatility function σ appearing in (1) is a constant, the model (1) turns out the B-S. In the B-S model for the pricing of options the influence of transaction costs was neglected and it was possible to construct a riskless portfolio that perfectly replicates the option pay-off. If transaction costs are taken into account perfect replication of the contingent claim is no longer possible and it has been shown in [20] that further restrictions have to be imposed in the model.

If bid-ask spreads and other transaction costs are also taken into account, each adjustment of the portfolio implies an additional cost and the replication property of the B-S hedge no longer holds. Making frequent adjustments to maintain the theoretical hedge can increase costs considerably. On the other hand, if only a few adjustments are made, the B-S exact hedge cannot be maintained due to movements in the price of the underlying asset between trades. Therefore, transaction costs cannot be ignored without incurring risk or loss. This is an important practice problem, especially in emerging markets where round-trip transaction costs of 1% and higher are not uncommon.

There are two competing approaches behind modeling transaction costs. One is the so called local in time methods [24] characterized by the assumption of hedging strategies taken place at discrete times. The second approach based on the work [13], the so called global in time methods, characterized by the use of optimality techniques and utility functions. Practical drawbacks of this last approach have motivated the search of asymptotic techniques to derive manageable option pricing formulae.

Starting with the local in time methods, Leland [16] introduced a theory for pricing a call option with transaction costs showing that the price of a call is given by the B-S formula with an augmented volatility

$$\sigma_A = \sigma_0\sqrt{1+A}, \quad A = \sqrt{\frac{2}{\pi}}\frac{k}{\sigma_0\sqrt{\Delta t}}. \qquad (2)$$

Here, σ_0 represents the volatility of the underlying security, k is the round-trip transaction cost (a percentage) and Δt is the time interval between successive adjustments of the portfolio. This time interval is considered fixed and is assumed to be much smaller than the time-to-expiration. We shall refer to the parameter A as the Leland number.

Boyle and Vorst [5] derived a similar option pricing formula using a binomial model. Both approaches are restricted to derivative securities with convex payoffs.

In practice, there are many derivative securities of interest which have non-convex payoffs. The foremost example is a portfolio of standard options combining short and long positions. In [14] Hoggard, Whalley and Wilmott proposed an extension of Leland's result with applies to derivative securities with arbitrary payoff functions. Their model involves the solution of an equation of type (1) where

$$\sigma = \sigma(V_{SS}) = \sigma_0 \sqrt{1 + A\ \text{sign}(V_{SS})}\ . \tag{3}$$

Note that if $A \geq 1$, then by (3), $\sigma^2 = \sigma_0^2(1 + A\ \text{sign}(V_{SS}))$ is nonpositive if $V_{SS} < 0$ what is unrealistic and ill-possed mathematically. From (2), $A \geq 1$ occurs when Δt is small enough or k is big enough. Avellaneda and Parás in [2] obtain the adjusted volatility (3) by proposing new hedging strategies to control effectively the hedging risk and transaction costs.

The global in time methods pioneered by Hodges and Neuberger, [13], and developed by Davis, Panas and Zariphopoulou [7], achieve an element of optimality, since they are based on the approach of utility maximization. Since there are no explicit solutions for the utility based hedging with transaction costs and the numerical methods are computationally hard, for practical applications a suitable alternative is the use of an asymptotic solution. In asymptotic analysis one studies the solution to a problem when some parameters in the problem assume large or small values.

Whalley and Wilmott [24] were the first to provide an asymptotic analysis of the model of Hodges and Neuberger, assuming that transaction costs are small. Barles and Soner [4] performed an alternative method assuming that both the transaction costs and the hedger's risk tolerance are small. While Whalley and Wilmott derive only an optimal form of the hedging bandwith which is centered about the B-S delta, Barles and Soner show a particular form of the hedging bandwith and provide a nonlinear adjusted volatility of the form

$$\sigma^2 = \sigma_0^2\big(1 + \Psi[\exp(r(T-t))a^2 S^2 V_{SS}]\big), \tag{4}$$

where μ is the proportional transaction cost; $a = \mu\sqrt{\gamma N}$, with risk aversion factor γ and the number N of options to be sold. The volatility correction function Ψ is the solution of the nonlinear initial-value problem

$$\Psi'(A) = \frac{\Psi(A) + 1}{2\sqrt{A\Psi(A)} - A}, \qquad A \neq 0, \quad \Psi(0) = 0. \tag{5}$$

In the mathematical literature, only a few results can be found on the numerical discretization of the B-S equation, mainly for linear B-S equations. The numerical approaches vary from finite element discretizations [10, 17], to finite difference approximations [8]. The numerical discretization of the B-S equation with the nonlinear volatility (3) has been performed using explicit finite-difference schemes in [2] and with the volatility given by (4) in [4]. However, explicit schemes have the disadvantage that restrictive conditions on the discretization parameters (for instance, the ratio of the time and the space step) are needed in order to obtain stable, convergent schemes [21]. Moreover, the order of convergence is only one in time and two in space. [9] combines high-order compact difference schemes derived by [18] and techniques to construct numerical solutions with frozen values of the nonlinear coefficient of the nonlinear B-S equation to make the formulation linear.

In this chapter we use a semidiscretization technique by using second-order and fourth-order difference approximations of the partial derivatives V_S and V_{SS} arising in the nonlinear B-S equation

$$V_t + \frac{1}{2}\big(\sigma(V_{SS})\big)^2 S^2 V_{SS} + rSV_S - rV = 0\,. \tag{6}$$

Then we achieve an ordinary system of nonlinear ordinary differential equations with respect to the time, that is solved numerically. Apart form (6) different terminal conditions and boundary conditions are considered depending on the option treated.

In order to compute the numerical solution, it is necessary to work in a bounded domain. Once this numerical domain has been chosen, the boundary conditions can be translated from the asymptotic condition, as it is done for instance in [4] or [9], or the boundary values must be found together with the solution and they are linked with the rest of the numerical solution in the interior of the numerical domain by using extrapolation techniques. This last approach is used in this chapter in accordance with the used scheme.

Apart from the selection of the bounded domain and the consideration of the boundary conditions, dealing with numerical solutions requires reliability guarantees in the sense that the numerical approximations must be close to the exact unknown solution. Reasonable strategies like the consideration of more discretization nodes near the maturity and the strike price, are not sufficient to guarantee accuracy of the numerical approximations. Stability and consistency of the numerical scheme are necessary.

Using the change of variable $\tau = T - t$, $U(S,\tau) = V(S,t)$ equation (6) together with the initial condition is transformed into

$$U_\tau - \frac{S^2}{2}\sigma^2 U_{SS} - rSU_S + rU = 0, \quad 0 < S < \infty, \quad 0 < \tau \le T, \tag{7}$$

$$U(S,0) = V(S,T)\,, \tag{8}$$

In this chapter we select a representative model of each of the two main approaches quoted above. With respect to the local in time approach we study the Hoggard, Whalley and Wilmott (H-W-W) and the global in time approach method we choose the Barles and Soner model.

This chapter is organized as follows. Section 2. deals with the semidiscretization technique and the numerical schemes construction of problem (7)-(8) for a general adjusted volatility function. Section 3. treats the consistency and stability of the proposed numerical scheme of the H-W-W model. Illustrative numerical examples are included paying attention to the binary option case. The binary option case is particularly interesting due to the convexity change behavior of the solution, see (3). Since for the Barles-Soner model the correction volatility function Ψ has not an explicit formula, see (5), Section 4. begins with the study of interesting properties of function Ψ including an implicit formula. Consistency and stability of the numerical scheme are studied using previously established properties of Ψ and illustrative examples for the vanilla option pricing case are included. The chapter concludes with the conclusions and references.

2. Numerical Schemes Construction

The computation of numerical solutions of the nonlinear models is necessary because an exact solution is not available. Among the more extended numerical techniques we should mention the finite-difference (FD) approach, [9, 22]. The numerical analysis of the computed solution using FD methods for nonlinear models uses to be difficult and difficulties are overcomed by means of linearization strategies that in some way falsify the model mainly near the maturity and the strike price. This fact motivates the search of an alternative numerical method which preserves the advantages of the FD method and that allows the full treatment of nonlinearities. The proposed method is the so called semidiscretization method (SD) or method of lines, [3, 19, 22]. By replacing the partial derivatives $U_S(S_i,\tau)$ and $U_{SS}(S_i,\tau)$ by finite-differences approximations [19,22,23], one gets multiple possible different schemes. In this chapter, we will use two different centered numerical schemes; the second order one that approximates both partial derivatives by second order centered FD approximations and the fourth order scheme that approximates both partial derivatives by fourth order centered FD approximations.

We begin this section with the treatment of second order centered schemes.

2.1. Second Order Schemes

Let us introduce the operators $\widetilde{\nabla}_i$ and $\widetilde{\Delta}_i$ defined by

$$\frac{\partial U}{\partial S}(S_i,\tau) = \widetilde{\nabla}_i(U(\tau)) + O(h^2), \tag{9}$$

$$\widetilde{\nabla}_i(U(\tau)) = \frac{U(S_{i+1},\tau) - U(S_{i-1},\tau)}{2h}, \tag{10}$$

$$\frac{\partial^2 U}{\partial S^2}(S_i,\tau) = \widetilde{\Delta}_i(U(\tau)) + O(h^2), \tag{11}$$

$$\widetilde{\Delta}_i(U(\tau)) = \frac{U(S_{i-1},\tau) - 2U(S_i,\tau) + U(S_{i+1},\tau)}{h^2}, \tag{12}$$

$$0 \le \tau \le T,$$

where $S_j = E - L + jh$, $1 \le j \le N-1$ are the nodes of the underlying asset interval $[E-L, E+L]$, E is the strike price and $L \le E$ is the radius of the neighborhood about E where the numerical solution is computed. Let $\widetilde{u}_i(\tau)$ be the approximation of theoretical value $U(S_i,\tau)$ and let us denote $\widetilde{u}(\tau) = [\widetilde{u}_1, \widetilde{u}_2, ..., \widetilde{u}_{N-1}]^\top$, then the partial differential equation (PDE) (7) is approximated by the system of ordinary differential equations

$$\frac{d\widetilde{u}(\tau)}{d\tau} = \widetilde{B}(\tau)\widetilde{u}(\tau) + \widetilde{w}(\tau), \tag{13}$$

where

$$\widetilde{B}(\tau) = \begin{bmatrix} \widetilde{\beta}_1 & \widetilde{\gamma}_1 & 0 & 0 & \cdots & 0 \\ \widetilde{\alpha}_2 & \widetilde{\beta}_2 & \widetilde{\gamma}_2 & 0 & \cdots & 0 \\ 0 & \widetilde{\alpha}_3 & \widetilde{\beta}_3 & \widetilde{\gamma}_3 & \cdots & 0 \\ \cdots & \cdots & \ddots & \ddots & \ddots & \cdots \\ 0 & \cdots & 0 & \widetilde{\alpha}_{N-2} & \widetilde{\beta}_{N-2} & \widetilde{\gamma}_{N-2} \\ 0 & \cdots & 0 & 0 & \widetilde{\alpha}_{N-1} & \widetilde{\beta}_{N-1} \end{bmatrix}, \tag{14}$$

$$\widetilde{w}(\tau) = \begin{bmatrix} \widetilde{\alpha}_1 \widetilde{u}_0 \\ 0 \\ \vdots \\ 0 \\ \widetilde{\gamma}_{N-1} \widetilde{u}_{N-1} \end{bmatrix}, \tag{15}$$

and the entries functions of τ are defined by

$$\begin{aligned} \widetilde{\alpha}_i &= \widetilde{\alpha}_i(\tau) = \frac{\widetilde{\sigma}_i(\tau)^2 S_i^2}{2h^2} - \frac{rS_i}{2h}, \\ \widetilde{\beta}_i &= \widetilde{\beta}_i(\tau) = -\frac{\widetilde{\sigma}_i(\tau)^2 S_i^2}{h^2} - r, \\ \widetilde{\gamma}_i &= \widetilde{\gamma}_i(\tau) = \frac{\widetilde{\sigma}_i(\tau)^2 S_i^2}{2h^2} + \frac{rS_i}{2h}, \end{aligned} \tag{16}$$

being $\widetilde{\sigma}_i(\tau)$ the adjusted volatility depending upon the model considered.

The boundary values $\widetilde{u}_0$ and $\widetilde{u}_N$ appearing in (15) are computed by using second order Lagrange interpolating polynomial passing throughout the two closest internal mesh points.

$$\begin{aligned} \widetilde{u}_0 &= 2\widetilde{u}_1 - \widetilde{u}_2 \ , \\ \widetilde{u}_N &= 2\widetilde{u}_{N-1} - \widetilde{u}_{N-2} \ . \end{aligned} \tag{17}$$

Taking into account (13)-(17), the semidiscretized difference approximating problem takes the form

$$\begin{aligned} &\frac{d\widetilde{u}}{d\tau} = \widetilde{M}(\tau)\widetilde{u}(\tau), \quad 0 < \tau \le T, \\ &\widetilde{u}(0) = (u_1(0), \ldots, u_{N-1}(0))^\top, \quad 1 \le i \le N-1, \end{aligned} \tag{18}$$

where $u_i(0)$ denotes the evaluation of the payoff function at S_i and $\widetilde{M}(\tau)$ is given by

$$\widetilde{M}(\tau) = \begin{bmatrix} \widetilde{\beta}_1 + 2\widetilde{\alpha}_1 & \widetilde{\gamma}_1 - \widetilde{\alpha}_1 & 0 & 0 & \cdots & 0 \\ \widetilde{\alpha}_2 & \widetilde{\beta}_2 & \widetilde{\gamma}_2 & 0 & \cdots & 0 \\ 0 & \widetilde{\alpha}_3 & \widetilde{\beta}_3 & \widetilde{\gamma}_3 & \cdots & 0 \\ \cdots & \cdots & \ddots & \ddots & \ddots & \cdots \\ 0 & \cdots & 0 & \widetilde{\alpha}_{N-2} & \widetilde{\beta}_{N-2} & \widetilde{\gamma}_{N-2} \\ 0 & \cdots & 0 & 0 & \widetilde{\alpha}_{N-1} - \widetilde{\gamma}_{N-1} & \widetilde{\beta}_{N-1} + 2\widetilde{\gamma}_{N-1} \end{bmatrix}. \tag{19}$$

Unlike the linear case where matrix $\widetilde{M}(\tau)$ is independent of τ, in general problem (18) has not an known solution and one has to perform a numerical solution of such problem. As well as in the S-discretization, there are various possible discretizations with respect to τ in (18). For the sake of simplicity we use forward τ-discretization, obtaining the Euler method

$$\widetilde{u}((n+1)k) = (I + k\widetilde{M}(nk))\widetilde{u}(nk), \ \ 0 \le n \le l-1, \ \ k = \triangle\tau, \tag{20}$$

whose scheme turns out by approximating in (18)

$$\frac{d\widetilde{u}}{d\tau}(nk) = \frac{\widetilde{u}^{n+1} - \widetilde{u}^{n}}{k} + O(k), \tag{21}$$

where $\widetilde{u}^n = \widetilde{u}(nk) = (\widetilde{u}_1(nk), ..., \widetilde{u}_{N-1}(nk))^\top = (\widetilde{u}_1^n, ..., \widetilde{u}_{N-1}^n)^\top$.

Solving (20) one gets for $\tau = lk,\ \ 0 < \tau \leq T$,

$$\widetilde{u}(\tau) = \left[\coprod_{n=l-1}^{n=0} (I + k\widetilde{M}(nk)) \right] \widetilde{u}(0). \tag{22}$$

Note that the τ-discretization of (18) described by (21) together with the S-discretization of [7] yields a full (S,τ)-discretization scheme defined by

$$\widetilde{F}(\widetilde{u}_j^n) = \frac{\widetilde{u}_j^{n+1} - \widetilde{u}_j^n}{k} - \widetilde{\alpha}_j^n\, \widetilde{u}_{j-1}^n - \widetilde{\beta}_j^n\, \widetilde{u}_j^n - \widetilde{\gamma}_j^n\, \widetilde{u}_{j+1}^n = 0, \quad 2 \leq j \leq N-2, \tag{23}$$

where $\widetilde{\alpha}_j^n = \widetilde{\alpha}_j(nk)$, $\widetilde{\beta}_j^n = \widetilde{\beta}_j(nk)$ and $\widetilde{\gamma}_j^n = \widetilde{\gamma}_j(nk)$ obtained taking $\tau = nk$ in (16) and $\widetilde{u}_j^n$ denotes the $j - th$ entry of vector $\widetilde{u}(nk)$.

For the node $j = 1$ the scheme takes the form

$$\widetilde{F}(\widetilde{u}_1^n) = \frac{\widetilde{u}_1^{n+1} - \widetilde{u}_1^n}{k} - (\widetilde{\beta}_1^n + 2\widetilde{\alpha}_1^n)\, \widetilde{u}_1^n - (\widetilde{\gamma}_1^n - \widetilde{\alpha}_1^n)\, \widetilde{u}_2^n = 0, \tag{24}$$

and for $j = N-1$,

$$\widetilde{F}(\widetilde{u}_{N-1}^n) = \frac{\widetilde{u}_{N-1}^{n+1} - \widetilde{u}_{N-1}^n}{k} - (\widetilde{\alpha}_{N-1}^n - \widetilde{\gamma}_{N-1}^n)\, \widetilde{u}_{N-2}^n - (\widetilde{\beta}_{N-1}^n + 2\widetilde{\gamma}_{N-1}^n)\, \widetilde{u}_{N-1}^n = 0. \tag{25}$$

For the sake of convenience in the further study of consistency and stability we can write also the scheme (23)-(25) starting from (10)-(12) in the form

$$\widetilde{F}(\widetilde{u}_j^n) = \frac{\widetilde{u}_j^{n+1} - \widetilde{u}_j^n}{k} - \frac{1}{2}(\widetilde{\sigma}_j^n)^2 S_j^2 \widetilde{\Delta}_j^n(\widetilde{u}) - rS_j \widetilde{\nabla}_j^n(\widetilde{u}) + r\widetilde{u}_j^n = 0, \tag{26}$$

$$1 \leq j \leq N-1,\ \ 0 \leq n \leq l-1,$$

where

$$\widetilde{\nabla}_j^n(\widetilde{u}) = \frac{\widetilde{u}_{j+1}^n - \widetilde{u}_{j-1}^n}{2h}, \tag{27}$$

$$\widetilde{\Delta}_j^n(\widetilde{u}) = \frac{\widetilde{u}_{j-1}^n - 2\widetilde{u}_j^n + \widetilde{u}_{j+1}^n}{h^2}, \tag{28}$$

and from (17),

$$\widetilde{u}_0^n = 2\widetilde{u}_1^n - \widetilde{u}_2^n, \quad \widetilde{u}_N^n = 2\widetilde{u}_{N-1}^n - \widetilde{u}_{N-2}^n. \tag{29}$$

2.2. Fourth Order Schemes

By replacing the partial derivatives $\frac{\partial U}{\partial S}(S_i, \tau)$ and $\frac{\partial^2 U}{\partial S^2}(S_i, \tau)$ by finite-differences approximations [19,22,23], one gets the operators ∇_i and $\triangle_i$ defined by

$$\frac{\partial U}{\partial S}(S_i, \tau) = \nabla_i(U(\tau)) + O(h^4), \tag{30}$$

$$\nabla_i(U(\tau)) = \frac{U(S_{i-2}, \tau) - 8U(S_{i-1}, \tau) + 8U(S_{i+1}, \tau) - U(S_{i+2}, \tau)}{12h}, \tag{31}$$

$$\frac{\partial^2 U}{\partial S^2}(S_i, \tau) = \triangle_i(U(\tau)) + O(h^4), \tag{32}$$

$$\begin{gathered}\triangle_i(U(\tau)) \\ = \frac{-U(S_{i-2}, \tau) + 16U(S_{i-1}, \tau) - 30U(S_i, \tau) + 16U(S_{i+1}, \tau) - U(S_{i+2}, \tau)}{12h^2}, \\ 0 \leq \tau \leq T,\end{gathered} \tag{33}$$

where $S_j = E - L + jh$, $1 \leq j \leq N-1$ are the nodes of the underlying asset interval $[E - L, E + L]$, E is the strike price and $L \leq E$ is the radius of the neighborhood about E where the numerical solution is computed. Let $u_i(\tau)$ be the approximation of theoretical value $U(S_i, \tau)$ and let us denote $u(\tau) = [u_1, u_2, ..., u_{N-1}]^\top$, then the PDE (7) is approximated by the system of ordinary differential equations

$$\frac{du(\tau)}{d\tau} = B(\tau)u(\tau) + w(\tau), \tag{34}$$

where

$$B(\tau) = \begin{bmatrix} \gamma_1 & \delta_1 & \xi_1 & 0 & 0 & 0 & 0 & \dots & 0 \\ \beta_2 & \gamma_2 & \delta_2 & \xi_2 & 0 & 0 & 0 & \dots & 0 \\ \alpha_3 & \beta_3 & \gamma_3 & \delta_3 & \xi_3 & 0 & 0 & \dots & 0 \\ 0 & \alpha_4 & \beta_4 & \gamma_4 & \delta_4 & \xi_4 & 0 & \dots & 0 \\ \dots & \dots & \dots & \dots & \dots & \dots & \dots & \dots & \dots \\ 0 & \dots & 0 & 0 & \alpha_{N-3} & \beta_{N-3} & \gamma_{N-3} & \delta_{N-3} & \xi_{N-3} \\ 0 & \dots & 0 & 0 & 0 & \alpha_{N-2} & \beta_{N-2} & \gamma_{N-2} & \delta_{N-2} \\ 0 & \dots & 0 & 0 & 0 & 0 & \alpha_{N-1} & \beta_{N-1} & \gamma_{N-1} \end{bmatrix}, \tag{35}$$

$$w(\tau) = \begin{bmatrix} \alpha_1 u_{-1} + \beta_1 u_0 \\ \alpha_2 u_0 \\ 0 \\ \vdots \\ 0 \\ \xi_{N-2} u_N \\ \delta_{N-1} u_N + \xi_{N-1} u_{N+1} \end{bmatrix}, \tag{36}$$

and the entries functions of τ are defined by

$$\begin{aligned}
\alpha_i = \alpha_i(\tau) &= -\frac{\sigma_i(\tau)^2 S_i^2}{24h^2} + \frac{rS_i}{12h}, \\
\beta_i = \beta_i(\tau) &= \frac{2\sigma_i(\tau)^2 S_i^2}{3h^2} - \frac{2rS_i}{3h}, \\
\gamma_i = \gamma_i(\tau) &= -\frac{5\sigma_i(\tau)^2 S_i^2}{4h^2} - r, \\
\delta_i = \delta_i(\tau) &= \frac{2\sigma_i(\tau)^2 S_i^2}{3h^2} + \frac{2rS_i}{3h}, \\
\xi_i = \xi_i(\tau) &= -\frac{\sigma_i(\tau)^2 S_i^2}{24h^2} - \frac{rS_i}{12h},
\end{aligned} \tag{37}$$

being $\sigma_i(\tau)$ the adjusted volatility depending of the model under consideration.

The boundary values u_{-1}, u_0, u_N and u_{N+1} appearing in (36) are computed by using fourth order Lagrange interpolating polynomial passing throughout the four closest internal mesh points:

$$\begin{aligned}
u_{-1} &= 10u_1 - 20u_2 + 15u_3 - 4u_4, \\
u_0 &= 4u_1 - 6u_2 + 4u_3 - u_4, \\
u_N &= 4u_{N-1} - 6u_{N-2} + 4u_{N-3} - u_{N-4}, \\
u_{N+1} &= 10u_{N-1} - 20u_{N-2} + 15u_{N-3} - 4u_{N-4}.
\end{aligned} \tag{38}$$

Taking into account (34)-(38), the semidiscretized approximating problem takes the form

$$\begin{aligned}
\frac{du}{d\tau} &= M(\tau)u(\tau), \quad 0 < \tau \leq T, \\
u(0) &= (u_1(0), \ldots, u_{N-1}(0))^\top, \quad 1 \leq i \leq N-1,
\end{aligned} \tag{39}$$

and $M(\tau)$ is given by

$$M(\tau) = \begin{bmatrix}
a_{11} & a_{12} & a_{13} & a_{14} & 0 & \cdots & \cdots & 0 \\
a_{21} & a_{22} & a_{23} & a_{24} & 0 & \ddots & \ddots & \vdots \\
\alpha_3 & \beta_3 & \gamma_3 & \delta_3 & \xi_3 & 0 & \ddots & \vdots \\
0 & \alpha_4 & \beta_4 & \gamma_4 & \delta_4 & \xi_4 & \ddots & \vdots \\
\vdots & \ddots & \ddots & \ddots & \ddots & \ddots & \ddots & 0 \\
\vdots & \cdots & 0 & \alpha_{N-3} & \beta_{N-3} & \gamma_{N-3} & \delta_{N-3} & \xi_{N-3} \\
\vdots & \cdots & \cdots & 0 & a_{N-2N-4} & a_{N-2N-3} & a_{N-2N-2} & a_{N-2N-1} \\
0 & \cdots & \cdots & 0 & a_{N-1N-4} & a_{N-1N-3} & a_{N-1N-2} & a_{N-1N-1}
\end{bmatrix}, \tag{40}$$

where the nonzero entries of the first, second, last but one and last rows are given by

$$
\begin{aligned}
a_{11} &= \gamma_1 + 10\alpha_1 + 4\beta_1 & a_{12} &= \delta_1 - 20\alpha_1 - 6\beta_1 \\
a_{13} &= \xi_1 + 15\alpha_1 + 4\beta_1 & a_{14} &= -4\alpha_1 - \beta_1 \\
a_{21} &= \beta_2 + 4\alpha_2 & a_{22} &= \gamma_2 - 6\alpha_2 \\
a_{23} &= \delta_2 + 4\alpha_2 & a_{24} &= \xi_2 - \alpha_2 \\
a_{N-2N-4} &= \alpha_{N-2} - \xi_{N-2} & a_{N-2N-3} &= \beta_{N-2} + 4\xi_{N-2} \\
a_{N-2N-2} &= \gamma_{N-2} - 6\xi_{N-2} & a_{N-2N-1} &= \delta_{N-2} + 4\xi_{N-2} \\
a_{N-1N-4} &= -\delta_{N-1} - 4\xi_{N-1} & a_{N-1N-3} &= \alpha_{N-1} + 4\delta_{N-1} + 15\xi_{N-1} \\
a_{N-1N-2} &= \beta_{N-1} - 6\delta_{N-1} - 20\xi_{N-1} & a_{N-1N-1} &= \gamma_{N-1} + 4\delta_{N-1} + 10\xi_{N-1}
\end{aligned}
$$

Using the Euler method

$$u((n+1)k) = (I + kM(nk))u(nk), \quad 0 \le n \le l-1, \tag{41}$$

the numerical solution of the vector initial value problem (39) takes the form

$$u(\tau) = \left[\prod_{n=l-1}^{n=0} (I + kM(nk)) \right] u(0), \tag{42}$$

where $k = \Delta\tau,\ lk = \tau$, and the entries of matrix $M(\tau)$ are given by (37).

From (41) one gets the FD scheme

$$F(u_j^n) = \frac{u_j^{n+1} - u_j^n}{k} - \frac{1}{2}(\sigma_j^n)^2 S_j^2 \Delta_j^n(u) - rS_j \nabla_j^n(u) + ru_j^n = 0, \tag{43}$$

$$1 \le j \le N-1, \ 0 \le n \le l-1,$$

where

$$\nabla_j^n(u) = \frac{u_{j-2}^n - 8u_{j-1}^n + 8u_{j+1}^n - u_{j+2}^n}{12h}, \tag{44}$$

$$\Delta_j^n(u) = \frac{-u_{j-2}^n + 16u_{j-1}^n - 30u_j^n + 16u_{j+1}^n - u_{j+2}^n}{12h^2}, \tag{45}$$

and the boundary values u_{-1}^n, u_0^n, u_N^n, u_{N+1}^n are obtained from (38).

3. Numerical Analysis about Local in Time Models

As we quoted in the introduction one of the most interesting local in time models is the H–W–W model, because it takes into account the Gamma of the option in the adjusted volatility function and it is applicable to arbitrary payoff functions.

In this section we study a second order scheme considered in Section 2.1. particularized for the H-W-W model whose vector form is given by (20) and with componentwise expression given by (26).

Consistency of a numerical scheme with a PDE means that the theoretical solution of the pde approximates well to the exact solution of the algebraic equation described by the scheme as the stepsize variables tend to zero [19].

In the next subsection we study the consistency of (26) with (7) with the volatility function given by (3) for a binary option payoff function (cash-or-nothing call) for values of the Leland number A, in $0 \leq A < 1$ to avoid unrealistic negative volatilities, cf. [8, p. 38],

$$U(S,0) = V(S,T) = \begin{cases} 0, & \text{if } S < E, \\ H, & \text{if } S > E, \end{cases} \tag{46}$$

for some $H > 0$.

This case is interesting because of the change of the convexity of the payoff function transmitted to the option price. The developed method is applicable to other local in time models and different payoff functions.

3.1. Consistency of the Numerical Scheme Related to the H-W-W Model

Let us represent equation (7) by $L(U) = 0$, and let $\widetilde{F}(\widetilde{u}_j^n) = 0$ represent the approximating difference equation defined by (26) with exact solution u. In accordance with ([19], p. 100) the FD scheme is consistent with (7) if

$$\widetilde{T}_j^n(U) = \widetilde{F}(U_j^n) - L(U_j^n) \to 0, \quad \text{as} \quad h = \triangle S \to 0,\ k = \triangle t \to 0, \tag{47}$$

where U_j^n denotes the theoretical solution of (7) evaluated at the point $(S_j,\ nk)$, i.e., $U_j^n = U(S_j, nk)$. If in (47) one has that $\widetilde{T}_j^n(U) = O(h^p) + O(k^q)$, then we say that the FD scheme is consistent of order (p,q).

Note that in order to show that (47) holds true, it is necessary to evaluate $\widetilde{F}(U_j^n)$. From (3) and (26), the quantity $\widetilde{\sigma}_j^n = \widetilde{\sigma}_j^n(U)$ takes the form

$$\widetilde{\sigma}_j^n(U)^2 = \sigma_0^2(1 + A\ \text{sign}(\widetilde{\triangle}_j^n(U)) \tag{48}$$

where σ_0 is the annualized volatility.

By applying the operators $\widetilde{\nabla}_j^n$ and $\widetilde{\triangle}_j^n$ defined by (27) and (28) respectively, to the theoretical solution U of equation (7), assuming that U is four times continuously partially differentiable with respect to S and using Taylor's expansion one gets

$$\begin{aligned}\widetilde{\nabla}_j^n(U) &= \frac{\partial U}{\partial S}(S_j, nk) + \frac{h^2}{12}\left\{\frac{\partial^3 U}{\partial S^3}(\xi_1, nk) - \frac{\partial^3 U}{\partial S^3}(\xi_{-1}, nk)\right\} \\ &= \frac{\partial U}{\partial S}(S_j, nk) + h^2 \widetilde{E}_j^n(1),\end{aligned} \tag{49}$$

where

$$S_j < \xi_1 < S_{j+1}\ ;\ S_{j-1} < \xi_{-1} < S_j\ ,$$

and

$$\mid \widetilde{E}_j^n(1) \mid \leq \frac{1}{6} \mid \widetilde{U}^n(1) \mid_{\max}, \tag{50}$$

with

$$\mid \widetilde{U}^n(1) \mid_{\max} = \max\left\{\left|\frac{\partial^3 U}{\partial S^3}(S, nk)\right|\ ;\ E - L \leq S \leq E + L\right\}, \tag{51}$$

$$\begin{aligned}\widetilde{\Delta}_j^n(U) &= \frac{\partial^2 U}{\partial S^2}(S_j, nk) + \frac{h^2}{12}\frac{\partial^4 U}{\partial S^4}(\eta, nk)\\ &= \frac{\partial^2 U}{\partial S^2}(S_j, nk) + h^2 \widetilde{E}_j^n(2),\end{aligned} \tag{52}$$

where $S_{j-1} < \eta < S_{j+1}$ and

$$|\widetilde{E}_j^n(2)| \leq \frac{1}{12}|\widetilde{U}^n(2)|_{\max}, \tag{53}$$

with

$$|\widetilde{U}^n(2)|_{\max} = \max\left\{\left|\frac{\partial^4 U}{\partial S^4}(S, nk)\right| \; ; \; E - L \leq S \leq E + L\right\}. \tag{54}$$

Assuming that U admits two times continuous partial derivatives with respect to time, we have

$$\frac{U_j^{n+1} - U_j^n}{k} = \frac{\partial U}{\partial \tau}(S_j, nk) + k\widetilde{E}_j^n(3), \tag{55}$$

with

$$\widetilde{E}_j^n(3) = \frac{1}{2}\frac{\partial^2 U}{\partial \tau^2}(S_j, \delta), \quad nk < \delta < (n+1)k, \tag{56}$$

$$|\widetilde{E}_j^n(3)| \leq \frac{1}{2}|\widetilde{U}_j(1)|_{\max} = \frac{1}{2}\max\left\{\left|\frac{\partial^2 U}{\partial \tau^2}(S_j, \tau)\right| \; ; \; nk \leq \tau \leq (n+1)k\right\}.$$

From (26), (48), (49), (52) and (55) one gets

$$\begin{aligned}\widetilde{T}_j^n(U) &= \widetilde{F}(U_j^n) - L(U_j^n)\\ &= k\widetilde{E}_j^n(3) - h^2 r S_j \widetilde{E}_j^n(1)\\ &\quad - \frac{S_j^2}{2}\sigma_0^2\Big\{ \left(1 + A \text{ sign}\left(\widetilde{\Delta}_j^n(U)\right)\widetilde{\Delta}_j^n(U)\right)\\ &\quad - \left(1 + A \text{ sign}\left(\frac{\partial^2 U}{\partial S^2}(S_j, nk)\right)\right)\frac{\partial^2 U}{\partial S^2}(S_j, nk)\Big\}.\end{aligned} \tag{57}$$

Let us introduce the expressions

$$B_j^n = \frac{\partial^2 U}{\partial S^2}(S_j, nk) \; ; \quad \triangle B_j^n = h^2 \widetilde{E}_j^n(2). \tag{58}$$

Taking into account (58) and that $x \,\text{sign}(x) = |x|$, (57) can be written in the form

$$\begin{aligned}\widetilde{T}_j^n(U) &= k\widetilde{E}_j^n(3) - h^2\left(rS_j\widetilde{E}_j^n(1) + \frac{S_j^2}{2}\sigma_0^2\widetilde{E}_j^n(2)\right)\\ &\quad - \frac{S_j^2}{2}\sigma_0^2 A\left(|B_j^n + \triangle B_j^n| - |B_j^n|\right).\end{aligned} \tag{59}$$

From (50), (53), (56), (57), (58) and (59) it follows that

$$|\widetilde{T}_j^n(U)| \leq \frac{h^2}{6}\left(r(E+L)|\widetilde{U}^n(1)|_{\max} + \frac{(E+L)^2}{4}\sigma_0^2(1+A)|\widetilde{U}^n(2)|_{\max}\right) + \frac{k}{2}|\widetilde{U}_j(1)|_{\max}.$$

Hence the consistency order of the numerical scheme (26) is given by

$$\widetilde{T}_j^n(U) = O(h^2) + O(k). \tag{60}$$

Summarizing the following result has been established:

Theorem 3.1 *Let* $S = E - L + jh,\ \ 0 < \tau = nk \leq T$ *with* $h = \triangle S,\ k = \triangle\tau$ *with* $2 \leq j \leq N-2$ *and* $2L = Nh$*. Then the numerical scheme (26) at the point* (S,τ) *is consistent of order* $(2,1)$ *with equation (7) and* σ *given by (3).*

3.2. Stability of the Numerical Scheme Related to the H-W-W Model

In this section we address the conditional time stability of the FD scheme (20) in the sense that for a fixed $h = \triangle S > 0$, the solution $\widetilde{u}(\tau)$ remains bounded as $k = \triangle\tau$ tends to zero, $l \to \infty$ but with $lk = \tau$. This is the concept of stability in the fixed station sense with respect to time. Note that as h is fixed the size of the matrix $\widetilde{M}(\tau)$ is also fixed. In other case, the size $N-1$ would increase to infinity as h tends to zero.

Let $\rho = nk$ with $0 \leq n \leq l-1$ and note that $\widetilde{M}(\rho)$ given by (19) can be expressed in the form

$$\widetilde{M}(\rho) = \frac{1}{2h^2}\widetilde{\Sigma}(\rho)D^2\widetilde{P} + \frac{r}{2h}D\widetilde{Q} - rI, \tag{61}$$

where D and $\widetilde{\Sigma}(\rho)$ are diagonal matrices, I is the identity matrix of size $N-1$ with

$$\widetilde{\Sigma}(\rho) = \sigma_0^2 \operatorname{diag}\left(1 + A \operatorname{sign}(\widetilde{\triangle}_1^n(\widetilde{u})), 1 + A \operatorname{sign}(\widetilde{\triangle}_2^n(\widetilde{u})), ..., 1 + A \operatorname{sign}(\widetilde{\triangle}_{N-1}^n(\widetilde{u}))\right), \tag{62}$$

$$D = \operatorname{diag}(S_1, S_2, ..., S_{N-1}), \quad S_i = E - L + ih, \quad 1 \leq i \leq N-1, \tag{63}$$

and $\widetilde{P},\ \widetilde{Q}$ are the constant matrices defined by

$$\widetilde{P} = \begin{bmatrix} 0 & 0 & 0 & 0 & 0 & \cdots & 0 \\ 1 & -2 & 1 & 0 & 0 & \cdots & 0 \\ 0 & 1 & -2 & 1 & 0 & \cdots & 0 \\ & \ddots & \ddots & \ddots & \ddots & \ddots & \\ 0 & \cdots & 0 & 1 & -2 & 1 & 0 \\ 0 & \cdots & 0 & 0 & 1 & -2 & 1 \\ 0 & \cdots & 0 & 0 & 0 & 0 & 0 \end{bmatrix}, \tag{64}$$

$$\widetilde{Q} = \begin{bmatrix} -2 & 2 & 0 & 0 & \cdots & 0 \\ -1 & 0 & 1 & 0 & \cdots & 0 \\ & \ddots & \ddots & \ddots & \ddots & \\ 0 & \cdots & 0 & -1 & 0 & 1 \\ 0 & \cdots & 0 & 0 & -2 & 2 \end{bmatrix}, \tag{65}$$

In order to show the stability of scheme (20) it is necessary to manage norms of vectors and matrices and for these goals it is convenient to introduce some concepts.

If A is a matrix in $\mathbb{R}^{p\times p}$ and $A^\top$ denotes its transposed matrix, we denote by $\|A\|$ the spectral norm of A defined as $\|A\| = \max\left\{\sqrt{\lambda};\ \lambda \text{ eigenvalue of } A^\top A\right\}$, cf. [11].

If q is an integer with $|q| \le p-1$, and A_q is a band matrix in $\mathbb{R}^{p\times p}$ such that $A_q = (a_{ij})$ with $a_{ij} = 0$ everywhere outside of the diagonal $j = i + q$, then it is easy to show that

$$\|A_q\| = \max\left\{|a_{i,i+q}|;\ \begin{array}{c} 1 \le i \le p-q \text{ if } q \ge 0 \\ 1-q \le i \le p \text{ if } q < 0 \end{array}\right\}. \tag{66}$$

Let $\widetilde{P}_0 = \widetilde{P}_0(0,-2,...,-2,0)$ be the diagonal matrix having the same principal diagonal as $\widetilde{P}$, let $\widetilde{P}_1 = \widetilde{P}_1(0,1,...,1)$ be the matrix having zeroes everywhere with the exception of the first upper diagonal that coincides with the first upper diagonal of $\widetilde{P}$. Let $\widetilde{P}_{-1} = \widetilde{P}_{-1}(1,...,1,0)$ be the matrix having zeroes everywhere with the exception of the first lower diagonal that coincides with the first lower diagonal of $\widetilde{P}$. Thus we may express $\widetilde{P}$ as the sum of three sparse matrices in $\mathbb{R}^{N-1\times N-1}$ having only one nonzero diagonal

$$\widetilde{P} = \widetilde{P}_{-1} + \widetilde{P}_0 + \widetilde{P}_1. \tag{67}$$

From (66) and (67) it follows that

$$\|\widetilde{P}\| \le \|\widetilde{P}_1\| + \|\widetilde{P}_0\| + \|\widetilde{P}_1\| = 1+2+1 = 4. \tag{68}$$

In an analogous way one can check that

$$\|\widetilde{Q}\| \le 6. \tag{69}$$

From (61), (68) and (69) one gets

$$\|\widetilde{M}(\rho)\| \le \frac{2}{h^2}\sigma_0^2(1+A)(E+L) + \frac{3r}{h}(E+L) + r = \widetilde{C}_1(h) \tag{70}$$

From (22) and (70) it follows that Euclidean norm of vector $u(\tau)$ is bounded by

$$\|u(\tau)\| \le \left(1 + k\widetilde{C}_1(h)\right)^l \|u(0)\|. \tag{71}$$

As the right hand side of (70) is quadratic in $1/h$, there exists h_0 small enough so that

$$\widetilde{C}_1(h) \le \frac{\widetilde{C}}{h^2},\quad 0 < h < h_0,\quad \widetilde{C} > 0. \tag{72}$$

From (71), (72) and Bernoulli's inequality it follows that

$$\|u(\tau)\| \le \exp\left(\frac{\widetilde{C}}{h^2}\tau\right)\|u(0)\|,\quad 0 < \tau \le T. \tag{73}$$

Summarizing we have proved the next theorem:

Theorem 3.2 *With previous notation, the numerical scheme (20) is stable in the fixed station sense with respect to time for the H-W-W model (3), (7).*

3.3. Cash-or-Nothing Call Options Examples

The first example shows the numerical values of the price option showing that this price increases with the Leland number. A remarkable fact is that the difference of the price option with the one obtained for the case $A = 0$, i.e., without transaction costs, is higher in the zone where the Gamma of the option price is negative. The exact option price without transaction costs, takes the form

$$N(S, \tau) = He^{-r\tau} N(d_2),$$

$$d_2 = \frac{log(S/E) + \tau(r - \frac{1}{2}\sigma_0^2)}{\sigma_0\sqrt{\tau}},$$

where $N(d_2)$ is the cumulative distribution function for the normal function evaluated at d_2, ([1], p.130), and H is given by (46).

3.3.1. Example 1

The following Figure 1 shows these facts for the data

$$\sigma_0 = 0.2, \quad r = 0.02, \quad E = 100\,€, \quad H = 50\,€,$$

$$\tau = 1 \text{ year}, \quad h = \triangle S = 2.4691, \quad k = 0.0025.$$

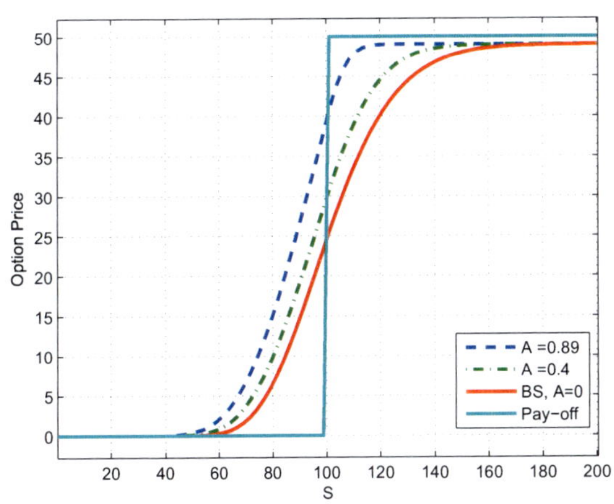

Figure 1. Option pricing for several Leland's number values.

3.3.2. Example 2

Theorem 3.2 guarantees a pointwise stability of the second order scheme (20) or (26) in the sense that given τ with $0 < \tau \leq T$ and a fixed value $h = \triangle S$, for any value of $k = \triangle\tau$ such that $kl = \tau$, the numerical solution remains bounded. However, this stability result is in some degree unsatisfactory because $h = \triangle S$ is fixed and we do not have a link between the stepsize $\triangle\tau = k$ and $h = \triangle S$ so that the numerical results are reliable. This example shows firstly that if $k = \triangle\tau$ is not small enough then spurious oscillations could appear for the data:

$$\sigma_0 = 0.2, \quad r = 0.02, \quad E = 100 \text{ €}, \quad H = 50 \text{ €},$$

$$\tau = 1 \text{ year}, \quad h = 3.9215686, \quad k = 0.0164.$$

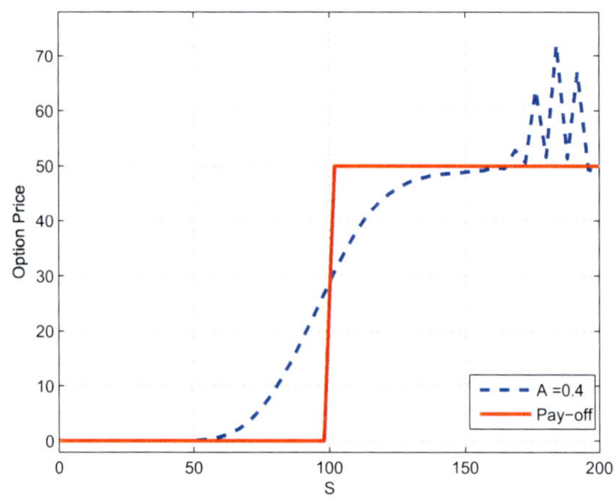

Figure 2. Spurious oscillations.

However, for the same data as in the previous situation with the same h but taking a smaller value of $k = 0.01$, the spurious oscillations disappear as it is shown in Figure 3.

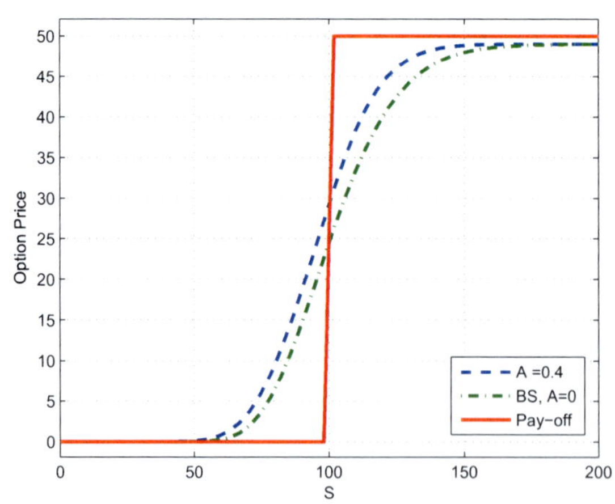

Figure 3. Recovered stability.

These facts suggest the suitable possibility of obtaining a relationship between the stepsizes $\triangle S = h$ and $\triangle\tau = k$ so that a stronger stable behavior could be granted. This task will be addressed in the next future.

4. Numerical Analysis about Global in Time Models

As we quoted in the introduction one of the most complete global in time models for option pricing with transaction costs is the Barles-Soner asymptotic model. Such model has the additional difficulty that its volatility correction function Ψ has not an explicit formula. We begin this section by studying useful properties of Ψ that will be used in the numerical analysis of the model.

4.1. Properties of the Volatility Correction Function

The first result deals with the behaviour of the volatility correction function including growth, concavity and an implicit expression. The proof of such result may be found in [6] and in a forthcoming paper.

Theorem 4.1 *The nonlinear volatility correction function Ψ, unique solution of (5) satisfies:*

(i) *Ψ is one to one increasing function mapping the real line onto the interval* $]-1,+\infty[$

(ii) $\Psi([0,\infty[) = [0,\infty[$ *is implicitly defined by*

$$A = \left(-\frac{\text{Arcsinh}\,\sqrt{\Psi}}{\sqrt{\Psi+1}} + \sqrt{\Psi} \right)^2, \quad if\ \Psi > 0\,, \tag{74}$$

and Ψ is a concave function in $[0,\infty[$.

(iii) $\Psi(]-\infty,0]) =]-1,0]$ *is implicitly defined by*

$$A = -\left(\frac{\arcsin\sqrt{-\Psi}}{\sqrt{\Psi+1}} - \sqrt{-\Psi} \right)^2, \quad if\ -1 < \Psi < 0\,, \tag{75}$$

and Ψ is a convex function in $]-\infty,0]$.

The Figure 4 shows graphically the above properties of Ψ.

The main difficulty in the numerical treatment of equation (6) comes out from the nonlinear term that involves the expression $V_{SS}(\sigma(V_{SS}))^2$ where σ is given by (4). This motivates the knowledge not only of $\Psi(A)$ but of $g(A) = A\Psi(A)$. The next result whose proof will appear in a forthcoming paper includes this information.

Theorem 4.2 . *Let $\Psi(A)$ be the nonlinear volatility correction function implicitly defined in theorem 4.1 and let $g(A) = A\,\Psi(A)$. Then*

(i)

$$|\Psi(A)| \le \max\{1, \Psi(|A|)\}, \quad A \in \mathbb{R}. \tag{76}$$

(ii) *If* $A > 0$, $A_2 = \left(\sinh 2 - \frac{2}{\sqrt{(\sinh\ 2)^2+1}}\right)^2 \simeq 9.58$, *and* $d_2 = \Psi(A_2) - \Psi'(A_2)\,A_2 \simeq 2.62$ *then*

$$|\Psi(A)| \le \Psi'(A_2)\,A + d_2. \tag{77}$$

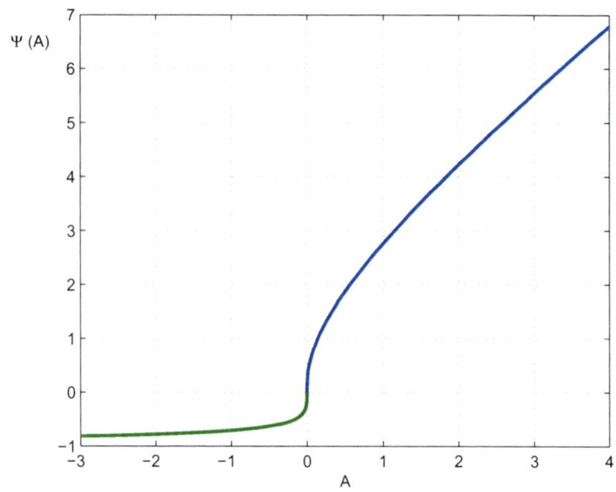

Figure 4. Graph of $\Psi(A)$

(iii) *If* $A < 0,\ \ A_1 = -\frac{(4\pi-3\sqrt{3})^2}{36} \simeq -1.51$, *and* $d_1 = \Psi'(A_1)\, A_1 - \Psi(A_1)$ $\simeq\ 0.64$ *then*

$$|\Psi(A)| \leq \Psi'(A_1)\, |A| \,+\, d_1,\ \ A < 0. \tag{78}$$

(iv) *Function* $g(A) = A\, \Psi(A)$ *is continuously differentiable at* $A = 0$

and

$$|g'(A)| \leq \max\{G,\ 2|A|\Psi'(A_2) + d_2\},\ \ A \in \mathbb{R} \tag{79}$$

where

$$G = \max\{|g'(A)|;\ \ A_1 \leq A \leq A_2. \tag{80}$$

4.2. Consistency of a Numerical Scheme for the Barles-Soner Model

In this subsection we address the study of the consistency of the fourth order scheme (43), in an analogous way to the development of section 3.1. associated to the second order scheme (26).

Let us represent equation (7) by $L(U) = 0$, and let $F(u_j^n) = 0$ represent the approximating difference equation defined by

$$F(u_j^n) = \frac{u_j^{n+1} - u_j^n}{k} - \frac{S_j^2}{2}\sigma_0^2(1 + \Psi_i^n(u))\triangle_j^n(u) - rS_j\nabla_j^n(u) + ru_j^n, \tag{81}$$

where $u_j^n = u(S_j, nk)$ with exact solution u. In accordance with ([19], p. 100) the FD scheme is consistent with (7) if

$$T_j^n(U) = F(U_j^n) - L(U_j^n) \to 0, \quad \text{as} \quad h = \triangle S \to 0,\ k = \triangle t \to 0, \tag{82}$$

where U_j^n denotes the theoretical solution of (7) evaluated at the point $(S_j,\ nk)$, i.e., $U_j^n = U(S_j, nk)$. If in (82) one has that $T_j^n(U) = O(h^p) + O(k^q)$, then we say that the FD scheme is consistent of order (p, q).

Taking into account (81), (31) and (33) for the internal points we have

$$F(U_j^n) = \frac{U_j^{n+1} - U_j^n}{k} - \frac{S_j^2}{2}\sigma_0^2(1 + \Psi_j^n(U))\triangle_j^n(U) - rS_j\nabla_j^n(U) + rU_j^n, \tag{83}$$

where

$$\Psi_j^n(U) = \Psi\left(e^{nkr}a^2S_j^2\triangle_j^n(U)\right) \tag{84}$$

Assuming that U admits enough continuous partial derivatives with respect to S and τ, and using Taylor expansion about (S_j, nk) one gets

$$\nabla_j^n(U) = \frac{\partial U}{\partial S}(S_j, nk) + h^4E_j^n(1), \tag{85}$$

$$\triangle_j^n(U) = \frac{\partial^2 U}{\partial S^2}(S_j, nk) + h^4E_j^n(2), \tag{86}$$

$$\frac{U_j^{n+1} - U_j^n}{k} = \frac{\partial U}{\partial \tau}(S_j, nk) + kE_j^n(3), \tag{87}$$

where $\nabla_j^n(U)$ and $\triangle_j^n(U)$ represent the action of operators ∇_j^n and $\triangle_j^n$ defined in (44)-(45), on the theoretical solution U of equation (7).

Let us introduce the notation

$$A_j^n = e^{nkr}a^2S_j^2\frac{\partial^2 U}{\partial S^2}(S_j, nk), \tag{88}$$

$$\triangle A_j^n = e^{nkr}a^2S_j^2E_j^n(2)h^4, \tag{89}$$

and note that using the function g introduced in theorem 4.2, and (86),(88) and (89), one gets

$$\begin{aligned} &\left\{\left(1 + \Psi_j^n(U)\right)\triangle_j^n(U) - \left(1 + \Psi\left(e^{nkr}a^2S_j^2\frac{\partial^2 U}{\partial S^2}(S_j, nk)\right)\right)\frac{\partial^2 U}{\partial S^2}(S_j, nk)\right\} \\ &= \left\{\left(1 + \Psi(A_j^n + \triangle A_j^n)\right)\left(A_j^n + \triangle A_j^n\right) - \left(1 + \Psi(A_j^n)\right)A_j^n\right\}\frac{1}{e^{nkr}a^2S_j^2} \\ &= \left\{g(A_j^n + \triangle A_j^n) - g(A_j^n) + \triangle A_j^n\right\}e^{-nkr}a^{-2}S_j^{-2}. \end{aligned} \tag{90}$$

By using the mean value theorem and theorem 4.2-(iv) it follows that

$$g(A_j^n + \triangle A_j^n) - g\left(A_j^n\right) = g'(A_j^n + \theta(\triangle A_j^n))\left(\triangle A_j^n\right),\ \ 0 < \theta < 1. \tag{91}$$

Hence, it is easy to show that the truncation error $T_j^n(U)$ is given by

$$T_j^n(U) = -\frac{S_j^2}{2}\sigma_0^2\left(1 + g'(A_j^n + \theta(\triangle A_j^n)\right)E_j^n(2)h^4 - rS_jE_j^n(1)h^4 + kE_j^n(3).$$

Summarizing, the FD scheme (43) is consistent of order (4,1), i.e.,

$$T_j^n(U) = O(h^4) + O(k), \tag{92}$$

and the following theorem can be established:

Theorem 4.3 *Let* $S = E - L + jh,\ \ 0 < \tau = nk \leq T$ *with* $h = \triangle S,\ \ k = \triangle\tau$ *with* $3 \leq j \leq N-3$ *and* $2L = Nh$. *Then the numerical scheme (43) at the point* (S,τ) *is consistent of order* $(4,1)$ *with equation (4),(7)*

4.3. Stability of the Numerical Scheme for the Barles-Soner Model

In this section we address the conditional time stability of the FD scheme (41).

Let $\rho = nk$ with $0 \leq n \leq l-1$ and note that $M(\rho)$ given by (40) can be expressed in the form

$$M(\rho) = \frac{1}{24h^2}\Sigma(\rho)D^2P + \frac{r}{12h}DQ - rI, \tag{93}$$

where D and $\Sigma(\rho)$ are diagonal matrices, I is the identity matrix of size $N-1$ with

$$\Sigma(\rho) = \sigma_0^2 \text{ diag}\left(1+\Psi_1^n(u), 1+\Psi_2^n(u), ..., 1+\Psi_{N-1}^n(u)\right), \tag{94}$$

$$D = \text{diag}\left(S_1, S_2, ..., S_{N-1}\right), \quad E - L < S_i < E + L, \quad 1 \leq i \leq N-1, \tag{95}$$

and $P,\ \ Q$ are the constant matrices defined by

$$P = \begin{bmatrix} 24 & -60 & 48 & -12 & 0 & \cdots & 0 \\ 12 & -24 & 12 & 0 & 0 & \cdots & 0 \\ -1 & 16 & -30 & 16 & -1 & \cdots & 0 \\ & \ddots & \ddots & \ddots & \ddots & \ddots & \\ 0 & \cdots & -1 & 16 & -30 & 16 & -1 \\ 0 & \cdots & 0 & 0 & 12 & -24 & 12 \\ 0 & \cdots & 0 & -12 & 48 & -60 & 24 \end{bmatrix}, \tag{96}$$

$$Q = \begin{bmatrix} -22 & 36 & -18 & 4 & 0 & \cdots & 0 \\ -4 & -6 & 12 & -2 & 0 & \cdots & 0 \\ 1 & -8 & 0 & 8 & -1 & \cdots & 0 \\ & \ddots & \ddots & \ddots & \ddots & \ddots & \\ 0 & \cdots & 1 & -8 & 0 & 8 & -1 \\ 0 & \cdots & 0 & 2 & -12 & 6 & 4 \\ 0 & \cdots & 0 & -4 & 18 & -36 & 22 \end{bmatrix}. \tag{97}$$

In an analogous way to the development of section 3.2., we can obtain explicit bounds of matrices P and Q, independent of h, or the number N of mesh points in the S-variable. In fact, it is easy to show that

$$\|P\| \leq 270\,, \quad \|Q\| \leq 138. \tag{98}$$

In order to find an appropriate bound of the norm of the matrix $\Sigma(\rho)$ given in (94), it is important to use parts (ii) and (iii) of theorem 4.2, obtaining

$$|\Psi_i^n(u)| \leq \Psi'(A_2)e^{nkr}a^2(E+L)^2\frac{12}{h^2}\max\left\{|u_j^n|;\ \ 0 \leq j \leq N\right\} + d_2. \tag{99}$$

Hence, for small enough values of h, one gets the existence of a positive constant $C_1(\rho)$ such that

$$\|\Sigma(\rho)\| \le \sigma_0^2 \left[1 + \left(\frac{C_1(\rho)}{h^2}\right) e^{nkr}\right], \quad \rho = nk. \tag{100}$$

From (93), (98) and (100), it easy to show that for some positive constant $C_2(\tau)$ and $C(\tau)$, one gets

$$\|M(\rho)\| \le \frac{45}{4h^2}\sigma_0^2 \left(1 + \frac{C_2(\tau)e^{r\tau}}{h^2}\right)(E+L)^2 + \frac{23r}{2h}(E+L) + r \le \frac{C(\tau)}{h^4}, \tag{101}$$

independent of n. Finally from (42) and (101) it follows that

$$\|u(\tau)\| \le e^{\frac{\tau C(\tau)}{h^4}} \|u(0)\|, \quad h = \Delta S \text{ fixed.} \tag{102}$$

Summarizing, the following result can be established:

Theorem 4.4 *With the previous notation, the numerical scheme (41) is stable in the fixed station sense with respect to time for the Barles-Soner model (4) and (7).*

4.4. Vanilla Call Option Pricing Example

4.4.1. Example 3

This example shows the numerical results obtained using the fourth order scheme (41) or (43) for the vanilla call option pricing with data

$$\sigma_0 = 0.2, \quad r = 0.02, \quad E = 100\text{ €}, \quad \tau = 1 \text{ year}, \quad h = \Delta S = 4, \quad k = 0.005.$$

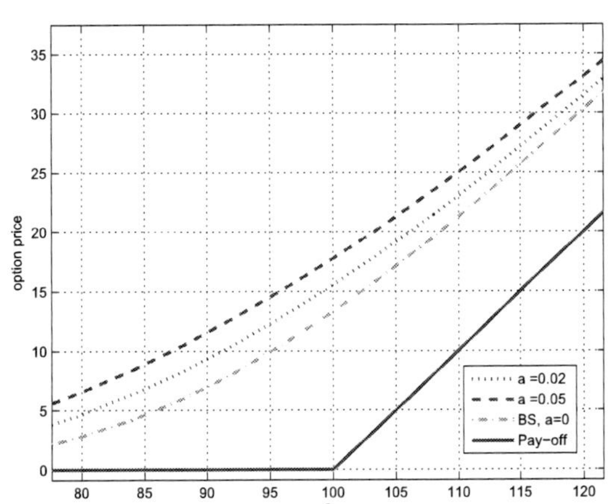

Figure 5. Numerical results for several values of the parameter a.

Note that the price option increases with the value of a, and that the difference with respect to the option price without transaction costs is maximal in the left proximities of the strike E.

Conclusion

This chapter deals with the numerical analysis and computing of nonlinear Black-Scholes equations modeling option pricing in market with transaction costs. We show that a good model can provide bad pricing if some cares in the numerical treatment of the problem are disregarded, such as consistency and stability of the underlying numerical scheme.

Acknowledgements

This work has been partially supported by the Spanish Ministery of Education and Research project TRA2007-68006-C02-02.

References

[1] M. Avellaneda and P. Laurence, *Quantitative modeling of derivative securities*, Chapman & Hall /CRC, New York, 2000.

[2] M. Avellaneda and A. Parás, Dynamic hedging portfolios for derivative securities in the presence of large transaction costs, *Appl. Math. Finance* **1**, 1994, 165–193.

[3] C. Ballester, R. Company and L. Jódar, *An efficient method for option pricing with discrete dividend payment*, Computer and Mathematics with Applications. In press.

[4] G. Barles and H.M. Soner, Option pricing with transaction costs and a nonlinear Black–Scholes equation, *Finance Stochast* **2** (1998), 369–397.

[5] P. Boyle and T. Vorst, Option replication in discrete time with transaction costs, *J. Finance* **47** (1973), 271–293.

[6] R. Company, E. Navarro, J.-R. Pintos and E. Ponsoda, *Numerical solution of linear and nonlinear Black–Scholes option pricing equations*, Computers and Mathematics with Applications. In press.

[7] M. Davis, V. Panis and T. Zariphopoulou, European option pricing with transaction fees, *SIAM J. Contr. Optim.* **31** (1993), 470–493.

[8] J. Dewynne, S. Howinson and P. Wilmott, Option pricing: mathematical models and computations, *Oxford Financial Press* **6**, Oxford, 2000.

[9] B. Düring, M. Fournier and A. Jüngel, Convergence of a high order compact finite difference scheme for a nonlinear Black–Scholes equation, Esaim–Mathematical Modelling and Numerical Analysis–Modelisation *Mathematique et Analyse Numerique,* **38** (2004), 359–369.

[10] P. Forsyth, K. Vetzal and R. Zvan, A finite element approach to the pricing of discrete lookbacks with stochastic volatility, *Appl. Math. Finance*, **6** (1999), 87–106.

[11] G. Golub and C. Van Loan, *Matrix computations*, The Johns Hopkins University Press, London, 1996.

[12] J.M. Harrison and S.R. Pliska, Martingales and stochastic integrals in the theory of continuous trading, *Stochastic Processes Appl.* **11** (1981), 215–260.

[13] S.D. Hodges and A. Neuberger, Optimal replication of contingent claims under transaction costs, *Rev. Futures Markets* **8** (1989) 222–239.

[14] T. Hoggard, A.E. Whalley, and P. Wilmott, Hedging Option Portfolios in the Presence of Transaction Costs, *Adv. Futures Options Research* **7** (1994) 21–35.

[15] J. Leitner, *Continoustime CAPM, price for risk and utility maximization. In Mathematical Finance*. Workshop of the Mathematical Finance Research Project, Konstanz, Germany, M. Kohlmanned et al., Birkhäuser, Basel, 2001.

[16] H.E. Leland, Option pricing and replication with transactions costs, *J. Finance* **40** (1985), 1283–1301.

[17] O. Pironneau and F. Hecht, Mesh adaption for the Black and Scholes equations, *East West J. Numer. Math.* **8** (2000), 25–35.

[18] A. Rigal, Numerical analisys of three-time-level finite difference schemes for unsteady diffusion-convection problems, *J. Numer. Meth. Engineering* **30** (1990), 307–330.

[19] G.D. Smith, *Numerical solution of partial differential equations: finite difference methods*, Clarendon Press, Oxford, 1985.

[20] H.M. Soner, S.E. Shreve and J. Cvitanic, There is non on trivial hedging portfolio for option pricing with transaction costs, *Ann. Appl. Probab.* **5** (1995), 327–355.

[21] J.C. Strikwerda, *Finite difference schemes and partial differential equations*, Wadsworth & Brooks/Cole, Mathematics Series, 1989.

[22] D. Tavella, C. Randall, *Pricing financial instruments. The finite difference method*, John Wiley & Sons, New York, 2000.

[23] J.W. Thomas, *Numerical partial differential equations: finite difference methods*, Springer, 1995.

[24] A.E. Whalley and P. Wilmott, An asymptotic analysis of an optimal hedging model for option pricing with transaction costs, *Math. Finance* **7** (1997), 307–324.

In: Nonlinear Models in Mathematical Finance...
Editor: Matthias Ehrhardt, pp. 173-218
ISBN 978-1-60456-931-5

Chapter 6

TRANSFORMATION METHODS FOR EVALUATING APPROXIMATIONS TO THE OPTIMAL EXERCISE BOUNDARY FOR LINEAR AND NONLINEAR BLACK-SCHOLES EQUATIONS

Daniel Ševčovič
Department of Applied Mathematics and Statistics
Faculty of Mathematics, Physics & Informatics, Comenius University
842 48 Bratislava, Slovak Republic

Abstract

The purpose of this survey chapter is to present a transformation technique that can be used in analysis and numerical computation of the early exercise boundary for an American style of vanilla options that can be modelled by class of generalized Black-Scholes equations. We analyze qualitatively and quantitatively the early exercise boundary for a linear as well as a class of nonlinear Black-Scholes equations with a volatility coefficient which can be a nonlinear function of the second derivative of the option price itself. A motivation for studying the nonlinear Black-Scholes equation with a nonlinear volatility arises from option pricing models taking into account e.g. nontrivial transaction costs, investor's preferences, feedback and illiquid markets effects and risk from a volatile (unprotected) portfolio. We present a method how to transform the free boundary problem for the early exercise boundary position into a solution of a time depending nonlinear nonlocal parabolic equation defined on a fixed domain. We furthermore propose an iterative numerical scheme that can be used in order to find an approximation of the free boundary. In the case of a linear Black-Scholes equation we are able to derive a nonlinear integral equation for the position of the free boundary. We present results of numerical approximation of the early exercise boundary for various types of linear and nonlinear Black-Scholes equations and we discuss dependence of the free boundary on model parameters. Finally, we discuss an application of the transformation method for the pricing equation for American type of Asian options.

1. Introduction

According to the classical theory due to Black, Scholes and Merton the price of an option in an idealized financial market can be computed from a solution to the well-known Black–Scholes linear parabolic equation derived by Black and Scholes in [7], and, independently by Merton (see also Kwok [38], Dewynne *et al.* [16], Hull [30], Wilmott *et al.* [53]). Recall that a European call (put) option is the right but not obligation to purchase (sell) an underlying asset at the expiration price E at the expiration time T. Assuming that the underlying asset S follows a geometric Brownian motion

$$dS = (\varrho - q)Sdt + \sigma S dW, \tag{1}$$

where ϱ is a drift, q is the asset dividend yield rate, σ is the volatility of the asset and W is the standard Wiener process (cf. [38]), one can derive a governing partial differential equation for the price of an option. We remind ourselves that the equation for option's price $V(S,t)$ is the following parabolic PDE:

$$\frac{\partial V}{\partial t} + (r-q)S\frac{\partial V}{\partial S} + \frac{\sigma^2}{2}S^2\frac{\partial^2 V}{\partial S^2} - rV = 0 \tag{2}$$

where σ is the volatility of the underlying asset price process, $r > 0$ is the interest rate of a zero-coupon bond, $q \geq 0$ is the dividend yield rate. A solution $V = V(S,t)$ represents the price of an option if the price of an underlying asset is $S > 0$ at time $t \in [0,T]$.

The case when the diffusion coefficient $\sigma > 0$ in (2) is constant represents a classical Black–Scholes equation originally derived by Black and Scholes in [7]. On the other hand, if we assume the volatility coefficient $\sigma > 0$ to be a function of the solution V itself then (2) with such a diffusion coefficient represents a nonlinear generalization of the Black–Scholes equation. It is a purpose of this chapter to focus our attention to the case when the diffusion coefficient σ^2 may depend on the time $T-t$ to expiry, the asset price S and the second derivative $\partial_S^2 V$ of the option price (hereafter referred to as Γ), i.e.

$$\sigma = \sigma(S^2\partial_S^2 V, S, T-t)\,. \tag{3}$$

A motivation for studying the nonlinear Black–Scholes equation (2) with a volatility σ having a general form (3) arises from option pricing models taking into account nontrivial transaction costs, market feedbacks and/or risk from a volatile (unprotected) portfolio. Recall that the linear Black–Scholes equation with constant σ has been derived under several restrictive assumptions like e.g. frictionless, liquid and complete markets, etc. We also recall that the linear Black–Scholes equation provides a perfectly replicated hedging portfolio. In the last decades some of these assumptions have been relaxed in order to model, for instance, the presence of transaction costs (see e.g. Leland [39], Hoggard *et al.* [29], Avellaneda and Paras [4]), feedback and illiquid market effects due to large traders choosing given stock-trading strategies (Frey and Patie [22], Frey and Stremme [23], During *et al.* [17], Schönbucher and Wilmott [49]), imperfect replication and investor's preferences (Barles and Soner [8]), risk from unprotected portfolio (Kratka [37], Jandačka and Ševčovič [32] or [47]). One of the first nonlinear models is the so-called Leland model (cf. [39]) for pricing call and put options under the presence of transaction costs. It has been generalized for more complex option strategies by Hoggard, Whaley and Wilmott in [29]. In this model the volatility σ is given by

$$\sigma^2(S^2\partial_S^2 V, S, \tau) = \hat{\sigma}^2(1 + \mathrm{Le}\,\mathrm{sgn}(\partial_S^2 V)) \tag{4}$$

where $\hat{\sigma} > 0$ is a constant historical volatility of the underlying asset price process and $\mathrm{Le} > 0$ is the so-called Leland constant given by $\mathrm{Le} = \sqrt{2/\pi}C/(\hat{\sigma}\sqrt{\Delta t})$ where $C > 0$ is a constant round trip transaction cost per unit dollar of transaction in the assets market and $\Delta t > 0$ is the time-lag between portfolio adjustments.

Notice that dependence of volatility adjustment on the second derivative of the price is quite natural. Indeed, in the idealized Black-Scholes theory, the optimal hedge is equal to $\pm\partial_S V$ and therefore one may expect more frequent transaction in regions with the high second derivative $\partial_S^2 V$ (cf. [7]).

A popular nonlinear generalization of the Black–Scholes equation has been proposed by Avellaneda, Levy, and Paras [5] for description of incomplete markets and uncertain but bounded volatility. In their model we have

$$\sigma^2(S^2\partial_S^2 V, S, \tau) = \begin{cases} \hat{\sigma}_1^2 & \text{if } \partial_S^2 V < 0, \\ \hat{\sigma}_2^2 & \text{if } \partial_S^2 V > 0, \end{cases} \tag{5}$$

where σ_1 and σ_2 represent a lower and upper a-priori bound on the otherwise unspecified asset price volatility.

If transaction costs are taken into account perfect replication of the contingent claim is no longer possible and further restrictions are needed in the model. By assuming that investor's preferences are characterized by an exponential utility function Barles and Soner (cf. [8]) derived a nonlinear Black–Scholes equation with the volatility σ given by

$$\sigma^2(S^2\partial_S^2 V, S, \tau) = \hat{\sigma}^2\left(1 + \Psi(a^2 e^{r\tau} S^2 \partial_S^2 V)\right) \tag{6}$$

where Ψ is a solution to the ODE: $\Psi'(x) = (\Psi(x) + 1)/(2\sqrt{x\Psi(x)} - x)$, $\Psi(0) = 0$, and $a > 0$ is a given constant representing risk aversion. Notice that $\Psi(x) = O(x^{\frac{1}{3}})$ for $x \to 0$ and $\Psi(x) = O(x)$ for $x \to \infty$.

Another popular model has been derived for the case when the asset dynamics takes into account the presence of feedback and illiquid market effects. Frey and Stremme (cf. [22, 23]) introduced directly the asset price dynamics in the case when a large trader chooses a given stock-trading strategy (see also [49]). The diffusion coefficient σ is again nonconstant and it can be expressed as:

$$\sigma^2(S^2\partial_S^2 V, S, \tau) = \hat{\sigma}^2\left(1 - \varrho\lambda(S)S\partial_S^2 V\right)^{-2} \tag{7}$$

where $\hat{\sigma}^2, \varrho > 0$ are constants and $\lambda(S)$ is a strictly convex function, $\lambda(S) \geq 1$. Interestingly enough, explicit solutions to the Black–Scholes equation with varying volatility as in (7) have been derived by Bordag and Chankova [9] and Bordag and Frey [10].

The last example of the Black–Scholes equation with a nonlinearly depending volatility is the so-called Risk Adjusted Pricing Methodology model proposed by Kratka in [37] and revisited by Jandačka and Ševčovič in [32]. In order to maintain (imperfect) replication of a portfolio by the delta hedge one has to make frequent portfolio adjustments leading to a substantial increase in transaction costs. On the other hand, rare portfolio adjustments may lead to an increase of the risk arising from a volatile (unprotected) portfolio. In the RAPM model the aim is to optimize the time-lag Δt between consecutive portfolio adjustments. By choosing $\Delta t > 0$ in such way that the sum of the rate of transaction costs and the rate of a risk from unprotected portfolio is minimal one can find the optimal time lag $\Delta t > 0$. In the RAPM model, it turns out that the volatility is again nonconstant and it has the following form:

$$\sigma^2(S^2\partial_S^2 V, S, \tau) = \hat{\sigma}^2 \left(1 + \mu(S\partial_S^2 V)^{\frac{1}{3}}\right). \tag{8}$$

Here $\hat{\sigma}^2 > 0$ is a constant historical volatility of the asset price returns and $\mu = 3(C^2R/2\pi)^{\frac{1}{3}}$ where $C, R \geq 0$ are nonnegative constant representing the transaction cost measure and the risk premium measure, resp. (see [32] for details).

Notice that all the above mentioned nonlinear models are consistent with the original Black–Scholes equation in the case the additional model parameters (e.g. Le, a, ϱ, μ) are vanishing. If plain call or put vanilla options are concerned then the function $V(S,t)$ is convex in S variable and therefore each of the above mentioned models has a diffusion coefficient strictly larger than $\hat{\sigma}^2$ leading to a larger values of computed option prices. They can be therefore identified with higher Ask option prices, i.e. offers to sell an option. Furthermore, these models have been considered and analyzed mostly for European style of options, i.e. options that can be exercised only at the maturity $t = T$. On the other hand, American options are much more common in financial markets as they allow for exercising of an option anytime before the expiry T. In the case of an American call option a solution to equation (2) is defined on a time dependent domain $0 < S < S_f(t)$, $0 < t < T$. It is subject to the boundary conditions

$$V(0,t) = 0\,, \quad V(S_f(t),t) = S_f(t) - E\,, \quad \partial_S V(S_f(t),t) = 1\,, \tag{9}$$

and terminal pay-off condition at expiry $t = T$

$$V(S,T) = \max(S - E, 0) \tag{10}$$

where $E > 0$ is a strike price (cf. [16, 38]). One of important problems in this field is the analysis of the early exercise boundary $S_f(t)$ and the optimal stopping time (an inverse function to $S_f(t)$) for American call (or put) options on stocks paying a continuous dividend yield with a rate $q > 0$ (or $q \geq 0$). However, an exact analytical expression for the free boundary profile is not even known for the case when the volatility σ is constant. Many authors have investigated various approximation models leading to approximate expressions for valuing American call and put options: analytic approximations (Barone–Adesi and Whaley [6], Kuske and Keller [36], Dewynne *et al.* [16], Geske *et al.* [24, 25], MacMillan [40], Mynemi [44]); methods of reduction to a nonlinear integral equation (Alobaidi [1], Kwok [38], Mallier *et al.* [41, 42], Ševčovič [46], Stamicar *et al.* [50]). In recent papers [55, 56], Zhu derived a closed form of the free boundary position in terms of on infinite series. We also refer to a recent survey paper by Chadam [12] focusing on free boundary problems in mathematical finance.

We remind ourselves that in the case of a constant volatility there are, in principle, two ways how to solve numerically the free boundary problem for the value of an American call resp. put option and the position of the early exercise boundary. The first class of algorithms is based on reformulation of the problem in terms of a variational inequality (see Kwok [38] and references therein). The variational inequality can be then solved numerically by the so-called Projected Super Over Relaxation method (PSOR for short). An advantage of this method is that it gives us immediately the value of a solution. A disadvantage is that one has to solve large systems of linear equations iteratively taking into account the obstacle for a solution, and, secondly, the free boundary position should be deduced from the solution a posteriori. Moreover, the PSOR method is not directly applicable for solving the problem (2)-(9) when the diffusion coefficient σ may depend on the second derivative of a solution itself. The second class of methods is based on derivation of a nonlinear integral equation for the position of the free boundary without the need of knowing the option price itself (see e.g. Evans, Kuske-Keller [21, 36], Mallier and Alobaidi [1, 41, 42], Ševčovič *et al.* [46, 50], Chadam *et al.* [12, 13]. In this approach an advantage is that only a single equation for the free boundary has to be solved provided that σ is constant; a disadvantage is that the method is based on integral transformation techniques and therefore the assumption σ is constant is crucial.

Higher order finite difference approximations of the free boundary problems for call or put options are discussed in recent papers by Ankudinova and Ehrhardt [2, 3], Ehrhardt and Mickens [18] and Zhao *et al.* [54]. Other interesting analytical and numerical methods for evaluating the early exercise boundary have been recently studied by e.g. Milstein *et al.* [43] (Monte–Carlo methods), Widdicks *et al.* [52] (singular pertubation techniques and asymptotic expansions), Cho *et al.* [14] (parameter estimation methods), Grandits and Schachinger *et al.* [26] (tracking of a discontinuity method), Imai *et al.* [31] (numerical method for generalized Lelend model).

In this survey we recall an iterative numerical algorithm for solving the free boundary problem for an American type of options in the case the volatility σ may depend on the option and asset values as well as on the time $T - t$ to expiry as well as for American type of Asian options. A key idea of this method consists in transformation of the free boundary problem into a semilinear parabolic equation defined on a fixed spatial domain coupled with a nonlocal algebraic constraint equation for the free boundary position. It has been proposed and analyzes by the author in a series of papers [46–48, 50]. Since the resulting parabolic equation contains a strong convective term we make use of the operator splitting method in order to overcome numerical difficulties. A full space-time discretization of the problem leads to a system of semi-linear algebraic equations that can be solved by an iterative procedure at each time level.

The rest of the chapter has the following organization: in the next section we recall the well known nonlinear generalization of Black–Scholes equations due to Frey and Stremme, Barles and Soner and Kratka, Jandačka and Ševčovič, resp. We also present qualitative and quantitative properties of the nonlinear models for pricing European style of options with special focus on the Risk adjusted pricing methodology (RAPM) due to Kratka [37], Jandačka and Ševčovič [32]. Section 3 is devoted to the free boundary problem for pricing American options by means of a linear Black–Scholes equation, i.e. $\sigma > 0$ is constant. This is important in order to understand important steps of the fixed domain transformation method. A resulting system of transformed equations consists of a nonlocal parabolic equation defined on a fixed domain with time depending coefficients and an algebraic constraint equation for the free boundary position. Since the volatility σ is constant in this case, by using Sine and Cosine Fourier transformations the system of transformed equations can be further simplified and reduced to a single nonlinear integral equation for the free boundary position. We show how this integral equation can be utilized in order to obtain qualitative properties of the free boundary position (early exercise behavior, long time behavior) as well as quantitative properties (a fast and stable numerical scheme for computing the free boundary function). In Section 4 we discuss a transformation method applied to a class of nonlinear Black–Scholes equations. We are able to derive a similar system of transformed parabolic–algebraic equation to the one from Section 3. However, as the volatility is no longer a constant and it may depend on the solution itself the resulting system of equations can not be reduced to a single integral equation for the free boundary and it has to be solved numerically. We propose a numerical method based on the finite difference approximation combined with an operator splitting technique for numerical approximation of the solution and computation of the free boundary position. Several numerical results for nonlinear Black–Scholes equations with volatility functions σ defined as in (6) and (8) are presented. We also compare our methodology with well-known methods for evaluation of approximation to the free boundary position in the case the volatility σ is constant. We analyze dependence of the free boundary position with respect to various parameters entering expressions (6) and (8). Finally, Section 5 is devoted to a recent application of the transformation method in the case of American style of Asian options in which the strike price is an arithmetical average of underlying asset prices. Although the volatility σ is assumed to be constant, due to a specific character of the Black–Scholes equation for pricing Asian option the resulting transformed system of equation cannot be further reduced and has to be solved numerically by a slight modification of the numerical method discussed in Section 4. We finish the last section by presentation of several illustrative examples describing the early exercise boundary for American style of Asian call options with floating strike.

2. Risk Adjusted Methodology Model

The aim of this section is to present one of nonlinear generalizations of the classical Black–Scholes equation with a volatility σ of the form (3) in a more detail. We focus on the so-called Risk adjusted pricing methodology model due to Kratka [37] and straightforward generalization by Jandačka and Ševčovič [32] (see also [47]). In this model both the risk arising from nontrivial transaction costs as well as the risk from unprotected volatile portfolio are taken into account. Their sum representing the total risk is subject of minimization. The original model was proposed by [37]. In [32] we modified Kratka's approach by considering a different measure for risk arising from unprotected portfolio in order to construct a model which is scale invariant and mathematically well posed. These two important features were missing in the original model of Kratka. The model is based on the Black–Scholes parabolic PDE in which transaction costs are described by the Hoggard, Whalley and Wilmott extension of the Leland model (cf. [29,30,38]) whereas the risk from a volatile portfolio is described by the average value of the variance of the synthesized portfolio. Transaction costs as well as the volatile portfolio risk depend on the time-lag between two consecutive transactions. We define the total risk premium as a sum of transaction costs and the risk cost from the unprotected volatile portfolio. By minimizing the total risk premium functional we obtain the optimal length of the hedge interval.

Concerning the dynamics of an underlying asset we will assume that the asset price $S = S(t), t \geq 0$, follows a geometric Brownian motion (1) with a drift ρ, standard deviation $\hat{\sigma} > 0$ and it may pay continuous dividends, i.e. $dS = (\rho - q)Sdt + \hat{\sigma}SdW$ where dW denotes the differential of the standard Wiener process and $q \geq 0$ is a continuous dividend yield rate. This assumption is usually made when deriving the classical Black–Scholes equation (see e.g. [30, 38]). Similarly as in the derivation of the classical Black–Scholes equation we construct a synthesized portfolio Π consisting of a one option with a price V and δ assets with a price S per one asset:

$$\Pi = V + \delta S\,. \tag{11}$$

We recall that the key idea in the Black–Scholes theory is to examine the differential $\Delta\Pi$ of equation (11). The right-hand side of (11) can be differentiated by using Itô's formula whereas portfolio's increment $\Delta\Pi(t) = \Pi(t + \Delta t) - \Pi(t)$ of the left-hand side can be expressed as follows:

$$\Delta\Pi = r\Pi\Delta t + \delta q S\Delta t \tag{12}$$

where $r > 0$ is a risk-free interest rate of a zero-coupon bond. In the real world, such a simplified assumption is not satisfied and a new term measuring the total risk should be added to (12). More precisely, the change of the portfolio Π is composed of two parts: the risk-free interest rate part $r\Pi\Delta t$ and the total risk premium: $r_R S\Delta t$ where r_R is a risk premium per unit asset price. We consider a short positioned call option. Therefore the writer of an option is exposed to this total risk. Hence we are going to price the higher Ask option price – an offer to sell an option. It means that $\Delta\Pi = r\Pi\Delta t - r_R S\Delta t$. The total risk premium r_R consists of the transaction risk premium r_{TC} and the portfolio volatility risk premium r_{VP}, i.e. $r_R = r_{TC} + r_{VP}$. Hence

$$\Delta\Pi = r\Pi\Delta t + \delta q S\Delta t - (r_{TC} + r_{VP})S\Delta t\,. \tag{13}$$

Our next goal is to show how these risk premium measures r_{TC}, r_{VP} depend on the time lag and other quantities, like e.g. $\hat{\sigma}$, S, V, and derivatives of V. The problem can be decomposed in two parts: modeling the transaction costs measure r_{TC} and volatile portfolio risk measure r_{VP}.

We begin with modeling transaction costs. In practice, we have to adjust our portfolio by frequent buying and selling of assets. In the presence of nontrivial transaction costs, continuous portfolio adjustments may lead to infinite total transaction costs. A natural way how to consider transaction costs within the frame of the Black–Scholes theory is to follow the well known Leland approach extended by Hoggard, Whalley and Wilmott (cf. [29,38]). We will recall an idea how to incorporate the effect of transaction costs into the governing equation. More precisely, we will derive the coefficient of transaction costs r_{TC} occurring in (13). Let us denote by C the round trip transaction cost per unit dollar of transaction. Then

$$C = (S_{ask} - S_{bid})/S \tag{14}$$

where S_{ask} and S_{bid} are the so-called Ask and Bid prices of the asset, i.e. the market price offers for selling and buying assets, resp. Here $S = (S_{ask} + S_{bid})/2$ denotes the mid value of the underlying asset price.

In order to derive the term r_{TC} in (13) measuring transaction costs we will assume, for a moment, that there is no risk from volatile portfolio, i.e. $r_{VP} = 0$. Then $\Delta V + \delta \Delta S = \Delta\Pi = r\Pi\Delta t + \delta q S \Delta t + r_{TC} S \Delta t$. Following Leland's approach (cf. [29]), using Itô's formula and assuming δ-hedging of a synthetised portfolio Π one can derive that the coefficient r_{TC} of transaction costs is given by the formula:

$$r_{TC} = \frac{C\hat{\sigma}S}{\sqrt{2\pi}} \left|\partial_S^2 V\right| \frac{1}{\sqrt{\Delta t}} \tag{15}$$

(see [29, Eq. (3)]). It leads to the well known Leland generalization of the Black–Scholes equation (2) in which the diffusion coefficient is given by (4) (see [29, 39] for details).

Next we focus our attention to the problem how to incorporate a risk from a volatile portfolio into the model. In the case when a portfolio consisting of options and assets is highly volatile, an investor usually asks for a price compensation. Notice that exposure to risk is higher when the time-lag between portfolio adjustments is higher. We shall propose a measure of such a risk based on the volatility of a fluctuating portfolio. It can be measured by the variance of relative increments of the replicating portfolio $\Pi = V + \delta S$, i.e. by the term $var((\Delta\Pi)/S)$. Hence it is reasonable to define the measure r_{VP} of the portfolio volatility risk as follows:

$$r_{VP} = R\frac{var\left(\frac{\Delta\Pi}{S}\right)}{\Delta t}. \tag{16}$$

In other words, r_{VP} is proportional to the variance of a relative change of a portfolio per time interval Δt. A constant R represents the so-called *risk premium coefficient*. It can be interpreted as the marginal value of investor's exposure to a risk. If we apply Itô's formula to the differential $\Delta\Pi = \Delta V + \delta\Delta S$ we obtain $\Delta\Pi = (\partial_S V + \delta)\,\hat{\sigma} S \Delta W + \frac{1}{2}\hat{\sigma}^2 S^2 \Gamma (\Delta W)^2 + \mathcal{G}$ where $\Gamma = \partial_S^2 V$ and $\mathcal{G} = (\partial_S V + \delta)\rho S \Delta t + \partial_t V \Delta t$ is a deterministic term, i.e $E(\mathcal{G}) = \mathcal{G}$ in the lowest order Δt - term approximation. Thus

$$\Delta\Pi - E(\Delta\Pi) = (\partial_S V + \delta)\,\hat{\sigma} S \phi\sqrt{\Delta t} + \frac{1}{2}\hat{\sigma}^2 S^2(\phi^2 - 1)\Gamma\Delta t$$

where ϕ is a random variable with the standard normal distribution such that $\Delta W = \phi\sqrt{\Delta t}$. Hence the variance of $\Delta\Pi$ can be computed as follows:

$$var(\Delta\Pi) = E\left([\Delta\Pi - E(\Delta\Pi)]^2\right) = E\left([(\partial_S V + \delta)\hat{\sigma} S\phi\sqrt{\Delta t} + \tfrac{1}{2}\hat{\sigma}^2 S^2 \Gamma\left(\phi^2 - 1\right)\Delta t]^2\right).$$

Similarly, as in the derivation of the transaction costs measure r_{TC} we assume the δ-hedging of portfolio adjustments, i.e. we choose $\delta = -\partial_S V$. Since $E((\phi^2 - 1)^2) = 2$ we obtain an expression for the risk premium r_{VP} in the form:

$$r_{VP} = \frac{1}{2} R \hat{\sigma}^4 S^2 \Gamma^2 \Delta t\,. \tag{17}$$

Notice that in our approach the increase in the time-lag Δt between consecutive transactions leads to a linear increase of the risk from a volatile portfolio where the coefficient of proportionality depends the asset price S, option's Gamma, $\Gamma = \partial_S^2 V$, as well as the constant historical volatility $\hat{\sigma}$ and the risk premium coefficient R.

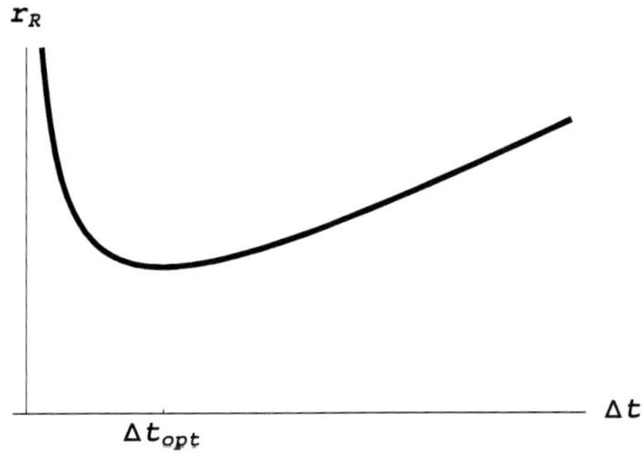

Figure 1. The total risk premium $r_R = r_{TC} + r_{VP}$ as a function of the time-lag Δt between two consecutive portfolio adjustments.

2.1. Risk Adjusted Black–Scholes Equation

The total risk premium $r_R = r_{TC} + r_{VP}$ consists of two parts: transaction costs premium r_{TC} and the risk from a volatile portfolio r_{VP} premium defined as in (15) and (17), resp. We assume that an investor is risk aversive and he/she wants to minimize the value of the total risk premium r_R. For this purpose one has to choose the optimal time-lag Δt between two consecutive portfolio adjustments. As both r_{TC} as well as r_{VP} depend on the time-lag Δt so does the total risk premium r_R. In order to find the optimal value of Δt we have to minimize the following function:

$$\Delta t \mapsto r_R = r_{TC} + r_{VP} = \frac{C|\Gamma|\hat{\sigma}S}{\sqrt{2\pi}}\frac{1}{\sqrt{\Delta t}} + \frac{1}{2}R\hat{\sigma}^4 S^2 \Gamma^2 \Delta t\,.$$

The unique minimum of the function $\Delta t \mapsto r_R(\Delta t)$ is attained at the time-lag $\Delta t_{opt} = K^2/(\hat{\sigma}^2|S\Gamma|^{\frac{2}{3}})$ where $K = (C/(R\sqrt{2\pi})^{\frac{1}{3}}$. For the minimal value of the function $\Delta t \mapsto r_R(\Delta t)$ we have

$$r_R(\Delta t_{opt}) = \frac{3}{2}\left(\frac{C^2 R}{2\pi}\right)^{\frac{1}{3}} \hat{\sigma}^2 |S\Gamma|^{\frac{4}{3}}\,. \tag{18}$$

Taking into account both transaction costs as well as risk from a volatile portfolio effects we have shown that the equation for the change $\Delta\Pi$ of a portfolio Π read as:

$$\Delta V + \delta \Delta S = \Delta\Pi\Delta t = r\Pi\Delta t + \delta q S \Delta t - r_R S \Delta t$$

where r_R represents the total risk premium, $r_R = r_{TC} + r_{VP}$. Applying Itô's lemma to a smooth function $V = V(S,t)$ and assuming the δ-hedging strategy for the portfolio adjustments we finally obtain the following generalization of the Black–Scholes equation for valuing options:

$$\frac{\partial V}{\partial t} + \frac{\hat{\sigma}^2}{2} S^2 \frac{\partial^2 V}{\partial S^2} + (r-q)S\frac{\partial V}{\partial S} - rV - r_R S = 0\,.$$

By taking the optimal value of the total risk coefficient r_R derived as in (18) the option price V is a solution to the following nonlinear parabolic equation:

(*Risk adjusted Black–Scholes equation*)

$$\frac{\partial V}{\partial t} + \frac{\hat{\sigma}^2}{2} S^2 \left(1 + \mu (S\partial_S^2 V)^{\frac{1}{3}}\right) \frac{\partial^2 V}{\partial S^2} + (r-q) S \frac{\partial V}{\partial S} - rV = 0\,, \tag{19}$$

where $\mu = 3\left(\frac{C^2 R}{2\pi}\right)^{\frac{1}{3}}$ and u^p with $u = S\partial_S^2 V$ and $p = 1/3$ stands for the signed power function, i.e. $u^p = |u|^{p-1}u$. In the case there are neither transaction costs ($C = 0$) nor the risk from a volatile portfolio ($R = 0$) we have $\mu = 0$. Then equation (19) reduces to the original Black–Scholes linear parabolic equation (2). We note that equation (19) is a backward parabolic PDE if and only if the function $\beta(H) = \frac{\hat{\sigma}^2}{2}(1 + \mu H^{\frac{1}{3}})H$ is an increasing function in the variable $H := S\Gamma = S\partial_S^2 V$. It is clearly satisfied if $\mu \ge 0$ and $H \ge 0$.

As it is usual in the classical Black-Scholes theory for European style of options (cf. [30,38]) we consider the change of independent variables: $x := \ln\left(\frac{S}{E}\right)\,,\ x \in \mathbb{R}\,,\quad \tau := T - t\,,\ \tau \in (0,T)$. As equation (19) contains the term $S\Gamma = S\partial_S^2 V$ it is convenient to introduce the following transformation: $H(x,\tau) := S\Gamma = S\partial_S^2 V(S,t)$. Now we are in position to derive an equation for the function H. It turns out that the function $H(x,\tau)$ is a solution to a nonlinear parabolic equation subject to the initial and boundary conditions. More precisely, by taking the second derivative of equation (19) with respect to x we obtain, after some calculations, that $H = H(x,\tau)$ is a solution to the quasilinear parabolic equation

$$\frac{\partial H}{\partial \tau} = \frac{\partial^2}{\partial x^2}\beta(H) + \frac{\partial}{\partial x}\beta(H) + r\frac{\partial H}{\partial x}\,, \tag{20}$$

$\tau \in (0,T), x \in \mathbb{R}$ (see [32]). Henceforth, we will refer to (20) as a Γ *equation*. A solution H to (20) is subjected to the initial condition at $\tau = 0$:

$$H(x,0) = \bar{H}(x)\,,\quad x \in \mathbb{R}\,, \tag{21}$$

where $\bar{H}(x)$ is the Dirac δ function $\bar{H}(x) = \delta(x)$. For the purpose of numerical approximation we approximate the initial Dirac delta function by $\bar{H}(x) = N'(d)/(\sigma\sqrt{\tau^*})$ where $\tau^* > 0$ is sufficiently small, $N(d)$ is the cumulative distribution function of the normal distribution, and $d = \left(x + (r - q - \sigma^2/2)\tau^* \big/ \sigma\sqrt{\tau^*}\right)$. It corresponds to the value $H = S\partial_S^2 V$ of a call (put) option valued by a linear Black–Scholes equation with a constant volatility $\sigma > 0$ at the time $T - \tau^*$ close to expiry T when the time parameter $0 < \tau^* \ll 1$ is sufficiently small. In the case of call or put options the function H is subjected to boundary conditions at $x = \pm\infty$,

$$H(-\infty,\tau) = H(\infty,\tau) = 0\,,\quad \tau \in (0,T)\,. \tag{22}$$

A numerical discretization scheme of the Γ equation (20) based on finite volume approximation has been discussed in [32] in more details.

2.2. Pricing of European Style of Options by the RAPM Model

Let us denote $V(S,t;C,\hat{\sigma},R)$ the value of a solution to (19) with parameters $C, \hat{\sigma}, R$. Suppose that the coefficient of transaction costs C is known from and is given by (14). In real

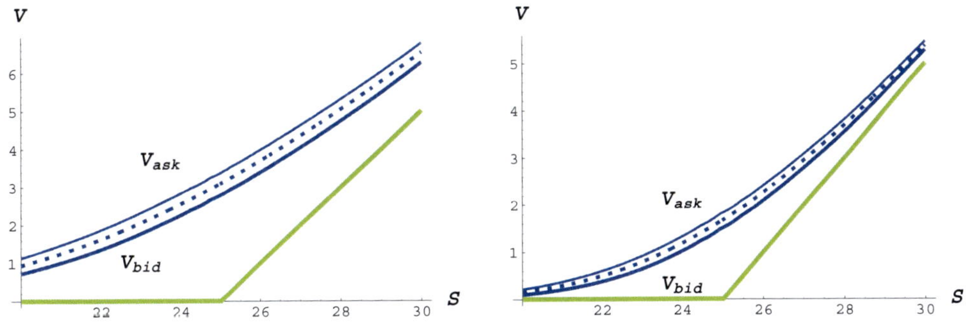

Figure 2. A comparison of Bid and Ask option prices computed by means of the RAPM model. The middle dotted line is the option price computed from the Black-Scholes equation. We chose $\sigma = 0.3, \mu = 0.2, r = 0.011, E = 25$ and $T = 1$ (left) and $T = 0.3$ (right).

option market data we can observe different Bid and Ask prices for an option, $V_{bid} < V_{ask}$, resp. Let us denote by V_{mid} the mid value, i.e. $V_{mid} = \frac{1}{2}(V_{bid} + V_{ask})$. By the RAPM model we are able to explain such a Bid-Ask spread in option prices. The higher Ask price corresponds to a solution to the RAPM model with some nontrivial risk premium $R > 0$ and $C > 0$ whereas the mid value V_{mid} corresponds to a solution $V(S,t)$ for vanishing risk premium $R = 0$, i.e. to a solution of the linear Black-Scholes equation (2). An illustrative example of Bid-Ask spreads captured by the RAPM model is shown in Fig. 2.

To calibrate the RAPM model we seek for a couple $(\hat{\sigma}_{RAPM}, R)$ of implied RAPM volatility σ_{RAPM} and risk premium R such that $V_{ask} = V(S,t;C,\hat{\sigma}_{RAPM},R)$ and $V_{mid} = V(S,t;C,\hat{\sigma}_{RAPM},0)$. Such a system of nonlinear equation for σ_{RAPM} and R can be solved by by means of the Newton-Kantorovich iterative method (cf. [32]).

As an example we considered sample data sets for call options on Microsoft stocks. We considered a flat interest rate $r = 0.02$, a constant transaction cost coefficient $C = 0.01$ estimated from (14), and we assumed that the underlying asset pays no dividends, i.e. $q = 0$. In Fig. 3 we present results of calibration of implied couple $(\hat{\sigma}_{RAPM}, R)$. Interestingly enough, two call options with higher strike prices $E = 25, 30$ had almost constant implied risk premium R. On the other the risk premium of an option with lowest $E = 23$ was fluctuating and it had highest average of R.

Finally, in Fig. 4 we present one week behavior of implied volatilities and risk premium coefficients for the Microsoft call option on $E = 25$ expiring at $T =$ April 19, 2003. In the beginning of the investigated period the risk premium coefficient R was rather high and fluctuating. On the other hand, it tends to a flat value of $R \approx 5$ at the end of the week. Interesting feature can be observed at the end of the second day when both stock and option prices went suddenly down. The time series analysis of the implied volatility $\hat{\sigma}_{RAPM}$ from first two days was unable to predict such a behavior. On the other, high fluctuation in the implied risk premium R during first two days can send a signal to an investor that sudden changes can be expected in the near future.

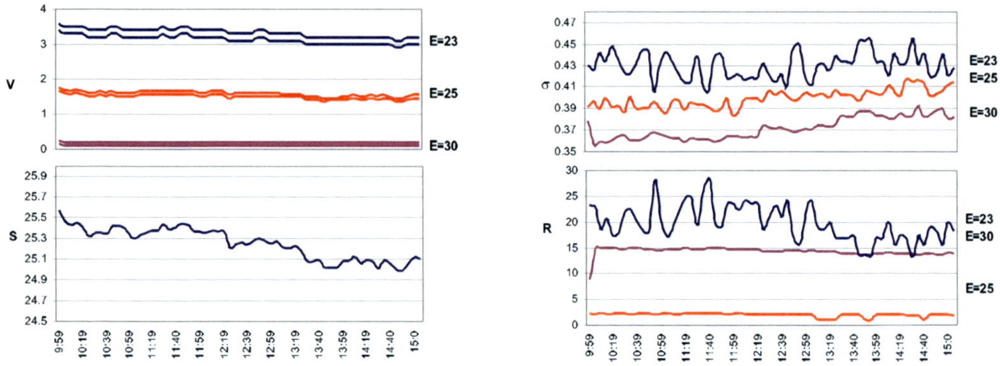

Figure 3. Intra-day behavior of Microsoft stocks (April 4, 2003) and shortly expiring call options with expiry date April 19, 2003. Computed implied volatilities $\hat{\sigma}_{RAPM}$ and risk premium coefficients R.

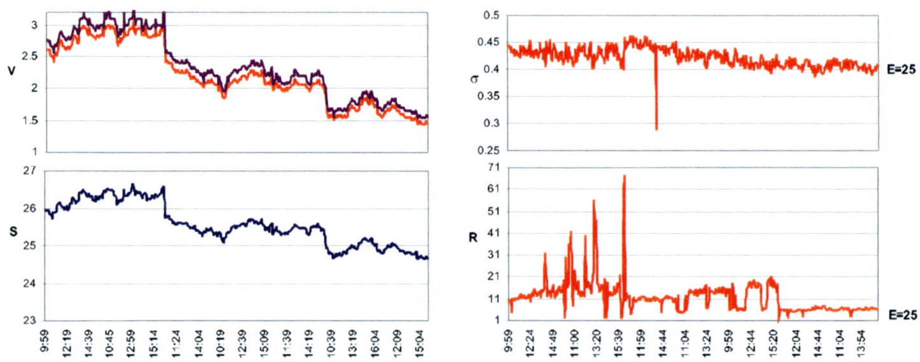

Figure 4. One week behavior of Microsoft stocks (March 20 - 27, 2003) and call options with expiration date April 19, 2003. Computed implied volatilities $\hat{\sigma}_{RAPM}$ and risk premiums R.

3. Transformation Method for a Linear Black–Scholes Equation

One of the interesting problems in this field is the analysis of the early exercise boundary and the optimal stopping time for American options on stocks paying a continuous dividend. It can be easily reduced to a problem of solving a certain free boundary problem for the Black–Scholes equation (cf. Black & Scholes [7]). However, the exact analytical expression for the free boundary profile is not known yet. Many authors have investigated various approximate models leading to approximate expressions for valuing American call and put options (see e.g. [24, 25, 33, 35, 36, 44, 45] and recent papers by Zhu [55, 56], Alobaidi *et al.* [1,41,42], Stamicar *et al.* [50] and the survey paper by Chadam [12] and other references therein). For the purpose of studying the free boundary profile near expiry, many different integral equations have been derived [6, 11, 30, 40].

Let us recall that the equation governing the time evolution of the price $V(S,t)$ of the American call option is the following parabolic PDE:

$$\begin{aligned} &\frac{\partial V}{\partial t} + (r-q)S\frac{\partial V}{\partial S} + \frac{\sigma^2}{2}S^2\frac{\partial^2 V}{\partial S^2} - rV = 0\,, \qquad 0 < S < S_f(t),\ 0 < t < T\,, \\ &V(0,t) = 0,\ V(S_f(t),t) = S_f(t) - E\,,\ \frac{\partial V}{\partial S}(S_f(t),t) = 1\,, \\ &V(S,T) = \max(S-E,0)\,, \end{aligned} \tag{23}$$

defined on a time-dependent domain $S \in (0, S_f(t))$, where $t \in (0,T)$. As usual $S > 0$ is the stock price, $E > 0$ is the exercise price, $r > 0$ is the risk-free rate, $q > 0$ is the continuous stock dividend rate and

$$\sigma \equiv const > 0$$

is a constant volatility of the underlying stock process.

The main purpose of this section is to present an alternative integral equation which will provide an accurate numerical method for calculating the early exercise boundary near expiry. The derivation of the nonlinear integral equation is based on the Fourier transform. A solution to this integral equation is the free boundary profile. The novelty of this approach consists in three steps:

1. The fixed domain transformation.
2. Derivation of a parabolic PDE for the so-called synthetic portfolio with a nonlinear nonlocal constraint between the free boundary position and the solution of this parabolic equation itself.
3. Construction of a solution by means of Fourier sine and cosine integral transforms.

Throughout this section we restrict our attention to the case when $r > q > 0$. It is well known that, for $r > q > 0$, the free boundary $\varrho(\tau) = S_f(T-\tau)$ starts at $\varrho(0) = rE/q$, whereas $\varrho(0) = E$ for the case $r \le q$ (cf. Dewynne *et al.* [16], Kwok [38]). Thus, the free boundary profile develops an initial jump in the case $r > q > 0$. Notice that the case $0 < r \le q$ can be also treated by other methods based on integral equations. Kwok [38] derived another integral equation which covers both cases $0 < r \le q$, as well as $r > q > 0$ (see equation (45) and Remark 3.2). However, in the latter case equation (45) becomes singular as $t \to T^-$, leading to numerical instabilities near expiry.

In the rest of this section to investigate the behavior of the free boundary $S_f(t)$. We present a method developed by Ševčovič in [46] of reducing the free boundary problem for (23) to a nonlinear integral equation with a singular kernel. Notice that our method of reducing the free boundary problem to a nonlinear integral equation can be also successfully adopted for valuing the American put option paying no dividends (cf. Stamicar *et al.* [50]).

3.1. Fixed Domain Transformation of the Free Boundary Problem

In this section we will perform a fixed domain transformation of the free boundary problem (23) into a parabolic equation defined on a fixed spatial domain. As it will be shown below, imposing of the free boundary condition will result in a nonlinear time-dependent term involved in the resulting equation. To transform equation (23) defined on a time dependent spatial domain $(0, S_f(t))$, we introduce the following change of variables:

$$\tau = T - t, \quad x = \ln\left(\frac{\varrho(\tau)}{S}\right) \qquad \text{where } \varrho(\tau) = S_f(T-\tau). \tag{24}$$

Clearly, $\tau \in (0,T)$ and $x \in (0,\infty)$ whenever $S \in (0, S_f(t))$. Let us furthermore define the auxiliary function $\Pi = \Pi(x,\tau)$ as follows:

$$\Pi(x,\tau) = V(S,t) - S\frac{\partial V}{\partial S}(S,t). \tag{25}$$

Notice that the quantity Π has an important financial meaning as it is a synthetic portfolio consisting of one long option and $\Delta = \frac{\partial V}{\partial S}$ underlying short positioned stocks. It follows from (24) that

$$\begin{aligned} \frac{\partial \Pi}{\partial x} &= S^2\frac{\partial^2 V}{\partial S^2}, \quad \frac{\partial^2 \Pi}{\partial x^2} + 2\frac{\partial \Pi}{\partial x} = -S^3\frac{\partial^3 V}{\partial S^3}, \\ \frac{\partial \Pi}{\partial \tau} + \frac{\dot\varrho}{\varrho}\frac{\partial \Pi}{\partial x} &= S\frac{\partial^2 V}{\partial S \partial t} - \frac{\partial V}{\partial t}, \end{aligned} \tag{26}$$

where $\dot\varrho = d\varrho/d\tau$. Now assuming that $V = V(S,t)$ is a sufficiently smooth solution of (23), we may differentiate (23) with respect to S. Plugging expressions (26) into (23), we finally obtain that the function $\Pi = \Pi(x,\tau)$ is a solution of the parabolic equation

$$\frac{\partial \Pi}{\partial \tau} + a(\tau)\frac{\partial \Pi}{\partial x} - \frac{\sigma^2}{2}\frac{\partial^2 \Pi}{\partial x^2} + r\Pi = 0, \tag{27}$$

$x \in (0,\infty)$, $\tau \in (0,T)$, with a time-dependent coefficient

$$a(\tau) = \frac{\dot\varrho(\tau)}{\varrho(\tau)} + r - q - \frac{\sigma^2}{2}.$$

It follows from the boundary condition $V(S_f(t),t) = S_f(t) - E$ and $V_S(S_f(t),t) = 1$ that

$$\Pi(0,\tau) = -E, \qquad \Pi(+\infty,\tau) = 0\,. \tag{28}$$

The initial condition $\Pi(x,0)$ can be deduced from the pay-off diagram for $V(S,T)$. We obtain

$$\Pi(x,0) = \begin{cases} -E & \text{for } x < \ln\left(\frac{\varrho(0)}{E}\right) \\ 0 & \text{otherwise.} \end{cases} \tag{29}$$

Notice that equation (27) is a parabolic PDE with a time-dependent coefficient $a(\tau)$. In what follows, we will show the function $a(\tau)$ depends upon a solution Π itself. This dependence is non-local in the spatial variable x. Moreover, the initial position of the interface $\varrho(0)$ enters the initial condition $\Pi(x,0)$. Therefore we have to determine the relationship between the solution $\Pi(x,\tau)$ and the free boundary function $\varrho(\tau)$ first. To this end, we make use of the boundary condition imposed on V at the interface $S = S_f(t)$. Since $S_f(t) - E = V(S_f(t), t)$ we have

$$\frac{d}{dt} S_f(t) = \frac{\partial V}{\partial S}(S_f(t), t) \frac{d}{dt} S_f(t) + \frac{\partial V}{\partial t}(S_f(t), t) \,.$$

As $\frac{\partial V}{\partial S}(S_f(t), t) = 1$ we obtain $\frac{\partial V}{\partial t}(S, t) = 0$ at $S = S_f(t)$. Assuming the function Π_x has a trace at $x = 0$, and taking into account (26), we may conclude that, for any $t = T - \tau \in [0, T)$,

$$S^2 \frac{\partial^2 V}{\partial S^2}(S, t) \to \frac{\partial \Pi}{\partial x}(0, \tau), \quad S \frac{\partial V}{\partial S}(S, t) \to \varrho(\tau) \text{ as } S \to S_f(t)^- \,.$$

If $\frac{\partial V}{\partial t}(S, t) \to \frac{\partial V}{\partial t}(S_f(t), t) = 0$ as $S \to S_f(t)^-$, then it follows from the Black–Scholes equation (23) that

$$\begin{aligned} &(r - q)\varrho(\tau) + \frac{\sigma^2}{2} \frac{\partial \Pi}{\partial x}(0, \tau) - r(\varrho(\tau) - E) \\ = &\lim_{S \to S_f(t)^-} \left(\frac{\partial V}{\partial t}(S, t) + (r - q) S \frac{\partial V}{\partial S}(S, t) + \frac{\sigma^2}{2} S^2 \frac{\partial^2 V}{\partial S^2}(S, t) - rV(S, t) \right) = 0. \end{aligned}$$

As a consequence we obtain a nonlocal algebraic constraint between the free boundary function $\varrho(\tau)$ and the boundary trace $\partial_x \Pi(0, \tau)$ of the derivative of the solution Π itself:

$$\varrho(\tau) = \frac{rE}{q} + \frac{\sigma^2}{2q} \frac{\partial \Pi}{\partial x}(0, \tau) \text{ for } 0 < \tau \le T. \tag{30}$$

It remains to determine the initial position of the interface $\varrho(0)$. According to Dewynne *et al.* [16] (see also Kwok [38]), the initial position $\varrho(0)$ of the free boundary is rE/q for the case $0 < q < r$. Alternatively, we can derive this condition from (27)–(29) by assuming the smoothness of Π in the x variable up to the boundary $x = 0$ uniformly for $\tau \to 0^+$. More precisely, we assume that

$$\lim_{\tau \to 0^+} \frac{\partial \Pi}{\partial x}(0, \tau) = \lim_{\tau \to 0^+, x \to 0^+} \frac{\partial \Pi}{\partial x}(x, \tau) = \lim_{x \to 0^+} \frac{\partial \Pi}{\partial x}(x, 0) = 0$$

because $\Pi(x, 0) = -E$ for x close to 0^+. By (30) we obtain

$$\varrho(0) = \frac{rE}{q}. \tag{31}$$

In summary, we have shown that, under suitable regularity assumptions imposed on a solution Π to (27), (28), (29), the free boundary problem (23) can be transformed into the initial boundary value problem for parabolic PDE

$$\begin{aligned} &\frac{\partial \Pi}{\partial \tau} + a(\tau) \frac{\partial \Pi}{\partial x} - \frac{\sigma^2}{2} \frac{\partial^2 \Pi}{\partial x^2} + r\Pi = 0, \\ &\Pi(0, \tau) = -E, \qquad \Pi(+\infty, \tau) = 0, \qquad x > 0, \tau \in (0, T), \\ &\Pi(x, 0) = \begin{cases} -E & \text{for } x < \ln(r/q), \\ 0 & \text{otherwise}, \end{cases} \end{aligned} \tag{32}$$

where $a(\tau) = \frac{\dot{\varrho}(\tau)}{\varrho(\tau)} + r - q - \frac{\sigma^2}{2}$ and

$$\varrho(\tau) = \frac{rE}{q} + \frac{\sigma^2}{2q}\frac{\partial \Pi}{\partial x}(0,\tau), \quad \varrho(0) = \frac{rE}{q}. \tag{33}$$

We emphasize that the problem (32) constitutes a nonlinear parabolic equation with a nonlocal constraint given by (33).

Remark 3.1 *In our derivation of the free boundary function $\varrho(\tau)$ and its initial condition $\varrho(0)$ we did not assume that the solution $V(S,t)$ is C^2 smooth up to the free boundary $S = S_f(t)$. Such an assumption would lead to an obvious contradiction $\Gamma := \partial^2 V/\partial S^2 = 0$ at $S = S_f(t)$. On the other hand, the jump in Γ at the free boundary is the only driving force for the evolution of the free boundary function ϱ (see (30)). Construction of a PDE for the synthetic portfolio function Π is a crucial step in our approach because the derivative $\partial_x \Pi$ admits a trace at the boundary $x = 0$ and the unknown free boundary function ϱ can be determined via (33)*

3.2. Reduction to a Nonlinear Integral Equation

The main purpose of this section is to show how the fully nonlinear nonlocal problem (32)–(33) can reduced to a single nonlinear integral equation for $\varrho(\tau)$ giving the explicit formula for the solution $\Pi(x,\tau)$ to (32). The idea is to apply the Fourier sine and cosine integral transforms (cf. Stein & Weiss [51]). Let us recall that for any Lebesgue integrable function $f \in L^1(\mathbb{R}^+)$ the sine and cosine transformations are defined as follows: $\mathcal{F}_S(f)(\omega) = \int_0^\infty f(x)\,\sin\omega x\,dx$, $\mathcal{F}_C(f)(\omega) = \int_0^\infty f(x)\,\cos\omega x\,dx$. Their inverse transforms are given by $\mathcal{F}_S^{-1}(g)(x) = \frac{2}{\pi}\int_0^\infty g(\omega)\,\sin\omega x\,d\omega$, $\mathcal{F}_C^{-1}(g)(x) = \frac{2}{\pi}\int_0^\infty g(\omega)\,\cos\omega x\,d\omega$. Now we suppose that the function $\varrho(\tau)$ and subsequently $a(\tau)$ are already know. Let $\Pi = \Pi(x,\tau)$ be a solution of (32) corresponding to a given function $a(\tau)$. Let us denote

$$p(\omega,\tau) = \mathcal{F}_S(\Pi(.,\tau))(\omega), \qquad q(\omega,\tau) = \mathcal{F}_C(\Pi(.,\tau))(\omega) \tag{34}$$

where $\omega \in \mathbb{R}^+, \tau \in (0,T)$. By applying the sine and cosine integral transforms, taking into account their basic properties and (33), we finally obtain a linear non-autonomous ω-parameterized system of ODEs

$$\begin{aligned}
&\frac{d}{d\tau}p(\omega,\tau) - a(\tau)\omega q(\omega,\tau) + \alpha(\omega)p(\omega,\tau) = -E\omega\frac{\sigma^2}{2},\\
&\frac{d}{d\tau}q(\omega,\tau) + a(\tau)\omega p(\omega,\tau) + \alpha(\omega)q(\omega,\tau) = -Ea(\tau) - q\varrho(\tau) + rE,
\end{aligned} \tag{35}$$

for the sine and cosine transforms of Π (see (34)) where

$$\alpha(\omega) = \frac{1}{2}(\sigma^2\omega^2 + 2r).$$

The system of equations (35) is subject to initial conditions at $\tau = 0$, $p(\omega,0) = \mathcal{F}_S(\Pi(.,0)(\omega)$, $q(\omega,0) = \mathcal{F}_C(\Pi(.,0)(\omega)$. In the case of the call option, we deduce from the initial condition for Π (see (32)) that

$$p(\omega,0) = \frac{E}{\omega}\left(\cos\left(\omega\ln\frac{r}{q}\right) - 1\right), \qquad q(\omega,0) = -\frac{E}{\omega}\sin\left(\omega\ln\frac{r}{q}\right). \tag{36}$$

Taking into account (36) and by using the variation of constants formula for solving linear non-autonomous ODEs, we obtain an explicit formula for $p(\omega,\tau) = -E\omega^{-1} + \tilde{p}(\omega,\tau)$ where

$$\tilde{p}(\omega,\tau) = \frac{E}{\omega} e^{-\alpha(\omega)\tau} \cos(\omega(A(\tau,0) + \ln(r/q)))$$
$$+ \int_0^\tau e^{-\alpha(\omega)(\tau-s)} \left[\frac{rE}{\omega} \cos(\omega A(\tau,s)) + (rE - q\varrho(s)) \sin(\omega A(\tau,s)) \right] ds. \tag{37}$$

Here we have denoted by A the function defined as

$$A(\tau,s) = \int_s^\tau a(\xi)\, d\xi = \ln \frac{\varrho(\tau)}{\varrho(s)} + \left(r - q - \frac{\sigma^2}{2}\right)(\tau - s) \, . \tag{38}$$

As $\mathcal{F}_S^{-1}(\omega^{-1}) = 1$ we have $\Pi(x,\tau) = \mathcal{F}_S^{-1}(p(\omega,\tau)) = -E + \frac{2}{\pi}\int_0^\infty \tilde{p}(\omega,\tau)\sin(\omega x)\, d\omega$. From (33) we conclude that the free boundary function ϱ satisfies the following equation:

$$\varrho(\tau) = \frac{rE}{q} + \frac{\sigma^2}{q\pi} \int_0^\infty \omega \tilde{p}(\omega,\tau)\, d\omega. \tag{39}$$

Taking into account (37) and (39) we end up with the following nonlinear singular integral equation for the free boundary function $\varrho(\tau)$:

$$\varrho(\tau) = \frac{rE}{q}\left(1 + \frac{\sigma}{r\sqrt{2\pi\tau}} \exp\left(-r\tau - \frac{(A(\tau,0)+\ln(r/q))^2}{2\sigma^2\tau}\right)\right. \tag{40}$$
$$\left. + \frac{1}{\sqrt{2\pi}} \int_0^\tau \left[\sigma + \frac{1}{\sigma}\left(1 - \frac{q\varrho(s)}{rE}\right)\frac{A(\tau,s)}{\tau - s}\right] \frac{\exp\left(-r(\tau-s) - \frac{A(\tau,s)^2}{2\sigma^2(\tau-s)}\right)}{\sqrt{\tau - s}}\, ds\right),$$

where the function A depends upon ϱ via equation (38). To simplify this integral equation, we introduce a new auxiliary function $H : [0, \sqrt{T}] \to \mathbb{R}$ as follows:

$$\varrho(\tau) = \frac{rE}{q}\left(1 + \sigma\sqrt{2}H(\sqrt{\tau})\right). \tag{41}$$

Using the change of variables $s = \xi^2\cos^2\theta$, one can rewrite the integral equation (39) in terms of the function H as follows:

$$H(\xi) = f_H(\xi) + \frac{1}{\sqrt{\pi}} \int_0^{\frac{\pi}{2}} [\xi\cos\theta - 2\mathrm{cotg}\,\theta\, H(\xi\cos\theta) g_H(\xi,\theta)]\, e^{-r\xi^2\sin^2\theta - g_H^2(\xi,\theta)}\, d\theta, \tag{42}$$

where

$$g_H(\xi,\theta) = \frac{1}{\sigma\sqrt{2}} \frac{1}{\xi\sin\theta} \ln\left(\frac{1+\sigma\sqrt{2}H(\xi)}{1+\sigma\sqrt{2}H(\xi\cos\theta)}\right) + \frac{\Lambda}{\sqrt{2}}\xi\sin\theta \tag{43}$$

for $\xi \in [0,\sqrt{T}]$, $\theta \in (0,\pi/2)$,

$$\Lambda = \frac{r-q}{\sigma} - \frac{\sigma}{2}$$

and

$$f_H(\xi) = \frac{1}{2r\sqrt{\pi}\xi}\, e^{-r\xi^2 - \left(g_H(\xi,\frac{\pi}{2}) + \frac{1}{\xi}\frac{1}{\sigma\sqrt{2}}\ln(r/q)\right)^2}. \tag{44}$$

Notice that equations (40) and (42) are integral equations with a singular kernel (cf. Gripenberg *et al.* [27]).

Remark 3.2 *Kwok [38] derived another integral equation for the early exercise boundary for the American call option on a stock paying continuous dividend. According to Kwok [38, Section 4.2.3], $\varrho(\tau)$ satisfies the integral equation*

$$\begin{aligned}\varrho(\tau) &= E + \varrho(\tau)e^{-q\tau}N(d) - E\mathrm{e}^{-r\tau}N(d-\sigma\sqrt{\tau}) \\ &+ \int_0^\tau q\varrho(\tau)e^{-q\xi}N(d_\xi) - rE\mathrm{e}^{-r\xi}N(d_\xi - \sigma\sqrt{\xi})d\xi \end{aligned} \tag{45}$$

where

$$d = \frac{1}{\sigma\sqrt{\tau}}\ln\left(\frac{\varrho(\tau)}{E}\right) + \Lambda\sqrt{2\tau}, \quad d_\xi = \frac{1}{\sigma\sqrt{\xi}}\ln\left(\frac{\varrho(\tau)}{\varrho(\tau-\xi)}\right) + \Lambda\sqrt{2\xi}$$

and $N(u)$ is the cumulative distribution function for the normal distribution. The above integral equation covers both cases: $r \le q$ as well as $r > q$. However, in the case $r > q$ this equation becomes singular as $\tau \to 0^+$.

In the rest of this section we derive a formula for pricing American call options based on the solution ϱ to the integral equation (42). With regard to (25), we have

$$\frac{\partial}{\partial S}\left(S^{-1}V(S,t)\right) = -S^{-2}\Pi\left(\ln\left(S^{-1}\varrho(T-t)\right), T-t\right).$$

Taking into account the boundary condition $V(S_f(t),t) = S_f(t) - E$ and integrating the above equation from S to $S_f(t) = \varrho(T-t)$, we obtain

$$V(S, T-\tau) = \frac{S}{\varrho(\tau)}\left(\varrho(\tau) - E + \int_0^{\ln\frac{\varrho(\tau)}{S}} \mathrm{e}^x\Pi(x,\tau)\,dx\right). \tag{46}$$

It is straightforward to verify that V given by (46 is indeed a solution to the free boundary problem (23). Inserting the expressions (37) and (39) (recall that $\Pi(x,\tau) = -E + \frac{2}{\pi}\int_0^\infty \tilde{p}(\omega,\tau)\sin\omega x\,d\omega$) into (46), we end up with the formula for pricing the American call option:

$$\begin{aligned} V(S,T-\tau) &= S - E + \frac{S}{\varrho(\tau)}E\,I_2(A(\tau,0) + \ln(r/q), \ln(\varrho(\tau)/S), \tau) \\ &+ \frac{S}{\varrho(\tau)}\int_0^\tau \Big[rE\,I_2(A(\tau,s), \ln(\varrho(\tau)/S), \tau - s) \\ &+ (rE - q\varrho(s))\,I_1(A(\tau,s), \ln(\varrho(\tau)/S), \tau - s)\Big]\,ds \end{aligned} \tag{47}$$

for any $S \in (0, S_f(t))$ and $t \in [0,T]$, where $A(\tau,s) = \ln\frac{\varrho(\tau)}{\varrho(s)} + (r - q - \frac{\sigma^2}{2})(\tau - s)$. Here

$$\begin{aligned} I_1(A,L) &= \frac{e^{-(r-\sigma^2/2)\tau}}{2}\left[e^A M\left(\frac{-A-\sigma^2\tau}{\sigma\sqrt{2\tau}}, \frac{L}{\sigma\sqrt{2\tau}}\right) - e^{-A}M\left(\frac{A-\sigma^2\tau}{\sigma\sqrt{2\tau}}, \frac{L}{\sigma\sqrt{2\tau}}\right)\right], \\ I_2(A,L) &= \frac{e^{-r\tau}e^L}{2}M\left(\frac{A-L}{\sigma\sqrt{2\tau}}, \frac{2L}{\sigma\sqrt{2\tau}}\right) \\ &-\frac{e^{-(r-\sigma^2/2)\tau}}{2}\left[e^A M\left(\frac{-A-\sigma^2\tau}{\sigma\sqrt{2\tau}}, \frac{L}{\sigma\sqrt{2\tau}}\right) + e^{-A}M\left(\frac{A-\sigma^2\tau}{\sigma\sqrt{2\tau}}, \frac{L}{\sigma\sqrt{2\tau}}\right)\right] \end{aligned} \tag{48}$$

and

$$M(x,y) = \text{erf}(x+y) - \text{erf}(x) = \frac{2}{\sqrt{\pi}} \int_x^{x+y} e^{-\xi^2}\, d\xi.$$

We will refer to (47) as the semi-explicit formula for pricing the American call option. We use the term 'semi-explicit' because (47) contains the free boundary function $\varrho(\tau) = S_f(T-\tau)$ which has to be determined first by solving the nonlinear integral equation (42).

3.3. Numerical Experiments

In this section we focus on numerical experiments. We will compute the free boundary profile

$$S_f(t) = \varrho(T-t) = \frac{rE}{q}\left(1 + \sigma\sqrt{2}H(\sqrt{T-t})\right) \tag{49}$$

(see (36)) by solving the nonlinear integral equation (42). We also present a comparison of the results obtained by our methods to those obtained by other known methods for solving the American call option problem.

The computation of a solution of the nonlinear integral equation is based on an iterative method. We will construct a sequence of approximate solutions to (42). Let H^0 be an initial approximation of a solution to (42). If we suppose $H^0(\xi) = h_1\xi$, then, plugging this ansatz into (42) yields the well-known first order approximation of a solution $H(\xi)$ in the form

$$H^0(\xi) = 0.451381\,\xi$$

i.e. $\varrho^0(\tau) = \frac{rE}{q}\,(1 + \,0.638349\,\sigma\sqrt{\tau})$ (cf. Ševčovič [46]). This asymptotics is in agreement with that of Dewynne *et al.* [16]. For $n = 0, 1, ...$ we will define the $n+1$ approximation H^{n+1} as follows:

$$H^{n+1}(\xi) =$$

$$f_{H^n}(\xi) + \frac{1}{\sqrt{\pi}}\int_0^{\frac{\pi}{2}} [\xi\cos\theta - 2\text{cotg}\,\theta\; H^n(\xi\cos\theta) g_{H^n}(\xi,\theta)]\, \mathrm{e}^{-r\xi^2\sin^2\theta - g^2_{H^n}(\xi,\theta)}\, d\theta \tag{50}$$

for $\xi \in [0, \sqrt{T}]$. With regard to (43) and (44), we have

$$g_{H^n}(\xi,\theta) = \frac{1}{\sigma\sqrt{2}}\frac{1}{\xi\sin\theta}\ln\left(\frac{1+\sigma\sqrt{2}H^n(\xi)}{1+\sigma\sqrt{2}H^n(\xi\cos\theta)}\right) + \frac{\Lambda}{\sqrt{2}}\xi\sin\theta$$

$$f_{H^n}(\xi) = \frac{1}{2r\sqrt{\pi}\xi}\,\mathrm{e}^{-r\xi^2 - \left(g_{H^n}(\xi,\frac{\pi}{2}) + \frac{1}{\xi}\frac{1}{\sigma\sqrt{2}}\ln\left(\frac{r}{q}\right)\right)^2}.$$

Notice that the function g_H is bounded provided that H is nonnegative and Lipschitz continuous on $[0, \sqrt{T}]$. Recall that we have assumed $r > q > 0$. Then the function $\xi \mapsto f_H(\xi)$ is bounded for $\xi \in [0, \sqrt{T}]$ and it vanishes at $\xi = 0$. Moreover, if H is smooth then f_H is a flat function at $\xi = 0$, i.e. $f_H(\xi) = o(\xi^n)$ as $\xi \to 0^+$ for all $n \in N$. From the numerical point of view such a flat function can be omitted from computations. For small values of θ we approximate the function $\text{cotg}\,\theta\, g_{H^n}(\xi,\theta)$ by its limit as $\theta \to 0^+$. It yields the approximation of the singular term in (50) in the form

$$\text{cotg}\,\theta g_H(\xi,\theta) \approx \frac{1}{2}\frac{H'(\xi)}{1+\sigma\sqrt{2}H(\xi)} + \frac{\Lambda}{\sqrt{2}}\xi \qquad \text{for } 0 < \theta \ll 1,$$

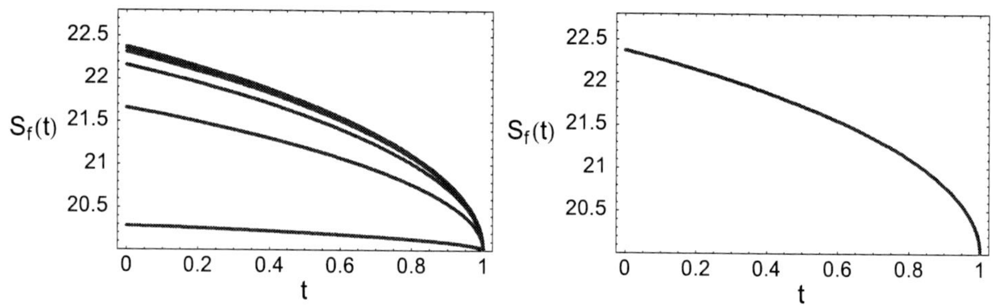

Figure 5. Five successive approximations of the free boundary $S_f(t)$ obtained from equation (42). The profile of the solution $S_f(t)$.

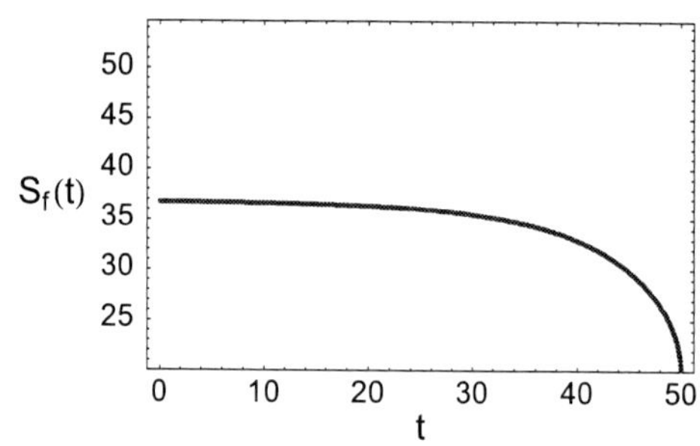

Figure 6. Long-time behavior of $S_f(t)$.

where $H = H^n$. In the following numerical simulations we chose $\varepsilon \approx 10^{-5}$.

In what follows, we present several computational examples. In Fig. 5 we show five iterates of the free boundary function $S_f(t)$, where the auxiliary function $H(\xi)$ is constructed by means of successive iterations of the nonlinear integral operator defined by the right-hand side of (42). This sequence converges to a fixed point of such a map, i.e. to a solution of (42). Parameter values were chosen as $E = 10, r = 0.1, \sigma = 0.2, q = 0.05, T = 1$. Fig. 5 depicts the final tenth iteration of approximation of the function $S_f(t)$. The mesh contained 100 grid points. One can observe very rapid convergence of iterates to a fixed point. In practice, no more than six iterates are sufficient to obtain the fixed point of (42). It is worth noting that in all our numerical simulations the convergence was monotone, i.e. the curve moves only up in the iteration process.

In Fig. 6 we show the long time behavior of the free boundary $S_f(t), t \in [0, T]$, for large values of the expiration time T. For the parameter values $T = 50, r = 0.1, q = 0.05, \sigma = 0.35$ and $E = 10$ the theoretical value of $S_f(+\infty)$ is 36.8179 (see Dewynne *et al.* [16]).

In Table 1 we show a comparison of results obtained by our method based on the semi-explicit formula (47) and those obtained by known methods based on trinomial trees (both with the depth of the tree equal to 100), finite difference approximation (with 200 spatial and time grids) and analytic approximation of Barone-Adesi & Whaley (cf. [6], [30, Ch.

Table 1. Comparison of the method based on formula (48) with other methods for the parameter values $E = 10, T = 1, \sigma = 0.2, r = 0.1, q = 0.05$. The position $S_f(0) = \varrho(T)$ of the free boundary at $t = 0$ (i.e. at $\tau = T$) was computed as $S_f(0) = \varrho(T) = 22.3754$

Method \ The asset value S	15	18	20	21	22.3754
Our method	5.15	8.09	10.03	11.01	12.37
Trinomial tree	5.15	8.09	10.03	11.01	12.37
Finite differences	5.49	8.48	10.48	11.48	12.48
Analytic approximation	5.23	8.10	10.04	11.02	12.38

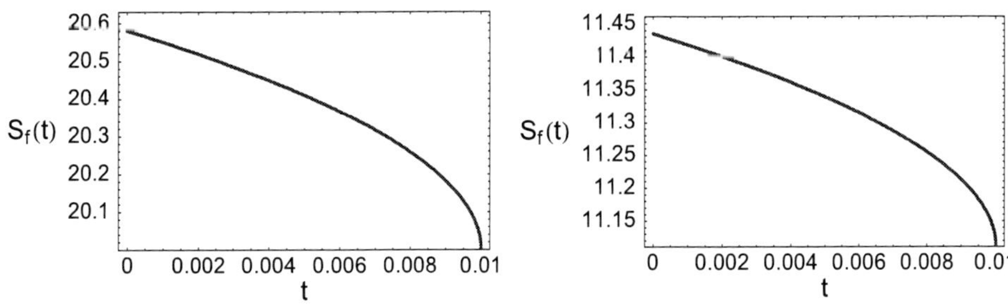

Figure 7. The early exercise boundary $S_f(t)$ for $r = 0.1, q = 0.05$. The early exercise boundary $S_f(t)$ for $r = 0.1, q = 0.09$.

15, p. 384]), resp. It also turned out that the method based on solving the integral equation (42) is 5-10 times faster than other methods based on trees or finite differences. The reason is that the computation of $V(S,t)$ for a wide range of values of S based on the semi-explicit pricing formula (47) is very fast provided that the free boundary function ϱ has already been computed.

In Fig. 7 the early exercise boundary $S_f(t)$ is computed for various values of the parameter q. In these computations we chose $E = 10, T = 0.01, \sigma = 0.45$. Of interest is the case where q is close to r ($q = 0.09$ and $r = 0.1$).

3.4. Early Exercise Boundary for an American Put Option

In this section we present results of transformation method applied to valuation of the early exercise boundary for American style of a put option. Recall that the early exercise boundary problem for American put option can be formulated as follows:

$$\begin{aligned}
&\frac{\partial V}{\partial t} + rS\frac{\partial V}{\partial S} + \frac{\sigma^2}{2}S^2\frac{\partial^2 V}{\partial S^2} - rV = 0\,, \qquad 0 < t < T,\ S_f(t) < S < \infty\,,\\
&V(+\infty, t) = 0,\ V(S_f(t), t) = E - S_f(t)\,,\ \frac{\partial V}{\partial S}(S_f(t), t) = -1\,, \qquad (51)\\
&V(S,T) = \max(E - S, 0)\,,
\end{aligned}$$

defined on a time-dependent domain $S \in (S_f(t), \infty)$, where $t \in (0,T)$ (cf. Kwok [38]). Again $S > 0$ stands for the stock price, $E > 0$ is the exercise price, $r > 0$ is the risk-free rate and $\sigma > 0$ is the volatility of the underlying stock process. We shall assume that the asset pay no dividends, i.e. $q = 0$. In order to perform a fixed domain transformation of the free boundary problem (52) we introduce the following change of variables

$$x = \ln\left(\frac{S}{\varrho(\tau)}\right), \quad \text{where } \tau = T - t, \varrho(\tau) = S_f(T-\tau).$$

Similarly as in the case of a call option we define a synthetised portfolio Π for the put option $\Pi(x,\tau) = V - S\frac{\partial V}{\partial S}$. Then it is easy to verify that Π is a solution to the following the parabolic equation

$$\begin{aligned}
&\frac{\partial \Pi}{\partial \tau} - a(\tau)\frac{\partial \Pi}{\partial x} - \frac{\sigma^2}{2}\frac{\partial^2 \Pi}{\partial x^2} + r\Pi = 0, \quad x > 0, \tau \in (0,T),\\
&\Pi(0,\tau) = E, \quad \Pi(\infty,\tau) = 0, \qquad (52)\\
&\frac{\sigma^2}{2}\frac{\partial \Pi}{\partial x}(0,\tau) = -rE, \quad \text{for } \tau \in (0,T), \qquad (53)\\
&\Pi(x,0) = 0 \quad \text{for} \quad x > 0,
\end{aligned}$$

where $a(\tau) = \frac{\dot\varrho(\tau)}{\varrho(\tau)} + r - \frac{\sigma^2}{2}$ (see Stamicar *et al.* [50]). Now, following the same methodology of applying the Fourier transform we are able to derive an integral equation for the free boundary position. It is just the equation $\frac{\sigma^2}{2}\frac{\partial \Pi}{\partial x}(0,\tau) = -rE$ in which the left hand side is expressed by the inverse Fourier transform of the solution Π as a weakly singular integral depending on the free boundary position ϱ (see [50] for details). We omit the technical details here. We just recall that the integral equation for the free boundary function $\varrho(\tau)$ yields the following expression for $\varrho(\tau)$

$$\varrho(\tau) = Ee^{-(r-\frac{\sigma^2}{2})\tau} e^{\sigma\sqrt{2\tau}\eta(\tau)}$$

in terms of a new auxiliary function $\eta(\tau)$ (see Stamicar *et al.* [50] for details). Further asymptotic analysis of the integral equation enables us to derive an asymptotic formula for $\eta(\tau)$ as $\tau \to 0$. Namely,

$$\eta(\tau) \sim -\sqrt{-\ln\left[\frac{2r}{\sigma}\sqrt{2\pi\tau}e^{r\tau}\right]} \quad \text{as } \tau \to 0^+. \qquad (54)$$

Next we examine how accurately our asymptotic approximation matches the data from the binomial method (cf. Kwok [38]). Near expiry at about one hour, the asymptotic approximation matches the data from the binomial method (see Fig. 8). With $\sigma = 0.25, r =$

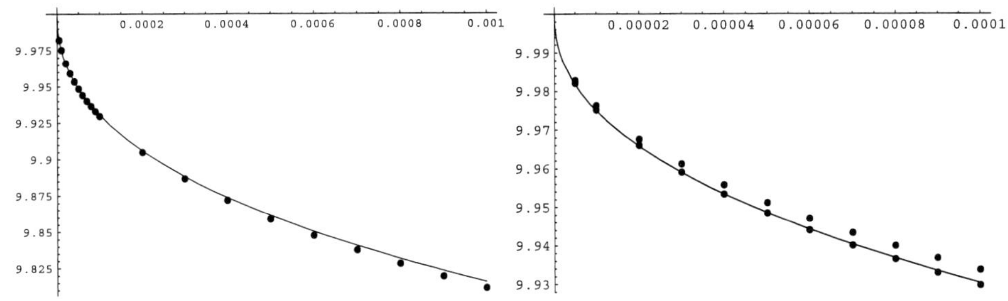

Figure 8. Left: asymptotic approximation vs. binomial method for $\sigma = 0.25, r = 0.1, E = 10$ and $T - t = 8.76$ hrs. (0.001 of a year), MBW approximation vs. the asymptotic solution (54) for $T - t = 0.876$ hrs (right).

$0.1, E = 10$ at 8.76 hours the approximation gives an overestimate but of only 0.4 cents (see [50]). We also compared our asymptotic solution with MacMillan, Barone-Adesi, and Whaley's [6], [4, 384–386], [40] numerical approximation of the American put free boundary (see Fig. 8). They apply a transformation that results in a Cauchy-Euler equation that can be solved analytically. For times very close to expiry, one can see that our approximation of the free boundary matches the data from the binomial and trinomial methods more accurately.

The numerical and analytical results obtained by the transformation method for solving free boundary problem for American put options are in agreement with those obtained recently by Zhu [55, 56], Mallier *et al.* [41, 42], Kuske and Keller [36], Knessl [34].

4. Transformation Method for a Nonlinear Black–Scholes Equation

The main goal of this section is to perform a fixed domain transformation of the free boundary problem for the nonlinear Black–Scholes equation (2) into a parabolic equation defined on a fixed spatial domain. For the sake of simplicity we will present a detailed derivation of an equation only for the case of an American call option. Derivation of the corresponding equation for the American put option is similar. Throughout this section we shall assume the volatility σ^2 appearing in the Black–Scholes equation to be a function of the option price S, time to expiry $T - t$ and $S^2 \partial_S^2 V$, i.e.

$$\sigma = \sigma(S^2 \partial_S^2 V, S, T - t)\,.$$

Following (24) we shall consider the following change of variables:

$$\tau = T - t, \quad x = \ln\left(\varrho(\tau)/S\right) \;\text{ where }\; \varrho(\tau) = S_f(T - \tau).$$

Then $\tau \in (0, T)$ and $x \in (0, \infty)$ iff $S \in (0, S_f(t))$. The boundary value $x = 0$ corresponds to the free boundary position $S = S_f(t)$ whereas $x \approx +\infty$ corresponds to the default value

$S = 0$ of the underlying asset. Following Stamicar *et al.* [50] and Ševčovič [46, 47] we construct the so-called synthetic portfolio function $\Pi = \Pi(x, \tau)$ defined as follows:

$$\Pi(x,\tau) = V(S,t) - S\frac{\partial V}{\partial S}(S,t)\,. \tag{55}$$

Again, it represents a synthetic portfolio consisting of one long positioned option and $\Delta = \partial_S V$ underlying short stocks. Similarly as in Section 3 we have

$$\frac{\partial \Pi}{\partial x} = S^2 \frac{\partial^2 V}{\partial S^2}, \quad \frac{\partial \Pi}{\partial \tau} + \frac{\dot{\varrho}}{\varrho}\frac{\partial \Pi}{\partial x} = -\frac{\partial}{\partial t}\left(V - S\frac{\partial V}{\partial S}\right)$$

where we have denoted $\dot{\varrho} = d\varrho/d\tau$. Assuming sufficient smoothness of a solution $V = V(S,t)$ to (2) we can deduce from (2) a parabolic equation for the synthetic portfolio function $\Pi = \Pi(x, \tau)$

$$\frac{\partial \Pi}{\partial \tau} + (b(\tau) - \frac{1}{2}\sigma^2)\frac{\partial \Pi}{\partial x} - \frac{1}{2}\frac{\partial}{\partial x}\left(\sigma^2 \frac{\partial \Pi}{\partial x}\right) + r\Pi = 0$$

defined on a fixed domain $x \in \mathbb{R}$, $t \in (0, T)$, with a time-dependent coefficient

$$b(\tau) = \frac{\dot{\varrho}(\tau)}{\varrho(\tau)} + r - q \tag{56}$$

and a diffusion coefficient given by: $\sigma^2 = \sigma^2(\partial_x \Pi(x,\tau), \varrho(\tau)e^{-x}, \tau)$ depending on τ, x and the gradient $\partial_x \Pi$ of a solution Π. Now the boundary conditions $V(0,t) = 0, V(S_f(t),t) = S_f(t) - E$ and $\partial_S V(S_f(t),t) = 1$ imply

$$\Pi(0,\tau) = -E, \quad \Pi(+\infty,\tau) = 0\,, \quad 0 < \tau < T\,, \tag{57}$$

and, from the terminal pay-off diagram for $V(S,T)$, we deduce

$$\Pi(x,0) = \begin{cases} -E & \text{for } x < \ln\left(\frac{\varrho(0)}{E}\right) \\ 0 & \text{otherwise.} \end{cases} \tag{58}$$

In order to close up the system of equations that determines the value of a synthetic portfolio Π we have to construct an equation for the free boundary position $\varrho(\tau)$. Indeed, both the coefficient b as well as the initial condition $\Pi(x,0)$ depend on the function $\varrho(\tau)$. Similarly as in the case of a constant volatility σ (see [46, 50]) we proceed as follows: since $S_f(t) - E = V(S_f(t),t)$ and $\partial_S V(S_f(t),t) = 1$ we have $\frac{d}{dt}S_f(t) = \partial_S V(S_f(t),t)\frac{d}{dt}S_f(t) + \partial_t V(S_f(t),t)$ and so $\partial_t V(S,t) = 0$ along the free boundary $S = S_f(t)$. Moreover, assuming $\partial_x \Pi$ is continuous up to the boundary $x = 0$ we obtain $S^2 \partial_S^2 V(S,t) \to \partial_x \Pi(0,\tau)$ and $S\partial_S V(S,t) \to \varrho(\tau)$ as $S \to S_f(t)^-$. Now, by taking the limit $S \to S_f(t)^-$ in the Black–Scholes equation (2) we obtain $(r-q)\varrho(\tau) + \frac{1}{2}\sigma^2 \partial_x \Pi(0,\tau) - r(\varrho(\tau) - E) = 0$. Therefore

$$\varrho(\tau) = \frac{rE}{q} + \frac{1}{2q}\sigma^2(\partial_x \Pi(0,\tau), \varrho(\tau), \tau)\frac{\partial \Pi}{\partial x}(0,\tau)$$

for $0 < \tau \le T$. The value of $\varrho(0)$ can be easily derived from the smoothness assumption made on $\partial_x \Pi$ at the origin $x = 0, \tau = 0$ under the structural assumption

$$0 < q < r \tag{59}$$

made on the interest and dividend yield rates r, q (cf. [46, 47]). Indeed, continuity of $\partial_x \Pi$ at the origin $(0,0)$ implies $\lim_{\tau \to 0^+} \partial_x \Pi(0,\tau) = \partial_x \Pi(0,0) = \lim_{x \to 0^+} \partial_x \Pi(x,0) = 0$ because $\Pi(x,0) = -E$ for x close to 0^+ provided $\ln(r/q) > 0$. From the above equation for $\varrho(\tau)$ we deduce $\varrho(0) = \frac{rE}{q}$ by taking the limit $\tau \to 0^+$. Putting all the above equations together we end up with a closed system of equations for $\Pi = \Pi(x,\tau)$ and $\varrho = \varrho(\tau)$

$$\begin{aligned}
&\frac{\partial \Pi}{\partial \tau} + \big(b(\tau) - \frac{\sigma^2}{2}\big)\frac{\partial \Pi}{\partial x} - \frac{1}{2}\frac{\partial}{\partial x}\big(\sigma^2 \frac{\partial \Pi}{\partial x}\big) + r\Pi = 0\,, \\
&\Pi(0,\tau) = -E\,, \ \ \Pi(+\infty,\tau) = 0\,, \ x > 0\,, \tau \in (0,T)\,, \\
&\Pi(x,0) = \begin{cases} -E & \text{for } x < \ln(r/q) \\ 0 & \text{otherwise}\,, \end{cases}
\end{aligned} \tag{60}$$

where $\sigma = \sigma(\partial_x \Pi(x,\tau), \varrho(\tau) e^{-x}, \tau)$, $b(\tau) = \frac{\dot\varrho(\tau)}{\varrho(\tau)} + r - q$ and the free boundary position $\varrho(\tau) = S_f(T - \tau)$ satisfies an implicit algebraic equation

$$\varrho(\tau) = \frac{rE}{q} + \frac{\sigma^2(\partial_x \Pi(0,\tau), \varrho(\tau), \tau)}{2q}\frac{\partial \Pi}{\partial x}(0,\tau)\,, \quad \text{with } \varrho(0) = \frac{rE}{q}, \tag{61}$$

where $\tau \in (0,T)$. Notice that, in order to guarantee parabolicity of equation (60) we have to assume that the function $p \mapsto \sigma^2(p, \varrho(\tau)e^{-x}, \tau)p$ is strictly increasing. More precisely, we shall assume that there exists a positive constant $\gamma > 0$ such that

$$\sigma^2(p,\xi,\tau) + p\partial_p \sigma^2(p,\xi,\tau) \ge \gamma > 0 \tag{62}$$

for any $\xi > 0, \tau \in (0,T)$ and $p \in \mathbb{R}$. Notice that condition (62) is satisfied for the RAPM model in which $\sigma^2 = \hat\sigma^2(1 + \mu p^{\frac{1}{3}} \xi^{-\frac{1}{3}})$ for any $\mu \ge 0$ and $p \ge 0$. Clearly $p = S^2 \partial_S^2 V > 0$ for the case of plain vanilla call or put options. As far as the Barles and Soner model is concerned, we have $\sigma^2 = \hat\sigma^2(1 + \Psi(a^2 e^{r\tau} p))$ and condition (62) is again satisfied because the function Ψ is a positive and increasing function in the Barles and Soner model.

Remark 4.1 *Following exactly the same argument as in (46) one can derive an explicit expression for the option price $V(S,t)$:*

$$V(S, T-\tau) = \frac{S}{\varrho(\tau)}\left(\varrho(\tau) - E + \int_0^{\ln\frac{\varrho(\tau)}{S}} e^x \Pi(x,\tau)\, dx\right). \tag{63}$$

4.1. An Iterative Algorithm for Approximation of the Early Exercise Boundary

The idea of the iterative numerical algorithm for solving the problem (60), (61) is rather simple: we use the backward Euler method of finite differences in order to discretize the parabolic equation (60) in time. In each time level we find a new approximation of a solution pair (Π, ϱ). First we determine a new position of ϱ from the algebraic equation (61). We remind ourselves that (even in the case σ is constant) the free boundary function $\varrho(\tau)$ behaves like $rE/q + O(\tau^{1/2})$ for $\tau \to 0^+$ (see e.g. [16] or [46]) and so $b(\tau) = O(\tau^{-1/2})$. Hence the convective term $b(\tau)\partial_x \Pi$ becomes a dominant part of equation (60) for small values of τ. In order to overcome this difficulty we employ the operator splitting technique for successive solving of the convective and diffusion parts of equation (60). Since the diffusion coefficient σ^2 depends on the derivative $\partial_x \Pi$ of a solution Π itself we make several micro-iterates to find a solution of a system of nonlinear algebraic equations.

Now we present our algorithm in more details. We restrict the spatial domain $x \in (0, \infty)$ to a finite interval of values $x \in (0, L)$ where $L > 0$ is sufficiently large. For practical purposes one can take $L \approx 3$ as it corresponds to the interval $S \in (S_f(t)e^{-L}, S_f(t))$ in the original asset price variable S. The value $S_f(t)e^{-L}$ is then could be a good approximation for the default value $S = 0$ if $L \approx 3$. Let us denote by $k > 0$ the time step, $k = T/m$, and, by $h > 0$ the spatial step, $h = L/n$ where $m, n \in N$ stand for the number of time and space discretization steps, resp. We denote by Π_i^j an approximation of $\Pi(x_i, \tau_j)$, $\varrho^j \approx \varrho(\tau_j)$, $b^j \approx b(\tau_j)$ where $x_i = ih, \tau_j = jk$. We approximate the value of the volatility σ at the node (x_i, τ_j) by finite difference as follows:

$$\sigma_i^j = \sigma_i^j(\varrho^j, \Pi^j) = \sigma((\Pi_{i+1}^j - \Pi_i^j)/h, \varrho^j e^{-x_i}, \tau_j) .$$

Then for the Euler backward in time finite difference approximation of equation (60) we have

$$\frac{\Pi^j - \Pi^{j-1}}{k} + \left(b^j - \frac{1}{2}(\sigma^j)^2\right)\partial_x \Pi^j - \frac{1}{2}\partial_x\left((\sigma^j)^2 \partial_x \Pi^j\right) + r\Pi^j = 0 \tag{64}$$

and the solution $\Pi^j = \Pi^j(x)$ is subject to Dirichlet boundary conditions at $x = 0$ and $x = L$. We set $\Pi^0(x) = \Pi(x, 0)$. Now we decompose the above problem into two parts - a convection part and a diffusive part by introducing an auxiliary intermediate step $\Pi^{j-\frac{1}{2}}$:

(*Convective part*)

$$\frac{\Pi^{j-\frac{1}{2}} - \Pi^{j-1}}{k} + b^j \partial_x \Pi^{j-\frac{1}{2}} = 0 , \tag{65}$$

(*Diffusive part*)

$$\frac{\Pi^j - \Pi^{j-\frac{1}{2}}}{k} - \frac{(\sigma^j)^2}{2}\partial_x \Pi^j - \frac{1}{2}\partial_x\left((\sigma^j)^2 \partial_x \Pi^j\right) + r\Pi^j = 0 . \tag{66}$$

The idea of the operator splitting technique consists in comparison the sum of solutions to convective and diffusion part to a solution of (64). Indeed, if $\partial_x \Pi^j \approx \partial_x \Pi^{j-\frac{1}{2}}$ then it is reasonable to assume that Π^j computed from the system (65)–(66) is a good approximation of the system (64).

The convective part can be approximated by an explicit solution to the transport equation:

$$\partial_\tau \tilde{\Pi} + b(\tau)\partial_x \tilde{\Pi} = 0 \qquad \text{for } x > 0,\ \tau \in (\tau_{j-1}, \tau_j] \tag{67}$$

subject to the boundary condition $\tilde{\Pi}(0, \tau) = -E$ and initial condition $\tilde{\Pi}(x, \tau_{j-1}) = \Pi^{j-1}(x)$. For American style of call option the free boundary $\varrho(\tau) = S_f(T - \tau)$ must be an increasing function in τ and we have assumed $0 < q < r$ we have $b(\tau) = \dot{\varrho}(\tau)/\varrho(\tau) + r - q > 0$ and so prescribing the in-flowing boundary condition $\tilde{\Pi}(0, \tau) = -E$ is consistent with the transport equation. Let us denote by $B(\tau)$ the primitive function to $b(\tau)$, i.e. $B(\tau) = \ln \varrho(\tau) + (r - q)\tau$. Equation (67) can be integrated to obtain its explicit solution:

$$\tilde{\Pi}(x, \tau) = \begin{cases} \Pi^{j-1}(x - B(\tau) + B(\tau_{j-1})) & \text{if } x - B(\tau) + B(\tau_{j-1}) > 0\,, \\ -E & \text{otherwise.} \end{cases} \tag{68}$$

Thus the spatial approximation $\Pi_i^{j-\frac{1}{2}}$ can be constructed from the formula

$$\Pi_i^{j-\frac{1}{2}} = \begin{cases} \Pi^{j-1}(\xi_i) & \text{if } \xi_i = x_i - \ln \varrho_j + \ln \varrho_{j-1} - (r-q)k > 0\,, \\ -E & \text{otherwise,} \end{cases} \tag{69}$$

where a linear approximation between discrete values $\Pi_i^{j-1}, i = 0, 1, ..., n$, is being used to compute the value $\Pi^{j-1}(x_i - \ln \varrho_j + \ln \varrho_{j-1} - (r - q)k)$.

The diffusive part can be solved numerically by means of finite differences. Using central finite difference for approximation of the derivative $\partial_x \Pi^j$ we obtain

$$\begin{aligned} \frac{\Pi_i^j - \Pi_i^{j-\frac{1}{2}}}{k} + r\Pi_i^j \quad &- \quad \frac{(\sigma_i^j)^2}{2} \frac{\Pi_{i+1}^j - \Pi_{i-1}^j}{2h} \\ &- \quad \frac{1}{2h}\left((\sigma_i^j)^2 \frac{\Pi_{i+1}^j - \Pi_i^j}{h} - (\sigma_{i-1}^j)^2 \frac{\Pi_i^j - \Pi_{i-1}^j}{h} \right) = 0\,. \end{aligned}$$

Hence, the vector of discrete values $\Pi^j = \{\Pi_i^j, i = 1, 2, ..., n\}$ at the time level $j \in \{1, 2, ..., m\}$ satisfies the tridiagonal system of equations

$$\alpha_i^j \Pi_{i-1}^j + \beta_i^j \Pi_i^j + \gamma_i^j \Pi_{i+1}^j = \Pi_i^{j-\frac{1}{2}}, \tag{70}$$

for $i = 1, 2, ..., n$, where

$$\begin{aligned} \alpha_i^j &\equiv \alpha_i^j(\varrho^j, \Pi^j) = -\frac{k}{2h^2}(\sigma_{i-1}^j)^2 + \frac{k}{2h}\frac{(\sigma_i^j)^2}{2}, \\ \gamma_i^j &\equiv \gamma_i^j(\varrho^j, \Pi^j) = -\frac{k}{2h^2}(\sigma_i^j)^2 - \frac{k}{2h}\frac{(\sigma_i^j)^2}{2}, \\ \beta_i^j &\equiv \beta_i^j(\varrho^j, \Pi^j) = 1 + rk - (\alpha_i^j + \gamma_i^j)\,. \end{aligned} \tag{71}$$

The initial and boundary conditions at $\tau = 0$ and $x = 0, L$, resp., can be approximated as follows:

$$\Pi_i^0 = \begin{cases} -E & \text{for } x_i < \ln(r/q)\,, \\ 0 & \text{for } x_i \geq \ln(r/q)\,, \end{cases}$$

for $i = 0, 1, ..., n$, and $\Pi_0^j = -E, \quad \Pi_n^j = 0$.

Next we proceed by approximation of equation (61) which introduces a nonlinear constraint condition between the early exercise boundary function $\varrho(\tau)$ and the trace of the solution Π at the boundary $x = 0$ ($S = S_f(t)$ in the original variable). Taking a finite difference approximation of $\partial_x \Pi$ at the origin $x = 0$ we obtain

$$\varrho^j = \frac{rE}{q} + \frac{1}{2q}\sigma^2\left((\Pi_1^j - \Pi_0^j)/h, \varrho^j, \tau_j\right)\frac{\Pi_1^j - \Pi_0^j}{h}. \tag{72}$$

Now, equations (69), (70) and (72) can be written in an abstract form as a system of nonlinear equations:

$$\begin{aligned} \varrho^j &= \mathcal{F}(\Pi^j, \varrho^j), \\ \Pi^{j-\frac{1}{2}} &= \mathcal{T}(\Pi^j, \varrho^j), \\ \mathcal{A}(\Pi^j, \varrho^j)\Pi^j &= \Pi^{j-\frac{1}{2}}, \end{aligned} \tag{73}$$

where $\mathcal{F}(\Pi^j, \varrho^j)$ is the right-hand side of the algebraic equation (72), $\mathcal{T}(\Pi^j, \varrho^j)$ is the transport equation solver given by the right-hand side of (69) and $\mathcal{A} = \mathcal{A}(\Pi^j, \varrho^j)$ is a tridiagonal matrix with coefficients given by (71). The system (73) can be approximately solved by means of successive iterates procedure. We define, for $j \geq 1$, $\Pi^{j,0} = \Pi^{j-1}$, $\varrho^{j,0} = \varrho^{j-1}$. Then the $(p+1)$-th approximation of Π^j and ϱ^j is obtained as a solution to the system:

$$\begin{aligned} \varrho^{j,p+1} &= \mathcal{F}(\Pi^{j,p}, \varrho^{j,p}), \\ \Pi^{j-\frac{1}{2},p+1} &= \mathcal{T}(\Pi^{j,p}, \varrho^{j,p+1}), \\ \mathcal{A}(\Pi^{j,p}, \varrho^{j,p+1})\Pi^{j,p+1} &= \Pi^{j-\frac{1}{2},p+1}. \end{aligned} \tag{74}$$

Notice that the last equation is a linear tridiagonal equation for the vector $\Pi^{j,p+1}$ whereas $\varrho^{j,p+1}$ and $\Pi^{j-\frac{1}{2},p+1}$ can be directly computed from (72) and (69), resp. If the sequence of approximate solutions $\{(\Pi^{j,p}, \varrho^{j,p})\}_{p=1}^{\infty}$ converges to some limiting value $(\Pi^{j,\infty}, \varrho^{j,\infty})$ as $p \to \infty$ then this limit is a solution to a nonlinear system of equations (73) at the time level j and we can proceed by computing the approximate solution the next time level $j+1$.

4.2. Numerical Approximations of the Early Exercise Boundary

In this section we focus on numerical experiments based on the iterative scheme described in the previous section. The main purpose is to compute the free boundary profile $S_f(t) = \varrho(T-t)$ for different (non)linear Black–Scholes models and for various model parameters. A solution (Π, ϱ) has been computed by our iterative algorithm for the following basic model parameters: $E = 10, T = 1$ (one year), $r = 0.1$ (10% p.a) , $q = 0.05$ (5% p.a.) and $\hat{\sigma} = 0.2$. We used $n = 750$ spatial points and $m = 225000$ time discretization steps. Such a time step $k = T/m$ corresponds to 140 seconds between consecutive time levels when expressed in real time scale. In average we needed $p \leq 6$ micro-iterates (74) in order to solve the nonlinear system (73) with the precision 10^{-7}.

4.2.1. Case of a Constant Volatility – Comparison Study

In our first numerical experiment we make attempt to compare our iterative approximation scheme for solving the free boundary problem for an American call option to known schemes in the case when the volatility $\sigma > 0$ is constant. We compare our solution

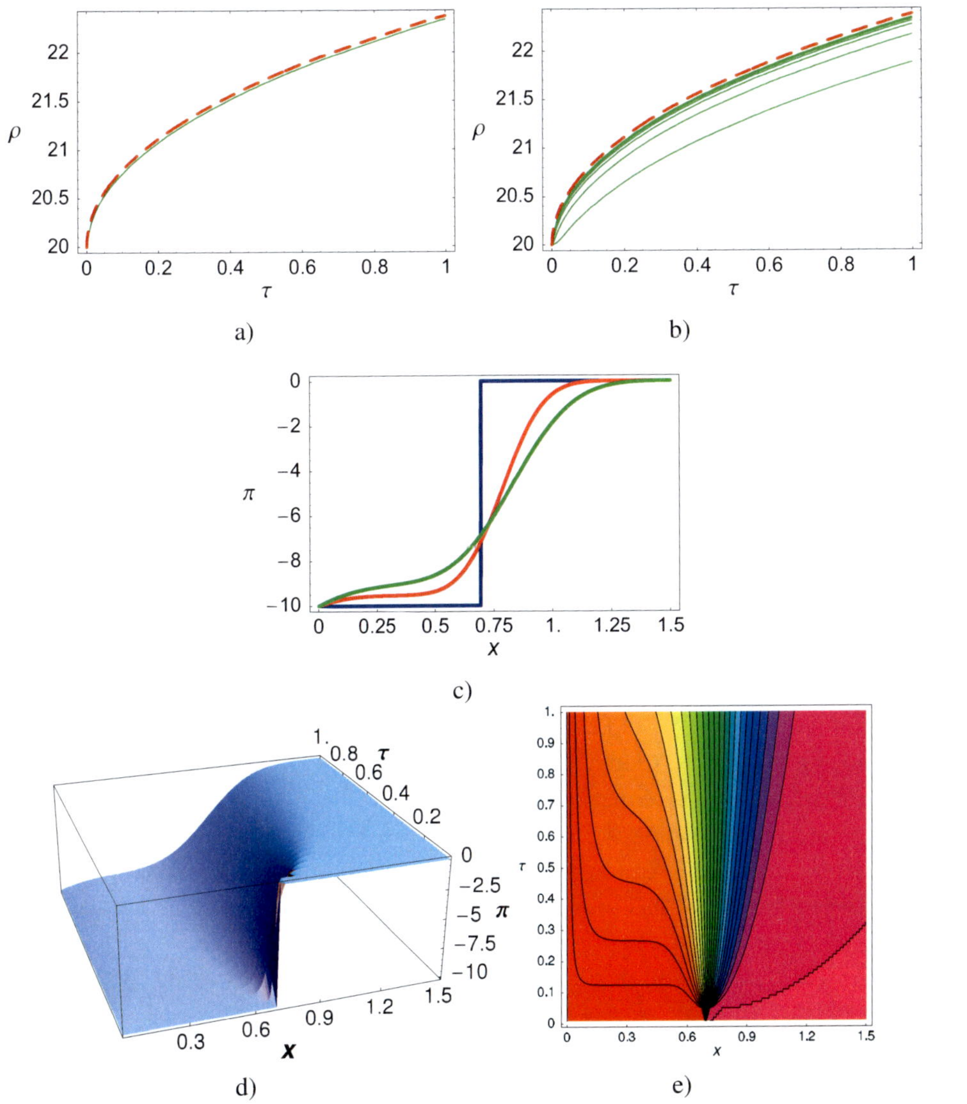

Figure 9. a) A comparison of the free boundary function $\varrho(\tau)$ computed by the iterative algorithm (green solid curve) to the integral equation based approximation (dashed red curve); b) free boundary positions computed for various mesh sizes; c) a solution profile $\Pi(x,\tau)$ for $\tau = 0$ (blue line), $\tau = T/2$ (red curve), $\tau = T$ (green curve); d) 3D plot and e) contour plot of the function $\Pi(x,\tau)$.

to the one computed by means of a solution to a nonlinear integral equation for $\varrho(\tau)$ (see also [46, 50]). This comparison can be also considered as a benchmark or test example for which we know a solution that can be computed by a another justified algorithm. In Fig. 9, part a), we show the function ϱ computed by our iterative algorithm for $E = 10, T = 1, r = 0.1, q = 0.05, \sigma = 0.2$. At the expiry $T = 1$ the value of $\varrho(T)$ was computed as: $\varrho(T) = 22.321$. The corresponding value $\varrho(T)$ computed from the integral

Table 2. Experimental order of convergence of the iterative algorithm for approximating the free boundary position.

h	err(L^∞)	eoc(L^∞)	err(L^2)	eoc(L^2)
0.03	0.5	-	0.808	-
0.012	0.215	0.92	0.227	1.39
0.006	0.111	0.96	0.0836	1.44
0.004	0.0747	0.97	0.0462	1.46
0.003	0.0563	0.98	0.0303	1.47
0.0024	0.0452	0.98	0.0218	1.48
0.002	0.0378	0.98	0.0166	1.48

equation (40) (cf. [46]) was $\varrho(T) = 22.375$. The relative error is less than 0.25%. In the part b) we present 7 approximations of the free boundary function $\varrho(\tau)$ computed for different mesh sizes h (see Tab. 2 for details). The sequence of approximate free boundaries $\varrho_h, h = h_1, h_2, ...,$ converges monotonically from below to the free boundary function ϱ as $h \downarrow 0$. The next part c) of Fig. 9 depicts various solution profiles of a function $\Pi(x,\tau)$. In order to achieve a reasonable approximation to equation (72) we need very accurate approximation of $\Pi(x,\tau)$ for x close to the origin 0. The parts d) and e) of Fig. 9 depict the contour and 3D plots of the function $\Pi(x,\tau)$.

In Tab. 2 we present the numerical error analysis for the distance $\|\varrho_h - \varrho\|_p$ measured in two different norms (L^∞ and L^2) of a computed free boundary position ϱ_h corresponding to the mesh size h and the solution ϱ computed from the integral equation described in (40) (cf. [46]). The time step k has been adjusted to the spatial mesh size h in order to satisfy CFL condition $\hat{\sigma}^2 k/h^2 \approx 1/2$. We also computed the experimental order of convergence eoc(L^p) for $p = 2, \infty$. Recall that the experimental order of convergence can be defined as the ratio:

$$\text{eoc}(L^p) = \frac{\ln(\|\varrho_{h_i} - \varrho\|_p) - \ln(\|\varrho_{h_{i-1}} - \varrho\|_p)}{\ln h_i - \ln h_{i-1}}\,.$$

It can be interpreted as an exponent $\alpha = \text{eoc}(L^p)$ for which we have $\|\varrho_h - \varrho\|_p = O(h^\alpha)$. It turns out from Tab. 2 that the conjecture on the order of convergence $\|\varrho_h - \varrho\|_\infty = O(h)$ whereas $\|\varrho_h - \varrho\|_2 = O(h^{3/2})$ as $h \to 0^+$ could be reasonable.

4.2.2. Risk Adjusted Pricing Methodology Model

In the next example we computed the position of the free boundary $\varrho(\tau)$ in the case of the Risk Adjusted Pricing Methodology model - a nonlinear Black–Scholes type model derived by Jandačka and Ševčovič in [32]. In this model the volatility σ is a nonlinear function of the asset price S and the second derivative $\partial_S^2 V$ of the option price, and it is given by formula (8). In Fig. 10 we present results of numerical approximation of the free boundary position $\varrho^R(\tau) = S_f^R(T-\tau)$ in the case when the coefficient of transaction costs $C = 0.01$ is fixed and the risk premium measure R varies from $R = 5, 15, 40, 70$, up to $R = 100$. We compare the position of the free boundary $\varrho^R(\tau)$ to the case when there are no transaction costs and no risk from volatile portfolio, i.e. we compare it with the free boundary position

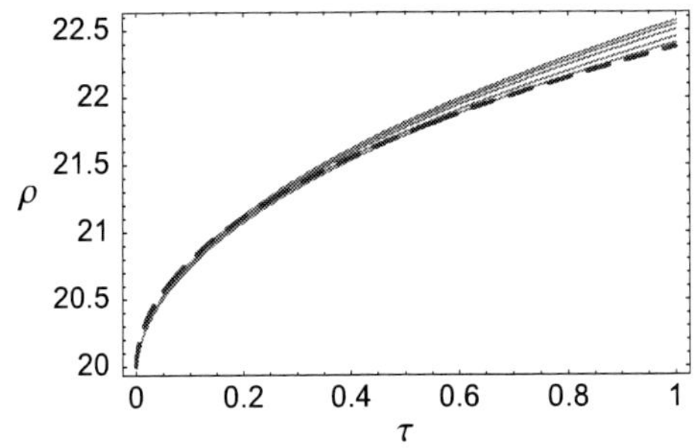

Figure 10. A comparison of the free boundary function $\varrho^R(\tau)$ computed for the Risk Adjusted Pricing Methodology model. Dashed red curve represents a solution corresponding to $R = 0$, whereas the green curves represent a solution $\varrho^R(\tau)$ for different values of the risk premium coefficients $R = 5, 15, 40, 70, 100$.

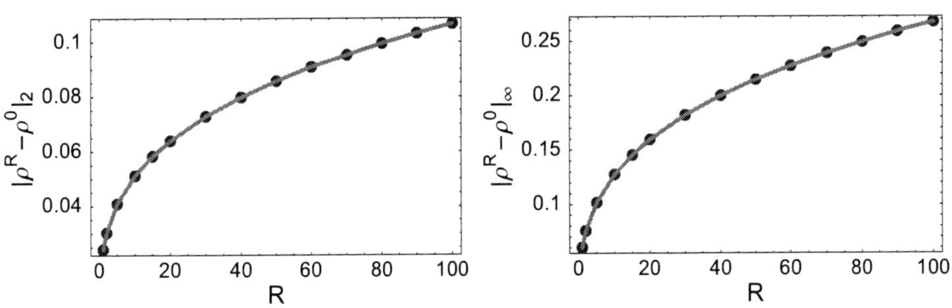

Figure 11. Dependence of the norms $\|\varrho^R - \varrho^0\|_p$ ($p = \infty, 2$) of the deviation of the free boundary $\varrho = \varrho^R(\tau)$ for the RAPM model on the risk premium coefficient R.

$\varrho^0(\tau)$ for the linear Black–Scholes equation (see Fig. 10). An increase in the risk premium coefficient R resulted in an increase of the free boundary position as it can be expected.

In Tab. 3 and Fig. 10 we summarize results of comparison of the free boundary position ϱ^R for various values of the risk premium coefficient to the reference position $\varrho = \varrho^0$ computed from the Black–Scholes model with a constant volatility $\sigma = \hat{\sigma}$, i.e. $R = 0$. The experimental order α_p of the distance function $\|\varrho^R - \varrho^0\|_p = O(R^{\alpha_p})$ has been computed for $p = 2, \infty$, as follows:

$$\alpha_p = \frac{\ln(\|\varrho^{R_i} - \varrho^0\|_p) - \ln(\|\varrho^{R_{i-1}} - \varrho^0\|_p)}{\ln R_i - \ln R_{i-1}}.$$

According to the values presented in Tab. 3 it turns out that a reasonable conjecture on the order of convergence is that $\|\varrho^R - \varrho^0\|_p = O(R^{1/3})$ for both norms $p = 2$ and $p = \infty$. Since the transaction cost coefficient C and risk premium measure R enter the expression for the RAPM volatility (8) only in the product $C^2 R$ we can conjecture that $\|\varrho^{R,C} - \varrho^{0,0}\|_p = O(C^{2/3} R^{1/3})$ as either $C \to 0^+$ or $R \to 0^+$.

Table 3. Distance $\|\varrho^R - \varrho^0\|_p$ ($p = 2, \infty$) of the free boundary position ϱ^R from the reference free boundary position ϱ^0 and experimental orders α_∞ and α_2 of convergence.

R	$\|\varrho^R - \varrho^0\|_\infty$	α_∞	$\|\varrho^R - \varrho^0\|_2$	α_2
1	0.0601	-	0.0241	-
2	0.0754	0.33	0.0303	0.328
5	0.102	0.33	0.0408	0.326
10	0.128	0.33	0.0511	0.324
15	0.145	0.32	0.0582	0.323
20	0.16	0.32	0.0639	0.322
30	0.182	0.32	0.0727	0.321
40	0.2	0.32	0.0798	0.32
50	0.214	0.32	0.0856	0.319
60	0.227	0.32	0.0907	0.318
70	0.239	0.32	0.0953	0.317
80	0.249	0.32	0.0994	0.317
90	0.259	0.32	0.103	0.316
100	0.268	0.32	0.107	0.316

4.2.3. Barles and Soner Model

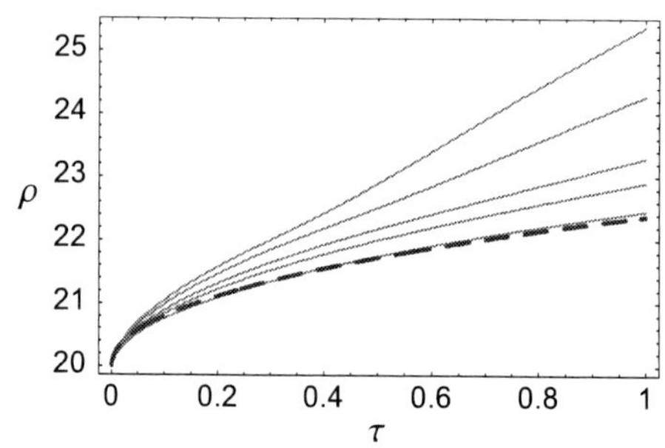

Figure 12. A comparison of the free boundary function $\varrho(\tau)$ computed for the Barles and Soner model. Dashed red curve represents a solution corresponding to $R = 0$, whereas the green curves represents a solution $\varrho(\tau)$ for different values of the risk aversion coefficient $a = 0.01, 0.07, 0.13, 0.25, 0.35$.

Our next example is devoted to the nonlinear Black–Scholes model due to Barles and Soner (see [8]). In this model the volatility is given by equation (6). Numerical results are depicted in Fig. 12. Choosing a larger value of the risk aversion coefficient $a > 0$ resulted in increase of the free boundary position $\varrho^a(\tau)$. The position of the early exercise boundary $\varrho^a(\tau)$ has considerably increased in comparison to the linear Black–Scholes equation with constant volatility $\sigma = \hat{\sigma}$. In contrast to the case of constant volatility as well as the RAPM model, there is, at least a numerical evidence (see Fig.12 and ϱ^a for the largest

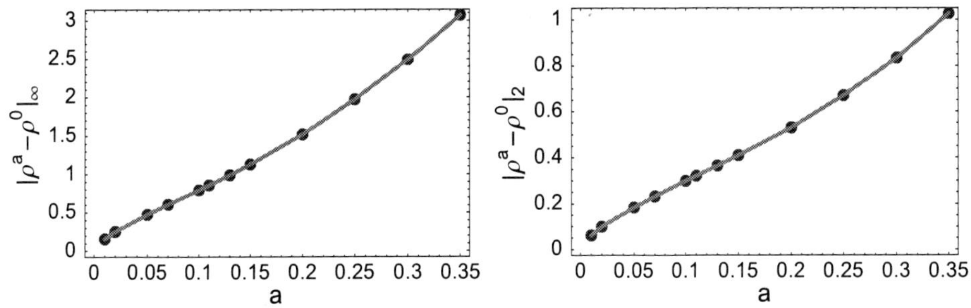

Figure 13. Dependence of the norms $\|\varrho^a - \varrho^0\|_p$ ($p = \infty, 2$) of the deviation of the free boundary $\varrho = \varrho^a(\tau)$ for the Barles-Soner model on the risk aversion parameter a.

Table 4. Distance $\|\varrho^a - \varrho^0\|_p$ ($p = 2, \infty$) of the free boundary position ϱ^a from the reference free boundary position ϱ^0 and experimental orders α_∞ and α_2 of convergence.

a	$\|\varrho^a - \varrho^0\|_\infty$	α_∞	$\|\varrho^a - \varrho^0\|_2$	α_2
0.01	0.156	-	0.0615	-
0.02	0.25	0.68	0.0985	0.68
0.05	0.472	0.69	0.184	0.679
0.07	0.602	0.72	0.232	0.69
0.1	0.793	0.77	0.298	0.712
0.11	0.857	0.82	0.32	0.74
0.13	0.99	0.86	0.364	0.766
0.15	1.13	0.92	0.409	0.807
0.2	1.52	1.	0.529	0.897
0.25	1.97	1.2	0.669	1.05
0.3	2.49	1.3	0.833	1.21
0.35	3.07	1.4	1.03	1.35

value $a = 0.35$) that the free boundary profile $\varrho^a(\tau)$ need not be necessarily convex. Recall that that convexity of the free boundary profile has been proved analytically by Ekström *et al.* and Chen *et al.* in a recent papers [13, 19, 20] in the case of a American put option and constant volatility $\sigma = \hat{\sigma}$.

Similarly as in the previous model we also investigated the dependence of the free boundary position $\varrho = \varrho^a(\tau)$ on the risk aversion parameter $a > 0$. In Tab. 4 and Fig. 13 we present results of comparison of the free boundary position ϱ^a for various values of the risk aversion coefficient a to the reference position $\varrho = \varrho^0$. Inspecting values α_p of the order of distance $\|\varrho^a - \varrho^0\|_p$ it can be conjectured that $\|\varrho^a - \varrho^0\|_p = O(a^{2/3})$ as $a \to 0$ for both norms $p = 2$ and $p = \infty$.

5. Transformation Methods for Asian Call Options

Path dependent options are options whose pay-off diagram depends on the path history of the underlying asset. Among path dependent options Asian options plays an important role as they are quite common in currency and commodity markets like e.g. oil industry (cf. [15,28]). Asian options may depend on the averaged path history in several ways. We shall restrict our attention to the so-called floating strike Asian call options. The floating strike price is assumed to be an arithmetic average of the underlying asset prices over the entire time interval $[0,T]$ where T is the expiration time.

Let us define the arithmetic average $A = A_t$ of the underlying asset $S = S_t$ by

$$A_t = \frac{1}{t}\int_0^t S_\tau \, d\tau \,.$$

For the case of the so-called Asian floating strike call option the pay-off diagram at expiry T reads as follows: $V(S,A,T) = \max(S-A,0)$. It means the price V of an option contract will depend not only on the underlying asset price S, time t, but also on the underlying asset path average A, i.e. $V = V(S,A,t)$.

5.1. Governing Equations for Asian Options

As it is usual in the option pricing theory, we shall describe the asset price dynamics by a geometric Brownian with drift ϱ, dividend yield $q \geq 0$ and volatility σ, i.e. $dS = (\varrho - q)Sdt + \sigma S dW$ where W is the standard Wiener process. If we apply Itô's formula to the function $V = V(S,A,t)$ we obtain

$$dV = \left(\frac{\partial V}{\partial t} + \frac{\sigma^2}{2}S^2\frac{\partial^2 V}{\partial S^2}\right) + \frac{\partial V}{\partial S}dS + \frac{\partial V}{\partial A}dA \,.$$

In the case of arithmetic averaging we have $dA = t^{-1}(S-A)dt$. Hence the differential dA is of the order of dt and this is why the above expression for dV indeed represents its lowest order approximation when taking into account stochastic character of the dynamics of the asset price S. Therefore, following standard arguments from the Black–Scholes theory one can derive the governing equation for pricing Asian option with arithmetic averaging in the form:

$$\frac{\partial V}{\partial t} + \frac{\sigma^2}{2}S^2\frac{\partial^2 V}{\partial S^2} + S(r-q)\frac{\partial V}{\partial S} + \frac{S-A}{t}\frac{\partial V}{\partial A} - rV = 0, \tag{75}$$

where $0 < t < T,\ S, A > 0$ (see e.g. [15]). For Asian call option the above equation is subject to the terminal pay-off condition

$$V(S,A,T) = \max(S-A,0), \quad S, A > 0 \,. \tag{76}$$

It is well known (see e.g. [15,38]) that for Asian options with floating strike we can achieve dimension reduction by introducing the following similarity variable:

$$x = \frac{S}{A}, \qquad W(x,\tau) = \frac{1}{A}V(S,A,t)$$

where $\tau = T - t$. It is straightforward to verify that $V(S,A,t) = W(S/A, T-t)A$ is a solution of (75) iff $W = W(x,\tau)$ is a solution to the following parabolic PDE:

$$\frac{\partial W}{\partial \tau} - \frac{\sigma^2}{2}x^2\frac{\partial^2 W}{\partial x^2} - (r-q)x\frac{\partial W}{\partial x} - \frac{x-1}{T-\tau}\left(W - x\frac{\partial W}{\partial x}\right) + rW = 0, \qquad (77)$$

where $x > 0$ and $0 < \tau < T$. The initial condition for W immediately follows from the terminal pay-off diagram for the call option,

$$W(x,0) = \max(x-1,0). \qquad (78)$$

5.2. American Style of Asian Options

Following Dai and Kwok [15], American style of Asian options is characterized by the exercise region

$$\mathcal{E} = \{(S,A,t) \in [0,\infty) \times [0,\infty) \times [0,T),\ V(S,A,t) = V(S,A,T)\}.$$

In the case of a call option this region can be described by an early exercise boundary function $S_f = S_f(A,t)$ such that

$$\mathcal{E} = \{(S,A,t) \in [0,\infty) \times [0,\infty) \times [0,T),\ S \geq S_f(A,t)\}.$$

For American style of an Asian call option we have to impose a homogeneous Dirichlet boundary condition $V(0,A,t) = 0$ at $S = 0$. According to [15] the C^1 continuity condition at the point $(S_f(A,t),A,t)$ of a contact of a solution V with its pay-off diagram implies the following boundary condition at the free boundary position $S_f(A,t)$:

$$\frac{\partial V}{\partial S}(S_f(A,t),A,t) = 1, \quad V(S_f(A,t),A,t) = S_f(A,t) - A \qquad (79)$$

for any $A > 0$ and $0 < t < T$. It is important to emphasize that the free boundary function S_f can be also reduced to a function of one variable by introducing a new state function $x_f(t)$ as follows:

$$S_f(A,t) = Ax_f(t).$$

The function $x_f = x_f(t)$ is a free boundary function for the transformed state variable $x = S/A$. For American style of Asian call options the spatial domain for the reduced equation (77) is given by

$$0 < x < \varrho(\tau),\ \ \tau \in (0,T), \quad \text{where } \varrho(\tau) = x_f(T-\tau).$$

Taking into account boundary conditions (79) for the option price V we end up with corresponding boundary conditions for the function W:

$$W(0,\tau) = 0, \qquad W(x,\tau) = x - 1, \quad \frac{\partial W}{\partial x}(x,\tau) = 1 \text{ at } x = \varrho(\tau) \qquad (80)$$

for any $0 < \tau < T$ and the initial condition

$$W(x,0) = \max(x-1,0) \qquad (81)$$

for any $x > 0$.

5.3. Fixed Domain Transformation for American Style of Asian Call Options

Similarly as in Section 3, in order to apply the fixed domain transformation for the free boundary problem (77), (80), (81) we introduce a new variable ξ and an auxiliary function $\Pi = \Pi(\xi, \tau)$ (again representing a synthetic portfolio) defined as follows:

$$\xi = \ln\left(\frac{\varrho(\tau)}{x}\right), \qquad \Pi(\xi,\tau) = W(x,\tau) - x\frac{\partial W}{\partial x}. \tag{82}$$

Clearly, $x \in (0, \varrho(\tau))$ iff $\xi \in (0, \infty)$ for $\tau \in (0, T)$. The value $\xi = \infty$ of the transformed variable corresponds to the value $x = 0$ ($S = 0$) expressed in the original variable. On the other hand, the value $\xi = 0$ corresponds to the free boundary position $x = \varrho(\tau)$ ($S = AS_f(A, t)$).

A straightforward calculation similar to that of Section 4 enables us to to justify that the function $\Pi = \Pi(\xi, \tau)$ is a solution to the following parabolic PDE

$$\frac{\partial \Pi}{\partial \tau} + a(\xi,\tau)\frac{\partial \Pi}{\partial \xi} - \frac{\sigma^2}{2}\frac{\partial^2 \Pi}{\partial \xi^2} + \left(r + \frac{1}{T-\tau}\right)\Pi = 0, \tag{83}$$

where the term $a(\xi, \tau)$ is given by

$$a(\xi,\tau) = \frac{\dot\rho(\tau)}{\rho(\tau)} + r - q - \frac{\sigma^2}{2} - \frac{\rho e^{-\xi} - 1}{T-\tau}. \tag{84}$$

Notice the spatial dependence of the coefficient $a = a(\xi, \tau)$ in comparison to the case of transformed equations for a linear or nonlinear Black–Scholes equations (see (32), (56)). The initial condition for the solution Π follows from (81)

$$\Pi(\xi, 0) = \begin{cases} -1 & \xi < \ln \rho(0), \\ 0 & \xi > \ln \rho(0). \end{cases}$$

Since $\partial_x W(x, \tau) = 1$ for $x = \varrho(\tau)$ and $W(0, \tau) = 0$ we conclude the Dirichlet boundary conditions for the function Π

$$\Pi(0,\tau) = -1, \qquad \Pi(\infty,\tau) = 0.$$

It remains to determine an algebraic constraint between the free boundary function $\varrho(\tau)$ and the solution Π. Similarly as in the case of a linear or nonlinear Black–Scholes equation we obtain, by differentiation the condition $W(\rho(\tau), \tau) = \rho(\tau) - 1$ with respect to τ, the following identity:

$$\frac{d}{d\tau}\rho(\tau) = \frac{\partial W}{\partial x}(\rho(\tau),\tau)\frac{d}{d\tau}\rho(\tau) + \frac{\partial W}{\partial \tau}(\rho(\tau),\tau).$$

Since $\frac{\partial W}{\partial x}(\rho(\tau), \tau) = 1$ we have $\frac{\partial W}{\partial \tau}(x, \tau) = 0$ at $x = \varrho(\tau)$. Assuming continuity of the function Π and its derivative Π_ξ up to the boundary $\xi = 0$ we have

$$x^2\frac{\partial^2 W}{\partial x^2}(x,\tau) \to \frac{\partial \Pi}{\partial \xi}(0,\tau), \quad x\frac{\partial W}{\partial x}(x,\tau) \to \rho(\tau)$$

as $x \to \rho(\tau)$. Passing to the limit $x \to \rho(\tau)$ in equation (77) we end up with the equation

$$-(r-q)\rho(\tau) - \frac{\sigma^2}{2}\frac{\partial \Pi}{\partial \xi}(0,\tau) + \frac{\rho(\tau)-1}{T-\tau} + r[\rho(\tau)-1] = 0.$$

It yields an algebraic nonlocal expression for the free boundary position $\varrho(\tau)$

$$\rho(\tau) = \frac{r + \frac{1}{T-\tau} + \frac{\sigma^2}{2}\frac{\partial \Pi}{\partial \xi}(0,\tau)}{q + \frac{1}{T-\tau}}. \qquad (85)$$

Next we determine the starting point of the free boundary function $\varrho(0)$. It means that we have to find the terminal value of the original state variable $x_f(T)$. We shall assume a structural assumption

$$r > q \geq 0 \qquad (86)$$

on the interest and dividend rates r, q. If $\varrho(0) > 1$ then $\ln \varrho(0) > 0$ and this is why the initial function $\Pi(\xi, 0)$ is equal to -1 in some right neighborhood of $\xi = 0$. Thus $\partial_\xi \Pi(\xi, 0) = 0$ for $0 < \xi \ll 1$. Again, assuming continuity of $\partial_\xi \Pi(\xi,\tau)$ at $(\xi,\tau) = (0,0)$ we obtain

$$\lim_{\tau\to 0^+} \frac{\partial \Pi}{\partial \xi}(0,\tau) = \lim_{\tau\to 0^+, x\to 0^+} \frac{\partial \Pi}{\partial \xi}(x,\tau) = \lim_{x\to 0^+} \frac{\partial \Pi}{\partial \xi}(x,0) = 0.$$

As a consequence we can conclude

$$\rho(0) = \frac{r + \frac{1}{T}}{q + \frac{1}{T}} = \frac{1+rT}{1+qT} > 1.$$

This initial condition for $\varrho(0)$ is exactly the same as the one derived recently by Dai and Kwok in [15]. They proved, for a general choice of $r, q \geq 0$ that the initial condition for the function ϱ is given by

$$\varrho(0) = \max\left(\frac{1+rT}{1+qT}, 1\right).$$

In summary, we have transformed the free boundary problem for pricing American style of Asian call option with floating strike price into the following nonlocal parabolic PDE with an algebraic constraint

$$\frac{\partial \Pi}{\partial \tau} + a(\xi,\tau)\frac{\partial \Pi}{\partial \xi} - \frac{\sigma^2}{2}\frac{\partial^2 \Pi}{\partial \xi^2} + \left(r + \frac{1}{T-\tau}\right)\Pi = 0, \quad 0 < \tau < T,\ \xi > 0,$$

$$\rho(\tau) = \frac{1 + r(T-\tau) + \frac{\sigma^2}{2}(T-\tau)\frac{\partial \Pi}{\partial \xi}(0,\tau)}{1 + q(T-\tau)}, \quad 0 < \tau < T,$$

subject to initial and boundary conditions

$$\begin{aligned} &\Pi(0,\tau) = -1, \qquad \Pi(\infty,\tau) = 0,\\ &\Pi(\xi,0) = \begin{cases} -1 & \xi < \ln((1+rT)/(1+qT)),\\ 0 & \xi > \ln((1+rT)/(1+qT)),\end{cases}\\ &\varrho(0) = (1+rT)/(1+qT), \end{aligned} \qquad (87)$$

where

$$a(\xi,\tau) = \frac{\dot\rho(\tau)}{\rho(\tau)} + r - q - \frac{\sigma^2}{2} - \frac{\rho e^{-\xi} - 1}{T-\tau}.$$

5.4. An Approximation Scheme for Pricing American Style of Asian Options

Similarly as in the case of a nonlinear Black–Scholes equation (see Section 4) we restrict the spatial domain $\xi \in (0,\infty)$ to a finite interval of values $\xi \in (0,L)$ where $L > 0$ is sufficiently large. Let $k > 0$ denote by the time step, $k = T/m$, and, by $h > 0$ the spatial step, $h = L/n$ where $m, n \in N$ again stand for the number of time and space discretization steps, resp. We denote by Π_i^j an approximation of $\Pi(\xi_i, \tau_j)$, $\varrho^j \approx \varrho(\tau_j)$ where $\xi_i = ih$ and $\tau_j = jk$. Then for the Euler backward in time finite difference approximation of equation (60) we have

$$\frac{\Pi^j - \Pi^{j-1}}{k} + b^j \frac{\partial \Pi^j}{\partial \xi} - \left(\frac{\sigma^2}{2} + \frac{\rho^j e^{-\xi} - 1}{T - \tau_j}\right)\frac{\partial \Pi^j}{\partial \xi} - \frac{\sigma^2}{2}\frac{\partial^2 \Pi^j}{\partial^2 \xi} + \left(r + \frac{1}{T - \tau_j}\right)\Pi^j = 0 \tag{88}$$

where b^j is an approximation of the value $b(\tau_j)$ where the function $b(\tau)$ is defined as in (56), i.e. $b(\tau) = \frac{\dot\varrho(\tau)}{\varrho(\tau)} + r - q$. The solution $\Pi^j = \Pi^j(x)$ is subject to Dirichlet boundary conditions at $\xi = 0$ and $\xi = L$. We set $\Pi^0(\xi) = \Pi(\xi, 0)$ (see (87). Again we split the above problem into a convection part and a diffusive part by introducing an auxiliary intermediate step $\Pi^{j-\frac{1}{2}}$:

(*Convective part*)

$$\frac{\Pi^{j-\frac{1}{2}} - \Pi^{j-1}}{k} + b^j \partial_x \Pi^{j-\frac{1}{2}} = 0\,, \tag{89}$$

(*Diffusive part*)

$$\frac{\Pi^j - \Pi^{j-\frac{1}{2}}}{k} - \left(\frac{\sigma^2}{2} + \frac{\rho^j e^{-\xi} - 1}{T - \tau_j}\right)\frac{\partial \Pi^j}{\partial \xi} - \frac{\sigma^2}{2}\frac{\partial^2 \Pi^j}{\partial^2 \xi} + \left(r + \frac{1}{T - \tau_j}\right)\Pi^j = 0. \tag{90}$$

The convective part can be approximated by an explicit solution to the transport equation $\partial_\tau \tilde\Pi + b(\tau)\partial_\xi \tilde\Pi = 0$ for $\xi > 0$ and $\tau \in (\tau_{j-1}, \tau_j]$ subject to the boundary condition $\tilde\Pi(0,\tau) = -1$ and the initial condition $\tilde\Pi(\xi, \tau_{j-1}) = \Pi^{j-1}(\xi)$. In contrast to classical plain vanilla call or put options the free boundary function $\varrho(\tau)$ need not be monotonically increasing (see e.g. [15] or [28]). Therefore depending on whether the term $b(\tau) = \frac{\dot\varrho(\tau)}{\varrho(\tau)} + r - q$ is positive or negative the boundary condition $\tilde\Pi(0,\tau) = -1$ at $\xi = 0$ is either in-flowing ($b > 0$) or out-flowing ($b < 0$). It means that the boundary condition $\Pi(0,\tau) = -1$ can be prescribed only if $b(\tau) \geq 0$. Let us denote by $B(\tau)$ the primitive function to $b(\tau)$, i.e. $B(\tau) = \ln \varrho(\tau) + (r - q)\tau$. Solving the equation $\partial_\tau \tilde\Pi + b(\tau)\partial_\xi \tilde\Pi = 0$ we obtain: $\tilde\Pi(\xi,\tau) = \Pi^{j-1}(\xi - B(\tau) + B(\tau_{j-1}))$ if $\xi - B(\tau) + B(\tau_{j-1}) > 0$ and $\tilde\Pi(\xi,\tau) = -1$ otherwise. Hence the full time-space approximation of the half-step solution $\Pi_i^{j-\frac{1}{2}}$ can be obtained from the formula

$$\Pi_i^{j-\frac{1}{2}} = \begin{cases} \Pi^{j-1}(\eta_i) & \text{if } \eta_i = \xi_i - \ln \varrho_j + \ln \varrho_{j-1} - (r-q)k > 0\,, \\ -1 & \text{otherwise.} \end{cases} \tag{91}$$

In order to compute the value $\Pi^{j-1}(\eta_i)$ we make use of a linear approximation between discrete values Π_i^{j-1}, $i = 0, 1, ..., n$.

Using central finite differences for approximation of the derivative $\partial_x \Pi^j$ we can approximate the diffusive part of a solution (90) as follows:

$$\frac{\Pi_i^j - \Pi_i^{j-\frac{1}{2}}}{k} + \left(r + \frac{1}{T-\tau_j}\right)\Pi_i^j \tag{92}$$
$$- \left(\frac{\sigma^2}{2} + \frac{\varrho^j e^{-\xi_i} - 1}{T-\tau_j}\right)\frac{\Pi_{i+1}^j - \Pi_{i-1}^j}{2h} - \frac{\sigma^2}{2}\frac{\Pi_{i+1}^j - 2\Pi_i^j + \Pi_{i-1}^j}{h^2} = 0 .$$

Therefore the vector of discrete values $\Pi^j = \{\Pi_i^j, i = 1, 2, ..., n\}$ at the time level $j \in \{1, 2, ..., m\}$ is a solution to a tridiagonal system of equations

$$\alpha_i^j \Pi_{i-1}^j + \beta_i^j \Pi_i^j + \gamma_i^j \Pi_{i+1}^j = \Pi_i^{j-\frac{1}{2}} \tag{93}$$

for $i = 1, 2, ..., n$, where

$$\begin{aligned}
\alpha_i^j &\equiv \alpha_i^j(\varrho^j) = \frac{k}{2h^2}\sigma^2 + \frac{k}{2h}\left(\frac{\sigma^2}{2} + \frac{\varrho^j e^{-\xi_i} - 1}{T-\tau_j}\right), \\
\gamma_i^j &\equiv \gamma_i^j(\varrho^j) = -\frac{k}{2h^2}\sigma^2 - \frac{k}{2h}\left(\frac{\sigma^2}{2} + \frac{\varrho^j e^{-\xi_i} - 1}{T-\tau_j}\right), \\
\beta_i^j &\equiv \beta_i^j(\varrho^j) = 1 + \left(r + \frac{1}{T-\tau_j}\right)k - (\alpha_i^j + \gamma_i^j) .
\end{aligned} \tag{94}$$

The initial and boundary conditions at $\tau = 0$ and $x = 0, L$, can be approximated as follows:

$$\Pi_i^0 = \begin{cases} -1 & \text{for } \xi_i < \ln\left((1+rT)/(1+qT)\right), \\ 0 & \text{for } \xi_i \geq \ln\left((1+rT)/(1+qT)\right), \end{cases}$$

for $i = 0, 1, ..., n$, and $\Pi_0^j = -E$, $\Pi_n^j = 0$. The equation for the free boundary position ϱ can be approximated by means of a finite difference approximation of $\partial_x \Pi$ at the origin $\xi = 0$ as follows:

$$\varrho^j = \frac{1 + r(T-\tau_j) + (T-\tau_j)\frac{\sigma^2}{2}\frac{\Pi_1^j - \Pi_0^j}{h}}{1 + q(T-\tau_j)} . \tag{95}$$

We formally rewrite discrete equations (91), (93) and (95) in the operator form:

$$\begin{aligned}
\varrho^j &= \mathcal{F}(\Pi^j), \\
\Pi^{j-\frac{1}{2}} &= \mathcal{T}(\Pi^j, \varrho^j), \\
\mathcal{A}(\varrho^j)\Pi^j &= \Pi^{j-\frac{1}{2}},
\end{aligned} \tag{96}$$

where $\mathcal{F}(\Pi^j)$ is the right-hand side of the algebraic equation (95), $\mathcal{T}(\Pi^j, \varrho^j)$ is the transport equation solver given by the right-hand side of (91) and $\mathcal{A} = \mathcal{A}(\varrho^j)$ is a tridiagonal matrix with coefficients given by (94). The system (96) can be approximately solved by means of successive iterates procedure. We define, for $j \geq 1$, $\Pi^{j,0} = \Pi^{j-1}$, $\varrho^{j,0} = \varrho^{j-1}$. Then the $(p+1)$-th approximation of Π^j and ϱ^j is obtained as a solution to the system:

$$\begin{aligned}
\varrho^{j,p+1} &= \mathcal{F}(\Pi^{j,p}), \\
\Pi^{j-\frac{1}{2},p+1} &= \mathcal{T}(\Pi^{j,p})\varrho^{j,p+1}), \\
\mathcal{A}(\varrho^{j,p+1})\Pi^{j,p+1} &= \Pi^{j-\frac{1}{2},p+1} .
\end{aligned} \tag{97}$$

Now, if the sequence of approximate discretized solutions $\{(\Pi^{j,p}, \varrho^{j,p})\}_{p=1}^{\infty}$ converges to some limiting value $(\Pi^{j,\infty}, \varrho^{j,\infty})$ as $p \to \infty$ then this limit is a solution to a nonlinear system of equations (96) at the time level j and we can proceed by computing the approximate solution in the next time level $j + 1$.

5.5. Computational Examples of the Free Boundary Approximation

We end this section with several computational examples documenting the capability of the new method for valuing early exercise boundary for American style of Asian call options with arithmetically averaged floating strike.

In Fig. 14 we show the behavior of the early exercise boundary function $\varrho(\tau)$ and the function $x_f(t) = \varrho(T - t)$. In these numerical experiments we chose $r = 0.06, q = 0.04, \sigma = 0.2$ and very long expiration time $T = 50$ years. These parameters correspond to the example presented in the preprint version of the paper [15] by Dai and Kwok. As far as other numerical parameters are concerned, we chose the mesh of $n = 100$ spatial grid points and we considered the number of time steps $m = 2 \times 10^6$ in other to achieve very fine time stepping corresponding to 13 minutes between consecutive time steps when expressed in the original time scale of the problem. In order to make a graphical comparison of the state variable $x = S/A$ to its reciprocal value A/S considered in [15] we also plot the function $1/x_f(t)$.

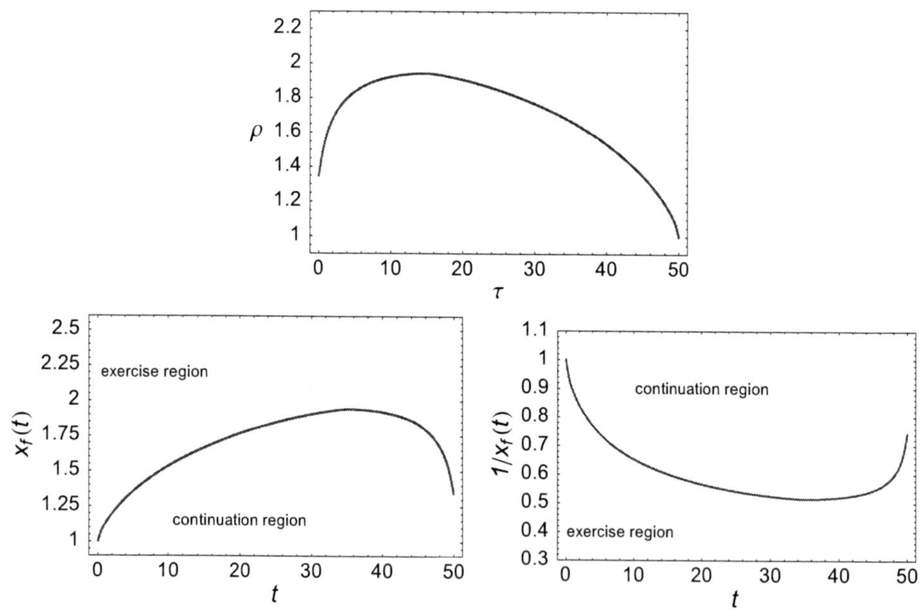

Figure 14. The function $\varrho(\tau)$ (above) and the free boundary position $x_f(t) = \varrho(T - t)$ (bottom-right) and the plot of the function $1/x_f(t) = 1/\varrho(T - t)$ (bottom-left).

In Fig. 15 we can see the behavior of the transformed function Π in both 3D as well as contour plot perspectives. Fig. 16 depicts the initial condition $\Pi(\xi, 0)$ and five time steps of the function $\xi \mapsto \Pi(\xi, \tau_j)$ for $\tau_j = 0.1, 1, 5, 25, 50$. A comparison of the free boundary

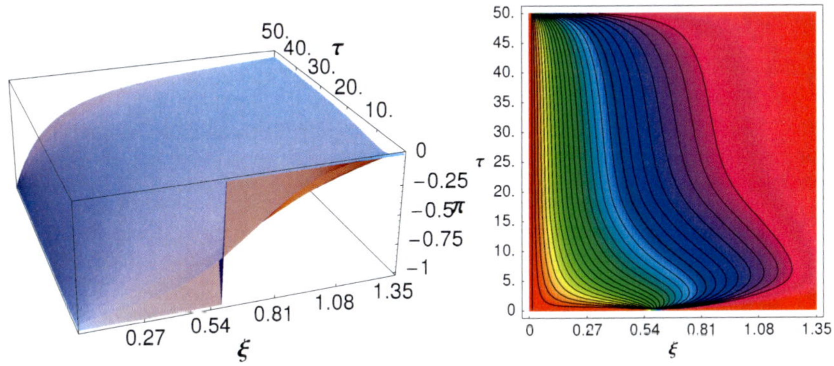

Figure 15. A 3D plot (left) and contour plot (right) of the function $\Pi(\xi, \tau)$.

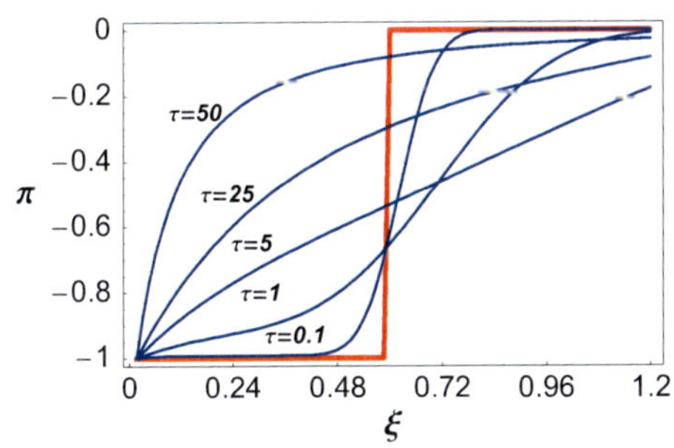

Figure 16. Profiles of the function $\Pi(\xi, \tau)$ for various times $\tau \in [0, T]$.

position $x_f(t) = \varrho(T-t)$ as well as of its reciprocal value $1/x_f(t) = 1/\varrho(T-t)$ obtained by our method (solid curve) and that of the projected successive over relaxation algorithm from [15] (dashed curve) is shown in Fig. 17. It is clear that our method and that of [15] give almost the same result in the one third of the time interval $[0, T]$ close to the expiration time $T = 50$. On the other hand, the long time behavior when $\tau \to T$ is quite different. Notice that $\lim_{\tau \to T} \varrho(\tau)$ can be analytically computed and is equal to 1 (see [28] or [48]). Hence the long time behavior of ϱ seems to be better approximated by our method. A comparison of early exercise profiles with respect to varying dividend rate q is shown in Fig. 18.

Finally, we present numerical experiments for shorter expiration times $T = 0.5833$ (seven months) and $T = 1$ (one year) with zero dividend rate $q = 0$ and $r = 0.05, \sigma = 0.2$.

Conclusion

In this survey chapter we presented recent developments in fixed domain transformation methods applied to evaluation of the early exercise boundary for American style of options. We discussed an iterative numerical scheme for approximating of the early exercise boundary for a class of Black–Scholes equations with a volatility which may depended on the asset price as well as the second derivative of the option price. The method consisted of transformation the free boundary problem for the early exercise boundary position into a so-

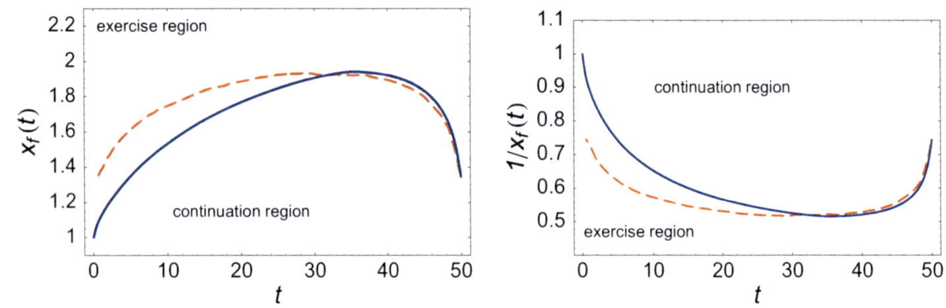

Figure 17. A comparison of the free boundary position $x_f(t) = \varrho(T-t)$ (left) and $1/x_f(t) = 1/\varrho(T-t)$ (right) obtained by our method (solid curve) and that of the projected successive over relaxation algorithm from [15] (dashed curve).

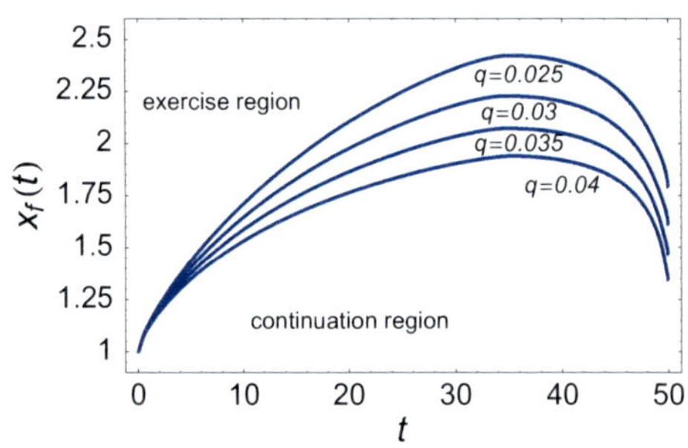

Figure 18. A comparison of the free boundary position $x_f(t) = \varrho(T-t)$ for various dividend yield rates $q = 0.04, 0.035, 0.3, 0.25$.

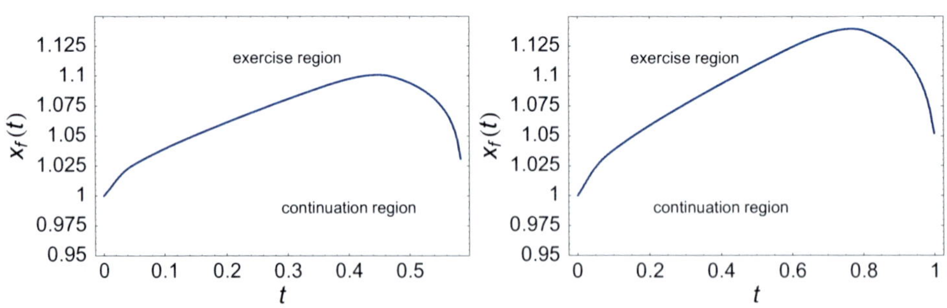

Figure 19. The free boundary position $x_f(t) = \varrho(T-t)$ for Asian call on assets paying no dividends ($q = 0$) with expiration time $T = 0.7$ (left) and $T = 1$ (right).

lution of a nonlinear parabolic equation and a nonlinear algebraic constraint equation. The transformed problem has been solved by means of operator splitting iterative technique. We also presented results of numerical approximation of the free boundary for several nonlinear Black–Scholes equation including, in particular, Barles and Soner model and the Risk adjusted pricing methodology model. The method of fixed domain transformation has been also applied for evaluation of early exercise boundary for American style of Asian option with arithmetically averaged strike price.

Acknowledgements

The author thanks Matthias Ehrhardt for fruitful discussions and his encouragement to write this survey chapter. I also appreciate help of my student B. Kucharčík with preparation of the last section. The work was supported by grants VEGA 1/3767/06 and DAAD-MSSR-11/2006.

References

[1] G. Alobaidi, R. Mallier and S. Deakin, Laplace transforms and installment options, *Math. Models & Methods in Appl. Science* **18** (2004), 1167–1189.

[2] J. Ankudinova and M. Ehrhardt, On the numerical solution of nonlinear Black-Scholes equations, to appear in: *Computers and Mathematics with Applications* (2008).

[3] J. Ankudinova and M. Ehrhardt, *Fixed domain transformations and highly accurate compact schemes for Nonlinear Black-Scholes equations for American Options*, in M. Ehrhardt (ed.), Nonlinear Models in Mathematical Finance: New Research Trends in Option Pricing, Nova Science Publishers, Inc., Hauppauge, to appear fall 2008.

[4] M. Avellaneda and A. Paras, Dynamic Hedging Portfolios for Derivative Securities in the Presence of Large Transaction Costs, *Applied Mathematical Finance,* **1** (1994) 165–193.

[5] M. Avellaneda, A. Levy and A. Paras, Pricing and hedging derivative securities in markets with uncertain volatilities, *Applied Mathematical Finance* **2** (1995) 73–88.

[6] B. Barone-Adesi and R.E. Whaley, Efficient analytic approximations of American option values, *J. Finance* **42** (1987) 301–320.

[7] F. Black and M. Scholes, The pricing of options and corporate liabilities, *J. Political Economy* **81** (1973) 637–654.

[8] G. Barles and H.M. Soner, Option Pricing with transaction costs and a nonlinear Black–Scholes equation, *Finance Stochast.,* **2** (1998) 369-397.

[9] L.A. Bordag and A.Y. Chmakova, Explicit solutions for a nonlinear model of financial derivatives, *International Journal of Theoretical and Applied finance* **10** (2007) 1–21.

[10] L.A. Bordag and R. Frey, *Nonlinear option pricing models for illiquid markets: scaling properties and explicit solutions*, arxiv.org/abs/0708.1568 (2007)

[11] P. Carr, R. Jarrow and R. Myneni, Alternative characterizations of American put options, *Mathematical Finance* **2** (1992) 87–105

[12] J. Chadam, *Free Boundary Problems in Mathematical Finance*, Progress in Industrial Mathematics at ECMI 2006, Vol 12, Springer Berlin Heidelberg, 2008.

[13] Xinfu Chen, J. Chadam, Lishang Jiang and Weian Zheng, Convexity of the Exercise Boundary of the American Put Option on a Zero Dividend Asset, *Mathematical Finance* **2008** (2008) 185–197.

[14] C.K. Cho, S. Kang, T. Kim and Y. Kwon, Parameter estimation approach to the free boundary for the pricing of an American call option, *Computers and Mathematics with applications* **51** (2006) 713–720.

[15] M. Dai, Y.K. Kwok, Characterization of optimal stopping regions of American Asian and lookback options, *Mathematical Finance* **16** (2006) 63–82.

[16] J.N. Dewynne, S.D. Howison, J. Rupf and P. Wilmott, Some mathematical results in the pricing of American options, *Euro. J. Appl. Math.* **4** (1993) 381–398.

[17] B. During, M. Fournier and A. Jungel, High order compact finite difference schemes for a nonlinear Black–Scholes equation, *Int. J. Appl. Theor. Finance* **7** (2003) 767–789.

[18] M. Ehrhardt and P. Mickens, A fast, stable and accurate numerical method for the Black-Scholes equation of American options, to appear in: *International Journal of Theoretical and Applied Finance.*

[19] E. Ekström and J. Tysk, The American put is log-concave in the log-price, *Journal of Mathematical Analysis and Appl.* **314**(2) (2006) 710–723.

[20] E. Ekström, Convexity of the optimal stopping boundary for the American put option, *Journal of Mathematical Analysis and Appl.* **299**(1) (2004) 147–156.

[21] J.D. Evans, R. Kuske and J.B. Keller, American options on assets with dividends near expiry, *Mathematical Finance* **12** (2002) 219–237.

[22] R. Frey and P. Patie, Risk Management for Derivatives in Illiquid Markets: A Simulation Study, *Advances in Finance and Stochastics*, Springer, Berlin, (2002) 137–159.

[23] R. Frey, and A. Stremme Market Volatility and Feedback Effects from Dynamic Hedging, *Mathematical Finance* **4** (1997) 351–374.

[24] R. Geske and H.E. Johnson, The American put option valued analytically, *J. Finance* **39** (1984) 1511–1524.

[25] R. Geske and R. Roll, On valuing American call options with the Black–Scholes European formula, *J. Finance* **89** (1984) 443–455.

[26] P. Grandits and W. Schachinger, Leland's approach to option pricing: the evolution of a discontinuity, *Mathematical Finance* **11**(3) (2001) 347–355.

[27] S. Gripenberg S.O. London and O. Staffans, *Volterra Integral and Functional Equations*, Cambridge University Press, 1990.

[28] A.T. Hansen and P.L. Jörgensen, Analytical Valuation of American-Style Asian Optionsl, *Management Science* **46** (2000) 1116–1136.

[29] T. Hoggard, A.E. Whalley and P. Wilmott, Hedging option portfolios in the presence of transaction costs, *Advances in Futures and Options Research* **7** (1994) 21–35.

[30] J. Hull, *Options, Futures and Other Derivative Securities*, third edition, Prentice-Hall, 1997.

[31] H. Imai, N. Ishimura, H. Sagakuchi, Computational technique for treating the nonlinear Black-Scholes equation with the effect of transactional costs, *Kybernetika* **43** 2007 807–816.

[32] M. Jandačka and D. Ševčovič, On the risk adjusted pricing methodology based valuation of vanilla options and explanation of the volatility smile, *Journal of Applied Mathematics* **3** (2005) 235–258.

[33] H. Johnson, An analytic approximation of the American put price, *J. Finan. Quant. Anal.* **18** (1983) 141–148.

[34] C. Knessl, A note on a moving boundary problem arising in the American put option, *Studies in Applied Mathematics* (107) (2001) 157–183.

[35] I. Karatzas, On the pricing American options, *Appl. Math. Optim.* **17** (1988) 37–60.

[36] R.A. Kuske and J.B. Keller, Optimal exercise boundary for an American put option, *Applied Mathematical Finance* **5** (1998) 107–116.

[37] M. Kratka, No Mystery Behind the Smile, *Risk* **9** (1998) 67–71.

[38] Y.K Kwok, *Mathematical Models of Financial Derivatives*, Springer-Verlag, 1998.

[39] H.E. Leland, Option pricing and replication with transaction costs, *Journal of Finance* **40** (1985) 1283–1301.

[40] L.W. MacMillan, Analytic approximation for the American put option, *Adv. in Futures Options Res.* **1** (1986) 119–134.

[41] R. Mallier, Evaluating approximations for the American put option, *Journal of Applied Mathematics* **2** (2002) 71–92.

[42] R. Mallier and G. Alobaidi, The American put option close to expiry, *Acta Mathematica Univ. Comenianae* **73** (2004) 161–174.

[43] G.N. Milstein, O. Reiss and J. Schoenmakers, A new Monte Carlo method for American Options, *International Journal of Theoretical and Applied Finance* **7** (2004) 591–614.

[44] R. Mynemi, The pricing of the American option, *Annal. Appl. Probab.* **2** (1992) 1–23.

[45] R. Roll, An analytic valuation formula for unprotected American call options on stock with known dividends, *J. Finan. Economy* **5** (1977) 251–258.

[46] D. Ševčovič, Analysis of the free boundary for the pricing of an American call option, *Euro. Journal on Applied Mathematics* **12** (2001) 25–37.

[47] D. Ševčovič, An iterative algorithm for evaluating approximations to the optimal exercise boundary for a nonlinear Black-Scholes equation, *Canad. Appl. Math. Quarterly* **15** (2007) 77–97.

[48] D. Ševčovič, *Early exercise boundary for Asian options*, in preparation.

[49] P. Schönbucher and P. Wilmott, The feedback-effect of hedging in illiquid markets, *SIAM Journal of Applied Mathematics* **61** (2000) 232–272.

[50] R. Stamicar, D. Ševčovič and J. Chadam, The early exercise boundary for the American put near expiry: numerical approximation, *Canad. Appl. Math. Quarterly* **7** (1999) 427–444.

[51] E.M. Stein and G. Weiss, *Introduction to Fourier Analysis on Euclidean Spaces*, Princeton University Press, 1971.

[52] M. Widdicks, P.W. Duck, A.D. Andricopoulos and D.P. Newton, The Black-Scholes equation revisited: Asyptotic expansions and singular perturbations, *Mathematical Finance* **15** (2005) 373–391.

[53] P. Wilmott, J. Dewynne and S.D. Howison, *Option Pricing: Mathematical Models and Computation*, UK: Oxford Financial Press, 1995.

[54] Jichao Zhao, R.M. Corless and M. Davison, Compact finite difference method for American option pricing, *Journal of Computational and Applied Mathematics* **206** (2007) 306–321.

[55] S.P. Zhu, A new analytical approximation formula for the optimal exercise boundary of American put options, *International Journal of Theoretical and Applied Finance* **9** (2006) 1141–1177.

[56] S.P. Zhu, Calculating the early exercise boundary of American put options with an approximation formula, *International Journal of Theoretical and Applied Finance* **10** (2007) 1203–1227.

In: Nonlinear Models in Mathematical Finance...
Editor: Matthias Ehrhardt, pp. 219-242
ISBN 978-1-60456-931-5

Chapter 7

Global in Space Numerical Computation for the Nonlinear Black-Scholes Equation

Naoyuki Ishimura[1] ***and Hitoshi Imai***[2]
[1]Graduate School of Economics, Hitotsubashi University,
Kunitachi, Tokyo 186-8601, Japan
[2] Institute of Technology and Science, The University of Tokushima,
Tokushima 770-8506, Japan

Abstract

Numerical treatment of the nonlinear Black-Scholes equation, which is global in the space variable, is discussed. Exploiting suitable transformation of variables, we do not make any truncation nor approximation of unbounded domains of the equation. Numerical implementation shows that our method and scheme are well formulated. As model problems we deal with the Black-Scholes equation with the effect of transaction costs, which is known to become nonlinear. Moreover, we apply our machinery to other singular nonlinear partial differential equation arising in the optimal investment problem.

1. Introduction

Like other areas of mathematical sciences, nonlinearities are known to be commonly observed also in mathematical finance. Just we recall several examples, although they are far from complete: the free boundary problem for the pricing of plain vanilla American option, various types of nonlinear Black-Scholes equations, a sophisticated modeling of optimal behaviors based on the stochastic control theory, and so on. To investigate these topics further in details, analytical manner of reasoning alone has a limitation; it is relatively exceptional to have analytically explicit solution such as the famous Black-Scholes pricing formula. Numerical treatment is therefore an important as well as an indispensable items; many excellent numerical schemes have been already introduced and much progress has been made so far. For basic materials in this aspect we refer for example to a nice monograph of Seydel [31] and/or to Part Seven of Wilmott [33].

Numerical computation cannot help approximations by nature; even to compute a differential quotient we must appeal to a discretized difference counterpart. Furthermore, to handle unboundedness, to be specific with an example, if the equation is defined on an unbounded domain, we must make certain truncation of the domain; on a bounded domain we should treat the equation, since machine can not realize the infinity. It is sure that in many cases these approximations can be carried out so that the resulting error is controlled within prescribed limit; indeed a large family of solutions to many kind of diffusion equations converge exponentially as the space variable tends to infinity. However, it is at the same time desirable to avoid unnecessary approximations. If there is a way to skip this process, we moderately think we had better advance such method and the outcomes is worth publishing.

In this article we are concerned with the numerical treatment of the nonlinear Black-Scholes equation as it is, which incorporates the effect of transaction costs. The formulation is then applied to the evolution equation for the risk preference in the optimal investment problem. These equations are defined on unbounded domains and solutions possibly become unbounded; as explained above we deal with such unboundedness directly and endeavor not to make any approximation in the numerical implementation with this respect. The word "global in space" in the title designates this principle. Of course there may arise objections against these attempts, we believe that our methodology has a reason to exist at least from the academic point of view. Moreover, additionally from the practical point of view, we point out that numerical examples indicate that our scheme is robust enough and works fairly well.

Most parts of ingredients are taken from our previous works with necessary modifications and arrangements [1, 17, 18, 19, 20, 21, 23]; on the standard criterion of research we do not claim the novelty of the current presentation. Analytical proofs of the theorems are omitted. Numerical examples, on the other hand, are newly replaced. We hope that the present article has an easy access to the readers.

The organization of this chapter is as follows. In §2 we recall background issues of the model on transaction costs which is selected according to our purposes. The numerical treatment of the nonlinear Black-Scholes equation in the presence of transaction costs is then exhibited in §3. The whole §4 is devotd to the optimal investment problem, with the stress on the evolution equation for the risk preference. The last section concludes.

2. Transaction Costs Model

2.1. Transaction Costs

In addition to the basic log-normal model for the asset price, it is well conceded that the celebrated Black-Scholes partial differential equation [4, 28] is derived upon several ideal assumptions. The list of these includes, for example [3, 15], the absence of arbitrage opportunities, the possibility of continuous self-financing trading of the underlying asset, the constant risk-less interest rate and the stock volatility, and so on. The transaction costs associated with trading is also excluded in the original Black-Scholes analysis, which, however, is invalid in general; the influence of transaction costs is actually very important for practitioners. Much attention has been paid so far on the effects of transaction costs and many substantial researches have been undertaken. We refer to [5, 8, 12, 22, 25, 32, 35] for instance and the references cited therein. Among others Leland [25] made a pioneering investigation, whose central idea is to modify the dynamic hedging so that transaction costs are considered to be charged to rehedge over a short discrete interval of time. The approach is nowadays classified into the so-called local in time hedging strategy.

In [20], we are concerned with the analytical solvability and a weak formulation of non-linear partial differential equations (PDEs) of Black-Scholes type of the next form, which interpretes the essence of Leland [25] in the PDE setting.

$$\begin{aligned} \frac{\partial V}{\partial t} + \frac{1}{2}\sigma^2 S^2 \frac{\partial^2 V}{\partial S^2} + rS\frac{\partial V}{\partial S} - rV = \kappa F\Big(S^2\Big|\frac{\partial^2 V}{\partial S^2}\Big|\Big), \\ V = V(S,t) \quad \text{in } (S,t) \in (0,\infty) \times (0,T) \\ V(S,T) = V_0(S) \qquad \text{for } S \geq 0, \end{aligned} \tag{1}$$

where $V(S,t)$ denotes the option price written on the underlying asset S. The constants r and σ stand for the risk-free interest rate and the asset volatility, respectively. The maturity data $V_0(S)$ (≥ 0) is assumed to enjoy $V_0'(S) \to \alpha$ exponentially as $S \to \infty$ with a nonnegative constant α. For simplicity we assume $V_0(0) = V(0,t) = 0$ hereafter. That is to say, we confine ourselves to treating the European call type options. The right hand side κF expresses the cost term where κ $(\kappa > 0)$ is a proportional constant. It is clear that if $\kappa F \equiv 0$ then (1) reduces to the original Black-Scholes equation. Given smooth function $F(Q)$ $(Q := S^2|\Gamma|)$, which depends on the absolute value of the option gamma $\Gamma := \partial^2 V/\partial S^2$ multiplied by the square of the asset price, is assumed to satisfy the following hypotheses.

(H1) $F(Q)$ is nonnegative, $F(0) = 0$, and uniformly C^3-class for $Q \in (0,\infty)$.
(H2) There exists a positive constant M such that

$$\Big|\frac{\partial F}{\partial Q}(Q)\Big| \leq M \qquad \text{for } Q \in (0,\infty).$$

Under the further assumption that $\kappa < \sigma^2/2M$ we have proved in [20] that (1) has a solution $V(S,t)$ with $V'(S,\cdot) \to \alpha$ exponentially as $S \to \infty$ both in strong and in weak sense. We merely recall the weak existence theorem. To do so we introduce Hilbert spaces.

$$\begin{aligned} E_0 &:= \{V \in L^1_{\text{loc}}(0,\infty) \mid \|V\|^2_{E_0} := \int_0^\infty V(S)^2 \frac{dS}{S} < \infty\} \\ E_1 &:= \{V \in E_0 \mid \|V\|^2_{E_1} := \|V\|^2_{E_0} + \int_0^\infty \Big(S\frac{\partial V(S)}{\partial S}\Big)^2 \frac{dS}{S} < \infty\} \\ E_{-1} &:= E_1^* \text{ the dual space of } E_1 \text{ with respect to the inner product of } E_0. \end{aligned}$$

Furthermore, we additionally introduce

$$E_1^0 := \{V \in L^1_{\text{loc}}(0,\infty) \mid S\frac{\partial V}{\partial S} \in E_0\}.$$

It is easy to see that E_0, E_1, E_{-1} are respectively equivalent to $L^2(\boldsymbol{R})$, $H^1(\boldsymbol{R})$, $H^{-1}(\boldsymbol{R})$ via the change of variable $u(x) = V(e^x) = V(S)$ $(x = \log S \in \boldsymbol{R})$.

In terms of these spaces we describe the definition of our weak solutions.

Definition 2.1 *We say $V \in C^1(0,T;E_1^0)$ a weak solution of (1) with maturity data $V_0 \in E_1^0$ if $V(S,t)$ is a solution of the linear partial differential equation*

$$\frac{\partial V}{\partial t} + \frac{1}{2}\sigma^2 S^2 \frac{\partial^2 V}{\partial S^2} + rS\frac{\partial V}{\partial S} - rV = \kappa F\Big(e^{-r(T-t)}\Big|S\frac{\partial W}{\partial S} - W\Big|\Big)$$
$$\text{in } (S,t) \in (0,\infty)\times(0,T)$$
$$V(S,T) = V_0(S) \quad \text{for } S \geq 0,$$

where $W(S,t)$ satisfies the next conditions.

(W1) $W \in L^\infty(0,T;E_0) \cap L^2(0,T;E_1)$, $\partial W/\partial t \in L^2(0,T;E_{-1})$

(W2) There holds for each $Z \in E_1$ and almost every $0 \leq t \leq T$ that

$$\int_0^\infty \Big(\frac{\partial W}{\partial t} + rS\frac{\partial W}{\partial S}\Big) Z \frac{dS}{S}$$
$$= \int_0^\infty \Big\{\frac{\sigma^2}{2}\Big(S\frac{\partial W}{\partial S} - W\Big) + \kappa e^{r(T-t)} F\Big(e^{-r(T-t)}\Big|S\frac{\partial W}{\partial S} - W\Big|\Big)\Big\} S\frac{\partial Z}{\partial S}\frac{dS}{S}.$$

(W3) $W(S,T) = S(\partial V_0/\partial S)(S)$ *in* E_0.

An inspection reveals that $W(S,t) = e^{r(T-t)}S(\partial V/\partial S)(S,t)$ should be the relevant relation.

One of the achievements of [20] is then addressed as follows.

Theorem 2.2 *Assume (H1)(H2) and suppose $\kappa < \sigma^2/2M$. Then for any $V_0 - \alpha S \in E_1^0$ $(\alpha \geq 0)$, there exists a solution $V(S,t)$ of (1) such that $V - \alpha S$ gives a weak solution in the sense of Definition 2.1.*

It is to be noted that for $V_0 \in E_1^0$ there holds $S(\partial V_0/\partial S) \to 0$ as $S \to \infty$, which apparently restricts the behavior of maturity data. However, since the linear function αS $(\alpha \in \boldsymbol{R})$ solves the equation (1), the existence of solutions should be understood up to the addition of linear functions. In particular, the maturity data $V_0(S) = \max\{S - E, 0\}$ of vanilla call options with exercise E enter our regime.

Typical example of (1) is proposed by Hoggard, Whalley, and Wilmott [13] in which $F(Q) = \sigma\sqrt{2/\pi\delta t}Q$. To be precise

$$\frac{\partial V}{\partial t} + \frac{1}{2}\sigma^2 S^2 \frac{\partial^2 V}{\partial S^2} + rS\frac{\partial V}{\partial S} - rV = \kappa\sigma S^2\sqrt{\frac{2}{\pi\delta t}}\left|\frac{\partial^2 V}{\partial S^2}\right| \quad \text{in } (S,t) \in (0,\infty)\times(0,T)$$
$$V(S,T) = V_0(S) \qquad \text{for } S \geq 0, \tag{2}$$

where δt is a non-infinitesimal fixed time-step not to be taken $\delta t \to 0$; the portfolio is considered to be revised every δt. This model claims that the transaction costs are charged proportionally to the value traded. In the next subsection we briefly review on the model and the derivation of this equation. We just mention that the equation (2) is claimed as one of the first nonlinear PDEs in finance.

2.2. Leland Model

Here we quickly make a sketch of the model of transaction costs whose idea originates from Leland [25] and of the derivation of corresponding partial differential equations. Although these are rather well known to the specialists, we reproduce the argument of [13, 33, 34], for completeness of our exposition and for the readers' convenience. We refer to above mentioned papers for further details.

We assume that the underlying asset price follows the random walk which is given in discrete time-step δt by

$$\delta S = \sigma S \phi \sqrt{\delta t} + \mu S \delta t,$$

where μ denotes the drift coefficient and ϕ is a random number drawn from the standardized normal distribution. Specifically we understand that when we take the expectation the probability density f is

$$f(\phi) = \frac{1}{\sqrt{2\pi}} \exp\Big[-\frac{\phi^2}{2}\Big]. \tag{3}$$

We further note that because of the discrete world, we return to the original Bachelier-Einstein law of the square root of time.

The reason why we introduce a non-infinitesimal fixed time-step δt is that if the costs associated with trading are independent of the time-scale of rehedging then infinite total transaction costs would be resulted in as we take the limit $\delta t \to 0$. Leland therefore has proposed a modification to the usual Black-Scholes continuous analysis so that the portfolio is now assumed to be revised every discrete δt.

Let $\Pi = V - \Delta S$ denote the value of the portfolio with typical delta hedging strategy; Δ denotes the number of shares S. We examine the change $\delta\Pi$ in Π over a discrete time-step δt. In the computation we need to involve the term which represents the effect of costs. For simplicity it takes the form $\kappa S|\nu|$, where ν is the number of shares which are bought ($\nu > 0$) or sold ($\nu < 0$) at a price S. This form means that a cost is proportional to the value traded with a constant κ depending on the individual investor. It is to be noted that the resulting partial differential equations will then be reduced to (2).

Now we derive

$$\delta\Pi = \sigma S\Big(\frac{\partial V}{\partial S} - \Delta\Big)\phi\sqrt{\delta t} + \Big(\frac{\partial V}{\partial t} + \frac{1}{2}\sigma^2 S^2 \frac{\partial^2 V}{\partial S^2}\phi^2 + \mu S \frac{\partial V}{\partial S} - \mu \Delta S\Big)\delta t - \kappa S|\nu|.$$

Taking into account that the transaction costs are always positive we have subtracted off $\kappa S|\nu|$ from $\delta\Pi$. Furthermore the square of the random variable ϕ should be saved in this discrete time world.

We follow the same hedging strategy as Black-Scholes concept and thus choose $\Delta = (\partial V/\partial S)(S,t)$, which has been evaluated at time t and asset S.

We compute the quantity ν. Since the number of shares held is provided by $\Delta(S,t)$, the number of shares traded after a time-step δt is $\Delta(S+\delta S, t+\delta t)$ and hence

$$\nu = \Delta(S+\delta S, t+\delta t) - \Delta(S,t) = \frac{\partial V}{\partial S}(S+\delta S, t+\delta t) - \frac{\partial V}{\partial S}(S,t).$$

We apply Taylor expansion to the right hand side for small δt and δS; it follows that

$$\nu = \frac{\partial^2 V}{\partial S^2}(S,t)\sigma S \phi \sqrt{\delta t} + O(\delta t).$$

To leading order the expected transaction cost in a time-step δt then turns out to be

$$E[\kappa S|\nu|] = \kappa\sigma S^2\sqrt{\frac{2}{\pi}}\left|\frac{\partial^2 V}{\partial S^2}\right|\sqrt{\delta t}.$$

Consequently with the above choice of Δ the expected change in the value of the portfolio becomes

$$E[\delta\Pi] = \left(\frac{\partial V}{\partial t} + \frac{1}{2}\sigma^2 S^2\frac{\partial^2 V}{\partial S^2} - \kappa\sigma S^2\sqrt{\frac{2}{\pi\delta t}}\left|\frac{\partial^2 V}{\partial S^2}\right|\right)\delta t,$$

which equals to the return from a risk-free deposit by the non-arbitrage principle. Precisely stated $E[\delta\Pi] = r(V - \Delta S)\delta t = r(V - S\partial V/\partial S)\delta t$. We thus obtain

$$\frac{\partial V}{\partial t} + \frac{1}{2}\sigma^2 S^2\frac{\partial^2 V}{\partial S^2} - \kappa\sigma S^2\sqrt{\frac{2}{\pi\delta t}}\left|\frac{\partial^2 V}{\partial S^2}\right| = r\left(V - S\frac{\partial V}{\partial S}\right).$$

This is the Hoggard-Whalley-Wilmott equation (2).

2.3. Generalized Models

The nonlinear function F in (1), which reflects the presence of transaction costs, is assumed to depend solely on $Q = S^2|\Gamma| = S^2|\partial^2 V/\partial S^2|$. There may be a concern that this is special. To reply such question we analyze other existing models, which at the same time shows that even slightly general form than (1) is also possible. We refer to [9, 33] for the background of these models. The first one is the so-called extended Leland model. This is

$$F(S,|\Gamma|) = \frac{\kappa_1}{\delta t} + \sigma(\kappa_2 + \kappa_3 S)S\sqrt{\frac{2}{\pi\delta t}}|\Gamma| = \frac{\kappa_1}{\delta t} + \sigma\left(\frac{\kappa_2}{S} + \kappa_3\right)\sqrt{\frac{2}{\pi\delta t}}Q, \tag{4}$$

where κ_1, κ_2 and κ_3 components describe a fixed cost component, a cost proportional to the number traded, and a cost proportional to the value traded, respectively. The apparent singularity $1/S$ appears if we extract Q variable. We remark that (4) verifies hypotheses (H1)(H2) provided $\kappa_2 = 0$. There seems no known existence result for the full model of (4). In recent work [26], Liu gives a nice numerical study on this extended model.

Another one is market practice model. This is

$$F(S,|\Gamma|) = \frac{\sigma^2 S^4}{H_0}\left(\kappa_1 + (\kappa_2 + \kappa_3 S)\frac{\sqrt{H_0}}{S}\right)|\Gamma|^2 = \frac{\sigma^2 Q^2}{H_0}\left(\kappa_1 + (\kappa_2 + \kappa_3 S)\frac{\sqrt{H_0}}{S}\right), \tag{5}$$

where H_0 is a function of S, t as well as V and its derivatives. The parameters κ_1, κ_2 and κ_3 have the same meaning as for (4). The fulfillment of hypotheses (H1)(H2) depends on the choice of H_0. If $H_0 \equiv$ constant, then (H1)(H2) is verified provided $\kappa_2 = 0$ and additionally Q stays bounded; the last condition is a restrictive requirement. If $H_0 \equiv \sigma^2\delta t S^4|\Gamma|^2$, then (5) reduces to (4) under an adjustment of parameters.

These two equations imply that the form (1) is not intended just for the mathematical sophistication. We further note that if F depends explicitly on S as in (4)(5) the mathematical procedure of handling the partial differential equation as in [20] would be harder.

Last but not least we consider a new model [23].

$$F(Q) = \kappa\sigma\sqrt{\frac{2}{\pi\delta t}}\sqrt{|S^2|\Gamma|(\beta - S^2|\Gamma|)|} = \kappa\sigma\sqrt{\frac{2}{\pi\delta t}}\sqrt{|Q(\beta - Q)|}. \tag{6}$$

That is, the transaction costs proportional to the value traded is of the so-called logistic type, which may cause the saturation effect.

3. Global in Space Computation

3.1. Scheme

Now we explain our scheme in order to numerically compute the PDE (1). To begin with we make the time inversion $t \mapsto T - t$ to consider the next initial boundary value problem.

$$\begin{gathered}\frac{\partial V_2}{\partial t} = \frac{1}{2}\sigma^2 S^2 \frac{\partial^2 V_2}{\partial S^2} - \kappa F\left(S^2 \left|\frac{\partial^2 V_2}{\partial S^2}\right|\right) + rS\frac{\partial V_2}{\partial S} - rV_2 \text{ in } (S,t) \in (0,\infty)\times(0,T) \\ V_2(S,0) = V_0(S) \quad \text{for } S \geq 0, \qquad V_2(0,t) = 0 \quad \text{for } 0 \leq t < T,\end{gathered} \tag{7}$$

where $V_0(S) \geq 0$ with $V_0(0) = 0$, and $V_0'(S) \simeq \alpha$ exponentially as $S \to \infty$ with a nonnegative constant α. The original $V(S,t)$ is recovered by the formula $V(S,t) = V_2(S, T-t)$.

To handle the unboundedness of the problem we perform transformations on (7), which are divided into two steps.

Step 1. Transformation of V_2.

We put

$$V_3(S,t) := V_2(S,t) - \alpha S.$$

It is easy to check that V_3 is bounded as $S \to \infty$ and satisfy

$$\begin{gathered}\frac{\partial V_3}{\partial t} = \frac{1}{2}\sigma^2 S^2 \frac{\partial^2 V_3}{\partial S^2} - \kappa F\left(S^2 \left|\frac{\partial^2 V_3}{\partial S^2}\right|\right) + rS\frac{\partial V_3}{\partial S} - rV_3 \text{ in } (S,t) \in (0,\infty)\times(0,T) \\ V_3(S,0) = V_0(S) - \alpha S \quad \text{for } S \geq 0, \qquad V_3(0,t) = 0 \quad \text{for } 0 \leq t < T.\end{gathered}$$

Step 2. Change of space variable.

We introduce $S = x/(1-x^2)$ (see Imai [16]). Then it is easy to see that $\{S \geq 0\}$ corresponds to $\{0 \leq x < 1\}$ and $x = 2S/(1+\sqrt{1+4S^2})$. We further define

$$u_3(x,t) := V_3(x/(1-x^2), t).$$

After a little tedious calculation we find that

$$\begin{aligned}
\frac{\partial u_3}{\partial t} &= \frac{1}{2}\sigma^2 \frac{x^2(1-x^2)^2}{(1+x^2)^2}\frac{\partial^2 u_3}{\partial x^2} \\
&- \kappa F\Big(\frac{x^2(1-x^2)}{(1+x^2)^3}\Big|(1-x^2)(1+x^2)\frac{\partial^2 u_3}{\partial x^2} - 2x(3+x^2)\frac{\partial u_3}{\partial x}\Big|\Big) \\
&+ \frac{x(1-x^2)}{(1+x^2)^3}\{r(1+x^2)^2 - \sigma^2 x^2(3+x^2)\}\frac{\partial u_3}{\partial x} - ru_3 \text{ in } (x,t) \in (0,1)\times(0,T) \\
u_3(x,0) &= V_0(x/(1-x^2)) - \frac{\alpha x}{1-x^2} \quad \text{for } 0 < x < 1, \\
u_3(0,t) &= u_3(1,t) = 0 \quad \text{for } 0 \le t < T.
\end{aligned} \tag{8}$$

This is our procedure.

A different type of transaformations is also possible. We introduce it below, which is valid only if $\alpha \neq 0$.

Step 1'. We put

$$V_4(S,t) := 1 - \frac{V_2(S,t)}{\alpha(\varepsilon + S)} \qquad \text{where } \varepsilon > 0.$$

It is easy to see that V_4 is also bounded as $S \to \infty$ and verify

$$\begin{aligned}
\frac{\partial V_4}{\partial t} &= \frac{1}{2}\sigma^2 S^2 \frac{\partial^2 V_4}{\partial S^2} + \frac{\kappa}{\alpha(\varepsilon+S)}F\Big(\alpha S^2\Big|(\varepsilon+S)\frac{\partial^2 V_4}{\partial S^2} + 2\frac{\partial V_4}{\partial S}\Big|\Big) \\
&+ \Big\{\frac{\sigma^2 S^2}{\varepsilon + S} + rS\Big\}\frac{\partial V_4}{\partial S} + \frac{r\varepsilon(1-V_4)}{\varepsilon+S} \quad \text{in } (S,t) \in (0,\infty)\times(0,T) \\
V_4(S,0) &= 1 - \frac{V_0(S)}{\alpha(\varepsilon+S)} \quad \text{for } S \ge 0, \qquad V_4(0,t) = 1 \quad \text{for } 0 \le t < T.
\end{aligned}$$

Step 2'. This step is the same. We define

$$u_4(x,t) := V_4(x/(1-x^2), t).$$

Then we discover

$$\begin{aligned}
\frac{\partial u_4}{\partial t} &= \frac{1}{2}\sigma^2\frac{x^2(1-x^2)^2}{(1+x^2)^2}\frac{\partial^2 u_4}{\partial x^2} + \frac{x(1-x^2)}{1+x^2}\Big\{r - \sigma^2\frac{x^2(3+x^2)}{(1+x^2)^2} + \frac{\sigma^2 x}{\varepsilon(1-x^2)+x}\Big\}\frac{\partial u_4}{\partial x} \\
&+ \frac{\kappa(1-x^2)}{\alpha(\varepsilon(1-x^2)+x)}F\Big(\frac{\alpha x^2(\varepsilon(1-x^2)+x)}{(1+x^2)^3}\Big|(1-x^2)(1+x^2)\frac{\partial^2 u_4}{\partial x^2} \\
&+ 2\Big\{\frac{(1+x^2)^2}{\varepsilon(1-x^2)+x} - x(3+x^2)\Big\}\frac{\partial u_4}{\partial x}\Big|\Big) \\
&+ \frac{r\varepsilon(1-x^2)(1-u_4)}{\varepsilon(1-x^2)+x} \quad \text{in } (x,t)\in(0,1)\times(0,T)
\end{aligned} \tag{9}$$

$$u_4(x,0) = 1 - \frac{1-x^2}{\alpha(\varepsilon(1-x^2)+x)} V_0(x/(1-x^2)) \quad \text{for } 0 < x < 1,$$

$$u_4(0,t) = \varepsilon, \quad u_4(1,t) = 0 \quad \text{for } 0 \le t < T.$$

Now our numerical experiments are manipulated on these PDEs, which are defined on a bounded interval and whose solution takes bounded values. We employ the explicit Euler method with the second order finite difference method in space; the computation is carried out in the double precision. To be more specific, for integers $N, M \gg 1$ we define $x_i := ih$ $(i = 0, 1, \cdots, N)$ with $h = 1/N$, $t_j := j\Delta t$ $(j = 0, 1, \cdots, M)$ with $\Delta t = T/M$, and $u^{i,j} := u_3(x_i, t_j)$. Here we use u_3 as an example. The case of u_4 is performed similarly. For $i = 0, 1, \cdots, N-1$ and $j = 0, 1, \cdots, M$ we obtain

$$\begin{aligned}
\frac{u^{i,j+1} - u^{i,j}}{\Delta t} &= \frac{1}{2}\sigma^2 \frac{x_i^2(1-x_i^2)^2}{(1+x_i^2)^2} \frac{u^{i+1,j} - 2u^{i,j} + u^{i-1,j}}{h^2} \\
&- \kappa F\Big(\frac{x_i^2(1-x_i^2)}{(1+x_i^2)^3} \Big| (1-x_i^2)(1+x_i^2) \frac{u^{i+1,j} - 2u^{i,j} + u^{i-1,j}}{h^2} \\
&- 2x_i(3+x_i^2) \frac{u^{i+1,j} - u^{i-1,j}}{2h} \Big| \Big) \\
&+ \frac{x_i(1-x_i^2)}{(1+x_i^2)^3} \{ r(1+x_i^2)^2 - \sigma^2 x_i^2 (3+x_i^2) \} \frac{u^{i+1,j} - u^{i-1,j}}{2h} - ru^{i,j} \\
u^{i,0} &= V_0(x_i/(1-x_i^2)) - \frac{\alpha x_i}{1-x_i^2}, \\
u^{0,j} &= u^{N,j} = 0 (= u^{-1,j} = u^{N+1,j}).
\end{aligned}$$

As is already mentioned, the absolute value and the nonlinear terms do not cause serious troubles in the numerical implementation so long as we use the explicit Euler method in time.

It is to be noted that a singularity $x = 1$ appeared in the transformation $S = x/(1-x^2)$ is just apparent. Indeed, as to the initial condition, we simply set $u(x_i, 0) = V_0(S_i)$ with $S_i := x_i/(1-x_i^2)$ for $0 \le i \le N-1$, and for $i = N$ it suffices to set $u(1,0) = 0$ by the Dirichlet condition. Moreover, this singularity is not involved in the PDEs (8)(9).

3.2. Numerical Implementation

Here we undertake the numerical computation. First we treat the Hoggard-Whalley-Wilmott equation (2). This is partly due to an intention to make sure the utility of our method.

To start with we ascertain the validity of our scheme; we check the error of computation. We take two ways. First since $V_0 \equiv S = x/(1-x^2)$ $(\alpha = 1)$ solves (8) with $u_3(x,t) \equiv 0$ and (9) $(\varepsilon = 1)$ with $u_4(x,t) \equiv (1-x^2)/(1-x^2+x)$, respectively, we estimate the $err(T) := \max_{0 \le i \le N-1} |u_3^{\text{num}}(x_i, T)|$ and $\max_{0 \le i \le N-1} |u_4^{\text{num}}(x_i, T) - u_4(x_i, T)|$, respectively, where $u_j^{\text{num}}(x,t)$ $(j = 3, 4)$ denote the numerical solution. The computation is performed under fixed $\Delta t = 10^{-5}$ with $T = 50$. The results are depicted in Figure 1, which shows that $err(T)$ decays approximately in polynomial order wih respect to N. We note that there always holds $u_3^{\text{num}} \equiv 0$. Although the nonlinear effect must be negligible in this case of $V_0 = S$, our numerical scheme is judged to be well constructed.

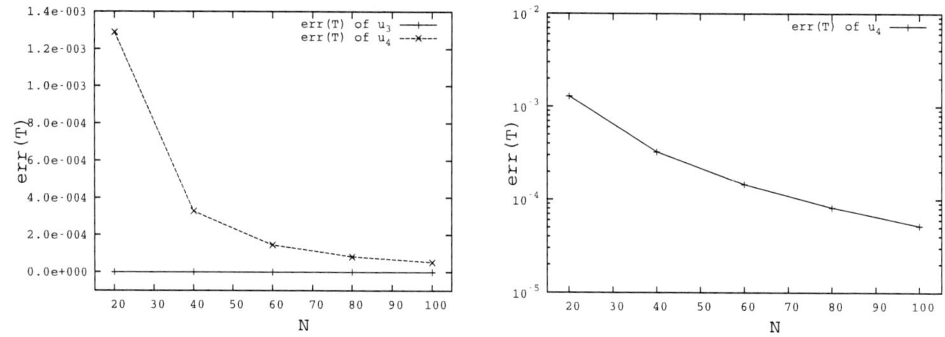

(a) linear scale (b) log-linear scale

Figure 1. Error analysis. Computation of the trivial steady state solution $V \equiv S$.

Table 1. Error analysis. Comparison with the Black-Scholes pricing formula.

S	Exact sol.	V_3	V_4
1.01	0.37696373	0.376966029	0.376965087
1.99	1.27427533	1.274260640	1.274269570
2.95	2.22344376	2.223446846	2.223462904
3.90	3.16628373	3.166340590	3.166322023
5.00	4.26532940	4.265534878	4.265381401

The second way to judge the validity of our scheme makes use of the Black-Scholes pricing formula. Put $\kappa = 0$ ($\sigma = 0.2, T = 10, r = 0.1$) and consider $V_0(S) = \max\{S - 2, 0\}$. We hasten to remark that in the case $\kappa = 0$, the equation (2) of course reduces to the ordinary Black-Scholes equation, and the exact solution is given by the famous Black-Scholes formula.

$$V(S,t) = SN(d_1) - 2e^{-0.1(10-t)}N(d_2),$$

where $N(x) := \int_{-\infty}^{x} f(\phi)d\phi$ with $f(\phi)$ being given in (3) and

$$d_1 := \frac{\log(S/2) + 0.12(10-t)}{\sqrt{10-t}}, \qquad d_2 := d_1 - \sqrt{10-t}.$$

The result is displayed in Table 1; the agreement of respective values may certify well formulation of our scheme.

Now we set $\Delta t = 10^{-5}$ ($M = 10^6$), $\Delta x = 1/200$ ($N = 200$), and various parameters values as follows.

$$\kappa = \frac{1}{2}, \quad \delta t = \frac{2}{\pi}, \quad T = 10, \quad r = 0.1, \quad \text{and } \varepsilon = 1 \text{ in (9).} \tag{10}$$

For comparison we employ three choices of σ; namely $\sigma = 1, 5, 10$, which are indicated in figures. As to the initial conditions we choose

$$\begin{aligned} V_0^1(S) &= 2S\tanh S \quad (\alpha = 2) \\ V_0^2(S) &= \max\{S-2,0\} - 2\max\{S-3,0\} + \max\{S-4,0\} \quad (\alpha = 0). \end{aligned} \tag{11}$$

It is easy to see that $V_0^1(S)$ is not convex nor concave and $V_0^2(S)$ gives rise to the linear combination of plain vanilla European call options. Since V_0^2 is bounded the transformation V_4, u_4 should be vacuous; only V_3, u_3 will be enough. Figure 2 summarizes the results. It may be seen that the larger σ smoothen the behavior of solutions more.

Next we consider a special case of the so-called extended Leland model [9]. See (4).

$$F(Q) = \frac{\kappa_1}{\delta t} + \sigma\sqrt{\frac{2}{\pi\delta t}}Q. \tag{12}$$

The parameters are set to be the same as (10) and $\kappa_1 = 2/\pi$. We similarly transform the equation and implement the computation. The initial condition is V_0^2. The result is shown in Figure 3. Since the constant term in (12) makes the option value negative, the pricing of this model resulted in an unrealistic situation.

Finally we are concerned with the logistic type costs term (6). The new parameter β is set to be $\beta = 10$. The result is depicted in Figure 4.

4. Optimal Investment Problem

4.1. Introduction

In this section we apply our computation procedure to other problems in the mathematical finance. To be precise we consider a singular PDE, which is introduced to describe the evolution of the risk preference in the optimal investment problem [1, 21]. The PDE is derived from the Hamilton-Jacobi-Bellman (HJB) equation for the value function. Our purpose is then to numerically investigate these singular PDEs.

We recall that the study of the optimal behavior in economics environment has been an intensive subject for researches. Various models have been introduced so far and much progress has been made. For instance optimal portfolio problems are discussed in [6, 7] after a pioneering work of Merton [27]. Applications to insurance are considered in [3, 10, 11, 14, 29]. We also refer to the references cited in these papers. We point out that the traditional way of investigation has been based on stochastic control and a number of authors have reduced the analysis to the treatment of the HJB equation for the value function. The resulting nonlinear equations, however, are typically hard to solve; it may be not an exaggeration to say that all that we can do is to merely guess a shape of solutions and manage to arrange the parameters. Observe the statement in [2]. Of course there are weak approaches to these equations and substantial success in mathematics was made. See for instance [24, 36] and the references cited there. The notion of weak solutions, the so-called viscosity solutions, however, is a little involved and does not seem to meet the wishes of practitioners. Additionally it is not so clear to numerically handle the weak solution. As a result the analysis of HJB equations has certainly stayed as principal difficulties to be surmounted.

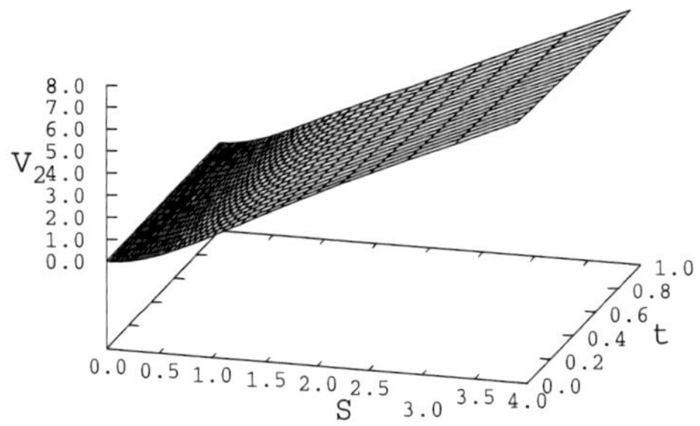

(a-1) $\sigma = 1$

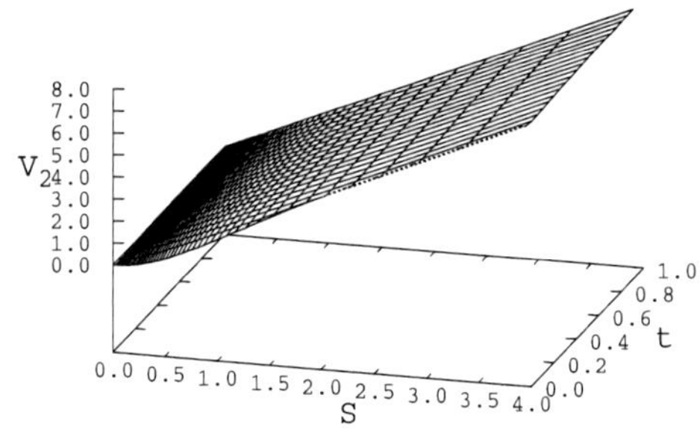

(a-2) $\sigma = 5$

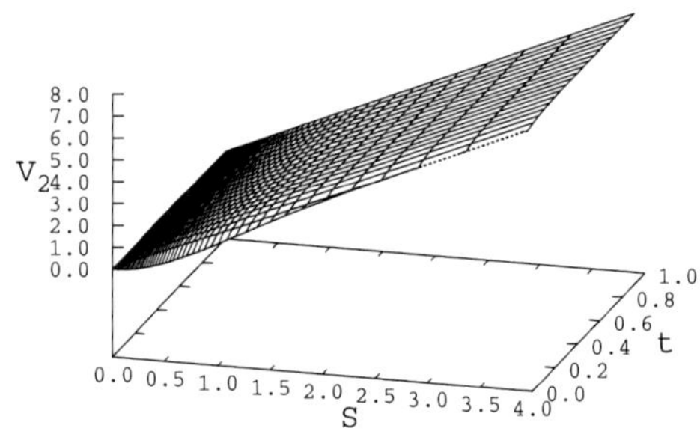

(a-3) $\sigma = 10$

(a) the initial condition $V_0^1(S)$.

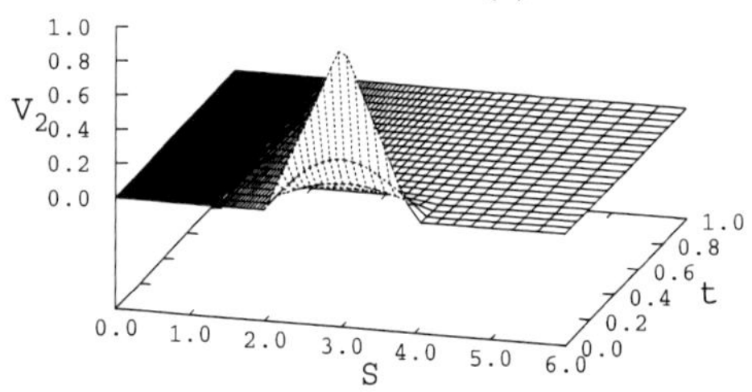

(b-1) $\sigma = 1$

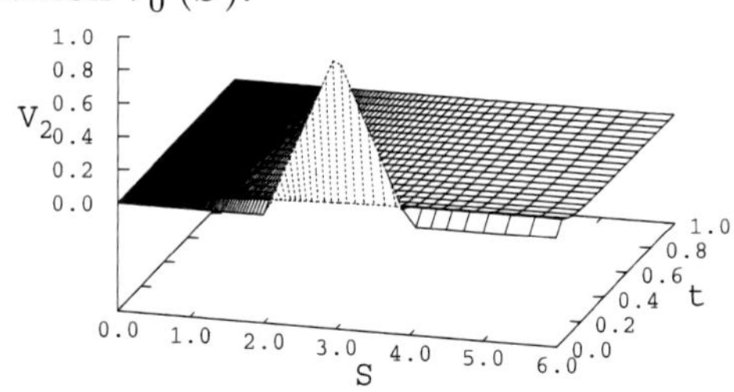

(b-2) $\sigma = 5$

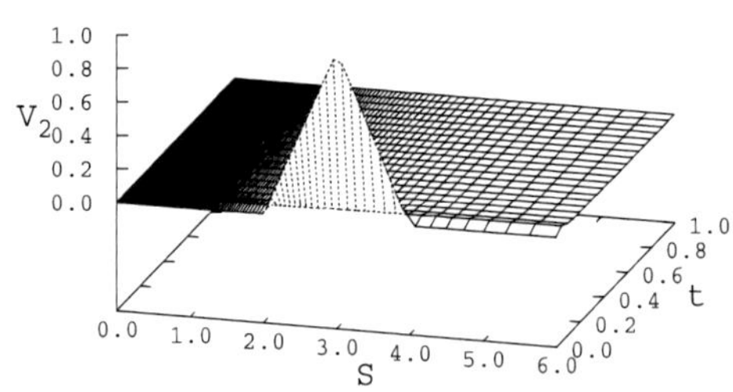

(b-3) $\sigma = 10$

(b) the initial condition $V_0^2(S)$.

Figure 2. Computation of the Hoggard-Whalley-Wilmott equation with the initial conditions (11). (See (10) for various parameter values.)

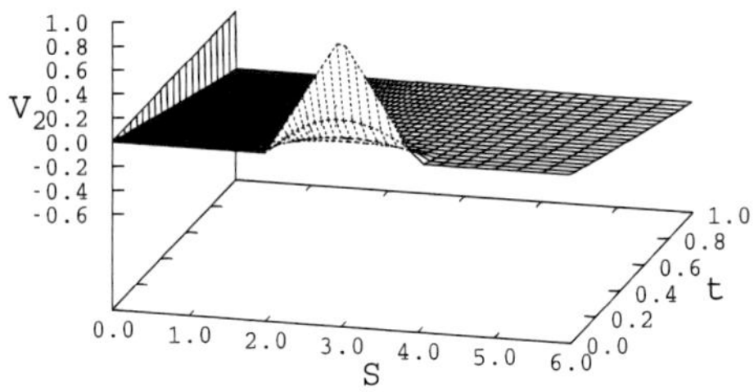

(1) $\sigma = 1$

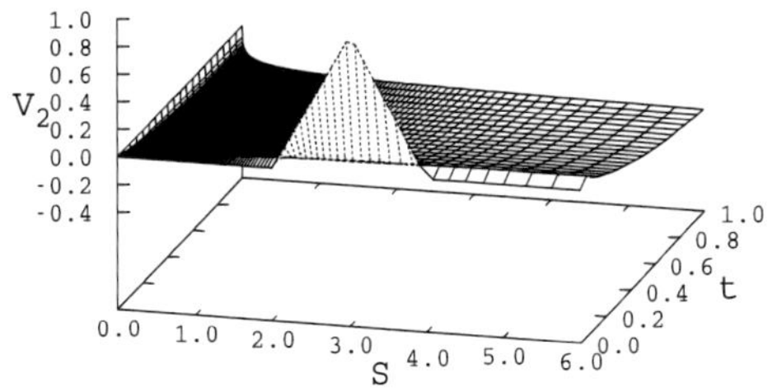

(2) $\sigma = 5$

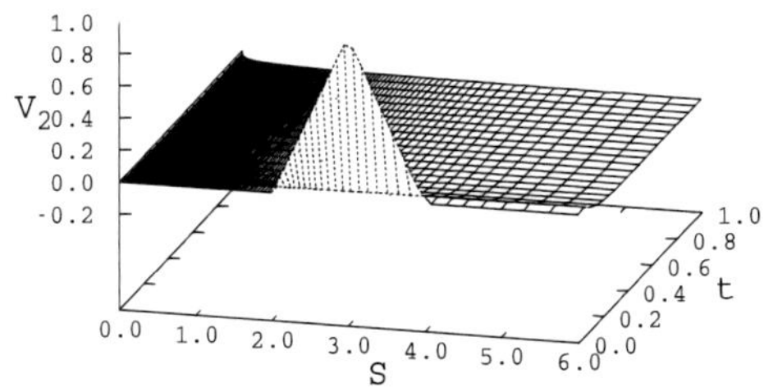

(3) $\sigma = 10$

Figure 3. Computation of the extended Leland model with the initial condition V_0^2.

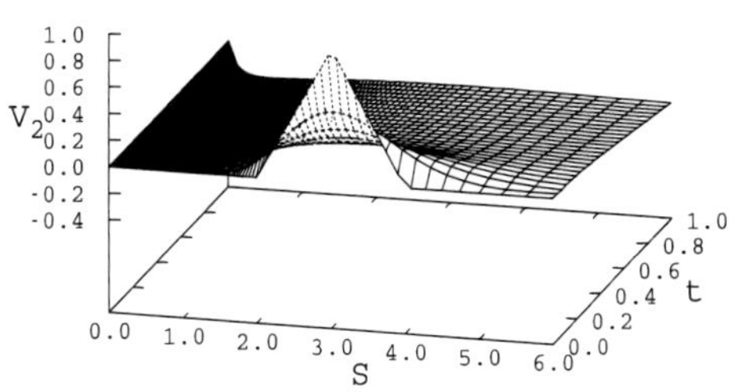

(1) $\sigma = 1$

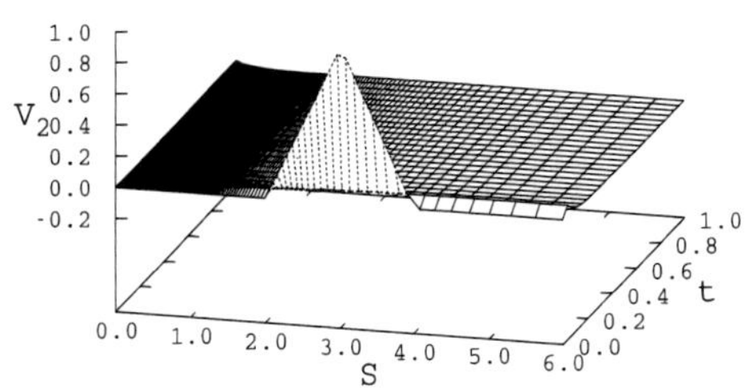

(2) $\sigma = 5$

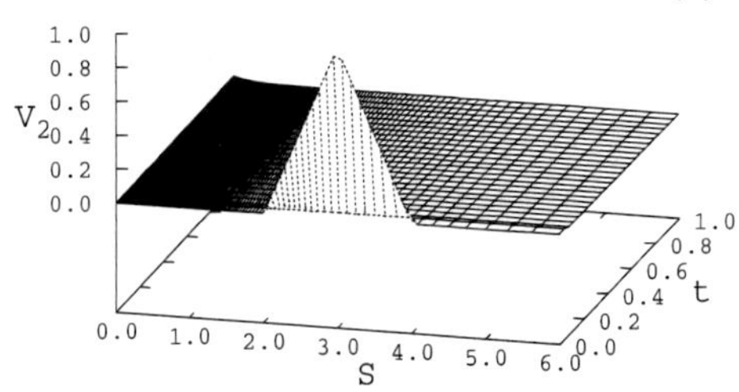

(3) $\sigma = 10$

Figure 4. Computation of the logistic type costs term with the initial condition V_0^2.

In [1, 21] we have derived a singular PDE in order to deal with such HJBs. See (19) below. Although essential difficulties are equivalent to those expressed by the HJB equation, this derived PDE has rather simple looking from the viewpoint of the theory of nonlinear PDEs. In addition, the unknown quantity is related to the Arrow-Pratt coefficient of absolute risk aversion [30] with respect to the optimal value function. In this sense our introduced PDE may be interpreted as the characteristic equation for the risk structure of the model. We do not insist that our PDE would replace the HJB itself but we at least believe that the study of this PDE is interesting as well as important and therefore worth investigating further. Here we undertake computational treatment.

4.2. Model

First we recall our basic model, which is essentially due to Browne [3]. See also [1, 21]. It is assumed that there is only one risky stock available for investment, whose prices P_t at time t is governed by the stochastic differential equation of Black-Scholes-Merton type

$$dP_t = P_t(\mu dt + \sigma dW_t^{(1)}),$$

where μ and σ are constants and $\{W_t^{(1)}\}_{t\geq 0}$ is a standard Brownian motion. There are also a risk process and a bond, whose price at time t are denoted by Y_t and B_t, respectively, and assumed to be evolved as

$$\begin{aligned} dY_t &= \alpha dt + \beta dW_t^{(2)}, \\ dB_t &= \gamma B_t, \end{aligned}$$

where α and β ($\beta > 0$) are constants and $\{W_t^{(2)}\}_{t\geq 0}$ is another standard Brownian motion. An interest rate $\gamma > 0$ is supposed to be constant and $\mu > \gamma$. It is allowed these two Brownian motions to be correlated with the correlation coefficient ρ. We prescribe $0 \leq \rho^2 < 1$ in the sequel.

The company invests in the risky stock under an investment policy f, where $f = \{f_t\}_{0\leq t\leq T}$ is a suitable, admissible adapted control process. T stands for the maturity date. Let X_t^f denote the wealth of the company at time t with $X_0 = x$, whose evolution process is given by

$$\begin{aligned} dX_t^f &= f_t \frac{dP_t}{P_t} + \gamma(X_t^f - f_t)dt + dY_t \\ &= (\gamma X_t^f + f_t(\mu - \gamma) + \alpha)dt + f_t dW_t^{(1)} + dW_t^{(2)}, \\ X_0 &= x. \end{aligned}$$

Suppose that the investor wants to maximize the utility $U(x)$ from his terminal wealth. The utility function $U(x)$ is customarily assumed to satisfy $U' > 0$ and $U'' < 0$. Let

$$V(x,t) := \sup_f E[e^{-\delta(T-t)}U(X_T^f) \mid X_t^f = x], \tag{13}$$

where δ stands for the rate at which consumption and terminal wealth are discounted. We remark that in the seminal work of Browne [3] the case of $\delta \equiv 0$ is treated.

Now the Bellman principle applied to (13) implies that the equation for V, which is called the Hamilton-Jacobi-Bellman (HJB) equation, becomes

$$\sup_f \{\mathcal{A}^f V(x,t)\} = -\delta V, \quad V(x,T) = U(x), \tag{14}$$

where $\mathcal{A}^f$ denotes the generator of the wealth process dX_t, which is expressed as

$$(\mathcal{A}^f g)(x,t) := \frac{\partial g}{\partial t} + (f(\mu-\gamma) + \gamma x + \alpha)\frac{\partial g}{\partial x} + \frac{1}{2}(f^2\sigma^2 + \beta^2 + 2\beta\sigma\rho f)\frac{\partial^2 g}{\partial x^2}.$$

Suppose that (14) has a classical solution V with $\partial V/\partial x > 0$, $\partial^2 V/\partial x^2 < 0$. We then infer that

$$f_t^* = -\frac{\mu-\gamma}{\sigma^2}\frac{\partial V/\partial x}{\partial^2 V/\partial x^2} - \frac{\beta\rho}{\sigma}, \tag{15}$$

where $\{f_t^*\}_{0\le t\le T}$ denotes the optimal policy. Placing (15) back into (14) we obtain

$$\begin{aligned} \frac{\partial V}{\partial t} + \Big(\gamma x + \alpha - \frac{\beta\rho(\mu-\gamma)}{\sigma}\Big)\frac{\partial V}{\partial x} - \frac{1}{2}\Big(\frac{\mu-\gamma}{\sigma}\Big)^2 \frac{(\partial V/\partial x)^2}{\partial^2 V/\partial x^2} & \\ + \frac{1}{2}\beta^2(1-\rho^2)\frac{\partial^2 V}{\partial x^2} &= -\delta V \qquad \text{for } 0 < t < T \\ V(T,x) &= U(x). \end{aligned} \tag{16}$$

Browne [3] shows that if $\delta = 0$ (16) possesses a solution in the case $U(x) = \lambda - (\nu/\theta)e^{-\theta x}$ with positive constants λ, ν, θ. This utility has constant absolute risk aversion parameter θ; precisely stated, $-U''(x)/U'(x) = \theta$. Here we proceed further in the analysis of (16) along the line of [1], where the case of $\gamma \equiv 0$ and $\delta \equiv 0$ is discussed.

Let $v(x,t)$ be defined by $V(x,t) = v(Ex, F(T-t))$, where

$$E := \sqrt{\frac{(\mu-\gamma)^2}{\beta^2(1-\rho^2)\sigma^2}}, \quad F := \frac{1}{2}\Big(\frac{\mu-\gamma}{\sigma}\Big)^2.$$

We further define

$$a := E\Big(\alpha - \frac{\beta\rho}{\sigma}(\mu-\gamma)\Big), \quad b := \frac{\delta}{F},$$

and write γ/F by the same γ with abuse of notation. It follows that after a calculation

$$\frac{\partial v}{\partial t} = \frac{\partial^2 v}{\partial x^2} - \frac{(\partial v/\partial x)^2}{\partial^2 v/\partial x^2} + (\gamma x + a)\frac{\partial v}{\partial x} - bv, \qquad v(x,0) = U(E^{-1}x). \tag{17}$$

Now we additionally introduce the next quantity.

$$r(x,t) := -\frac{\partial^2 v/\partial x^2}{\partial v/\partial x} = -\frac{\partial}{\partial x}\log\Big|\frac{\partial v}{\partial x}(x,t)\Big|. \tag{18}$$

It should be noted that (18) is related to the coefficient of absolute risk aversion. A little tedious computation then leads us to the following equation.

$$\frac{\partial r}{\partial t} = \frac{\partial}{\partial x}\Big\{\Big(1+\frac{1}{r^2}\Big)\frac{\partial r}{\partial x} - r^2 + (\gamma x + a)r\Big\}, \quad r = r(x,t) \quad \text{in } (x,t) \in \Omega_T := \boldsymbol{R}^+ \times (0,T) \tag{19}$$

where $T > 0$ and $\boldsymbol{R}^+ = \{x > 0\}$.

This is the PDE we wish to propose. Compared to the equation (17), which is in itself worth investigating further, the equation (19) is quasilinear and has divergence form; (19) is rather popular types in the PDE world, although it is singular at the same time. We also remark that in [1] the equation (19) with $\gamma \equiv 0$ and $a \equiv 0$ is derived.

The conditions imposed on (19) are as follows. Since r is related to the Arrow-Pratt coefficient of absolute risk aversion, it is desirable to be nonnegative and nonincreasing. Moreover, $r(x, \cdot)$ should tend to zero as $x \to \infty$. The boundary condition at $x = 0$ is Neumann type, which is somewhat by mathematical simplicity. In summary we understand that (19) is supplemented by the next condition.

$$\frac{\partial r}{\partial x}(0,t) = 0, \qquad \frac{\partial r}{\partial x}(x,t) \leq 0, \qquad r(x,t) \to 0 \quad \text{as } x \to \infty. \tag{20}$$

We notice that in particular the domain is not bounded but the half line.

4.3. Analytical Results

Here we recall analytical results established in [1], where $\gamma = a = 0$ is treated.

$$\begin{aligned} &\frac{\partial r}{\partial t} = \frac{\partial}{\partial x}\Big\{\Big(1 + \frac{1}{r^2}\Big)\frac{\partial r}{\partial x} - r^2\Big\}, \qquad r = r(x,t) \quad \text{in } (x,t) \in \Omega_T := \boldsymbol{R}^+ \times (0,T) \\ &r \geq 0, \qquad \frac{\partial r}{\partial x}(0,t) = 0, \qquad \frac{\partial r}{\partial x}(x,t) \leq 0, \\ &r(x,t) \to \alpha \quad \text{as } x \to \infty \\ &r(x,0) = r_0(x) \quad \text{on } x \in \boldsymbol{R}^+. \end{aligned} \tag{21}$$

Here α denotes nonnegative constant and the non-increasing initial datum r_0 belongs to $H^1(\boldsymbol{R}^+)$ and satisfies the required compatibility condition. Despite the harmony in accordance with (20), we slightly lessen the limitation so that α is allowed to be nonnegative. This is partly due to the imperfectness of our mathmematical theory.

The solutions is discussed in a weak sense. To clarify the notion of weak solutions the next definition is in order.

Definition 4.1 *We say r a weak solution of (21) if the following conditions are fulfilled.*

1. $r - \alpha \in L^\infty(0,T; L^2(\boldsymbol{R}^+)) \cap L^2(0,T; H^1(\boldsymbol{R}^+))$, $\partial r/\partial t \in L^2(0,T; H^{-1}(\boldsymbol{R}^+))$.
2. *There holds for each* $\varphi \in H^1(\boldsymbol{R}^+)$ *and almost every* $0 \leq t \leq T$

$$\int_{\boldsymbol{R}^+} \frac{\partial r}{\partial t}(x,t)\varphi(x)dx = -\int_{\boldsymbol{R}^+} \Big\{\Big(1 + \frac{1}{r^2}\Big)\frac{\partial r}{\partial x}\frac{\partial \varphi}{\partial x} + 2r\frac{\partial r}{\partial x}\varphi\Big\}dx. \tag{22}$$

3. $r(x,0) = r_0(x)$ *in* $L^2(\boldsymbol{R}^+)$.

Our analytical existence theorem is then stated as follows [1].

Theorem 4.2 *For any positive and non-increasing intitial data $r_0 \in H^1(\boldsymbol{R}^+)$ satisfying $(\partial r_0/\partial x)(0) = 0$ and $r_0(x) \to \alpha$ as $x \to \infty$ with $\alpha > 0$, there corresponds a finite $T = T(r_0)$ such that there exists a positive solution r for (21), which is non-increasing in x, in the sense of Definition 4.1.*

The proof involves a fixed point argument and a convergence procedure. See [1].

To make familiar with the nature of the equation (21), we analyze the structure of steady state solutions; that is, we determine the set of solutions for

$$\left(1 + \frac{1}{r^2}\right)\frac{\partial r}{\partial x} - r^2 = C, \qquad r = r(x),$$

where C denotes a constant independent of x and t. There are three possibilities according to the sign of C. We note that, however, the first two cases are meaningless for the economics because r takes a negative value.

If $C > 0$ then we write $C = M^2$ to obtain

$$\begin{aligned} -\frac{1/M^2}{r(x)} + \left(\frac{1}{M} - \frac{1}{M^3}\right)\tan^{-1}\frac{r(x)}{M} \\ = x - \frac{1/M^2}{r(0)} + \left(\frac{1}{M} - \frac{1}{M^3}\right)\tan^{-1}\frac{r(0)}{M}. \end{aligned} \tag{23}$$

It follows that $r(x) \sim -M^{-2}x^{-1}$ as $x \to \infty$.

If $C = 0$ then we have

$$-\frac{1}{r(x)} - \frac{1}{3r(x)^3} = x - \frac{1}{r(0)} - \frac{1}{3r(0)^3}. \tag{24}$$

It follows that $r(x) \sim -(3x)^{-1/3}$ as $x \to \infty$.

Consequently if $C \geq 0$ then there is no steady state solution suitable to the finance. Only the next last case fits appropriately to mathematical finance.

If $C < 0$ then we write $C = -M^2$ to obtain

$$\begin{aligned} \frac{1/M^2}{r(x)} + \frac{1+M^2}{2M^3}\log\left|\frac{r(x)-M}{r(x)+M}\right| \\ = x + \frac{1/M^2}{r(0)} + \frac{1+M^2}{2M^3}\log\left|\frac{r(0)-M}{r(0)+M}\right|, \end{aligned} \tag{25}$$

provided $r(x) \neq M$. In this case $r(x) \sim M^{-2}x^{-1}$ as $x \to \infty$. It is also clear that $r(x) \equiv M$ gives one of steady state solutions, which has a character of constant absolute risk aversion. It should be noted that the last steady state solutions correspond to those presented in [3].

4.4. Numerical Scheme

In this subsection we formulate our numerical scheme. First we make the next change of variables, which is now familiar to us.

$$s = \frac{2x}{1 + \sqrt{4x^2+1}} \quad \left(x = \frac{s}{1-s^2}\right), \qquad r(s,t) = r(x,t)$$

The half line $[0, \infty)$ then corresponds bijectively to $[0, 1)$. Furthermore the equation (19) is transformed into

$$\begin{aligned}
\frac{\partial r}{\partial t} &= \Big(1 + \frac{1}{r^2}\Big) \frac{(1-s^2)^4}{(1+s^2)^2} \frac{\partial^2 r}{\partial s^2} - \frac{2(1-s^2)^4}{r^3(1+s^2)^2} \Big(\frac{\partial r}{\partial s}\Big)^2 \\
&\quad - 2\Big\{\Big(1 + \frac{1}{r^2}\Big) \frac{s(1-s^2)(3+s^2)}{(1+s^2)^3} \\
&\quad + ((r-a)(1-s^2) - \gamma s) \frac{1-s^2}{1+s^2}\Big\} \frac{\partial r}{\partial s} + \gamma r \\
r &= r(s,t) \qquad \text{in } (s,t) \in (0,1) \times (0,T) \\
r(s,0) &= r_0\Big(\frac{s}{1-s^2}\Big) \qquad \text{on } s \in [0,1].
\end{aligned}$$

As to the first and the third conditions of (20) we simply put

$$\frac{\partial r}{\partial s}(0,t) = 0, \quad r(1,t) = \alpha \qquad \text{for } t \in [0,T]. \tag{26}$$

Our discretization is now performed on this transformed equation. For integers $N, M \gg 1$ we define $s_i := ih$ $(i = 0, 1, 2, \cdots, N)$ with $h = 1/N$, $t_j := j\Delta t$ $(j = 0, 1, 2, \cdots, M)$ with $\Delta t = T/M$, and $r^{i,j} := r(s_i, t_j)$. We obtain for $i = 0, 1, 2, \cdots, N-1$ and $j = 0, 1, 2, \cdots, M$,

$$\begin{aligned}
\frac{r^{i,j+1} - r^{i,j}}{\Delta t} &= \Big(1 + \frac{1}{(r^{i,j})^2}\Big) \frac{(1-s_i^2)^4}{(1+s_i^2)^2} \frac{r^{i+1,j} - 2r^{i,j} + r^{i-1,j}}{h^2} \\
&\quad - \frac{2(1-s_i^2)^4}{(r^{i,j})^3(1+s_i^2)^2} \frac{(r^{i+1,j} - r^{i-1,j})^2}{4h^2} \\
&\quad - 2\Big\{\Big(1 + \frac{1}{(r^{i,j})^2}\Big) \frac{s_i(1-s_i^2)^3(3+s_i^2)}{(1+s_i^2)^3} \\
&\quad + ((r^{i,j} - a)(1-s_i^2) - \gamma s_i) \frac{1-s_i^2}{1+s_i^2}\Big\} \frac{r^{i+1,j} - r^{i-1,j}}{2h} + \gamma r^{i,j} \\
\frac{-r^{2,j} + 4r^{1,j} - 3r^{0,j}}{2h} &= 0, \qquad r^{N,j} = 0.
\end{aligned} \tag{27}$$

This is our scheme. The last two equations come from boundary conditions. We remark once again that the condition (26), namely the point $s = 1$, does not produce a singularity.

4.5. Numerical Implementation

We carry out our numerical experiment based on the scheme (27). We here consider three examples. To keep our situation simple enough we assume $a = 0$ here. In both examples we set $N = 10^3$, $M = 10^7$ and $T = 1$.

Example 1. $\alpha \geq 0$ and we take $r_0(x) = 2e^{-x^2} + \alpha$. The results are shown in Figures 5, 6 and 7. Figures 5, 6 deal with $\alpha = 1$ and $\gamma = 0$, $\gamma = 1$ while Figure 7 with $\alpha = 10$, $\gamma = 1$.

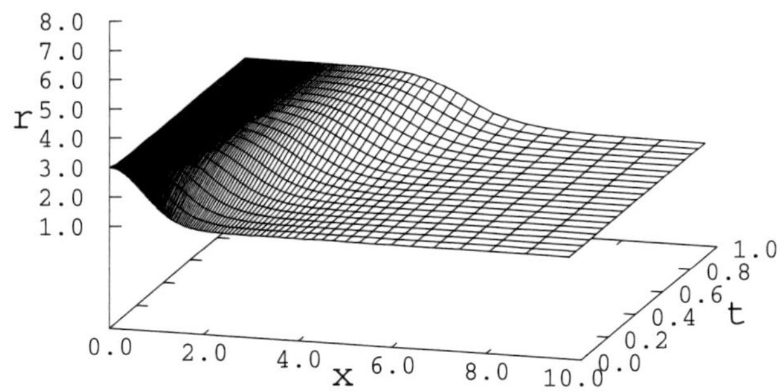

Figure 5. $\gamma = 0$ and the initial condition $r_0(x) = 2e^{-x^2} + 1$.

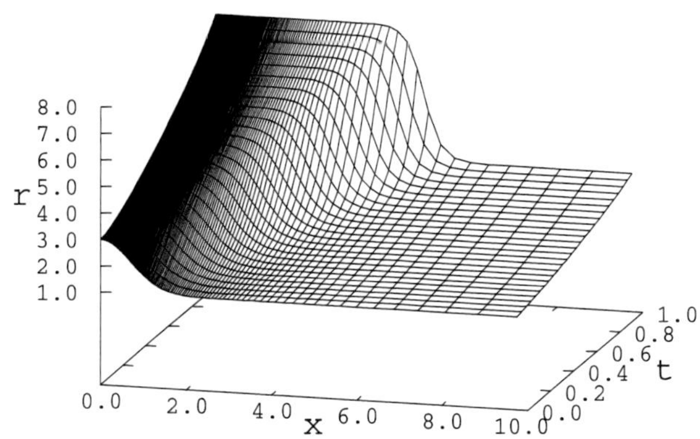

Figure 6. $\gamma = 1$ and the initial condition $r_0(x) = 2e^{-x^2} + 1$.

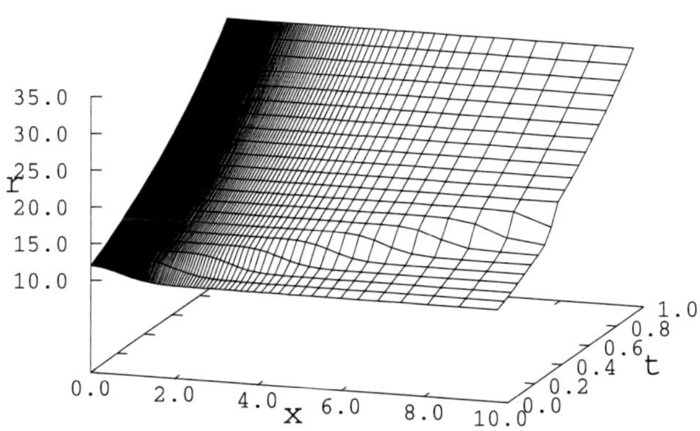

Figure 7. $\gamma = 0$ and the initial condition $r_0(x) = 2e^{-x^2} + 10$.

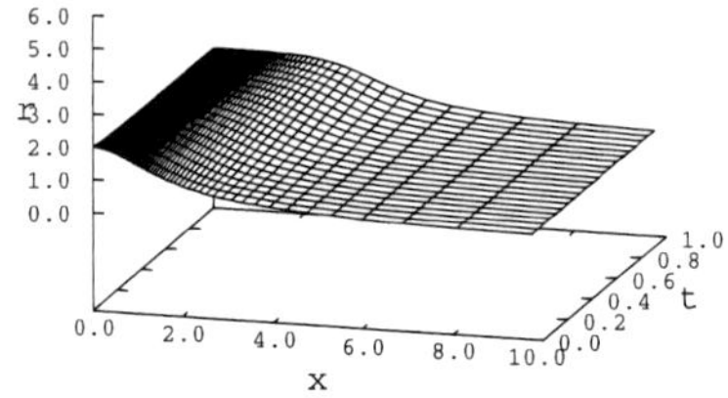

(a) $\gamma = 0$

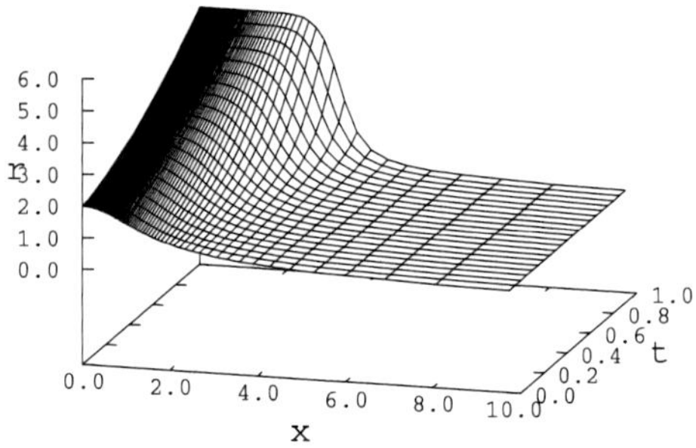

(b) $\gamma = 1$

Figure 8. Computation under the initial condition $r_0(x) = 2x^{-1} \tanh x$.

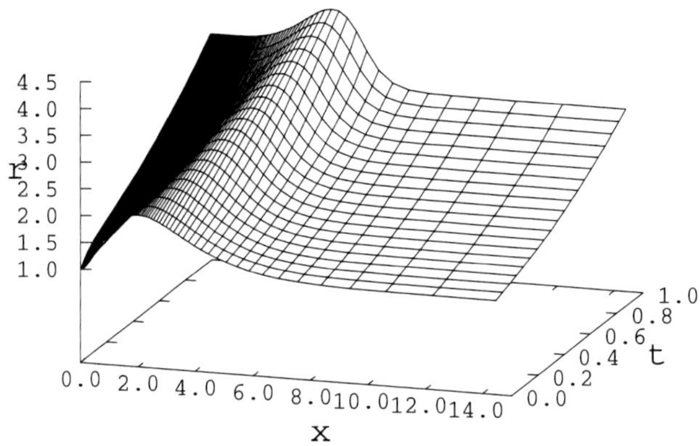

Figure 9. Computation under the initial condition $r_0(x) = 2x^2 e^{-x} + 1$ and $\gamma = 1$.

The function $r_0(x)$ is monotone decreasing. It is seen that the presence of the interest rate γ makes the solution grow; in terms of the original setting, we can say that the interest rate calms the risk preference as time comes near to the maturity.

Example 2. We take $r_0(x) = 2x^{-1} \tanh x$. The results are shown in Figure 8. Figure 8(a) deals with $\gamma = 0$ while Figure 8(b) with $\gamma = 1$. The latter case corresponds to the economy with positive interest rate.

The function $r_0(x)$ is not monotone decreasing and attains the local maximum at the point $2 = \sinh(2x)/x$. It is observed that the shape of solutions is smoothened as time proceeds. The positive γ makes the solution grow rapidly.

Example 3. $\alpha \geq 0$ and we take $r_0(x) = 2x^2 e^{-x} + \alpha$. The results are shown in Figures 9 and 10. Figure 9 deals with $\alpha = 1$ while Figure 10 with $\alpha = 5$. In both cases we put $\gamma = 1$.

The function $r_0(x)$ is not monotone decreasing and attains the local maximum at $x = 2$. It is seen that the point of local maximum grows as time proceeds.

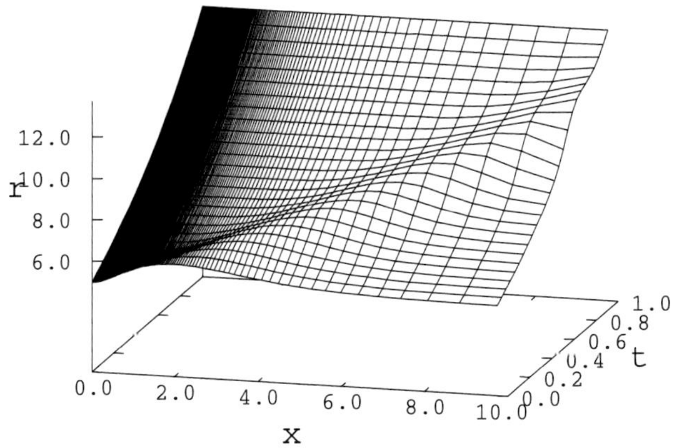

Figure 10. Computation under the initial condition $r_0(x) = 2x^2e^{-x} + 5$ and $\gamma = 1$.

Conclusion

We have proposed a numerical scheme to effectively compute the nonlinear partial differential equations (PDEs) as they are, especially when equations are defined on unbounded domains and their solutions generally grow infinitely large; we are forced to be confronted with unboundedness. We overcome these difficulties by exploiting suitable transformations so that we eventually discuss the PDE on a bounded interval whose solution stays bounded. We hope that our strategy is accessible to large part of the researchers and easy to understand. We additionally point out that numerical implementation shows that our scheme is robust enough and well works.

We discuss two problems as examples of our procedure. One is the nonlinear PDE of Black-Scholes type which incorporates the effects of transaction costs; the influence of transaction costs makes the PDE nonlinear. We directly argue the PDE rather than simulate the underlying stochastic processes. It can be seen that transaction costs certainly make the value lower; in the presence of fixed cost component the value even becomes negative, which will be meaningless in real trading. In any case, however, transaction costs smoothen off the behavior of solutions.

The other example is the nonlinear evolution equation for the risk preference in the optimal investment problem. The unknown quantity is related to the Arrow-Pratt coefficient of absolute risk aversion with respect to the optimal value function. The proposed equation is therefore interpreted as to describe the evolution of the risk preference of the system. The domain where the evolution takes place is half line and the solution may grow. From the results of numerical implementation, it is indicated in particular that the interest rate makes the solution dissipate as time advances to the maturity. In a sense the presence of the interest rate calms the risk preference.

The models in mathematical finance are generically expressed by stochastic terminology; the possibility is in principle restricted that the model is ultimately realized by an explicit PDE. Nevertheless we believe that our method may be applicable to such PDE obtained in finances, which will be our next topics for researches.

Acknowledgements

We are grateful to Professor Masahiro Kushida for great help in numerical computation and to Professor Hideo Sakaguchi for assistance. Thanks are also due to Professor Shigeaki Koike for informing us about [24, 36]. This work is partially supported by Grants-in-Aids for Scientific Research (Nos. 18340045, 18654023) from the Japan Society for Promotion of Sciences. The first author (N.I.) as well as supported by the Inamori Foundation through the Inamori Grants in 2006–2007.

References

[1] R. Abe and N. Ishimura, Existence of solutions for the nonlinear partial differential equation arising in the optimal investment problem, *Proc. Japan Acad., Ser. A.*, **84** (2008), 11–14.

[2] T. Björk, *Arbitrage Theory in Continuous Time*, 2nd ed., Oxford University Press, Oxford, 2004.

[3] S. Browne, Optimal investment policies for a firm with a random risk process: exponential utility and minimizing the probability of ruin, *Math. Operations Research,* **20** (1995), 937–958.

[4] F. Black and M. Scholes, The pricing of options and corporate liabilities, *J. Political Economy,* **81** (1973), 637–659.

[5] P.P. Boyle and T. Vorst, Option replication in discrete time with transaction costs, *J. Finance,* **47** (1992), 271–293.

[6] S. Browne, Survival and growth with a liability: optimal portfolio strategies in continuous time, *Math. Operations Research,* **22** (1997), 458–493.

[7] J.C. Cox and C.-f. Huang, *A variational problem arising in financial economics* , J. Math. Economics, **20** (1991), 465–487.

[8] M.H.A. Davis, V.G. Panes and T. Zariphopoulou, European option pricing with transaction costs, *SIAM J. Control and Optim.*, **31** (1993), 470–493.

[9] J.N. Dewynne, A.E. Whalley and P. Wilmott, *Path-dependent options and transaction costs*, in "Mathematical Models in Finance," (S.D. Howison, F.P. Kelly and P. Wilmott, eds.), Chapman and Hall, London, (1995), pp. 67–79.

[10] C. Hipp, Stochastic control with application in insurance, in "Stochastic Methods in Finance," (M. Frittelli and W. Runggaldier, eds.), *Springer Lecture Notes in Mathematics* **1856**, Springer, Berlin, 2004, pp. 127–164.

[11] C. Hipp and M. Plum, Optimal investment for insures, *Insurance: Mathematics and Economics,* **27** (2000), 215–228.

[12] S.D. Hodges and A. Neuberger, Optimal replication of contingent claims under transaction costs, *Rev. Futures Markets,* **8** (1989), 222–239.

[13] T. Hoggard, A.E. Whalley and P. Wilmott, Hedging option portfolios in the presence of transaction costs, *Advances in Futures and Options Res.* , **7** (1994), 21–35.

[14] B. Højgaard and M. Taksar, Optimal proportional reinsurance policies for diffusion models, *Scand. Actuarial J.* **1998** (2), 166–180.

[15] J. Hull, *Options, Futures, and Other Derivatives*, 4th edition, Prentice-Hall, New Jersey, 2000.

[16] H. Imai, *Some methods for removing singularities and infinity in numerical simulations*, in "The Proceedings of Third Polish-Japanese Days on Mathematical Approach to Nonlinear Phenomena: Modelling, Analysis and Simulations" (T. Aiki, N. Kenmochi, M. Niezgódka and M. Ôtani, eds.), Gakuto, Tokyo, 2005, pp. 103–118.

[17] H. Imai, N. Ishimura and M. Kushida, *Numerical treatment of a singular nonlinear partial differential equation arising in the optimal investment* , to appear in Thai J. Math.

[18] H. Imai, N. Ishimura, I. Mottate and M.A. Nakamura, On the Hoggard-Whalley-Wilmott equation for the pricing of options with transaction costs, *Asia-Pacific Financial Markets,* **13** (2007), 315–326.

[19] H. Imai, N. Ishimura and H. Sakaguchi, Computational technique for treating the nonlinear Black-Scholes equation with the effect of transaction costs, *Kybernetika,* **43** (2007), 807–815.

[20] N. Ishimura, *Remarks on the nonlinear partial differential equations of Black-Scholes type with transaction costs*, preprint, submitted.

[21] N. Ishimura and K. Murao, *Nonlinear evolution equations for the risk preference in the optimal investment problem*, Paper presented at AsianFA/NFA 2008 International Conference in Yokohama, to be available at http://fs.ics.hit-u.ac.jp/nfa-net/

[22] M. Jandačka and D. Ševčovič, On the risk-adjusted pricing-methodology-based valuation of vanilla options and explanation of the volatility smile, *J. Appl. Math.*, **3** (2005), 235–258.

[23] Z. Jin, H. Sakaguchi, N. Ishimura and H. Imai, Global in space simulation for the Black-Scholes equation incorporating transaction costs, *Theoretical and Applied Mechanics Japan,* **56** (2007), 445–450.

[24] S. Koike and H. Morimoto, Optimal consumption and portfolio choice with stopping, *Funkcialaj Ekvacioj,* **48** (2005), 183–202.

[25] H.E. Leland, Option pricing and replication with transaction costs, *J. Finance,* **40** (1985), 1283–1301.

[26] H.-K. Liu, An approximation and properties for valuating European options in presence of the transaction costs, *Working paper* (2007).

[27] R.C. Merton, Optimal consumption and portfolio rules in a continuous time model, *J. Economic Theory,* **3** (1971), 373–413.

[28] R.C. Merton, Theory of rational option pricing, *Bell J. Econ. Manag. Sci.*, **4** (1973), 141–183.

[29] K.S. Moore and V.R. Young, Optimal insurance in a continuous-time model, *Insurance: Mathematics and Economics,* **39** (2006), 47–68.

[30] J.W. Pratt, Risk aversion in the small and in the large, *Econometrica,* **32** (1964), 122–136.

[31] R. Seydel, *Tools for Computational Finance*, 3rd edition, Springer, Berlin, 2006.

[32] A.E. Whalley and P. Wilmott, Optimal hedging of options with small but arbitrary transaction cost structure, *Euro. J. Appl. Math.,* **10** (1999), 117–139.

[33] P. Wilmott, *Paul Wilmott on Quantitative Finance*, Vol. I, II, John Wiley & Sons, Ltd., New York, 2000.

[34] P. Wilmott, S. Howison and J. Dewynne, *The Mathematics of Financial Derivatives*, Cambridge University Press, Cambridge, 1995.

[35] S. Yokoya, Option hedging strategy with transaction costs, *Theoretical and Applied Mechanics Japan*, **53** (2004), 291–295.

[36] T. Zariphopolou, Consumption-investment models with constraints, *SIAM J. Control Optim.,* **32** (1994), 59–85.

In: Nonlinear Models in Mathematical Finance... ISBN 978-1-60456-931-5
Editor: Matthias Ehrhardt, pp. 243-273

Chapter 8

FIXED DOMAIN TRANSFORMATIONS AND SPLIT–STEP FINITE DIFFERENCE SCHEMES FOR NONLINEAR BLACK–SCHOLES EQUATIONS FOR AMERICAN OPTIONS

***Julia Ankudinova*[1] *and Matthias Ehrhardt*[2]**
[1]Institut für Mathematik, TU Berlin, Strasse des 17. Juni 136,
10623 Berlin, Germany
[2] Weierstraß–Institut für Angewandte Analysis und Stochastik,
Mohrenstr. 39, 10117 Berlin, Germany

Abstract

Due to transaction costs, illiquid markets, large investors or risks from an unprotected portfolio the assumptions in the classical Black–Scholes model become unrealistic and the model results in strongly or fully nonlinear, possibly degenerate, parabolic diffusion–convection equations, where the stock price, volatility, trend and option price may depend on the time, the stock price or the option price itself.

In this chapter we will be concerned with several models from the most relevant class of nonlinear Black–Scholes equations for American options with a volatility depending on different factors, such as the stock price, the time, the option price and its derivatives.

We will analytically approach the option price by following the ideas proposed by Ševčovič and transforming the free boundary problem into a fully nonlinear nonlocal parabolic equation defined on a fixed, but unbounded domain. Finally, we will present the results of a split–step finite difference scheme for various volatility models including the Leland model, the Barles and Soner model and the Risk adjusted pricing methodology model.

1. Introduction

The strong interest in pricing *financial derivatives* – among them in pricing options – arises from the fact that financial derivatives, also called contingent claims, can be used to minimize losses caused by price fluctuations of the underlying assets. This process of protection is called *hedging*. There is a variety of financial products on the market, such as futures, forwards, swaps and options. In this chapter we will focus on American Call options.

We recall that an *American Call* option is a contract where at any time before a prescribed time in the future, known as the *expiry date* T, the owner of the option, known as the *holder*, may purchase a prescribed asset, known as the *underlying asset* $S(t)$, for a prescribed amount, known as the exercise or *strike price* K. The opposite party, or the *writer*, has the obligation to sell the asset if the holder chooses to excercise his right. The value of the American Call option at the time of execution, known as the *pay-off function*, is

$$V(S,t) = (S-K)^+.$$

Option pricing theory has made a great leap forward since the development of the Black–Scholes option pricing model by Black and Scholes [6] in 1973 and previously by Merton [44]. The solution of the famous *(linear) Black–Scholes equation*

$$0 = V_t + \frac{1}{2}\sigma^2 S^2 V_{SS} + rSV_S - rV, \quad 0 < S < S_f(t),\, t \in (0,T), \tag{1}$$

where V denotes the value of the option and r the riskless interest rate, provides both an option pricing formula for an American Call option and a hedging portfolio that replicates the contingent claim. This is true under the assumption that the market is *complete*, which means that any derivative and any asset can be replicated or hedged with a portfolio of other assets in the market, cf. [61].

However, this assumption of a complete market is never fulfilled in reality. Due to transaction costs [4, 8, 41], large investor preferences [23, 24, 50] and incomplete markets [55] the classical model results in strongly or fully nonlinear, possibly degenerate, parabolic convection–diffusion equations, where both the volatility σ and the drift μ can depend on the time t, the stock price S or the derivatives of the option price V itself. Here, we will be concerned with several transaction cost models from the most relevant class of nonlinear Black–Scholes equations for American options with a constant drift μ and a nonconstant *modified volatility function*

$$\tilde{\sigma}^2 := \tilde{\sigma}^2(t, S, V_S, V_{SS}).$$

Under these circumstances (1) becomes the following *nonlinear Black–Scholes equation*:

$$0 = V_t + \frac{1}{2}\tilde{\sigma}^2(t, S, V_S, V_{SS}) S^2 V_{SS} + rSV_S - rV, \quad S > 0,\, t \in (0,T) \tag{2}$$

Studying (1) for an American Call option would be redundant, since the value of an American Call option equals the value of a European Call option if no dividends are paid and the volatility is constant. In order to make the model more realistic, we will consider a modification of (2) for American options, where S pays out a *continuous dividend* $qSdt$ in a time step dt:

$$0 = V_t + \frac{1}{2}\tilde{\sigma}^2(t, S, V_S, V_{SS}) S^2 V_{SS} + (r-q)SV_S - rV, \quad S > 0,\, t \in (0,T), \tag{3}$$

where the dividend yield q is constant.

Remark 1.1 *Most dividend payments on an index — such as the Dow Jones Industrial Average (DJIA) or the Standard and Poor's 500 (S&P500) — are so frequent that they can be modeled as a continuous payment, which is the case in* (3). *However, if companies only make two or four dividend payments per year, then one has to treat the dividend payments discretely and the question of how to incorporate* discrete dividend payments *into the Black–Scholes equation arises.*

Even though in this work we will focus on the case of continuous dividend payments, we briefly review the results for discrete dividend payments from [63] in the sequel.

We assume that there is only one dividend payment of the dividend yield q during the lifetime of the option at the dividend date t_q. *Neglecting other factors, such as taxes, the asset price S must decrease exactly by the amount of the dividend payment q at time t_q. Thus we have the* jump condition

$$S(t_q^+) = (1-q)S(t_q^-),$$

where t_q^-, t_q^+ denote the moments just before and after the dividend date t_q. This leads to the following effect on the option price:

$$V(S,t_q^-) = V((1-q)S,t_q^+), \tag{4}$$

i.e. the value of the option at S and time t_q^- is the same as the value immediately after the dividend date t_q but at the asset value $(1-q)S$. In order to calculate the value of a Call option with one dividend payment we solve the Black–Scholes equation from expiry $t = T$ until $t = t_q^+$ and use the relation (4) *to compute the values at $t = t_q^-$. Finally, we continue to solve the Black–Scholes equation backwards starting at $t = t_q^-$ using these values as the initial data. The boundary conditions, that are discussed in the next section, do not need to be modified for this case.*

In the mathematical sense equations (2) and (3) are called convection–diffusion equations. The second-order term $\frac{1}{2}\tilde{\sigma}^2(t,S,V_S,V_{SS})S^2V_{SS}$ is responsible for the *diffusion*, the first-order term rSV_S or $(r-q)SV_S$ is the *convection* term and $-rV$ can be interpreted as the *reaction* term (cf. [53, 62]).

In the financial sense, the partial derivatives indicate the sensitivity of the option price V to the corresponding parameter and are called *Greeks*. The option delta is denoted by $\Delta = V_S$, the option gamma by $\Gamma = V_{SS}$ and the option theta by $\theta = V_t$ [33].

Since American options can be exercised at any time before expiry, we need to find the optimal time t of exercise, known as the *optimal exercise time*. At this time, which mathematically is a *stopping time*, the asset price reaches the *optimal exercise price* or *optimal exercise boundary* $S_f(t)$.

This leads to the formulation of the problem for American options by dividing the domain $[0,\infty[\times[0,T]$ of (3) into two parts along the curve $S_f(t)$ and analyzing each of them (see Fig. 1(a)). Since $S_f(t)$ is not known in advance but has to be determined in the process of the solution, the problem is called *free boundary value problem* (see e.g. [67]).

For different numerical approaches, the free boundary problem for American options can be reformulated into a *linear complementary problem*, a *variational inequality* and a *minimization problem* [26]. Here, we will only consider the formulation as a free boundary problem.

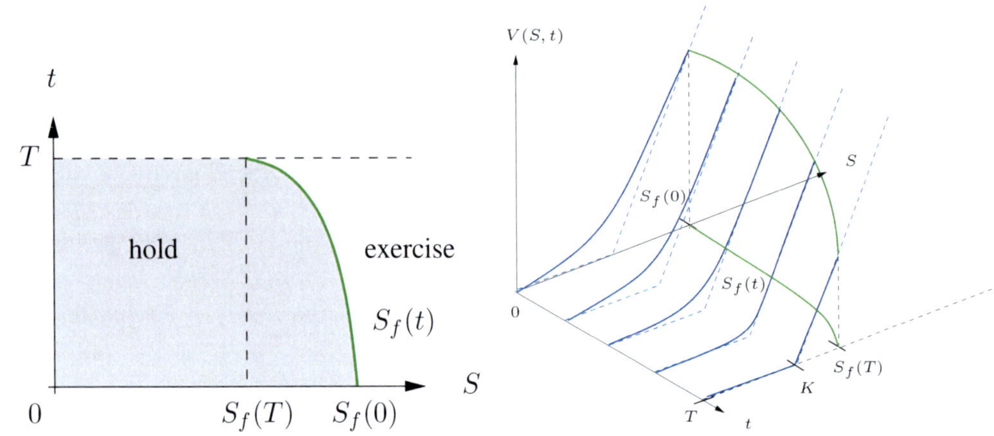

(a) Exercising and holding regions.

(b) Schematical value $V(S,t)$.

Figure 1. American Call.

For the American Call option the *spatial* domain is divided into two regions by the *free boundary* $S_f(t)$, the *stopping region* $S_f(t) < S < \infty$, $0 \le t \le T$, where the option is exercised or dead with $V(S,t) = S - K$ and the *continuation region* $0 \le S \le S_f(t)$, $0 \le t \le T$, where the option is held or stays alive and equation (3) is valid under the following terminal and boundary conditions:

$$\begin{aligned} V(S,T) &= (S-K)^+ && \text{for } 0 \le S \le S_f(T) \\ V(0,t) &= 0 && \text{for } 0 \le t \le T \\ V(S_f(t),t) &= S_f(t) - K && \text{for } 0 \le t \le T \\ V_S(S_f(t),t) &= 1 && \text{for } 0 \le t \le T \\ S_f(T) &= \max(K, rK/q). \end{aligned} \tag{5}$$

For the sake of simplicity we will assume $r > q$ in this chapter, and therefore we have $S_f(T) = rK/q$ for the American Call.

The structure of the value of an American Call can be seen Fig. 1(b), where we notice that the free boundary $S_f(t)$ determines the position of the exercise.

For American options, in general, analytic valuation formulae are not available, except for a few special types, which we are not going to address in this chapter. Those types are Calls on an asset that pays discrete dividends and *perpetual* Calls — meaning Calls with an infinite time to expiry [40]. For the other types, there are various kinds of analytical and numerical approximations that will be discussed in this chapter.

The structure of this chapter is as follows. In Section 2. several nonconstant volatility models that lead to the nonlinearity of the Black–Scholes equation will be introduced. The focus of this chapter is the solution of the resulting nonlinear problems for American Call options. Since in general, a closed–form solution to the nonlinear Black–Scholes equation for American options does not exist (even in the linear case), we have to solve the problems numerically. The numerical solution and the comparison study for American options will be achieved by the transformation of the free boundary problem (3) subject to (5) into a forward-in-time parabolic equation defined on a fixed (but unbounded) spatial domain (Section 3.). This new problem will be numerically solved by the method of finite differences using an operator splitting technique (Section 4.). It will then be evaluated and concisely discussed in Section 5. thereafter.

2. Volatility Models

The essential parameter of the standard Black–Scholes model, that is not directly observable and is assumed to be constant, is the volatility σ. There have been many approaches to improve the model by treating the volatility in different ways and using a *modified volatility function* $\widetilde{\sigma}(\cdot)$ to model the effects of transaction costs, illiquid markets and large traders, which is the reason for the nonlinearity of (2) and (3). In this section we will first give a brief overview of several volatility models and then focus on the volatility models of transaction costs.

- The constant volatility σ in the standard Black–Scholes model can be replaced by the estimated volatility from the former values of the underlying. This volatility is known as the *historical volatility* [26].

- If the price of the option and the other parameters are known, which is e.g. the case for the European Call and Put options, then the *implied volatility* can be calculated from those Black–Scholes formulae. The implied volatility is the value σ, for which the Black–Scholes equation. is true compared to the real market data. It can be calculated implicitly via the difference between the observed option price V (from the market data) and the Black–Scholes formulae, where all the parameters - except for the implied volatility σ - are taken from the market data (the stock price S, the time t, the expiration date T, the strike price K, the interest rate r the dividend rate q).

  Considering options with different strike prices K but otherwise identical parameters, we see that the implicit volatility changes depending on the strike price. If the implicit volatility for a certain strike price K is less than the implicit volatility for both the strike price greater and less than K, this effect is called *volatility smile* (see e.g. [39]).

- Replacing the constant volatility with the observed implicit volatilities at each stock price and time leads to the term of the *local volatility* $\widetilde{\sigma} := \widetilde{\sigma}(S,t)$. Dupire [16] examines the dependencies and expresses the local volatility as a function of implicit volatilities.

- Hull and White [32] and Heston [28] develop a model, in which the volatility follows the dynamics of a stochastic process. This is known as the *stochastic volatility*.

- The assumption, that each security is available at any time and any size, or that individual trading will not influence the price, is not always true. Therefore, illiquid markets and large trader effects have been modeled by several authors. In [23] Frey and Stremme and later Frey and Patie [24] consider these effects on the price and come up with the result

$$\widetilde{\sigma} = \frac{\sigma}{1 - \rho\lambda(S)SV_{SS}}, \tag{6}$$

where σ is the historical volatility, ρ is constant, $\lambda(S)$ is a strictly convex function and $\lambda(S) \geq 1$. The function $\lambda(S)$ depends on the pay-off function of the financial derivative. For the European Call option, Frey and Patie show that $\lambda(S)$ is a smooth, slightly increasing function for $S \geq K$. Bordag and Chmakova [7] assume that $\lambda(S)$ is constant and solve the problem (2) with the modified volatility (6) explicitly using Lie-group theory (see also [12]).

The main scope of this chapter is the numerical solution of the nonlinear Black–Scholes equation for the American Call option, where the nonlinearity results from transaction costs. Therefore, after this general overview, we devote our attention to a more detailed description of several transaction cost models.

2.1. Transaction Costs

The Black–Scholes model requires a continuous portfolio adjustment in order to hedge the position without any risk. In the presence of transaction costs it is likely that this adjustment easily becomes expensive, since an infinite number of transactions is needed [40]. Thus, the hedger needs to find the balance between the transaction costs that are required to rebalance the portfolio and the implied costs of hedging errors. As a result to this "imperfect" hedging, the option might be over- or underpriced up to the extent where the riskless profit obtained by the arbitrageur is offset by the transaction costs, so that there is no single equilibrium price but a range of feasible prices. It has been shown that in a market with transaction costs there is no replicating portfolio for the European Call option and the portfolio is required to dominate rather than replicate the value of the option (see [4]). Soner, Shreve and Cvitanič prove in [54] that the minimal hedging portfolio that dominates a European Call is the trivial one (hence holding one share of the stock that the Call is written on), so that efforts have been made to find an alternate relaxation of the hedging conditions to better replicate the pay-offs of derivative securities.

2.1.1. The Model of Leland

Leland's idea of relaxing the hedging conditions is to trade at discrete times [41], which promises to reduce the expenses of the portfolio adjustment. He assumes that the transaction cost $\kappa|\Delta|S/2$, where κ denotes the round trip transaction cost per unit dollar of the transaction and Δ the number of assets bought ($\Delta > 0$) or sold ($\Delta < 0$) at price S, is proportional to the monetary value of the assets bought or sold. Leland derives the relation

$$rB\delta t - \frac{\kappa}{2}|\delta\Delta|S = (V_t + \frac{\sigma^2}{2}S^2V_{SS})\delta t, \tag{7}$$

where B is the bond and r the riskless interest rate, and shows that

$$\frac{\kappa}{2}|\delta\Delta|S = \frac{\sigma^2}{2}\,\mathrm{Le}\,S^2|V_{SS}|\delta t. \tag{8}$$

Here, Le denotes the *Leland number*, which is given by

$$\text{Le} = \sqrt{\frac{2}{\pi}}\left(\frac{\kappa}{\sigma\sqrt{\delta t}}\right), \tag{9}$$

with δt being the transaction frequency (interval between successive revisions of the portfolio) and κ the round trip transaction cost per unit dollar of the transaction. Plugging (8) and $B = \Pi - \Delta S = V - SV_S$ into the equation (7) becomes

$$rV - rSV_S - \frac{\sigma^2}{2} LeS^2|V_{SS}| = V_t + \frac{\sigma^2}{2} S^2 V_{SS}. \tag{10}$$

Therefore, Leland deduces that the option price is the solution of the nonlinear Black–Scholes equation

$$0 = V_t + \frac{1}{2}\tilde{\sigma}^2 S^2 V_{SS} + rSV_S - rV,$$

with the *modified volatility*

$$\tilde{\sigma}^2 = \sigma^2\Big(1 + \text{Le}\,\text{sign}(V_{SS})\Big), \tag{11}$$

where σ represents the historical volatility and Le the Leland number. It follows from the definition of the Leland number (9) that the more frequent the rebalancing (δt smaller), the higher the transaction cost and the greater the value of V.
Leland's model has played a significant role in financial mathematics, even though it has been partly criticized by e.g. Kabanov and Safarian in [37], who prove that Leland's result has a hedging error. The restriction of his model is the convexity of the resulting option price V (hence $V_{SS} > 0$) and the possibility to only consider one option in the portfolio. Hoggard, Whalley and Wilmott study equation (2) with the modified volatility (11) for several underlyings in [30]. An extension to this approach to general pay–offs is obtained by Avellaneda and Parás in [3].

2.1.2. The Model of Barles and Soner

In [4] Barles and Soner derive a more complicated model by following the *utility function approach* of Hodges and Neuberger [29].

Supposing that the proportional transaction cost κ is equal to $a\sqrt{\varepsilon}$ for some constant $a > 0$, they prove that as ε and κ go to 0, V is the unique (viscosity) solution of the nonlinear Black–Scholes equation

$$0 = V_t + \frac{1}{2}\tilde{\sigma}^2 S^2 V_{SS}^2 + rSV_S - rV,$$

where

$$\tilde{\sigma}^2 = \sigma^2\Big(1 + \Psi(e^{r(T-t)}a^2S^2V_{SS})\Big). \tag{12}$$

Here σ denotes the historical volatility, $a = \kappa/\sqrt{\varepsilon}$ and $\Psi(x)$ is the solution to the following nonlinear ordinary differential equation (ODE)

$$\Psi'(x) = \frac{\Psi(x)+1}{2\sqrt{x\Psi(x)} - x}, \quad x \neq 0, \tag{13a}$$

with the initial condition

$$\Psi(0) = 0. \tag{13b}$$

The analysis of this ODE (13) by Barles and Soner in [4] implies that

$$\lim_{x\to\infty} \frac{\Psi(x)}{x} = 1 \quad \text{and} \quad \lim_{x\to-\infty} \Psi(x) = -1. \tag{14}$$

The property (14) encourages to treat the function $\Psi(\cdot)$ as the identity for large arguments and therefore to simplify the calculations. In this case the volatility becomes

$$\tilde{\sigma}^2 = \sigma^2(1 + e^{r(T-t)}a^2 S^2 V_{SS}). \tag{15}$$

The existence of a viscosity solution to (2) for European options with the volatility given by (12) is proved by Barles and Soner in [4] and their numerical results indicate an economically significant price difference between the standard Black–Scholes model and the nonlinear model with transaction costs.

2.1.3. Risk Adjusted Pricing Methodology

In this model, proposed by Kratka in [39] and improved by Jandačka and Ševčovič in [35], the optimal time-lag δt between the transactions is found to minimize the sum of the rate of the transaction costs and the rate of the risk from an unprotected portfolio. That way the portfolio is still well protected with the Risk Adjusted Pricing Methodology (RAPM) and the *modified volatility* is now of the form

$$\tilde{\sigma}^2 = \sigma^2\left(1 + 3\left(\frac{C^2 M}{2\pi} S V_{SS}\right)^{\frac{1}{3}}\right), \tag{16}$$

where $M \geq 0$ is the transaction cost measure and $C \geq 0$ the risk premium measure.

It is worth mentioning that these nonlinear transaction cost models that are described above are all consistent with the linear model if the additional parameters for transaction costs are equal to zero and vanish (Le, $\Psi(\cdot)$, M). We will study these models – more precisely equations (2) and (3) where the volatility is given by the equations (11), (12), (15) and (16) – for American Call options.

In general, an exact analytical solution leading to a closed expression is not known for American options in a market with transaction costs. In the next section we will analytically approach the solution of (3) by a transformation, that facilitates the numerical solution in the section thereafter. We will compare and evaluate the results in the section thereafter.

3. The Fixed Domain Transformation

The equation (3) subject to (5) is a backward-in-time free boundary problem. In order to ease the numerical solution of (3) (5) for American Call options, we transform the problem into a problem posed on a a fixed (unbounded) domain additionally to the forward transformation in time. Hence, the domain does not depend on the free boundary $S_f(t)$ anymore and we simply calculate an algebraic constraint equation for the position of the free boundary. Following the idea of Ševčovič [52] we change the variables to:

$$\tau = T - t, \quad x = \ln\left(\frac{\varrho(\tau)}{S}\right) \Leftrightarrow S = e^{-x}\varrho(\tau), \quad \varrho(\tau) = S_f(T-\tau),$$

such that $x \in \mathbb{R}^+$ and $\tau \in [0, T]$.
Then, we construct a portfolio

$$\Pi(x,\tau) = V(S,t) - SV_S(S,t)$$

by buying $\Delta = V_S$ shares S and selling an option V. Differentiating Π with respect to x and τ gives us

$$\Pi_x = V_S S_x - S_x V_S - SV_{SS}S_x = S^2 V_{SS}$$

and

$$\begin{aligned}\Pi_\tau &= V_S S_\tau + V_t t_\tau - S_\tau V_S - S(V_{SS}S_\tau + V_{St}t_\tau)\\ &= -V_t - \frac{\varrho'(\tau)}{\varrho(\tau)}S^2 V_{SS} + SV_{St}\\ &= -V_t - \frac{\varrho'(\tau)}{\varrho(\tau)}\Pi_x - S\partial_S(-V_t).\end{aligned} \tag{17}$$

Substituting

$$-V_t = \frac{\widetilde{\sigma}^2}{2}S^2 V_{SS} - r(V - SV_S) - qSV_S = \frac{\widetilde{\sigma}^2}{2}\Pi_x - r\Pi - qSV_S$$

from (3) into (17) and using the fact that $-S\partial_S = \partial_x$, we get

$$\begin{aligned}\Pi_\tau &= \frac{\widetilde{\sigma}^2}{2}\Pi_x - r\Pi - qSV_S - \frac{\varrho'(\tau)}{\varrho(\tau)}\Pi_x + \partial_x\left(\frac{\widetilde{\sigma}^2}{2}\Pi_x - r\Pi\right) + S\partial_S(qSV_S)\\ &= \left(\frac{\widetilde{\sigma}^2}{2} - \frac{\varrho'(\tau)}{\varrho(\tau)} - r + q\right)\Pi_x - r\Pi + \frac{1}{2}\partial_x(\widetilde{\sigma}^2\Pi_x).\end{aligned}$$

We therefore obtain

$$0 = \Pi_\tau + \big(b(\tau) - \frac{\widetilde{\sigma}^2}{2}\big)\Pi_x - \frac{1}{2}\partial_x(\widetilde{\sigma}^2\Pi_x) + r\Pi, \tag{18}$$

defined on $x \in \mathbb{R}^+$, $0 \le \tau \le T$, where the coefficient $b(\tau)$ is

$$b(\tau) = \frac{\varrho'}{\varrho}(\tau) + r - q.$$

The terminal condition from (5) in the original variables (S,T) becomes the initial condition in the new variables $(x,0)$:

$$\begin{aligned}\Pi(x,0) &= V(S,T) - SV_S(S,T)\\ &= \begin{cases} -K & \text{for } S > K \Leftrightarrow x < \ln\frac{\varrho(0)}{K} \\ 0 & \text{otherwise}\end{cases}.\end{aligned} \tag{19a}$$

and the boundary conditions from (5) transform to

$$\Pi(x,\tau) = 0 \qquad \text{as } x \to \infty,\ 0 \le \tau \le T, \tag{19b}$$
$$\Pi(0,\tau) = -K \qquad \text{for } 0 \le \tau \le T. \tag{19c}$$

To complete the system of equations that enables the computation of the portfolio Π we need to use the last two conditions of (5) to obtain an expression at the free boundary position $\varrho(\tau)$. Differentiating and evaluating $V(S_f(t),t) = S_f(t) - K$ at the free boundary gives us

$$V_S(S_f(t),t)S_f'(t) + V_t(S_f(t),t) = S_f'(t).$$

Using (5), we conclude that

$$V_t(S_f(t),t) = 0 \text{ for } 0 \le \tau \le T.$$

Computing (3) at the point $(S_f(t),t)$ or at $(0,\tau)$ in the transformed variables yields:

$$\begin{aligned} 0 &= V_t(S_f(t),t) + \frac{1}{2}\widetilde{\sigma}^2\Pi_x(0,\tau) + (r-q)S_f(t)V_S(S_f(t),t) - rV(S_f(t),t) \\ &= \frac{1}{2}\widetilde{\sigma}^2\Pi_x(0,\tau) + rK - q\varrho(\tau). \end{aligned}$$

We remind the reader that we have assumed $r \ge q$ and therefore we obtain the last condition:

$$\varrho(\tau) = \frac{1}{2q}\widetilde{\sigma}^2\Pi_x(0,\tau) + \frac{rK}{q} \quad \text{with} \quad \varrho(0) = \frac{rK}{q}, \tag{19d}$$

where $0 \le \tau \le T$ and $\widetilde{\sigma}^2$ depends on the volatility model we choose. The volatility (11) from Leland's model becomes

$$\widetilde{\sigma}^2 = \sigma^2\Big(1 + Le\,\mathrm{sign}(\Pi_x)\Big), \tag{20a}$$

for Barles' and Soner's model (12) we get

$$\widetilde{\sigma}^2 = \sigma^2(1 + \Psi(e^{r\tau}a^2\Pi_x)), \tag{20b}$$

for the identity model (15) we obtain

$$\widetilde{\sigma}^2 = \sigma^2(1 + e^{r\tau}a^2\Pi_x), \tag{20c}$$

and for the Risk Adjusted Princing Methodology (16) we derive

$$\widetilde{\sigma}^2 = \sigma^2\left(1 + 3\big(\frac{C^2M}{2\pi}\Pi_x\varrho(\tau)e^{-x}\big)^{\frac{1}{3}}\right). \tag{20d}$$

This transformed problem (18) subject to (19) with the corresponding volatilities (20) is solved by the split-step finite-difference method proposed by Ševčovič [52]

Once we have numerically solved the transformed problem by calculating the solution to our portfolio $\Pi(x,\tau)$ and the free boundary $\varrho(\tau)$, we calculate the value of the American Call $V(S,t)$ option by transforming

$$\Pi(x,\tau) = V(S,t) - SV_S(S,t)$$

back to the original variables. Since we know that

$$\frac{\Pi(x,\tau)}{S^2} = \frac{V(S,t)}{S^2} - \frac{V_S(S,t)}{S} = \partial_S\Big(- \frac{V(S,t)}{S}\Big),$$

we integrate the above equation from S to $S_f(t)$, take into account the boundary condition $V(S_f(t),t) = S_f(t) - K$ and obtain:

$$\int_S^{S_f(t)} \frac{\Pi(\ln(\varrho(\tau)/S),\tau)}{S^2} dS = \int_S^{S_f(t)} \partial_S\Big(- \frac{V(S,t)}{S}\Big) dS$$

$$\int_{\ln\frac{\varrho(\tau)}{S}}^{\ln\frac{\varrho(\tau)}{S_f(t)}} \frac{\Pi(x,\tau)}{S^2}(-S)dx = -\frac{V(S_f(t),t)}{S_f(t)} + \frac{V(S,t)}{S}$$

$$S\int_0^{\ln\frac{\varrho(\tau)}{S}} \frac{\Pi(x,\tau)}{e^{-x}\varrho(\tau)}dx = -S\frac{\varrho(\tau)-K}{\varrho(\tau)} + V(S,t)$$

$$V(S,T-\tau) = \frac{S}{\varrho(\tau)}\Big(\varrho(\tau) - K + \int_0^{\ln\frac{\varrho(\tau)}{S}} e^x\Pi(x,\tau)dx\Big). \tag{21}$$

Therefore, (21) yields the price of the American Call option $V(S,t)$ in the presence (and absence) of transaction costs.

4. Numerical Solution

Due to the lack of general closed–form solutions to the Black–Scholes equations, there are various numerical methods for solving Black–Scholes equations for American options.

For European Call and Put options, the Black–Scholes formulae provide the correct answer, but for more complicated contracts in more general settings analytical formulae are seldom available and numerical methods have to be used to solve the problem. These vary from lattice methods (including binomial and trinomial approximations [14]), Monte-Carlo methods using the least-square techniques [34], analytical approximations [5, 11, 42], finite-element discretizations [26] to finite-difference methods [2, 9, 13].

There are numerous other methods for pricing American options including the method of lines [45], front-tracking algorithms [64], penalty methods [68] and many others. One of the standard approaches for solving the Black–Scholes equation for American options consists of the transformation of the original equation into the heat equation posed on a semi–unbounded domain with a free boundary $S_f(t)$ [53, 63]. For a new alternative direct method using the Mellin transformation we refer to [36, 47].

Up to now, an exact analytical formula for the free boundary profile $S_f(t)$ in (3) subject to (5) is not known, but several authors derived approximate expressions to evaluate American Call and Put options in the linear case [25]. Recently, in a promising approach [51], Ševčovič obtained a semi–explicit formula for an American Call in the case of $r > q$. By transforming the linear Black–Scholes equation for the American Call option into a nonlinear parabolic equation on a fixed domain and applying Fourier sine and cosine transformations, he derives a nonlinear singular integral equation determining the shape of the free boundary. This integral equation can be solved effectively by the means of successive iterations.

Another standard method consists of the reformulation of the free boundary problem into a linear complementary problem (LCP) and the solution by the *Projected Successive Over Relaxation (PSOR)* method of Cryer [15]. Alternatively, penalty and front–fixing methods are developed (e.g. in [22, 46]). A disadvantage of these methods is the change of the underlying model.

A different approach [31] is based on a recursive calculation of the early exercise boundary, estimating the boundary only at some points and then approximating the whole boundary by Richardson extrapolation. Explicit boundary tracking algorithms are e.g. a *finite-difference bisection scheme* [38] or the *front–tracking strategy* of Han and Wu [27].

This emphasis of this chapter is on finite-difference schemes, thus other methods will not be further elaborated on here. For more information on numerical methods we refer the reader to [48, 49, 66] and the references therein.

4.1. American Call Option

Now we want to solve the transformed problem from the previous section.

$$0 = \Pi_\tau + \big(b(\tau) - \frac{\widetilde{\sigma}^2}{2}\big)\Pi_x - \frac{1}{2}\partial_x(\widetilde{\sigma}^2\Pi_x) + r\Pi, \quad x \in \mathbb{R}^+, \quad 0 \le \tau \le T \tag{22}$$

with the corresponding volatilities (20) subject to the conditions

$$\begin{aligned} \Pi(x,0) &= \begin{cases} -K & \text{for } x < \ln\frac{\varrho(0)}{K} \\ 0 & \text{otherwise} \end{cases} \\ \Pi(x,\tau) &= 0 \quad \text{as } x \to \infty,\ 0 \le \tau \le T, \\ \Pi(0,\tau) &= -K \quad \text{for } 0 \le \tau \le T, \end{aligned} \tag{23}$$

and the constraint

$$\varrho(\tau) = \frac{1}{2q}\widetilde{\sigma}^2\Pi_x(0,\tau) + \frac{rK}{q} \quad \text{with} \quad \varrho(0) = \frac{rK}{q}. \tag{24}$$

We therefore first describe the solution of (22) subject to (23) and (24) with the corresponding volatilities (20) by finite-difference schemes and then present the numerical results.

4.2. Finite-Difference Schemes

There have been many approaches to calculate the value of an American option numerically by compact finite-difference schemes in the absence of transaction costs. Recently, Tangman et al. [59, 60] introduced a compact scheme of order $(2, 4)$. Two other compact schemes, known as the *Numerov-type* (see [58, 66]) and the *Crandall-Douglas scheme* (see [43]), are analyzed for linear Black–Scholes equations. However, these schemes are not directly transferable to the model in the presence transaction costs.

In order to find a solution for the nonlinear Black–Scholes equation (22) subject to (23) with the corresponding volatilities (20) and the constraint (24), Ševčovič suggests to combine two approaches that solve the problem for the American Call with a constant volatility numerically [52]. One of them is the transformation of the problem into a variational inequality and its solution by the PSOR algorithm [26, 53]. The other one is the derivation of a nonlinear integral equation for the position of the free boundary without the knowledge of the price itself [40, 64].

Even though these methods are not directly applicable, since they require a constant volatility σ, this approach is successful when it is combined with an operator splitting technique. The idea is to discretize (22) in time, to split the equation into a convective and a diffusive part and to find an approximation for the solution pair (Π, ϱ) at each time level. The detailed derivation is given in the sequel.

4.2.1. Grid

We discretize the problem (22) subject to the conditions (23) with the corresponding volatilities (20) by confining the unbounded domain $x \in \mathbb{R}^+$ and $\tau \in [0, T]$ to $x \in (0, R)$ with $R > 0$ sufficiently large (see [52]). For the calculation Ševčovič chooses to take $R = 3$, since this is equivalent to $S \in (S_f(t)e^{-R}, S_f(t))$ and yields a good approximation for $S \in (0, S_f(t))$ (as the transformation was $S = S_f(t)e^{-x}$). In the sequel we refer to $h > 0$ as the spatial step and to $k > 0$ as the time step, $x_i = ih$, $i \in [0, N]$, $R = Nh$ and $\tau_n = nk$, $n \in [0, M]$, $T = Mk$ (see Fig. 2).

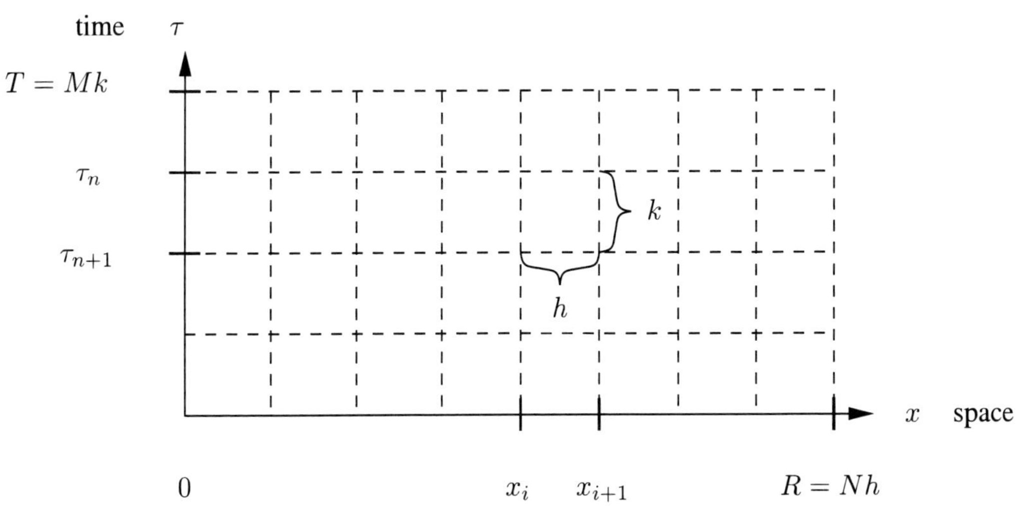

Figure 2. Uniform grid for an American Call option.

The approximate solution of (22) in x_i at time τ_n is denoted by $\Pi_i^n := \Pi(x_i, \tau_n)$, the value of the free boundary at time τ_n by $\varrho^n := \varrho(\tau_n)$ and the value of the coefficient $b(\tau)$ at τ_n by $b^n := b(\tau_n)$.

We treat the initial and boundary conditions (23) in the following way:

$$\begin{aligned} \Pi_i^0 &= \Pi(x_i, 0) = \begin{cases} -K & \text{for } x_i < \ln \frac{\varrho(0)}{K} = \ln \frac{r}{q} \\ 0 & \text{otherwise} \end{cases}, \\ \Pi_0^n &= -K, \\ \Pi_N^n &= 0. \end{aligned} \tag{25}$$

4.2.2. Difference Quotients

We denote the forward difference quotient with respect to the spatial variable in x_i at time τ_n with the spatial step size h by:

$$D_h^+ \Pi_i^n := \frac{\Pi_{i+1}^n - \Pi_i^n}{h} \approx \Pi_x(x_i, \tau_n),$$

the backward difference quotient by:

$$D_h^- \Pi_i^n := \frac{\Pi_i^n - \Pi_{i-1}^n}{h} \approx \Pi_x(x_i, \tau_n)$$

and the central difference quotient by

$$D_h^0 \Pi_i^n := \frac{\Pi_{i+1}^n - \Pi_{i-1}^n}{2h} \approx \Pi_x(x_i, \tau_n),$$

omitting the truncation error $\mathrm{O}(h)$, $\mathrm{O}(h)$ and $\mathrm{O}(h^2)$, respectively. For the second spatial derivative we introduce the standard difference quotient

$$D_h^2 \Pi_i^n := \frac{\Pi_{i+1}^n - 2\Pi_i^n + \Pi_{i-1}^n}{h^2} \approx \Pi_{xx}(x_i, \tau_n),$$

with the error term $\mathrm{O}(h^2)$.

4.3. Volatility Functions

The volatilities (20) can all be written in the form

$$\left(\widetilde{\sigma}_i^n\right)^2 = \sigma^2(1 + s_i^n),$$

where s_i^n denotes the volatility correction in x_i at time τ_n. We choose forward differences to approximate Π_x in the volatility formulae, so that for Leland's model with the volatility (20a) our volatility correction becomes

$$s_i^n = Le \,\mathrm{sign}(D_h^+ \Pi_i^n), \tag{26a}$$

for the volatility correction in Barles' and Soner's model with the volatility (20b) we get

$$s_i^n = \Psi\left(e^{r\tau_n} a^2 D_h^+ \Pi_i^n\right), \tag{26b}$$

for the volatility correction in case of treating $\Psi(\cdot)$ as the identity with the original volatility (20c) we obtain

$$s_i^n = e^{r\tau_n} a^2 D_h^+ \Pi_i^n, \tag{26c}$$

and for the volatility (20d) in the Risk Adjusted Pricing Methodology (RAPM) the volatility correction is

$$s_i^n = 3\left(\frac{C^2 M}{2\pi} D_h^+ \Pi_i^n \varrho^n e^{-x_i}\right)^{\frac{1}{3}}. \tag{26d}$$

4.4. The Treatment of the Free Boundary

We discretize the free boundary (24) by approximating the spatial derivative at the origin $x = 0$ by forward differences and obtain:

$$\varrho^n = \frac{1}{2q}\sigma^2(1+s_0^n)D_h^+\Pi_0^n + \frac{rK}{q} \quad \text{with} \quad \varrho^0 = \frac{rK}{q}, \tag{27}$$

where s_0^n denotes (26) at $x = 0$ depending on the volatility model.

Note, that in case of the RAPM, where the volatility correction is given by equation (26d), s_0^n depends on ϱ^n and therefore ϱ^n in (27) is expressed by a fixed point equation.

Remark 4.1 *For the American Call option (in contrast to the American Put option) it is possible to derive a series for the location of the optimal exercise boundary close to expiry using standard asymptotic analysis [1, 63]. This local analysis of the free boundary $S_f(t)$ yields*

$$S_f(t) \sim S_f(T)\left(1 + \xi_0\sqrt{\frac{1}{2}\sigma^2(T-t)} + \dots\right), \qquad \text{as } t \to T, \tag{28}$$

where $\xi_0 = 0.9034\dots$ is a universal constant *of Call option pricing. Equation* (28) *can be rewritten as*

$$\varrho(\tau) \sim \varrho(0)\left(1 + \xi_0\sqrt{\frac{1}{2}\sigma^2(\tau)} + \dots\right), \qquad \text{as } \tau \to 0. \tag{29}$$

With only very few terms we get a fairly accurate result for the free boundary and thus equation (29) *will serve us as a check for the case of a constant volatility $\widetilde{\sigma}^2 = \sigma^2$ (see Fig. 3). Note that this result is especially useful in the first time levels of a numerical calculation where rapid changes in $\varrho(\tau)$ influence the whole solution region.*

4.5. The Splitting in Time Method

We approximate the time derivative of (22) by backward differences $D_k^-\Pi_i^n$, the first and second spatial derivatives by central differences $D_h^0\Pi_i^n$ and $D_h^2\Pi_i^n$. Then, (22) becomes:

$$0 = D_k^-\Pi_i^n + \left(b^n - \frac{\sigma^2}{2}(1+s_i^n)\right)D_h^0\Pi_i^n - \frac{1}{2}\partial_x\left(\sigma^2(1+s_i^n)D_h^0\Pi_i^n\right) + r\Pi_i^n \tag{30}$$

subject to the Dirichlet conditions (25). We introduce an intermediate step at time $\tau_{n-\frac{1}{2}}$, such that

$$D_k^-\Pi_i^n = \frac{\Pi_i^n - \Pi_i^{n-1}}{k} = \frac{\Pi_i^n - \Pi_i^{n-\frac{1}{2}} + \Pi_i^{n-\frac{1}{2}} - \Pi_i^{n-1}}{k},$$

and then split the problem (30) into a *convective part* with the linear first-order term $b^n D_h^0\Pi_i^n$:

$$0 = \frac{\Pi_i^{n-\frac{1}{2}} - \Pi_i^{n-1}}{k} + b^n D_h^0\Pi_i^n \tag{31}$$

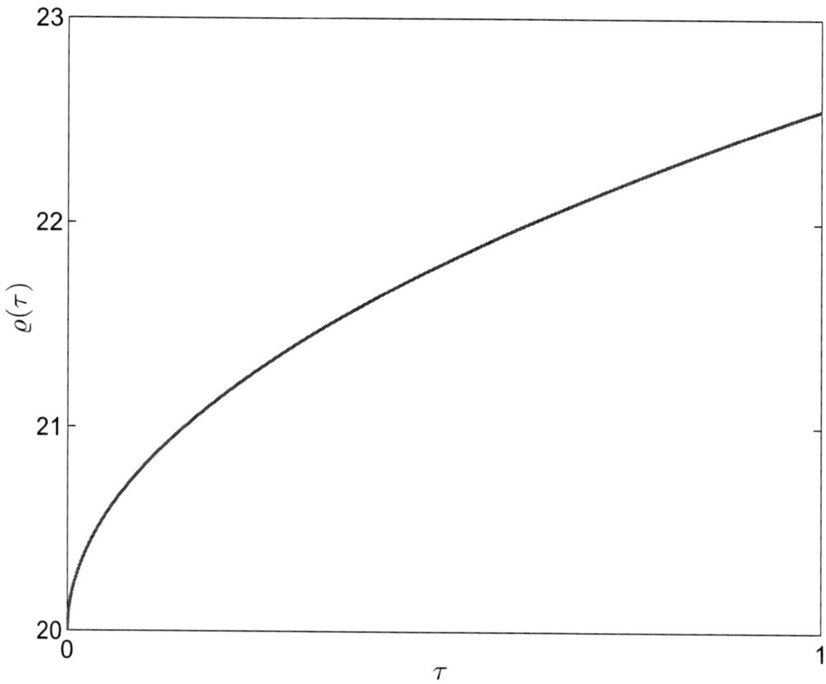

Figure 3. Asymptotic solution for the free boundary $\varrho(\tau)$ with $T = 1$, $K = 10$, $\sigma = 0.2$, $r = 0.1$, $q = 0.05$.

and a *diffusive part* with the nonlinear first- and second-order terms $\sigma^2/2(1 + s_i^n)D_h^0\Pi_i^n$ and $-\partial_x\big(\sigma^2/2(1 + s_i^n)D_h^0\Pi_i^n\big)$:

$$0 = \frac{\Pi_i^n - \Pi_i^{n-\frac{1}{2}}}{k} - \frac{\sigma^2}{2}(1 + s_i^n)D_h^0\Pi_i^n - \frac{1}{2}\partial_x\big(\sigma^2(1 + s_i^n)D_h^0\Pi_i^n\big) + r\Pi_i^n. \tag{32}$$

Assuming that $D_h^0\Pi_i^n \approx D_h^0\Pi_i^{n-\frac{1}{2}}$, which is reasonable for small time steps k, we can approximate the convective part (31) as

$$0 = \frac{\Pi_i^{n-\frac{1}{2}} - \Pi_i^{n-1}}{k} + b^n D_h^0\Pi_i^{n-\frac{1}{2}}. \tag{33}$$

Now the solution to (32)-(33) gives a good approximation to the solution of (30) (see [52]). This decomposition of the problem is called *Lie-Splitting* and is a spitting of order 1 in time.

4.5.1. Convective Part

First, we solve the convective part (33), which can be approximated by an explicit solution to the *transport equation*

$$\Pi_\tau + b(\tau)\Pi_x = 0, \tag{34}$$

for $(x,\tau) \in \mathbb{R} \times [0,T]$, subject to the boundary and initial conditions

$$\Pi(0,\tau) = -K,$$
$$\Pi(x,0) = \begin{cases} -K & \text{for } x < \ln\frac{r}{q} \\ 0 & \text{otherwise} \end{cases} = \Pi^0(x). \tag{35}$$

We then know by the theory of partial differential equations (see e.g. [20]) that the solution for this problem (34)–(35) is

$$\Pi(x,\tau) = \Pi(x - \int_0^\tau b(s)\,ds, 0) = \Pi^0(x - \int_0^\tau b(s)\,ds) \tag{36}$$

with the primitive function $\int b(s)\,ds = B(\tau) + c = \ln \varrho(\tau) + (r-q)\tau + c$. Hence, considering the problem (34) for $(x_i,\tau_j) \in \mathbb{R} \times [\tau_{n-1},\tau_n]$ subject to the boundary and initial conditions

$$\Pi(0,\tau_j) = -K,$$
$$\Pi(x_i,\tau_{n-1}) - \Pi^{n-1}(x_i), \tag{37}$$

we know that the solution is given by

$$\Pi(x_i,\tau_j) = \Pi\Big(x_i - \int_{\tau_{n-1}}^{\tau_j} b(s)\,ds, \tau_{n-1}\Big)$$
$$= \begin{cases} \Pi(\xi_i^j,\tau_{n-1}) & \text{for } \xi_i^j > 0 \\ -K & \text{otherwise,} \end{cases} \tag{38}$$

where we set $\xi_i^j = x_i - B(\tau_j) + B(\tau_{n-1}) = x_i - \ln\frac{\varrho^j}{\varrho^{n-1}} - (\tau_j - \tau_{n-1})(r-q)$. Then we can write

$$\Pi_i^{n-\frac{1}{2}} = \begin{cases} \Pi(\xi_i^n,\tau_{n-1}) & \xi_i^n = x_i - \ln\frac{\varrho^n}{\varrho^{n-1}} - k(r-q) > 0 \\ -K & \text{otherwise.} \end{cases} \tag{39}$$

Here, we use a linear approximation between the discrete values $\Pi(x_i,\tau_{n-1})$, $i \in \mathbb{N}$ in order to compute the value of $\Pi(\xi_i^n,\tau_{n-1})$.
Hence, (39) is the solution to the convective part (33) of the problem (30).

4.5.2. Diffusive Part

We solve the diffusive part (32) of the problem (30) by the finite-difference method. We approximate the second spatial derivative by central differences $D_h^2\Pi_i^n$ and the first spatial derivative by both central $D_h^0\Pi_i^n$ and backward differences $D_h^-\Pi_i^n$. Then, (32) becomes:

$$\begin{aligned} 0 = {} & \frac{\Pi_i^n - \Pi_i^{n-\frac{1}{2}}}{k} - \frac{\sigma^2}{2}(1+s_i^n)\frac{\Pi_{i+1}^n - \Pi_{i-1}^n}{2h} + r\Pi_i^n \\ & - \frac{\sigma^2}{2}\left((1+s_i^n)\frac{\Pi_{i+1}^n - 2\Pi_i^n + \Pi_{i-1}^n}{h^2} + \frac{(1+s_i^n)-(1+s_{i-1}^n)}{h}\frac{\Pi_i^n - \Pi_{i-1}^n}{h}\right) \\ = {} & \frac{\Pi_i^n - \Pi_i^{n-\frac{1}{2}}}{k} - \frac{\sigma^2}{2}(1+s_i^n)\frac{\Pi_{i+1}^n - \Pi_{i-1}^n}{2h} + r\Pi_i^n \\ & - \frac{\sigma^2}{2}\left((1+s_i^n)\frac{\Pi_{i+1}^n - \Pi_i^n}{h^2} - (1+s_{i-1}^n)\frac{\Pi_i^n - \Pi_{i-1}^n}{h^2}\right). \end{aligned}$$

Rearranging leads to a tridiagonal system of equations

$$\Pi_i^{n-\frac{1}{2}} = a_i^n \Pi_{i-1}^n - b_i^n \Pi_i^n + c_i^n \Pi_{i+1}^n, \tag{40}$$

with the coefficients

$$\begin{aligned} a_i^n &= \frac{\sigma^2}{2}(1+s_i^n)\frac{k}{2h} - \frac{\sigma^2}{2}(1+s_{i-1}^n)\frac{k}{h^2}, \\ b_i^n &= 1 + kr + \frac{\sigma^2}{2}(1+s_i^n)\frac{k}{h^2} + \frac{\sigma^2}{2}(1+s_{i-1}^n)\frac{k}{h^2}, \\ c_i^n &= -\frac{\sigma^2}{2}(1+s_i^n)\frac{k}{2h} - \frac{\sigma^2}{2}(1+s_i^n)\frac{k}{h^2}. \end{aligned}$$

Equation (40) can be written in the form of matrices:

$$\Pi^{n-\frac{1}{2}} = A^n \Pi^n + d^n, \tag{41}$$

where

$$\Pi^n = \left(\Pi_1^n, \cdots, \Pi_{N-1}^n\right)^\top \in \mathbb{R}^{N-1},$$

$$A^n = \begin{pmatrix} b_1^n & c_1^n & 0 & \cdots & 0 \\ a_2^n & b_2^n & \ddots & \ddots & \vdots \\ 0 & \ddots & \ddots & \ddots & 0 \\ \vdots & \ddots & \ddots & b_{N-2}^n & c_{N-2}^n \\ 0 & \cdots & 0 & a_{N-1}^n & b_{N-1}^n \end{pmatrix} \in \mathbb{R}^{(N-1)\times(N-1)},$$

and

$$d^n = \left(a_1^n \Pi_0^n, \quad 0, \quad \cdots, \quad 0, \quad c_{N-1}^n \Pi_N^n\right)^\top \in \mathbb{R}^{N-1}.$$

Therefore, (41) solves the diffusive part (32) of the problem (30).

Now, we have a set of nonlinear equations (26), (27), (39) and (41) that delivers the solution to our portfolio $\Pi(x,\tau)$ and to the free boundary $\varrho(\tau)$, from which we can calculate the value of the American Call option $V(S,t)$ with equation (21).

In order to see the dependencies of the equations, we rewrite them in the following abstract form:

$$\begin{aligned} s^n &= \mathcal{D}(\Pi^n, \varrho^n), \\ \varrho^n &= \mathcal{F}(\Pi^n, s^n) = \mathcal{F}(\Pi^n, \varrho^n), \\ \Pi^{n-\frac{1}{2}} &= \mathcal{G}(\Pi^{n-1}, \varrho^n, \varrho^{n-1}) = \mathcal{G}(\Pi^{n-1}, \varrho^n), \\ A(s^n)\Pi^n &= A(\Pi^n, \varrho^n)\Pi^n = \Pi^{n-\frac{1}{2}} - d(s^n), \end{aligned} \tag{42}$$

where

$$s^n = \left(s_0^n, \quad \cdots, \quad s_N^n \right)^\top \in \mathbb{R}^{N+1},$$

$\mathcal{D}(\cdot)$ denotes the right-hand side of (26), $\mathcal{F}(\cdot)$ is the right-hand side of (27), $\mathcal{G}(\cdot)$ is the right-hand side of the transport equation (39), $A(\cdot)$ is the tridiagonal matrix and $d(s^n)$ the vector as defined in (41).

As we can see by this notation (42), both ϱ^n and Π^n are given in terms of themselves, hence each is given in terms of ϱ^n and Π^n. This problem can be approximately solved by a successive fixed point iteration over $p \in \mathbb{N}$ at each time level n.

Following Ševčovič [52] we define for $n \geq 1$: $\Pi^{n,0} = \Pi^{n-1}$, $\varrho^{n,0} = \varrho^{n-1}$ and $s^{n,0} = s^{n-1}$. Then the (p+1)-th approximation of Π^n, ϱ^n and s^n is obtained as the solution of the system:

$$\begin{aligned} s^{n,p+1} &= \mathcal{D}(\Pi^{n,p}, \varrho^{n,p}), \\ \varrho^{n,p+1} &= \mathcal{F}\big(\Pi^{n,p}, s^{n,p+1}\big), \\ \Pi^{n-\frac{1}{2},p+1} &= \mathcal{G}\big(\Pi^{n-1,p}, \varrho^{n,p+1}\big), \\ A\big(s^{n,p+1}\big)\Pi^{n,p+1} &= \Pi^{n-\frac{1}{2},p+1} - d\big(s^{n,p+1}\big). \end{aligned} \tag{43}$$

Both the volatility correction $s_i^{n,p+1}$, the free boundary $\varrho^{n,p+1}$ and the solution $\Pi^{n-\frac{1}{2},p+1}$ to the convective part (31) can be directly computed from (26), (27) and (39) respectively. The solution $\Pi^{n,p+1}$ to the diffusive part (32) has to be calculated from the system of equations (41).

Assuming that the system (43) converges to some limiting values $s^{n,p_{max}}$, $\varrho^{n,p_{max}}$, $\Pi^{n-\frac{1}{2},p_{max}}$ and $\Pi^{n,p_{max}}$ at each time level n [52], we can calculate $V(S_i, t_n) = V(e^{-x_i}\varrho^n, T - \tau_n)$ with these values and proceed to the next time level $n+1$.

From (21) we then know that:

$$V(S_i, t_n) = e^{-x_i}\big(\varrho^n - K + \mathcal{I}_i\big), \tag{44}$$

where

$$\begin{aligned} \mathcal{I}_i &= \sum_{j=0}^{i-1} \mathcal{I}_k + \int_{x_{i-1}}^{x_i} e^x \Pi(x,\tau)dx \\ &= \sum_{j=0}^{i-1} \mathcal{I}_k + \frac{x_i - x_{i-1}}{2}\left(e^{x_{i-1}}\Pi_{i-1}^n + e^{x_i}\Pi_i^n\right). \end{aligned}$$

Here, we use the *trapezoidal rule* in order to approximate the integral in equation (21).

We summarize the calculation of the price $V(S,t)$ for the American Call option in the presence or absence of transaction costs by the Algorithm 1 given in the appendix.

5. Comparison Study

Based on the iterative algorithm described in the previous section (Algorithm 1), we solve the transformed Black–Scholes equation (22) with the corresponding volatilities (20) for the American Call option and finally transform $\Pi(x,\tau)$ back to the original option price $V(S,t)$.

The main purpose of this section is to compare the resulting option value $V(S,t)$ and the free boundary $S_f(T-t) = \varrho(\tau)$ for the four different transaction cost models (26) to the linear model (σ constant) and to each other.

We choose $p_{max} = 5$ for the successive iteration over p in our algorithm in order to solve the system (42) with the precision of 10^{-7} [52]. We use the following parameters to calculate $\Pi(x,\tau)$ and $\varrho(\tau)$:

$$r = 0.1, \quad \sigma = 0.2, \quad K = 10, \quad T = 1 \text{ (one year)}, \quad R = 3.$$

We start by comparing the free boundary $\varrho(\tau)$ computed with Algorithm 1 to the asymptotic solution (29) from Remark 4.1 for the linear case ($s_i^n = 0$). In Fig. 4 we observe that for smaller spatial steps $h \to 0$ the free boundary computed by the iterative algorithm converges monotonically towards the asymptotic solution (29) from below.

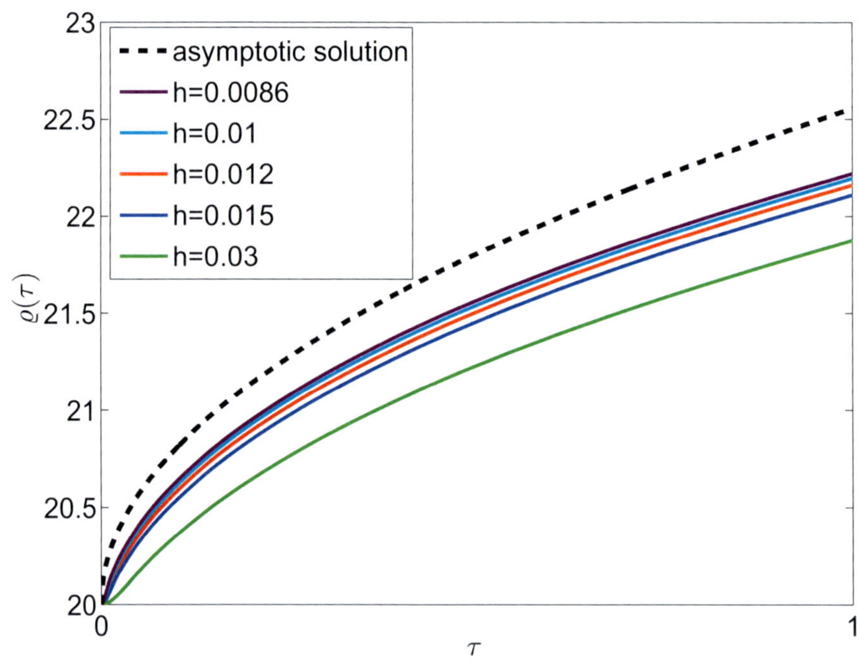

Figure 4. Free boundary for various spatial steps h with a constant time step $k = 0.0008$ and a constant volatility σ^2 computed by Algorithm 1 vs. the asymptotic solution of (29).

We keep the time step $k = 0.0008$ constant and see that for $h = 0.0086$ (purple line) the free boundary at T is computed by our algorithm as $\varrho(T) \approx 22.2201$. The asymptotic solution at T is $\varrho(T) \approx 22.5552$, which means a relative error of 1.49%. The free boundary values for the other spatial steps can be seen in Table 1.

Table 1. Values of the free boundary position for various spatial steps h with a constant time step $k = 0.0008$ and a constant volatility σ^2.

h	0.03	0.015	0.012	0.01	0.0086
$\varrho(T)$	21.8764	22.1111	22.1619	22.1955	22.2201

Since the asymptotic solution of (29) is only an approximation, we are satisfied by our results and take the free boundary $\varrho(T) \approx 22.1111$ for $k = 0.0008$, $h = 0.015$ (blue line in Fig. 4) as our reference solution in the absence of transaction costs for the sake of the computational time.

Fig. 5 shows the structure of the price for the American Call option $V(S,t)$ without transaction costs with $k = 0.0008$ and $h = 0.015$. It is computed with the iterative algorithm described in the previous sections and the parameters above.

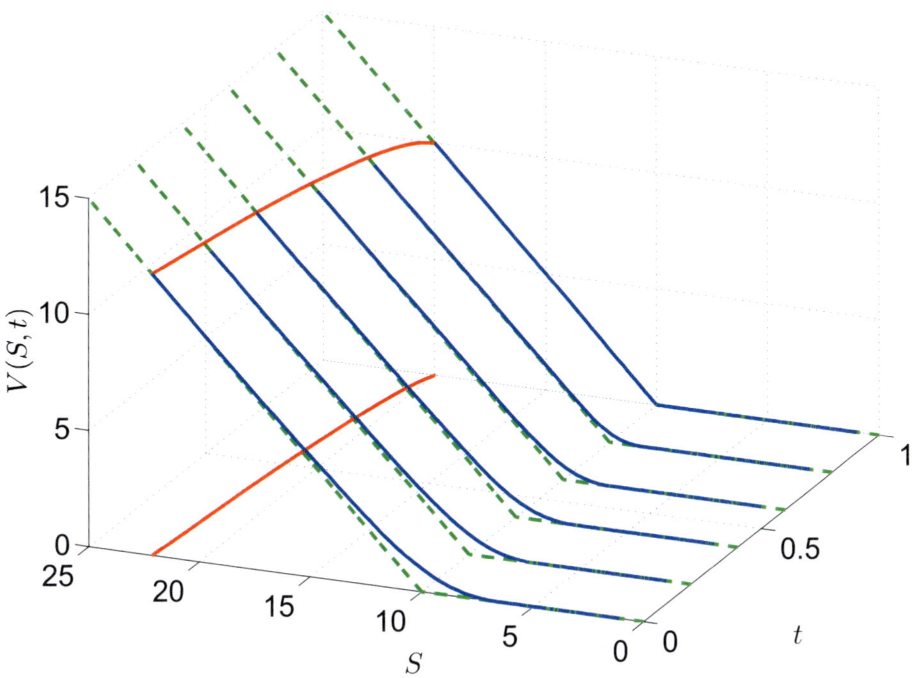

Figure 5. Value of an American Call option $V(S,t)$ in the absence transaction costs computed with Algorithm 1 determined by the free boundary (red line).

The corresponding synthetic portfolio $\Pi(x,\tau)$ in the absence of transaction costs is illustrated in Fig. 6. Note, that we include rounding and discretization errors when transforming $\Pi(x,\tau)$ back into $V(S,t)$, since equation (44) involves an integral approximation. However, the analysis of $V(S,t)$ is more interesting for us and we therefore assume that these errors are sufficiently small due to the chosen mesh.

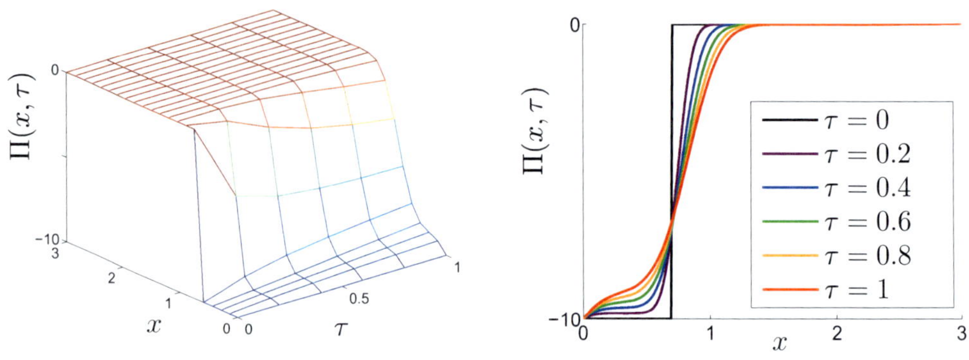

(a) 3-D profile. (b) Profile at different time points.

Figure 6. Value of the synthetic portfolio $\Pi(x, \tau)$ in the absence of transaction costs computed with Algorithm 1.

We now compare the price $V(S, 0)$ computed by Algorithm 1 to the price $V_{PSOR}(S, 0)$ computed by the PSOR algorithm in the linear case $s_i^n = 0$. Fig. 7 shows that with the given mesh size $k = 0.0008$ and $h = 0.015$ the price computed by our algorithm (Fig. 7(a)) only slightly differs from the price computed by the PSOR algorithm (Fig. 7(b)).

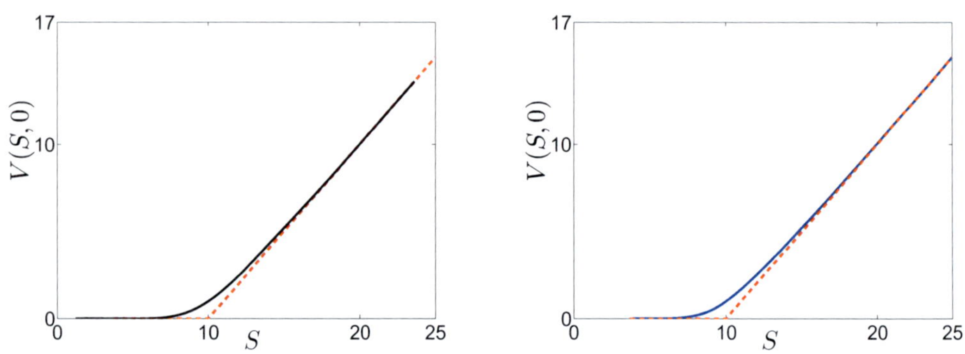

(a) Computed with Algorithm 1. (b) Computed with PSOR.

Figure 7. Price of an American Call option $V(S, 0)$ in the absence of transaction costs and the pay-off $V(S, T)$ (red dotted line).

We calculate the error of accuracy of our computation one year to expiry at $t = 0$, denoted by the ℓ^2-error

$$\mathrm{err}_2(0) = \left(h \sum_{i=0}^{N} |V_{\mathrm{PSOR}}(S_i, 0) - V_i^0|^2 \right)^{\frac{1}{2}},$$

where $V_{\mathrm{PSOR}}(S_i, 0)$ denotes the solution computed by the PSOR algorithm at $S_i = e^{-ih} \varrho(T)$ and $\varrho(T)$ depends on the step size h. For this purpose, we interpolate the solution computed by the PSOR algorithm by the MATLAB routines `spline` and `ppval`.

For V_i^0 we use our corresponding solution, where $k = 0.0008$. The error can be seen in Table 2, which reveals that it is reasonable to assume the accuracy $\mathrm{O}(h)$.

Table 2. ℓ^2-error of accuracy of Algorithm 1 compared to the PSOR algorithm in the absence of transaction costs.

h	0.03	0.015	0.012	0.01	0.0086
ℓ^2-error	0.0365	0.0162	0.0257	0.0084	0.0167

We further compute the free boundary profiles for the four different transaction cost models (26) by Algorithm 1 and compare them to the profile of the free boundary in the absence transaction costs. For our computations we take $k = 0.0008$ and $h = 0.015$.

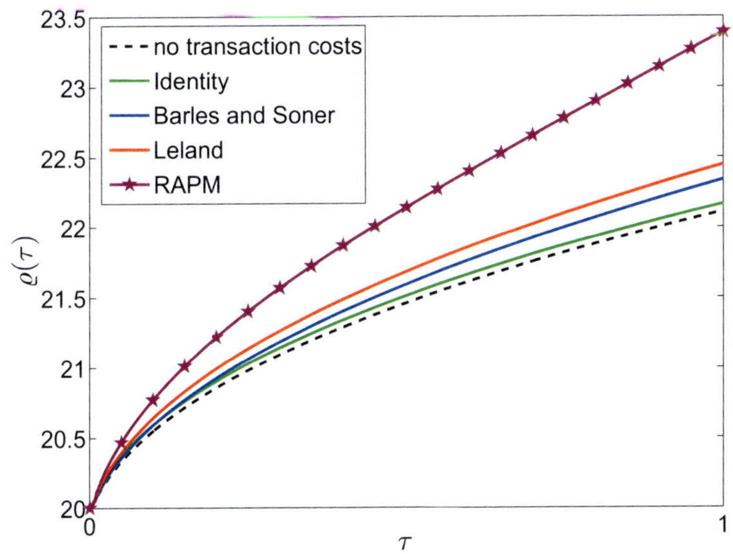

Figure 8. Free boundary positions for various transaction cost models vs. the free boundary profile in the absence of transaction costs.

As expected, we see that for all the transaction cost models the free boundary values are greater than in the case without transaction costs (Fig. 8). With the given parameters the free boundary in the absence of transaction costs is $\varrho(T) \approx 22.11$, followed by the identity model with $a = 0.02$ ($\varrho(T) \approx 22.16$), Barles' and Soner's model with $a = 0.02$ ($\varrho(T) \approx 22.34$), Leland's model with $\delta t = 0.1$, $\kappa = 0.02$ ($\varrho(T) \approx 22.44$) and finally the RAPM with $C = 0.01$, $R = 30$ ($\varrho(T) \approx 23.39$).

Furthermore, we compute the corresponding values $V(S,t)$ for the American Call option by Algorithm 1 and check the price difference between the American Call option with transaction costs and the American Call option without transaction costs

$$V_{\text{nonlinear}}(S,t) - V_{\text{linear}}(S,t).$$

The influence of transaction costs for the four models can be seen in Fig. 9. We notice that the difference is maximal one year to expiry at $t = 0$ and $S \approx 9.5$. The difference is not symmetric, but decreases towards the expiry. This seems plausible, since towards expiry the portfolio can not be adjusted as often at it could be adjusted before. Hence, the transaction costs and the value of the American Call option with transaction costs decrease towards $t = 1$.

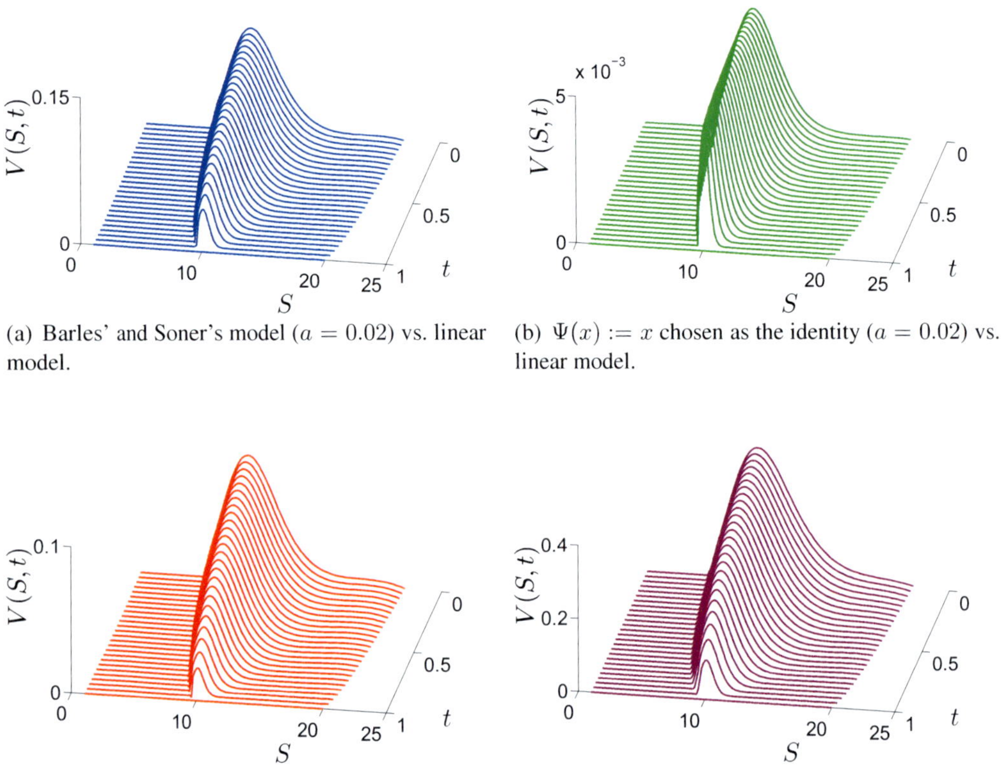

(a) Barles' and Soner's model ($a = 0.02$) vs. linear model.

(b) $\Psi(x) := x$ chosen as the identity ($a = 0.02$) vs. linear model.

(c) Leland's model ($\delta t = 0.1$, $\kappa = 0.02$) vs. linear model.

(d) RAPM model ($M = 0.01$, $C = 30$) vs. linear model.

Figure 9. The influence of transaction costs $V_{\text{nonlinear}}(S,t) - V_{\text{linear}}(S,t)$.

The corresponding prices $V(S,0)$ in the presence of transaction costs can be seen in Fig. 10. At $S \approx 9.5$ with the parameters as indicated above and $k = 0.0008$, $h = 0.015$ the price of the American Call option evaluated with the RAPM transaction cost model is the highest (≈ 1.06). It is followed by Barles' and Soner's model (≈ 0.82), Leland's model (≈ 0.78), the identity model (≈ 0.74) and finally the model in the absence of transaction costs (≈ 0.71). As already shown in Table 2, the linear price computed by our algorithm (light blue solid line in Fig. 10) only slightly deviates from the price computed by the PSOR algorithm (black dotted line in Fig. 10).

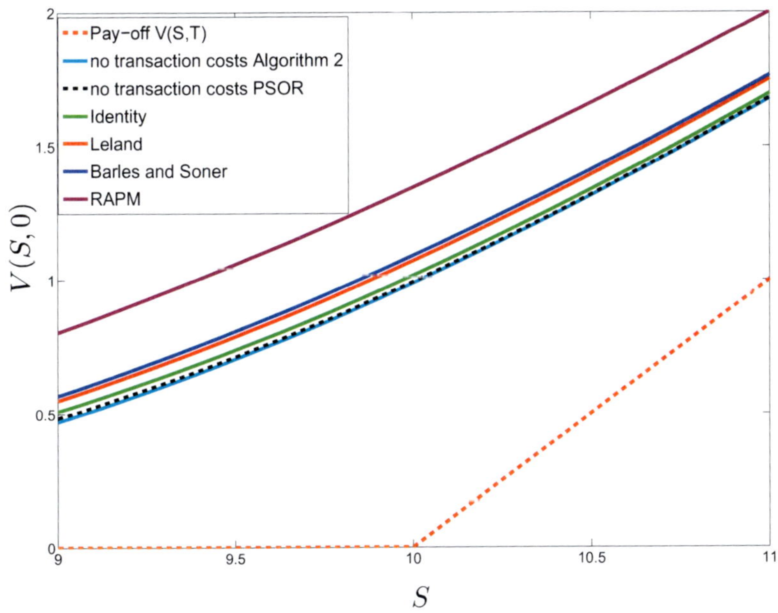

Figure 10. Price of an American Call option $V(S,0)$ for different transaction cost models vs. the price without transaction costs.

For other numerical experiments in the future is recommendable to use rather `C` or `C++` in order to reduce the computational time which is relatively high in `MATLAB`.

Conclusion

In this chapter we solved the nonlinear Black–Scholes equation for American options in the presence of transaction costs. Summing up, our numerical results showed a considerable price difference between linear and nonlinear prices for American Call options.

While we focused in this chapter on standard options (known as *plain–vanilla options*) of American type, our future work will deal with extensions: forward and future contracts, options on futures, more general pay–off functions (e.g. 'cash–or–nothing call') with transaction costs and instalment options.

Moreover, we will consider a higher-order splitting in time, e.g. the well–known *Strang–Splitting* [56] and combine this with modern compact finite difference of high spatial order, like the *Crandall–Douglas Scheme* [43] which is fourth-order accurate in 'space' (i.e. asset price) or the high–order compact methods proposed in [59], [60], [66]. Especially, the method of [60] is promising, since it is already an improvement of the Han and Wu method [27] with a higher order interior scheme and more accurate tracking of the free boundary.

Acknowledgements

The authors acknowledge very stimulating discussions with Daniel Ševčovič in the framework of a bilateral German–Slovakian DAAD project *Fin-Diff-Fin: Finite differences for Financial derivative models.*

Algorithm 1 Computation of the price $V(S,t)$ for the American Call

Appendix

Require: R, T, h, k, M, N, r, K, D, σ, Le, a, C, M

solve the ODE (13) required for the volatility model of Barles and Soner and interpolate the solution
initialize Π^0
initialize the free boundary $\varrho^0 = rK/q$
transform Π^0 into V^0
set $\pi = \Pi^0$ and $v = V^0$
set $\Pi^{1,0} = \Pi^0$ and $\varrho^{1,0} = \varrho^0$
calculate Π^n, ϱ^n at each time level
for $n = 1 : M$ **do**
  calculate $s^{n,p}$, $\varrho^{n,p}$, $\Pi^{n-1/2,p}$ and $\Pi^{n,p}$ in the successive loop over p
  for $p = 1 : p_{max}$ **do**
    calculate the volatility correction $s^{n,p}$ depending on the volatility model using $\Pi^{n,p-1}$ and $\varrho^{n,p-1}$ (in the case of Barles' and Soner's model use the interpolated solution of (13), in the case without transaction costs $s^{n,p} = (0,\cdots,0)^\top \in \mathbb{R}^{N+1}$)
    calculate $\varrho^{n,p}$ using $\Pi^{n,p-1}$ and $s^{n,p}$
    calculate $\Pi^{n-1/2,p}$ using Π^{n-1} and $\varrho^{n,p}$
    fill the matrix $A^{n,p}$ and the vector $d^{n,p}$ with the corresponding coefficients using $s^{n,p}$
    L-R-decompose $A^{n,p} = L^{n,p}R^{n,p}$
    solve $L^{n,p}y^{n,p} = \Pi^{n-1/2,p} - d^{n,p}$ for $y^{n,p}$
    solve $R^{n,p}\Pi^{n,p} = y^{n,p}$ for $\Pi^{n,p}$
    start over with the loop over p
  end for
  set $\Pi^n = \Pi^{n,p}$ and $\varrho^n = \varrho^{n,p}$
  transform Π^n into V^n
  save the solution in the transformed variables in the array
  $\pi = \begin{bmatrix}\pi & [-K; \Pi^n; 0]\end{bmatrix}$
  save the solution in the original variables in the array
  $v = \begin{bmatrix}v & [\varrho^n - K; V^n; 0]\end{bmatrix}$
  start over with the loop over n
end for
plot v at each time level and each stock price, plot ϱ at each time level

References

[1] G. Alobaidi, *American options and their strategies,* Ph.D. thesis, University of Western Ontario, London, Canada, 2000.

[2] J. Ankudinova and M. Ehrhardt, The numerical solution of nonlinear Black–Scholes equations, *Comput. Math. Appl.* **56** (2008), 799–812.

[3] M. Avellaneda and A. Parás, Dynamic hedging portfolios for derivative securities in the presence of large transaction costs, *Appl. Math. Finance* **1** (1994), 165–193.

[4] G. Barles and H.M. Soner, Option pricing with transaction costs and a nonlinear Black–Scholes equation, *Finance Stoch.* **2** (1998), 369–397.

[5] G. Barone-Adesi and R. Whaley, Efficient Analytic Approximation of American Option Values, *J. Finance* **42** (1987), 302–320.

[6] F. Black and M. Scholes, The pricing of options and corporate liabilities, *J. Polit. Econ.* **81** (1973), 637–659.

[7] L. Bordag and A. Chmakova, Explicit solutions for a nonlinear model of financial derivatives, *Int. J. Theor. Appl. Finance* **10** (2007), 1–21.

[8] P. Boyle and T. Vorst, Option Replication in Discrete Time with Transaction Costs, *J. Finance* **47** (1992), 271–293.

[9] M. Brennan and E. Schwartz, The valuation of American put options, *J. Finance* **32** (1977), 449–462.

[10] M. Broadie and J. Detemple, American Option valuation: New bounds, approximations, and a comparison of existing methods, *Rev. Fin. Stud.* **9** (1996), 1211–1250.

[11] P. Carr, R. Jarrow and R. Myneni, Alternative Characterizations of American Put Options, *Math. Finance* **2** (1992), 87–105.

[12] A. Chmakova, *Symmetriereduktionen und explizite Lösungen für ein nichtlineares Modell eines Preisbildungsprozesses in illiquiden Märkten*, Dissertation, Technische Universität Cottbus, 2005 (in german).

[13] G. Courtadon, A More Accurate Finite Difference Approximation for the Valuation of Options, *J. Fin. Quant. Anal.* **15** (1982), 697–703.

[14] J. Cox and S. Ross and M. Rubinstein, Option pricing: A simplified approach, *J. Fin. Econ.* **7** (1979), 229–263.

[15] E. Cryer, The solution of a quadratic programming problem using systematic overrelaxation, *SIAM J. Control* **9** (1971), 385–392.

[16] B. Dupire, Pricing with a Smile, *Risk* **7** (1994), 18–20.

[17] M. Ehrhardt, Discrete Transparent Boundary Conditions for Parabolic Equations, *Z. Angew. Math. Mech.* **77** (1997) S2, 543–544.

[18] M. Ehrhardt, *Discrete Artificial Boundary Conditions*, Ph.D. Thesis, Technische Universität Berlin, 2001.

[19] M. Ehrhardt and R.E. Mickens, A fast, stable and accurate numerical method for the Black–Scholes equation of American options, *Int. J. Theoret. Appl. Finance* **11** (2008), 471–501.

[20] L.C. Evans, *Partial Differential Equations*, American Mathematical Society, Providence, 1998.

[21] G. Ferreyra, The Mathematics Behind the 1997 Nobel Prize in Economics, *e-MATH, What's New in Mathematics*, (electr. magazine of the Amer. Math. Soc.), December 1997.

[22] P.A. Forsyth and K.R. Vetzal, Quadratic Convergence for Valuing American Options using a Penalty Method, *SIAM J. Sci. Comput.* **23** (2002), 2095–2122.

[23] R. Frey and A. Stremme, Market Volatility and Feedback Effects from Dynamic Hedging, *Math. Finance* **4** (1997), 351–374.

[24] R. Frey and P. Patie, *Risk Management for Derivatives in Illiquid Markets: A Simulation Study*, Springer, Berlin, 2002.

[25] R. Geske and R. Roll, On valuing American call options with the Black–Scholes European formula, *J. Finance* **89** (1984), 443–455.

[26] M. Günther and A. Jüngel, *Finanzderivate mit MATLAB*, Vieweg, 2003 (in german).

[27] H. Han and X. Wu, A fast numerical method for the Black–Scholes equation of American options, *SIAM J. Numer. Anal.* **41** (2003), 2081–2095.

[28] S. Heston, A Closed–Form Solution for Options with Stochastic Volatility with Applications to Bond and Currency Options, *Rev. Fin. Studies* **6** (1993), 327–343.

[29] S. Hodges and A. Neuberger, Optimal replication of contingent claims under transaction costs, *Review of Futures Markets* **8** (1989), 222–239.

[30] T. Hoggard and A. E. Whalley and P. Wilmott, Hedging option portfolios in the presence of transaction costs, *Adv. Fut. Opt. Res.* **7** (1994), 21–35.

[31] J.-Z. Huang, M.G. Subrahmanyam and G.G. Yu, Pricing and Hedging American Options: A Recursive Integration Method, *Rev. Fin. Studies* **9** (1996), 277–300.

[32] J. Hull and A. White, The Pricing of Options on Assets with Stochastic Volatilities, *J. Finance* **42** (1987), 281–300.

[33] J. Hull, *Options, Futures and Other Derivatives*, 5th Edition, Prentice Hall, 2002.

[34] P. Jäckel, *Monte Carlo methods in finance*, John Wiley & Sons, Inc., London, 2002.

[35] M. Jandačka and D. Ševčovič, On the risk-adjusted pricing-methodology-based valuation of vanilla options and explanation of the volatility smile, *J. Appl. Math.* **3** (2005), 235–258.

[36] L. Jódar, P. Sevilla–Peris, J.C. Cortés, and R. Sala, A new direct method for solving the Black–Scholes equation, *Appl. Math. Lett.* **18** (2005), 29–32.

[37] Y. Kabanov and M. Safarian, On Leland's Strategy of Option Pricing with Transactions Costs, *Finance Stoch.* **1** (1997), 239–250.

[38] T. Kim and Y. Kwon, *Finite–Difference Bisection Algorithms for free boundaries of American Call and Put Options*, Preprint Pohang University of Science and Technology, 2002.

[39] M. Kratka, No Mystery Behind the Smile, *Risk* **9** (1998), 67–71.

[40] Y.-K. Kwok, *Mathematical Models of Financial Derivatives*, Springer Finance, 1998.

[41] H.E. Leland, Option pricing and replication with transactions costs, *J. Finance* **40** (1985), 1283–1301.

[42] L. MacMillan, Analytic Approximation for the American Put Option, *Adv. Fut. Opt. Res.* **1** (1986), 119–139.

[43] B.J. McCartin and S.M. Labadie, Accurate and efficient pricing of vanilla stock options via the Crandall–Douglas Scheme, *Appl. Math. Comp.* **143** (2003), 39–60.

[44] R.C. Merton, Theory of rational option pricing, *Bell J. Econ. Manag. Sci.* **4** (1973), 141–183.

[45] G. Meyer and J. van der Hoek, The valuation of American options with the method of lines, *Adv. Fut. Opt. Res.* **9** (1997), 265–286.

[46] B.F. Nielsen, O. Skavhaug and A. Tveito, Penalty and front–fixing methods for the numerical solution of American option problems, *J. Comp. Finance* **5** (2002), 69–97.

[47] R. Panini and R. P. Srivastav, Option Pricing with Mellin Transforms, *Math. Comput. Modelling* **40** (2004), 43–56.

[48] O. Pauly, *Numerical simulation of American Options*, Diploma thesis, Universität Ulm, 2004.

[49] J. Persson, *Accurate Finite Difference Methods for Option Pricing*, Dissertation, Uppsala University, 2006.

[50] P. Schönbucher and P. Wilmott, The feedback–effect of hedging in illiquid markets, *SIAM J. Appl.Math.* **61** (2000), 232–272.

[51] D. Ševčovič, Analysis of the free boundary for the pricing of an American call option, *Euro J. Appl. Math* **12** (2001), 25–37.

[52] D. Ševčovič, An iterative algorithm for evaluating approximations to the optimal exercise boundary for a nonlinear Black–Scholes equation, *Canad. Appl. Math. Quarterly* **15** (2007), 77–79.

[53] R. Seydel, *Tools for Computational Finance*, Second ed., Springer, 2004.

[54] H.M. Soner, S.E. Shreve and J. Cvitanič, There is No Nontrivial Hedging Portfolio for Option Pricing with Transaction Costs, *Ann. Appl. Probab.* **5** (1995), 327–355.

[55] H.M. Soner and N. Touzi, Superreplication Under Gamma Constraints, *SIAM J. Contr. Optim.* **39** (2001), 73–96.

[56] G. Strang, On the Construction and Comparison of Difference Schemes, *SIAM J. Numer. Anal.* **5** (1968), 506–517.

[57] J.C. Strikwerda, *Finite Difference Schemes and Partial Differential Equations*, Wadsworth & Brooks/Cole, 1989.

[58] H. Sun and J. Zhang, A high order compact boundary value method for solving one dimensional heat equations, *Tech. Rep.* No. 333–02, University of Kentucky, 2002.

[59] D.Y. Tangman, A. Gopaul and M. Bhuruth, Numerical pricing of options using high–order compact finite difference schemes, in press: *J. Comput. Appl. Math.* (2007).

[60] D.Y. Tangman, A. Gopaul and M. Bhuruth, A Fast High–Order Finite Difference Algorithm for Pricing American Options, submitted to: *J. Comput. Appl. Math.*

[61] D. Tavella and C. Randall, Pricing financial instruments: The finite difference method, John Wiley & Sons, 2000.

[62] P. Wilmott, J. Dewynne and S. Howison, *Option Pricing: Mathematical Models and Computation*, Oxford, Financial Press, 1993.

[63] P. Wilmott, S. Howison, and J. Dewynne, *The Mathematics of Financial Derivatives,* A Student Introduction, Cambridge University Press, 2002.

[64] L. Wu and Y. Kwok, A front-fixing finite difference method for the valuation of American options, *J. Fin. Engin.* **6** (1997), 83–97.

[65] X. Wu and Z.-Z. Sun, Convergence of difference scheme for heat equation in unbounded domains using artificial boundary conditions, *Appl. Numer. Math.* **50** (2004), 261–277.

[66] J. Zhao, M. Davison and R.M. Corless, Compact finite difference method for American option pricing, *J. Comput. Appl. Math.* **206** (2007), 306–321.

[67] Y. Zhu, X. Wu and I. Chern, *Derivative securities and difference methods*, Springer Finance, New York, 2004.

[68] R. Zvan, P. Forsyth and K. Vetzal, A penalty method for American options with stochastic volatility, *J. Comp. Appl. Math.* **91** (1998), 199–218.

In: Nonlinear Models in Mathematical Finance... ISBN 978-1-60456-931-5
Editor: Matthias Ehrhardt, pp. 273-301

Chapter 9

PRICING HYDROELECTRIC POWER PLANTS WITH/WITHOUT OPERATIONAL RESTRICTIONS: A STOCHASTIC CONTROL APPROACH

Zhuliang Chen and Peter A. Forsyth
David R. Cheriton School of Computer Science,
University of Waterloo, Waterloo, ON, Canada N2L 3G1

Abstract

In this chapter, we value hydroelectric power plant cash flows under a stochastic control framework, taking into consideration the implication of operational constraints such as ramping and minimum flow rate constraints for the purpose of environmental protection. The power plant valuation problem under a ramping constraint is characterized as a bounded stochastic control problem, resulting in a Hamilton-Jacobi-Bellman (HJB) partial integrodifferential equation (PIDE). The valuation problem without the ramping restriction is characterized as an unbounded stochastic control problem; we propose an impulse control formulation, resulting in an HJB variational inequality, for the valuation problem under this scenario. We develop a consistent numerical scheme for solving both the HJB PIDE for the bounded control problem and the HJB variational inequality for the unbounded control problem. We prove the convergence of the numerical scheme to the viscosity solution of each pricing equation, provided a strong comparison result holds. Numerical results indicate that failing to consider operational constraints may considerably overestimate the value of hydroelectric power plant cashflows.

1. Introduction

A hydroelectric power plant generates electricity by releasing water from a reservoir through the turbine located downstream of the dam. The electricity power produced relies on the rate of water flowing through the turbine, which is controlled by the ramping rate (the rate of change of flow through the turbine) chosen by the operator. In order to gain profits, the operator of a hydroelectric power plant must determine the appropriate ramping rate in response to stochastic electricity prices. Furthermore, the operator is required to conform to various operational restrictions, such as ramping and minimum flow rate constraints, imposed by the government for the purpose of environmental protection [14]. The valuation of hydroelectric power plants is characterized as a stochastic control problem with the ramping rate (which can be either bounded or unbounded) as the control variable.

In this chapter, we focus on valuing hydroelectric power plant cash flows by solving the corresponding stochastic control problem, and studying the implication of the operational constraints on the value of power plants.

Following [7, 8, 24], we can formulate the hydroelectric power plant pricing problem (i.e., the stochastic control problem) as a Hamilton-Jacobi-Bellman (HJB) partial integrodifferential equation (PIDE) and an HJB variational inequality respectively for the cases with and without a ramping constraint. Then we can use PDE based approaches to solve the HJB equation/variational inequality numerically.

In general, the solution to an HJB equation/variational inequality may not be unique. According to [15, 23], the viscosity solution to the HJB equation/variational inequality is normally identical to the value of the corresponding stochastic control problem. Therefore, as noted in [2, 16], it is important to ensure that a numerical scheme converges to the viscosity solution of the equation, which is the appropriate solution of the corresponding stochastic control problem.

Based on the work in [7, 8], we develop a consistent numerical scheme to solve both the HJB equation and the HJB variational inequality. In both cases, the numerical technique can be shown to converge to the viscosity solution of the pricing equation. Our main results are the following:

- We propose an HJB PIDE for the valuation of a hydroelectric power plant under a ramping constraint. We use a one-factor model for the electricity spot price that is able to capture major features of electricity prices such as mean-reverting dynamics, daily price trends, and price spikes.

- We present a semi-Lagrangian scheme, based on the scheme proposed in [8], to solve the HJB PIDE. The timestepping scheme is an implicit-explicit scheme with the integral terms treated explicitly. As discussed in [12], this scheme is unconditionally stable. The two integral terms in the PIDE are evaluated using a Fast Fourier Transform (FFT). Provided a strong comparison result holds, we prove the scheme converges to the unique viscosity solution of the pricing PIDE.

- We also consider a scenario with no ramping constraint. Formally, this case can be considered to be the limit obtained by allowing the ramping rate to become unbounded. We propose an impulse control formulation for this situation, resulting in an HJB variational inequality.

- We derive a simple extension of the the discretization scheme for the HJB PIDE under ramping constraints to solve the HJB variational inequality with unbounded ramping rates. Provided a strong comparison result holds, we prove the scheme converges to the unique viscosity solution of the HJB variational inequality.

- We study the implication of operational restrictions on the optimal control strategies and the value of the power plant. Through an example, our numerical results indicate that the operational constraints can reduce the value of the power plant cash flows by more than 37% compared to the case with no ramping constraints and no minimum flow requirements. Therefore, it is important to take these constraints into consideration in order to accurately value the plant revenues.

1.1. Previous Work

Pricing and scheduling hydroelectric and thermal power generation assets is a popular research topic (see, e.g., [14, 18, 19, 21, 24] for the work on hydroelectric power generators and [10, 24, 25, 28] for the research on thermal power generators).

The authors of [14, 18, 19] study the short-term scheduling of hydroelectric power generation systems with various operational constraints including a ramping constraint. Their research focuses on scheduling a number of power systems under the assumption of known electricity prices with the main effort placed on solving nonlinear optimization problems, while in this chapter we consider the operation of a single power plant under electricity price uncertainty.

The authors of [21] consider the long and medium term production planning of a hydropower system under price and inflow uncertainty. However, they assume a parametric form for the control strategies in order to solve the valuation problem. As such, the resulting control strategies may not be optimal, and hence the method tends to undervalue the power system. Furthermore, no operational constraints are considered in [21].

The research in [24] focuses on the value of a pump-storage facility in the PDE framework. The pump storage facility allows water to be pumped to the reservoir when the electricity price is low and to be released from the storage to generate electricity when the electricity price is high, and thus resulting in profit. Their work, however, does not consider operational constraints. Moreover, the pricing equation is solved numerically using an explicit finite differencing method, which is known to suffer from timestep restrictions due to stability considerations.

2. Hydroelectric Power Plant Valuation Problem

This section defines the short-term hydroelectric power plant valuation problem and proposes the pricing equations for the value of a hydroelectric power plant. The section is arranged as follows: first, we define some notation for the problem; we then present a one-factor model for electricity spot prices that is able to capture main features of the market electricity prices. Based on the spot price model, we propose two pricing equations for the problem: one incorporates the regulatory operational constraints, the other does not. We also provide the boundary conditions for the pricing equations to completely specify the hydroelectric power plant valuation problem.

2.1. Problem notation

We use the following notation for the hydroelectric power plant valuation problem:

- t: time in hours.

- P: current electricity spot price.

- h: head of the water above the turbine (i.e., the difference in elevation between water levels upstream and downstream of a dam). We assume that h can be any value lying within the domain $[h_{\min}, h_{\max}]$, where $h_{\min}, h_{\max}$ are positive constants specified in the management plan.

- c: rate of the water flowing through the turbine (outflow rate). We assume c can be any value residing within the domain $[c_{\min}, c_{\max}]$, where $c_{\min} \geq 0$ and $c_{\max}$ is determined by the turbine's head due to Bernoulli's law [24]. For a hydroelectric power plant, the change on the head is usually much smaller (e.g., within meters) compared with the head itself (e.g., a hundred meters). As a result, instead of treating $c_{\max}$ as a function of the head h, in this chapter, we regard $c_{\max}$ as a positive constant.

- T: Cash flow horizon. We will typically choose $T = 1$ week. This will determine the risk neutral revenues obtained through optimal operation of the plant in a typical weekly cycle.

- $V(P, c, h, \tau)$: value of the hydroelectric power plant as a function of electricity price P, outflow rate c, head h and current time to the cash flow horizon $\tau = T - t$, where t denotes the current time.

- z: control variable that represents the ramping rate imposed by the power plant. That is, we have
$$\frac{dc}{dt} = z. \tag{1}$$
According to equation (1), $z > 0$ represents ramp-up (i.e., increasing the outflow rate c), $z < 0$ represents ramp-down (i.e., decreasing the outflow rate c), while $z = 0$ represents keeping the current outflow rate unchanged.

- f: rate of the water flowing into the dam (inflow rate). In this chapter, we assume that f is a positive constant. It is straightforward to extend f to be a deterministic function of t or to follow a stochastic process [24].

- a: surface area of the hydroelectric reservoir. We assume that the volume of the reservoir is $a \cdot \bar{h}$, where $\bar{h}$ is the depth of the reservoir.

- $H(c, h)$: amount of power produced by the power generator as a function of outflow rate c and head h. According to [24], in theory, the power generated by a hydroelectric power plant is given by
$$H_m(c, h) = g\rho ch, \tag{2}$$
where g is the acceleration of gravity, ρ is the density of water. Let $\eta(c, h)$ be a nonlinear function of c and h representing the efficiency of the turbine. Then the actual power generated by the turbine is represented by
$$H(c, h) = H_m(c, h)\eta(c, h). \tag{3}$$
Following [24], we use the following quadratic function to represent $\eta(c, h)$:
$$\eta(c, h) = -k_1 \left(\frac{H_m(c, h)}{k_2} - 1 \right)^2 + k_1, \tag{4}$$

where $0 < k_1 < 1$ and $k_2 > H_m(c_{\max}, h_{\max})/2$ so that $0 < \eta(c, h) < 1$. The above equation has the physical meaning that the turbine achieves the maximum efficiency k_1 when it generates a power of k_2. According to (2), $H_m(c, h)$ will achieve the maximum at $(c, h) = (c_{\max}, h_{\max})$. Note that the maximum efficiency η in (4) may not correspond to $(c_{\max}, h_{\max})$. In other words, if $H_m(c_{\max}, h_{\max}) \neq k_2$ then $\eta(c_{\max}, h_{\max}) < k_1$.

Given the above notation, when h is away from the extreme values $h_{\min}$ or $h_{\max}$, the rate of change of the head satisfies

$$\frac{dh}{dt} = \frac{f - c}{a}. \tag{5}$$

Equation (5) implies that if the outflow rate c is larger than the inflow rate f, then the head decreases. Similarly, the head increases if the outflow rate is smaller than the inflow rate. Nevertheless, equation (5) cannot be generalized to the case when either when $h \to h_{\min}$ and $c > f$, or when $h \to h_{\max}$ and $c < f$. As $h \to h_{\min}$ and $c > f$, the head h can fall below $h_{\min}$ according to equation (5), which is not allowed. As $h \to h_{\max}$ and $c < f$, equation (5) implies that the head can increase above $h_{\max}$, the highest head permitted. To handle these cases, we generalize equation (5) to

$$\frac{dh}{dt} = G(c, h) \cdot \frac{f - c}{a} \tag{6}$$

for any $h \in [h_{\min}, h_{\max}]$ and any $c \in [c_{\min}, c_{\max}]$, where G is a smooth function of c and h satisfying

$$\begin{array}{ll} G(c, h) \to 0 & \text{if } c > f \text{ and } h \to h_{\min}, \\ G(c, h) \to 0 & \text{if } c < f \text{ and } h \to h_{\max}, \\ G(c, h) = 1 & \text{otherwise.} \end{array} \tag{7}$$

Consequently, equations (6-7) ensure that $h \in [h_{\min}, h_{\max}]$. For future reference, note that equations (6-7) imply that

$$G(c, h) \cdot \frac{f - c}{a} \geq 0 \Leftrightarrow \frac{dh}{dt} \geq 0 \ ; \ \forall c \in [c_{\min}, c_{\max}] \ , \ h = h_{\min} \ , \tag{8}$$

$$G(c, h) \cdot \frac{f - c}{a} \leq 0 \Leftrightarrow \frac{dh}{dt} \leq 0 \ ; \ \forall c \in [c_{\min}, c_{\max}] \ , \ h = h_{\max} \ . \tag{9}$$

In order to satisfy (7), we assume that function $G(c, h)$ has the following property

$$G(c, h) = \begin{cases} O\left((h - h_{\min})^{\nu}\right) & \text{if } c > f \text{ and } h \to h_{\min}, \\ O\left((h_{\max} - h)^{\nu}\right) & \text{if } c < f \text{ and } h \to h_{\max}, \\ 1 & \text{otherwise,} \end{cases} \tag{10}$$

where ν is any small positive constant.

As outflow rate $c \to c_{\min}$, condition

$$\frac{dc}{dt} \geq 0 \tag{11}$$

needs to be satisfied so that the outflow rate does not decrease below the minimum flow rate $c_{\min}$. Similarly, condition

$$\frac{dc}{dt} \leq 0 \tag{12}$$

must be satisfied when $c \to c_{\max}$ so that the outflow rate does not increase above the maximum flow rate $c_{\max}$. From equations (1) and (11-12), the above analysis results in the following conditions on the ramping rate z:

$$z \geq 0\,, \quad c = c_{\min}, \tag{13}$$

$$z \leq 0\,, \quad c = c_{\max}. \tag{14}$$

In order to satisfy conditions (13-14), we will further assume that

$$\min\{z\} = O\Big((c - c_{\min})^{\theta}\Big)\,, \quad \text{if } c \to c_{\min}, \tag{15}$$

$$\max\{z\} = O\Big((c_{\max} - c)^{\theta}\Big)\,, \quad \text{if } c \to c_{\max}, \tag{16}$$

where θ is any positive constant. Condition (15) has the physical meaning that the operator of the power plant has to gradually decrease the maximum ramp-down rate $\min(z)$ towards zero when c is close to $c_{\min}$, instead of suddenly switching it to zero at $c = c_{\min}$. A similar explanation applies to the condition (16).

In the following, we only require that $\theta, \nu > 0$. If $\theta \ll 1$, then according to (15-16), for all practical purposes, the operator can switch ramping rates virtually instantaneously to zero at $c_{\min}$ and $c_{\max}$. If $\nu \ll 1$, then (7) implies that the change of head can be reduced to zero virtually instantaneously at $h_{\min}$ and $h_{\max}$. We require $\theta, \nu > 0$ to simplify some of the proofs in subsequent sections.

At a point (P, c, h), the rate of cash flow (instantaneous revenue) received by the operator of the power plant is $H(c, h) \cdot P$, obtained from selling the generated electricity to the spot market at price P. Similar to the above analysis, we need to pay attention to two cases when $h \to h_{\min}$, $c > f$ and $h \to h_{\max}$, $c < f$. Since these two cases can result in potential decrease/increase of the head below/above the lower/upper bound, we will impose a penalty on the revenue function in these cases to keep the head and the outflow rate from reaching these states (since there are penalties imposed, it would not be optimal to reach these states). Specifically, we will set the rate of cash flow as $G(c, h)H(c, h)P$, where the function G is given in (7), so that the revenue rate decreases to zero as $h \to h_{\min}$, $c > f$ or $h \to h_{\max}$, $c < f$, and otherwise remains at $H(c, h)P$.

2.2. Electricity Spot Price Model

Electricity spot prices mainly exhibit three features: mean-reverting dynamics, seasonality trends on daily, weekly and annual time scales, and occasional price spikes caused by short-term disparities between supply and demand. A price spike consists of a cluster of up-jumps of relatively large size with respect to normal fluctuations, shortly followed by a return to normal price levels [9, 17]. Since we are interested in considering short-term (e.g., one week) valuation and optimal operation of a hydroelectric generation asset, we will model hourly electricity spot price and ignore the weekly and annual price trends.

In this section, we specify a one-factor hourly electricity spot price model similar to the model given in [24] which is capable of qualitatively capturing the realistic electricity spot price dynamics. We will consider directly the risk adjusted (or risk neutral) price process with parameters given under the $\mathbb{Q}$ measure since we will value the cash flows under the risk neutral measure.

We assume that the risk adjusted hourly spot price is modeled by the a stochastic differential equation (SDE) given by

$$\begin{aligned} dP &= [\alpha(K(t) - P) - \lambda_1 \kappa_1 P - \lambda_2(P)\kappa_2 P]dt + \sigma P dZ \\ &\quad +(J_1 - 1)P dq_1 + (J_2 - 1)P dq_2 \,, \end{aligned} \tag{17}$$

$$K(t) = K_0 + \beta \sin\left(\frac{2\pi(t - t_0)}{24}\right) , \tag{18}$$

where

- $\alpha > 0$ is the mean-reversion rate,
- $K(t) \geq 0$ is the long-term equilibrium price that incorporates the daily price cycle,
- σ is the volatility,
- dZ is an increment of the standard Gauss-Wiener process,
- $K_0 \geq 0$ is the equilibrium price without incorporating a daily price fluctuation,
- β is the parameter for the daily price trend,
- t_0 is the centering parameter, representing the time of daily peak of the equilibrium price,
- dq_1 is the independent Poisson process representing the up-jump of a price spike event, that is, $dq_1 = \begin{cases} 0 & \text{with probability } 1 - \lambda_1 dt \,, \\ 1 & \text{with probability } \lambda_1 dt \,. \end{cases}$
- λ_1 is the constant jump intensity representing the mean arrival time of the up-jump event,
- J_1, $J_1 \geq 1$, is a random variable representing the up-jump size of electricity price. When $dq_1 = 1$, price jumps from P up to PJ_1. We assume that J_1 follows a probability density function $g_1(J_1)$.
- dq_2 is the independent Poisson process representing the down-jump of a price spike event, that is, $dq_2 = \begin{cases} 0 & \text{with probability } 1 - \lambda_2(P)dt \,, \\ 1 & \text{with probability } \lambda_2(P)dt \,. \end{cases}$
- $\lambda_2(P)$ is the jump intensity as a function of P, representing the mean arrival time of the down-jump event,
- J_2, $0 < J_2 \leq 1$, is a random variable representing the down-jump size of electricity price. When $dq_2 = 1$, price jumps from P down to PJ_2. We assume that J_2 follows a probability density function $g_2(J_2)$.

- κ_1 (or κ_2) is $E[J_1 - 1]$ (or $E[J_2 - 1]$), where $E[\cdot]$ is the expectation operator.

We assume that the two Poisson processes are uncorrelated, i.e., $Cov(dq_1, dq_2) = 0$.

Following [24], we use parameters λ_1 and J_1 to simulate the upward jumps, and use $\lambda_2(P)$ and J_2 to simulate the downward jumps. Let $P_0 > 0$ be a constant representing the threshold value for the spike state. In other words, with high probability, we regard any spot price $P > P_0$ as a price resulting after up-jumps, but before down-jumps, during a price spike event. According to [24], we set

$$\lambda_2(P) = \begin{cases} \bar{\lambda}_2 & \text{if } P \geq P_0 \,, \\ 0 & \text{if } P < P_0 \,, \end{cases} \tag{19}$$

where $\bar{\lambda}_2$ is a positive constant. Consequently, if the price resides in the spike state after upward jumps (i.e., $P \geq P_0$), then with high probability a down-jump will occur, bringing the price back to the normal state; conversely, if the price lies in the normal state (i.e., $P < P_0$), the down-jump never occurs. Note that in the spike state another up-jump of the spot price can still occur before dropping back to the normal state, which will form a cluster of up-jumps, as frequently observed in the electricity spot market [9, 17].

2.3. Pricing Equations for Hydroelectric Power Plant Valuation

The valuation of power generation assets is characterized as a stochastic control problem resulting in Hamilton-Jacobi-Bellman (HJB) equations. In [24], a pricing PIDE is proposed for the pump-storage facility valuation. The pricing equation in [24] does not consider environmental and regulatory operational constraints for hydroelectric power generators such as ramping and minimum flow rate constraints [14, 18, 19]. As we demonstrate in a later section, ignoring these constraints will lead to over-valuing the assets.

In this section, we first give the pricing equation, an HJB PIDE, with a ramping constraint imposed. We then present the pricing equation, an HJB variational inequality, for the case without imposing a ramping constraint.

Incorporating the Ramping Constraint

Under a ramping constraint, the ramping rate z must satisfy

$$z_{\min} \leq z \leq z_{\max} \,, \tag{20}$$

where $z_{\min} \leq 0$, $z_{\max} \geq 0$ are bounded constants. We denote by $Z(c)$ the set of admissible controls that satisfy constraint (20) as well as conditions (15-16). As a result, we have $Z(c) \subseteq [z_{\min}, z_{\max}]$.

Based on the standard hedging arguments in the financial valuation literature [24], the value of a hydroelectric power plant $V(P, c, h, \tau)$, assuming that the risk adjusted electricity spot price follows the stochastic process defined in (17-18), is given by the following HJB PIDE:

$$V_\tau = \mathcal{C}V + \mathcal{B}V + \frac{1}{a}(f - c)G(c, h)V_h + \sup_{z \in Z(c)} \left(zV_c\right) + G(c, h)H(c, h)P, \tag{21}$$

where the operators $\mathcal{C}$ and $\mathcal{B}$ are

$$
\begin{aligned}
\mathcal{C}V &= \frac{1}{2}\sigma^2 P^2 V_{PP} + \big[\alpha(K(t) - P) - \lambda_1\kappa_1 P - \lambda_2(P)\kappa_2 P\big]V_P \\
&\quad -\big[r + \lambda_1 + \lambda_2(P)\big]V \qquad (22)\\
\mathcal{B}V &= \lambda_1 \int_{-\infty}^{\infty} V(J_1P)g_1(J_1)dJ_1 + \lambda_2(P)\int_{-\infty}^{\infty} V(J_2P)g_2(J_2)dJ_2. \qquad (23)
\end{aligned}
$$

Here r is the riskless interest rate.

Remark 2.1. *In PIDE (21), if $V_c > 0$, then the supremum is achieved at $z = \max\{Z(c)\}$ (note that $Z(c)$ is a closed region). Similarly, if $V_c < 0$, then the supremum is achieved at $z = \min\{Z(c)\}$; if $V_c = 0$, then for any $z \in Z(c)$ the supremum is zero.*

Relaxing the Ramping Constraint

In the absence of a ramping constraint, the ramping rate z is unbounded, that is, $z \in [-\infty, \infty]$. According to equation (1), this means that the outflow rate c can instantaneously switch from one state to another. In this case, we follow the steps in [7] and formulate the valuation problem as an impulse control problem by introducing positive fixed costs d_- and d_+ for control values $z = -\infty$ and $z = +\infty$, respectively (d_- and d_+ can be infinitesimally small). The optimal ramping rate will either be $-\infty$, 0, or $+\infty$.

At a point (P, c, h, τ), one of the following three scenarios occurs:

- The optimal control satisfies $z = 0$. Substituting $z = 0$ into the PIDE (21) gives

$$
V_\tau - \mathcal{C}V - \mathcal{B}V - \frac{1}{a}(f - c)G(c,h)V_h - G(c,h)H(c,h)P = 0 \qquad (24)
$$

- The optimal control satisfies $z = -\infty$, which corresponds to switching the outflow rate c to a smaller value instantaneously. In this case, the value of V satisfies the following no-arbitrage jump condition:

$$
V - \sup_{\delta \in [c_{\min} - c, 0)} \big[V(P, c+\delta, h, \tau) - d_-\big] = 0. \qquad (25)
$$

- The optimal control satisfies $z = +\infty$, which corresponds to switching the outflow rate c to a higher value instantaneously. In this case, V satisfies the following jump condition:

$$
V - \sup_{\delta \in (0, c_{\max} - c]} \big[V(P, c+\delta, h, \tau) - d_+\big] = 0. \qquad (26)
$$

More generally, at each point (P, c, h, τ), we have

$$
\begin{aligned}
V_\tau - \frac{1}{a}(f - c)G(c,h)V_h - \mathcal{C}V - \mathcal{B}V - G(c,h)H(c,h)P &\ge 0 \\
V - \sup_{\delta \in [c_{\min} - c, 0)} \big[V(P, c+\delta, h, \tau) - d_-\big] &\ge 0 \qquad (27)\\
V - \sup_{\delta \in (0, c_{\max} - c]} \big[V(P, c+\delta, h, \tau) - d_+\big] &\ge 0
\end{aligned}
$$

with at least one of these inequalities holding with equality. Therefore, the value function V satisfies the following HJB variational inequality

$$\begin{aligned}\min\Big\{ & V_\tau - \mathcal{C}V - \mathcal{B}V - \frac{1}{a}(f-c)G(c,h)V_h - G(c,h)H(c,h)P, \\ & V - \sup_{\delta\in[c_{\min}-c,0)} \big[V(P,c+\delta,h,\tau) - d_-\big], \\ & V - \sup_{\delta\in(0,c_{\max}-c]} \big[V(P,c+\delta,h,\tau) - d_+\big]\Big\} = 0.\end{aligned} \tag{28}$$

2.4. Boundary Conditions for Pricing Equations

In order to completely specify the power plant valuation problem, we need to provide boundary conditions. As for the terminal boundary conditions, we use the following zero payoff as specified in [24]:

$$V(P,c,h,\tau=0) = 0. \tag{29}$$

The domain for pricing equations (21) and (28) is $(P,c,h) \in [0,\infty] \times [c_{\min}, c_{\max}] \times [h_{\min}, h_{\max}]$. For computational purposes, we need to solve the equations in a finite computational domain $[0, P_{\max}] \times [c_{\min}, c_{\max}] \times [h_{\min}, h_{\max}]$.

Boundary Equations for PIDE (21)

As $h \to h_{\min}$, from inequality (8), the characteristics are outgoing (or zero) in the h direction at $h = h_{\min}$, and we simply solve PIDE (21) along the $h = h_{\min}$ boundary, no further information is needed. Similarly, as $h \to h_{\max}$, inequality (9) implies that the characteristics are outgoing (or zero) in the h direction at $h = h_{\max}$. We can simply solve the pricing equation along the $h = h_{\max}$ boundary, no further information is needed.

Conditions (13-14) respectively imply that the characteristics are outgoing in the c direction at $c = c_{\min}$ and $c = c_{\max}$. As such, we can solve the PIDEs along the $c = c_{\min}$ and $c = c_{\max}$ boundaries without requiring further information.

Taking the limit of equation (21) as $P \to 0$, we obtain the boundary PDE

$$V_\tau = \mathcal{C}_0 V + \frac{1}{a}(f-c)G(c,h)V_h + \sup_{z\in Z(c)} \big(zV_c\big)\ ;\ \ P \to 0 \tag{30}$$

$$\text{with} \quad \mathcal{C}_0 V = \alpha K(t) V_P - rV. \tag{31}$$

Note that the integral terms disappear as we take the limit $P \to 0$ and interchange the integrals and the limit. Since $\alpha K(t) \geq 0$ in $\mathcal{C}_0 V$, we can solve (30) without requiring additional boundary conditions, as we do not need information from outside the computational domain $[0, P_{\max}]$.

As $P \to \infty$, we need to deal with two major issues. The first problem is that there is no obvious Dirichlet-type condition that can be imposed for P large. The second issue is that for any finite domain $[0, P_{\max}]$, the two integral terms require information from outside the computational domain.

We can resolve the issues using the localization strategy presented in [12, 13]. Specifically, we apply the commonly used boundary condition $V_{PP} \to 0$ [24, 27], which implies that

$$V \simeq x(h, c, \tau)P + y(h, c, \tau) \tag{32}$$

in the region $P \in [P_{\max} - \chi, P_{\max}]$ for a sufficiently large $P_{\max}$, where functions x and y are independent of P and χ is a positive constant. Substituting equation (32) into the pricing PIDE (21) results in

$$\begin{aligned} V_\tau = \mathcal{C}_1 V + \frac{1}{a}(f - c)G(c, h)V_h + \sup_{z \in Z(c)} (zV_c) + G(c, h)H(c, h)P \\ P \in [P_{\max} - \chi, P_{\max}], \end{aligned} \tag{33}$$

with

$$\mathcal{C}_1 V = \begin{cases} \frac{1}{2}\sigma^2 P^2 V_{PP} + \alpha(K(t) - P)V_P - rV & \text{if } P \in [P_{\max} - \chi, P_{\max}), \\ \alpha(K(t) - P)V_P - rV & \text{if } P = P_{\max}\,. \end{cases} \tag{34}$$

We leave the V_{PP} term in the operator $\mathcal{C}_1$, though we have assumed that $V_{PP} = 0$ as $P \to \infty$, so that the differential operator is formally parabolic. Note that the integral terms vanish in equation (33).

Following [13], we solve PIDE (21) in the region $P \in (0, P_{\max} - \chi)$, where the integrals in the PIDE are evaluated using a Fast Fourier Transform (FFT); we solve PDE (30) at $P = 0$ boundary; and solve PDE (33) in the region $P \in [P_{\max} - \chi, P_{\max}]$. By carefully choosing χ using the method presented in [13], we can provide sufficient data for the computation of integral terms and at the same time reduce the effect of FFT wrap-around. Note that we will choose $P_{\max}$ sufficiently large so that $P_{\max} \gg K(t)$, hence equation (33) can be solved at $P = P_{\max}$ without additional information.

Boundary Conditions for Equation (28)

Following a similar analysis as above, we can solve the HJB variational inequality (28) directly along the $h = h_{\min}$ and $h = h_{\max}$ boundaries without requiring further information.

Taking the limit of equation (28) as $c \to c_{\min}$, we obtain the boundary equation

$$\begin{aligned} \min\Bigg\{ & V_\tau - \mathcal{C}V - \mathcal{B}V - \frac{1}{a}(f - c)G(c, h)V_h - G(c, h)H(c, h)P, \\ & V - \sup_{\delta \in (0, c_{\max} - c]} \left[V(P, c + \delta, h, \tau) - d_+ \right] \Bigg\} = 0 \;\; ; \;\; c \to c_{\min}. \end{aligned} \tag{35}$$

Taking the limit of equation (28) as $c \to c_{\max}$, we obtain the boundary equation

$$\begin{aligned} \min\Bigg\{ & V_\tau - \mathcal{C}V - \mathcal{B}V - \frac{1}{a}(f - c)G(c, h)V_h - G(c, h)H(c, h)P, \\ & V - \sup_{\delta \in [c_{\min} - c, 0)} \left[V(P, c + \delta, h, \tau) - d_- \right] \Bigg\} = 0 \;\; ; \;\; c \to c_{\max}. \end{aligned} \tag{36}$$

Equations (35-36) can be solved directly without using further information from outside the computational domain.

Taking the limit of equation (28) as $P \to 0$, we obtain the boundary equation

$$\min\left\{V_\tau - \mathcal{C}_0 V - \frac{1}{a}(f-c)G(c,h)V_h,\ V - \sup_{\delta\in[c_{\min}-c,0)}\left[V(P,c+\delta,h,\tau)-d_-\right],\right.$$
$$\left. V - \sup_{\delta\in(0,c_{\max}-c]}\left[V(P,c+\delta,h,\tau)-d_+\right]\right\} = 0;\ \ P\to 0. \tag{37}$$

Since $\alpha K(t) \geq 0$ in $\mathcal{C}_0 V$ given in (31), we can solve (37) without requiring additional boundary conditions.

As $P \to P_{\max}$, we also assume that V satisfies the linear form (32) in the region $P \in [P_{\max} - \chi, P_{\max}]$. Substituting (32) into the pricing equation (28) gives

$$\begin{aligned}\min\Big\{ & V_\tau - \mathcal{C}_1 V - \frac{1}{a}(f-c)G(c,h)V_h - G(c,h)H(c,h)P,\\ & V - \sup_{\delta\in[c_{\min}-c,0)}\left[V(P,c+\delta,h,\tau)-d_-\right],\\ & V - \sup_{\delta\in(0,c_{\max}-c]}\left[V(P,c+\delta,h,\tau)-d_+\right]\Big\} = 0; \qquad P\in[P_{\max}-\chi, P_{\max}].\end{aligned} \tag{38}$$

We will choose $P_{\max}$ sufficiently large so that $P_{\max} \gg K(t)$, hence equation (38) can be solved without requiring additional information. Similar to the way we handle the boundary conditions for PIDE (21), we solve equation (28) in the region $P \in (0, P_{\max} - \chi)$; solve equation (37) at $P = 0$ boundary; and solve equation (38) in the region $P \in [P_{\max} - \chi, P_{\max}]$ with χ chosen to reduce the effect of FFT wrap-around for the computation of integral terms.

3. Numerical Algorithms

In [8], we develop a numerical scheme based on a semi-Lagrangian approach for solving the gas storage valuation problem—an optimal stochastic control problem with a bounded control. The scheme has advantage that it is more efficient than the existing numerical methods and satisfies sufficient conditions for convergence to the viscosity solution of the gas storage equation. In [7], we extend the scheme in [8] to price variable annuities with a guaranteed minimum withdrawal benefit—an optimal stochastic control problem with an unbounded control. Since the power plant valuation problem is characterized as a stochastic control problem with either a bounded or an unbounded control, respectively for the case when the ramping constraint is imposed or not, we can solve the power plant valuation problem using the schemes proposed in [8] and [7]. We will directly present the schemes in this section. Refer to [7, 8] for more motivation for the schemes.

Prior to introducing the numerical schemes, we introduce the following notation. We use an unequally spaced grid in P direction for the PDE discretization, represented by $[P_0, P_1, \ldots, P_{i_{\max}}]$ with $P_0 = 0$ and $P_{i_{\max}} = P_{\max}$. Similarly, we use unequally spaced grids in c and h directions, respectively denoted by $[c_0, c_1, \ldots, c_{j_{\max}}]$ and $[h_0, h_1, \ldots, h_{k_{\max}}]$ with $c_0 = c_{\min}$, $c_{j_{\max}} = c_{\max}$, $h_0 = h_{\min}$, and $h_{k_{\max}} = h_{\max}$. We denote by $0 = \Delta\tau < \ldots < N\Delta\tau = T$ the discrete timesteps. Let $\tau^n = n\Delta\tau$ denote the nth timestep. Let $V(P_i, c_j, h_k, \tau^n)$ denote the exact solution of the pricing equation when the electricity spot price is P_i, the outflow rate is c_j, the head is h_k and discrete time is τ^n. Let $V^n_{i,j,k}$ denote an approximation of the exact solution $V(P_i, c_j, h_k, \tau^n)$.

It will be convenient to define $\Delta P_{\max} = \max_i(P_{i+1} - P_i)$, $\Delta P_{\min} = \min_i(P_{i+1} - P_i)$, $\Delta c_{\max} = \max_j(c_{j+1} - c_j)$, $\Delta c_{\min} = \min_j(c_{j+1} - c_j)$, $\Delta h_{\max} = \max_k(h_{k+1} - h_k)$, $\Delta h_{\min} = \min_k(h_{k+1} - h_k)$. We assume that there is a mesh size/timestep parameter ε such that

$$\begin{aligned} &\Delta P_{\max} = C_1\varepsilon\,, \quad \Delta c_{\max} = C_2\varepsilon\,, \quad \Delta h_{\max} = C_3\varepsilon\,, \quad \Delta\tau = C_4\varepsilon\,, \\ &\Delta P_{\min} = C_1'\varepsilon\,, \quad \Delta c_{\min} = C_2'\varepsilon\,, \quad \Delta h_{\min} = C_3'\varepsilon\,, \end{aligned} \tag{39}$$

where C_1, C_1', C_2, C_2', C_3, C_3', C_4 are constants independent of ε.

We use standard finite difference methods to discretize the operator $\mathcal{C}_0V$, $\mathcal{C}V$ and $\mathcal{C}_1V$ as given in (31), (22) and (34). Let $(\mathcal{C}_\varepsilon V)^n_{i,j,k}$ denote the discrete value of the differential operators $\mathcal{C}_0V$, $\mathcal{C}V$, or $\mathcal{C}_1V$ at a node (P_i, c_j, h_k, τ^n) so that $(\mathcal{C}_\varepsilon V)^n_{i,j,k}$ is an approximation for $(\mathcal{C}_0V)^n_{i,j,k}$ if $P_i = 0$, an approximation for $(\mathcal{C}V)^n_{i,j,k}$ if $P_i \in (0, P_{\max} - \chi)$ and for $(\mathcal{C}_1V)^n_{i,j,k}$ if $P_i \in [P_{\max} - \chi, P_{\max}]$. The operators (31), (22) and (34) can be discretized using central, forward, or backward differencing in the P direction to give

$$\begin{aligned} &(\mathcal{C}_\varepsilon V)^n_{i,j,k} \\ &= \begin{cases} \gamma_i^n V^n_{i-1,j,k} + \beta_i^n V^n_{i+1,j,k} & \\ \qquad - \big(\gamma_i^n + \beta_i^n + r + \lambda_1 + \lambda_2(P_i)\big)V^n_{i,j,k} & \text{if } P_i \in (0, P_{\max} - \chi), \\ \gamma_i^n V^n_{i-1,j,k} + \beta_i^n V^n_{i+1,j,k} - (\gamma_i^n + \beta_i^n + r)V^n_{i,j,k} & \text{if } P_i \in [P_{\max} - \chi, P_{\max}), \\ \beta_i^n V^n_{i+1,j,k} - (\beta_i^n + r)V^n_{i,j,k} & \text{if } P_i = 0, \\ \gamma_i^n V^n_{i-1,j,k} - (\gamma_i^n + r)V^n_{i,j,k} & \text{if } P_i = P_{\max}, \end{cases} \end{aligned} \tag{40}$$

where γ_i^n and β_i^n are determined using an algorithm in [8]. The algorithm guarantees γ_i^n and β_i^n satisfy the following positive coefficient conditions:

$$\gamma_i^n \geq 0\,, \quad \beta_i^n \geq 0 \qquad\qquad i = 0, \ldots, i_{\max}\,, \quad n = 1, \ldots, N. \tag{41}$$

Let $(\mathcal{B}_\varepsilon V)^n_{i,j,k}$ be an approximation of the operator $\mathcal{B}V$ at a mesh node (P_i, c_j, h_k, τ^n). We compute two integrals in $(\mathcal{B}_\varepsilon V)^n_{i,j,k}$ using the FFT approach described in [11,13]: transforming discrete values $V^n_{i,j,k}$ to an equally spaced $\log P$ grid, carrying out an FFT on the data, computing the correlations resulting from approximation of the integrals using a Trapezoidal rule, and then transforming back to P coordinates.

Following [11,13], we use linear interpolation to transform discrete solution values from equally spaced $\log P$ grid to unequally spaced P grid (and vice versa), which introduces a second-order discretization error. In other words, if $\phi(P, c, h, \tau)$ is a smooth function on (P, c, h, τ) with $\phi^n_{i,j,k} = \phi(P_i, c_j, h_k, \tau^n)$, then we have

$$(\mathcal{B}_\varepsilon \phi)^n_{i,j,k} = (\mathcal{B}\phi)^n_{i,j,k} + O\big(\Delta P^2_{\max}\big). \tag{42}$$

See [11, 13] for more details of discretizing and computing the integral terms. Effectively, we can approximate $(\mathcal{B}V)^n_{i,j,k}$ by

$$(\mathcal{B}_\varepsilon V)^n_{i,j,k} = \lambda_1 \sum_l b^1_{i,l} V^n_{l,k,j} + \lambda_2(P_i) \sum_l b^2_{i,l} V^n_{l,k,j}$$
$$\text{with } 0 \le b^1_{i,l} \le 1\,,\ \sum_l b^1_{i,l} \le 1\,,\ 0 \le b^2_{i,l} \le 1\,,\ \sum_l b^2_{i,l} \le 1. \tag{43}$$

Note that $b^1_{i,l}$, $b^2_{i,l}$ satisfy

$$b^1_{i,l} = 0\ ,\ \ b^2_{i,l} = 0 \qquad \forall l,\ \forall P_i \in \{0\} \cup [P_{\max} - \chi, P_{\max}] \tag{44}$$

since boundary equations (33) and (38) imply that the integral terms vanish in the boundary regions $P = 0$ and $P \in [P_{\max} - \chi, P_{\max}]$.

3.1. Numerical Scheme for Pricing Equation (21)

Following [8], we discretize the terms

$$\frac{DV}{D\tau} \equiv V_\tau - \frac{1}{a}(f - c)G(c,h)V_h - zV_c \tag{45}$$

in PIDE (21) using a semi-Lagrangian timestepping. Let $\zeta^{n+1}_{i,j,k}$ denote the value of the control variable z at the mesh node $(P_i, c_j, h_k, \tau^{n+1})$. Then we can approximate the value of $\frac{DV}{D\tau}$ at $(P_i, c_j, h_k, \tau^{n+1})$ by the following:

$$\left(\frac{DV}{D\tau}\right)^{n+1}_{i,j,k} = \frac{1}{\Delta\tau}\left(V^{n+1}_{i,j,k} - V^n_{i,\hat{j},\hat{k}}\right) + \text{truncation error}, \tag{46}$$

where $V^n_{i,\hat{j},\hat{k}}$ is an approximation of $V\left(P_i, c^n_{\hat{j}}, h^n_{\hat{k}}, \tau^n\right)$ obtained by linear interpolation with $c^n_{\hat{j}}$ and $h^n_{\hat{k}}$ given by

$$c^n_{\hat{j}} = c_j + \zeta^{n+1}_{i,j,k}\Delta\tau, \tag{47}$$

$$h^n_{\hat{k}} = \min\left[\max\left[h_k + \frac{f - c_j}{a}G(c_j, h_k)\Delta\tau,\ h_{\min}\right], h_{\max}\right]. \tag{48}$$

Following [8], the control $\zeta^{n+1}_{i,j,k}$ must satisfy the constraint $\zeta^{n+1}_{i,j,k} \in Z(c_j)$, where $Z(c_j)$ is the set of controls satisfying constraint (20) and conditions (15-16). Moreover, to prevent the value of $c^n_{\hat{j}}$ from going outside of the domain $[c_{\min}, c_{\max}]$, we need to impose further constraints on $\zeta^{n+1}_{i,j,k}$. Let $Z_j \subseteq Z(c_j)$ denote the set of values of $\zeta^{n+1}_{i,j,k} \in Z(c_j)$ such that the resulting $c^n_{\hat{j}}$ computed from (47) is bounded within $[c_{\min}, c_{\max}]$. We regard all elements in Z_j as admissible controls. Note that Z_j is a closed region and is determined only by the grid node c_j.

Equation (48) guarantees that the value of $h^n_{\hat{k}}$ will never go outside of the domain $[h_{\min}, h_{\max}]$. As shown in Lemma 4.4, when the mesh size/timestep parameter ε is sufficiently small, (48) reduces to $h^n_{\hat{k}} = h_k + \frac{f-c_j}{a}G(c_j, h_k)\Delta\tau$.

Given the notation above, at any discrete mesh node $(P_i, c_j, h_k, \tau^{n+1})$, $n \geq 0$, PIDE (21) and the associated boundary equations (30) and (33) can be discretized as

$$V_{i,j,k}^{n+1} = \sup_{\zeta_{i,j,k}^{n+1} \in Z_j} V_{i,\hat{j},\hat{k}}^{n} + \Delta\tau (\mathcal{C}_\varepsilon V)_{i,j,k}^{n+1} + \Delta\tau (\mathcal{B}_\varepsilon V)_{i,j,k}^{n} + \Delta\tau G(c_j, h_k) H(c_j, h_k) P_i. \quad (49)$$

We can rewrite the discrete equation (49) at a node $(P_i, c_j, h_k, \tau^{n+1})$, $n \geq 0$, as

$$\begin{aligned} &\mathcal{G}_{i,j,k}^{n+1}\left(\varepsilon, V_{i,j,k}^{n+1}, \{V_{l,j,k}^{n+1}\}_{l\neq i}, \{V_{i,j,k}^{n}\}\right) \\ &\equiv \inf_{\zeta_{i,j,k}^{n+1} \in Z_j} \left[\frac{V_{i,j,k}^{n+1} - V_{i,\hat{j},\hat{k}}^{n}}{\Delta\tau} \right] - (\mathcal{C}_\varepsilon V)_{i,j,k}^{n+1} - (\mathcal{B}_\varepsilon V)_{i,j,k}^{n} - G(c_j, h_k) H(c_j, h_k) P_i \quad (50) \\ &= 0, \end{aligned}$$

where $\{V_{l,j,k}^{n+1}\}_{l\neq i}$ is the set of values $V_{l,j,k}^{n+1}$, $l \neq i$, $l = 0, \ldots, i_{\max}$, and $\{V_{i,j,k}^{n}\}$ is the set of values $V_{i,j,k}^{n}$, $i = 0, \ldots, i_{\max}$, $j = 0, \ldots, j_{\max}$, $k = 0, \ldots, k_{\max}$.

Remark 3.1. *Based on our assumption of the mesh size/timestep parameters (39), conditions (15-16) and (20) imply that*

$$\min\{\zeta_{i,j,k}^{n+1}\} = \begin{cases} 0 & \text{if } c_j = c_{\min}, \\ O(x^\theta) & \text{if } c_j > c_{\min} \text{ and} \\ & \quad c_j - c_{\min} = O(x),\ x \ll 1, \\ z_{\min} & \text{otherwise.} \end{cases} \quad (51)$$

and

$$\max\{\zeta_{i,j,k}^{n+1}\} = \begin{cases} 0 & \text{if } c_j = c_{\max}, \\ O(x^\theta) & \text{if } c_j < c_{\max} \text{ and} \\ & \quad c_{\max} - c_j = O(x),\ x \ll 1, \\ z_{\max} & \text{otherwise.} \end{cases} \quad (52)$$

Since θ is any positive constant, we can set θ such that $\theta|\log\varepsilon| \ll 1$ for all values of ε chosen for practical purposes, which implies that $\varepsilon^\theta \approx 1$. This means that the numerical implementation assuming that control z has the behavior in (51-52) is, for all practical purposes, the same as an implementation assuming that

$$\begin{aligned} Z(c_j) &= [0, z_{\max}] \ ; \ c_j = c_{\min} \\ &= [z_{\min}, z_{\max}] \ ; \ c_{\min} < c_j < c_{\max} \\ &= [z_{\min}, 0] \ ; \ c_j = c_{\max} \ . \end{aligned}$$

Note that the set of admissible values $Z_j \subseteq Z(c_j)$ in equation (50) ensures that $c_{\hat{j}}^n \in [c_{\min}, c_{\max}]$.

Following a discussion similar to the above, we can show that from a practical point of view, the implementation assuming that function $G(c, h)$ satisfies (10) is the same as an implementation assuming that $G(c_j, h_k) = 0$ if $c_j > f$, $h_k = h_{\min}$ or if $c_j < f$ and $h_k = h_{\max}$, and $G(c_j, h_k) = 1$ at all other mesh nodes.

Remark 3.2. *In (49), we evaluate the integral terms $\mathcal{B}V$ explicitly at timestep τ^n, instead of implicitly at timestep τ^{n+1}, so that no Policy-type iteration is required to solve the linear system resulting from the scheme (49) at each timestep. As shown in [12] and Section 4., the scheme is still unconditionally stable and monotone. Such an explicit evaluation results in a first order error in time, which is asymptotically identical to the error in space generated by linear interpolating $V^n_{i,\hat{j},\hat{k}}$. In the following scheme (56) for the pricing equation (28), we also explicitly evaluate the integral terms.*

Numerical tests show that there is no advantage in terms of convergence as $\varepsilon \to 0$ if we use an implicit discretization of the integral terms.

3.2. Numerical Scheme for Pricing Equation (28)

In order to develop a discretization scheme for the impulse control case, we will proceed in a heuristic fashion. In a later section, we will show rigorously that the resulting discretization is consistent with the impulse control problem (28).

We can consider the impulse control problem to be the limiting case where the ramping rate $\zeta^{n+1}_{i,j,k}$ in (47) is unbounded, hence $c^n_{\hat{j}}$ in (47) can be any value between $c_{\min}$ and $c_{\max}$. Therefore, we rewrite (47) as

$$c^n_{\hat{j}} = c_j + \delta^{n+1}_{i,j,k}, \tag{53}$$

where the control variable $\delta^{n+1}_{i,j,k} \in [c_{\min} - c_j, c_{\max} - c_j]$ so that $c^n_{\hat{j}} \in [c_{\min}, c_{\max}]$. We denote by Δ_j the admissible control set

$$\Delta_j = [c_{\min} - c_j, c_{\max} - c_j]. \tag{54}$$

According to (28), if $\delta^{n+1}_{i,j,k} < 0$, then a fixed cost d_- is charged; if $\delta^{n+1}_{i,j,k} > 0$, then a fixed cost d_+ is charged. Therefore, we use $d(\delta)$ to represent the fixed cost as a function of δ given by

$$d(\delta) = \begin{cases} d_- & \text{if } \delta < 0, \\ 0 & \text{if } \delta = 0, \\ d_+ & \text{if } \delta > 0. \end{cases} \tag{55}$$

Using (53-55), we can generalize the scheme (49) to the following discretization for the impulse control case

$$\begin{aligned} V^{n+1}_{i,j,k} = & \sup_{\delta^{n+1}_{i,j,k} \in \Delta_j} \left[V^n_{i,\hat{j},\hat{k}} - d(\delta^{n+1}_{i,j,k}) \right] + \Delta\tau (\mathcal{C}_\varepsilon V)^{n+1}_{i,j,k} + \Delta\tau (\mathcal{B}_\varepsilon V)^n_{i,j,k} \\ & + \Delta\tau G(c_j, h_k) H(c_j, h_k) P_i, \end{aligned} \tag{56}$$

where $V^n_{i,\hat{j},\hat{k}}$ is an approximation of $V\left(P_i, c^n_{\hat{j}}, h^n_{\hat{k}}, \tau^n\right)$ obtained by linear interpolation with $c^n_{\hat{j}}$ and $h^n_{\hat{k}}$ given by (53) and (48).

For the purpose of proving consistency of the scheme, following [7], we separate the control region Δ_j in (54) into three subregions: $\Delta_j = [c_{\min} - c_j, 0) \cup \{0\} \cup (0, c_{\max} - c_j]$, where we adopt the convention that $(\alpha, \beta] = \emptyset$ and $[\alpha, \beta) = \emptyset$ if $\alpha = \beta$. We will write equation (56) in terms of these three subregions. Let us define

$$\mathcal{H}_{i,j,k}^{n+1}\left(\varepsilon, V_{i,j,k}^{n+1}, \{V_{l,j,k}^{n+1}\}_{l\neq i}, \{V_{i,j,k}^{n}\}\right) = \frac{V_{i,j,k}^{n+1} - V_{i,j,\hat{k}}^{n}}{\Delta\tau} - (\mathcal{C}_\varepsilon V)_{i,j,k}^{n+1} - (\mathcal{B}_\varepsilon V)_{i,j,k}^{n} - G(c_j, h_k)H(c_j, h_k)P_i\ , \tag{57}$$

where $V_{i,j,\hat{k}}^{n}$ is an approximation of $V(P_i, c_j, h_{\hat{k}}^{n}, \tau^n)$. We denote (assuming $c_j > c_{\min}$)

$$\begin{aligned}&\mathcal{I}_{i,j,k}^{n+1}\left(\varepsilon, V_{i,j,k}^{n+1}, \{V_{l,j,k}^{n+1}\}_{l\neq i}, \{V_{i,j,k}^{n}\}\right)\\ &= V_{i,j,k}^{n+1} - \sup_{\delta_{i,j,k}^{n+1}\in[c_{\min}-c_j,0)}\left[V_{i,\hat{j},\hat{k}}^{n} - d_-\right] - \Delta\tau(\mathcal{C}_\varepsilon V)_{i,j,k}^{n+1}\\ &\quad - \Delta\tau(\mathcal{B}_\varepsilon V)_{i,j,k}^{n} - \Delta\tau G(c_j, h_k)H(c_j, h_k)P_i\end{aligned} \tag{58}$$

where $V_{i,\hat{j},\hat{k}}^{n}$ is an approximation of $V(P_i, c_{\hat{j}}^{n}, h_{\hat{k}}^{n}, \tau^n)$, and (assuming $c_j < c_{\max}$)

$$\begin{aligned}&\mathcal{J}_{i,j,k}^{n+1}\left(\varepsilon, V_{i,j,k}^{n+1}, \{V_{l,j,k}^{n+1}\}_{l\neq i}, \{V_{i,j,k}^{n}\}\right)\\ &= V_{i,j,k}^{n+1} - \sup_{\delta_{i,j,k}^{n+1}\in(0,c_{\max}-c_j]}\left[V_{i,\hat{j},\hat{k}}^{n} - d_+\right] - \Delta\tau(\mathcal{C}_\varepsilon V)_{i,j,k}^{n+1}\\ &\quad - \Delta\tau(\mathcal{B}_\varepsilon V)_{i,j,k}^{n} - \Delta\tau G(c_j, h_k)H(c_j, h_k)P_i.\end{aligned} \tag{59}$$

Note that within (57-59), the fixed cost term $d(\delta_{i,j,k}^{n+1})$ in (56) is replaced by the representation given in (55) based on the subregion where the control $\delta_{i,j,k}^{n+1}$ resides. Given the definitions of $\mathcal{H}, \mathcal{I}, \mathcal{J}$, we can write scheme (56) in an equivalent way at a node $(P_i, c_j, h_k, \tau^{n+1})$, $n \geq 0$, as

$$\mathcal{G}_{i,j,k}^{n+1}\left(\varepsilon, V_{i,j,k}^{n+1}, \{V_{l,j,k}^{n+1}\}_{l\neq i}, \{V_{i,j,k}^{n}\}\right) \equiv \begin{cases} \min\left\{\mathcal{H}_{i,j,k}^{n+1}, \mathcal{I}_{i,j,k}^{n+1}, \mathcal{J}_{i,j,k}^{n+1}\right\} & \text{if } c_{\min} < c_j < c_{\max},\\ \min\left\{\mathcal{H}_{i,j,k}^{n+1}, \mathcal{I}_{i,j,k}^{n+1}\right\} & \text{if } c_j = c_{\max},\\ \min\left\{\mathcal{H}_{i,j,k}^{n+1}, \mathcal{J}_{i,j,k}^{n+1}\right\} & \text{if } c_j = c_{\min},\end{cases}$$
$$= 0. \tag{60}$$

Remark 3.3. *In the impulse control case, although the ramping rate is of bang-bang type, i.e., $\zeta_{i,j,k}^{n+1} \in \{-\infty, 0, \infty\}$, our numerical results indicate that the outflow rate $c_{\hat{j}}^{n}$ resulting from (53) is not of bang-bang type, i.e., $c_{\hat{j}}^{n}$ can be any value between $c_{\min}$ and $c_{\max}$. This is due to the nonlinear revenue structure resulting from the nonlinear function $H(c, h)$ in (3).*

3.3. Solving the Local Optimization Problems

In scheme (49) we need to solve a discrete local optimization problem

$$\sup_{\zeta_{i,j,k}^{n+1}\in Z_j} V_{i,\hat{j},\hat{k}}^{n} \tag{61}$$

at a mesh node $(P_i, c_j, h_k, \tau^{n+1})$. We solve problem (61) using an approach similar to that given in [7], which we briefly describe as follows. If we fix a mesh node $(P_i, c_j, h_k, \tau^{n+1})$, then from (47-48), $h^n_{\hat{k}}$ is fixed and $c^n_{\hat{k}}$ varies according to different values of $\zeta^{n+1}_{i,j,k}$, where all the values of $c^n_{\hat{k}}$ form a closed region, denoted by

$$C_j = \left\{ c^n_{\hat{j}} \mid c^n_{\hat{j}} = c_j + \zeta^{n+1}_{i,j,k} \Delta\tau, \ \forall \zeta^{n+1}_{i,j,k} \in Z_j \right\}. \tag{62}$$

We first sample a sequence of values from the region C_j, denoted by $\hat{C}_j$, where $\hat{C}_j$ includes the lower and upper bounds of region C_j as well as all the discrete grid nodes in the c direction residing within the region. We then evaluate $V^n_{i,\hat{j},\hat{k}}$ for all elements in the sequence $\hat{C}_j$ and return as output the maximum among the set of computed values. In other words, we solve an alternative problem

$$\sup_{c^n_{\hat{j}} \in \hat{C}_j} V^n_{i,\hat{j},\hat{k}}. \tag{63}$$

Following [7], we can prove the following results, showing that the solutions to problems (61) and (63) are consistent.

Proposition 3.4. *Let $\phi(P, c, h, \tau)$ be a smooth function with $\phi^n_{i,j,k} = \phi(P_i, c_j, h_k, \tau^n)$. Then the optimization procedure introduced above results in*

$$\begin{aligned} \sup_{c^n_{\hat{j}} \in \hat{C}_j} \phi^n_{i,\hat{j},\hat{k}} &= \sup_{\zeta^{n+1}_{i,j,k} \in Z_j} \phi^n_{i,\hat{j},\hat{k}} + O(\varepsilon^2) \\ &= \sup_{\zeta^{n+1}_{i,j,k} \in Z_j} \phi(P_i, \hat{c}^n_j, \hat{h}^n_j, \tau^n) + O(\varepsilon^2). \end{aligned} \tag{64}$$

We can also use the approach described above to solve the local optimization problem

$$\sup_{\delta^{n+1}_{i,j,k} \in \Delta_j} \left[V^n_{i,\hat{j},\hat{k}} - d(\delta^{n+1}_{i,j,k}) \right] \tag{65}$$

for scheme (56).

Remark 3.5. *Note that the discretizations (49) and (56) are virtually identical. As discussed above, in both cases, we reduce the local optimization problems to a grid search using a set of values $\hat{C}_j$. Hence, we use essentially the same discretization for either the finite control case or the impulse control case. Only the set of admissible controls and the inclusion/exclusion of the fixed cost term is different in each case.*

Remark 3.6. *Since $\zeta^{n+1}_{i,j,k}$ in discretization (49) is bounded, then according to (47), at each mesh node (P_i, c_j, h_k, τ^n) we need to perform only a finite number of linear interpolations to solve problem (63). More precisely, in view of Remark 2.1, we need only examine the upper and lower bounds of Z_j and $c^n_{\hat{j}} = c_j$.*

Moreover, since Z_j is independent of the timestep τ^n, we can precompute the interpolation weights to avoid binary searches for linear interpolations at each timestep.

As for problem (56), however, we need to perform $O(1/\varepsilon)$ interpolations at each mesh node since we have to examine $O(1/\varepsilon)$ grid nodes in the c direction according to (53). Nevertheless, we can still save the cost for binary searches at each timestep by precomputing the interpolation weights.

4. Properties of the Numerical Schemes

Provided a strong comparison result for the pricing PDE/PIDE applies, [2, 5] demonstrate that a numerical scheme will converge to the viscosity solution of the equation if it is l_∞ stable, monotone, and consistent. In this section, we will prove the convergence of our numerical schemes (49) and (56) (or equivalently, schemes (50) and (60)) to the viscosity solution of the pricing equations (21) and (28) respectively by verifying these three properties.

4.1. l_∞ Stability

Definition 4.1 (l_∞ stability). *Discretizations (49) and (56) are l_∞ stable if*

$$\|V^{n+1}\|_\infty \leq C_5 \tag{66}$$

for $0 \leq n \leq N-1$ as $\Delta\tau \to 0$, $\Delta P_{\min} \to 0$, $\Delta c_{\min} \to 0$, $\Delta h_{\min} \to 0$, where C_5 is a constant independent of $\Delta\tau$, $\Delta P_{\min}$, $\Delta c_{\min}$, $\Delta h_{\min}$. Here $\|V^{n+1}\|_\infty = \max_{i,j,k} |V^{n+1}_{i,j,k}|$.

Lemma 4.2 (l_∞ stability). *If discretizations (49) and (56) satisfy the positive coefficient condition (41) and conditions (43-44) and if linear interpolation is used to compute $V^n_{i,\hat{j},\hat{k}}$, then schemes (49) and (56) satisfy*

$$\|V^{n+1}\|_\infty \leq \|V^0\|_\infty + T \cdot P_{\max} \cdot \|H\|, \tag{67}$$

where

$$\|H\| = \max_{j,k} |H(c_j, h_k)|. \tag{68}$$

Therefore, discretizations (49) and (56) are l_∞ stable according to Definition 4.1.

Proof. The proof directly follows from applying the maximum principle to the discrete equations (49) and (56). We omit the details here. Readers can refer to [12, Theorem 5.5] and [16] for complete stability proofs of the semi-Lagrangian fully implicit scheme for American Asian options and that of finite difference schemes for controlled HJB equations, respectively. □

4.2. Consistency

Let us define a vector $\mathbf{x} = (P, c, h, \tau)$ and let $DV(\mathbf{x})$ and $D^2V(\mathbf{x})$ be the first and second order derivatives of $V(\mathbf{x})$, respectively. Let $IV(\mathbf{x})$ be the integral terms in the pricing equation. Let $\bar{\Omega} = [0, P_{\max}] \times [c_{\min}, c_{\max}] \times [h_{\min}, h_{\max}] \times [0, T]$ be the closed domain in which our problem is defined. Then following the steps in [7], we can rewrite the pricing problem (21), (29), (30), (33) or the problem (28), (29), (35-38) into one equation as follows:

$$F\big(D^2V(\mathbf{x}), DV(\mathbf{x}), IV(\mathbf{x}), V(\mathbf{x}), \mathbf{x}\big) = 0 \quad \text{for all } \mathbf{x} = (P, c, h, \tau) \in \bar{\Omega}\,. \tag{69}$$

The authors of [2, 5] define the consistency of a numerical scheme to a possible discontinuous viscosity solution of a nonlinear PDE, which in our case is given as follows.

Definition 4.3 (Consistency). *The schemes $\mathcal{G}_{i,j,k}^{n+1}$ given in (50) and (60) are consistent with the pricing problem (21), (29), (30), (33) and the pricing problem (28), (29), (35-38), respectively, if for all $\hat{\mathbf{x}} = (\hat{P}, \hat{c}, \hat{h}, \hat{\tau}) \in \bar{\Omega}$ and any function $\phi(P, c, h, \tau)$ having bounded derivatives of all orders in $(P, c, h, \tau) \in \bar{\Omega}$ with $\phi_{i,j,k}^{n+1} = \phi(P_i, c_j, h_k, \tau^{n+1})$ and $\mathbf{x} = (P_i, c_j, h_k, \tau^{n+1})$, we have*

$$\limsup_{\substack{\varepsilon \to 0 \\ \mathbf{x} \to \hat{\mathbf{x}} \\ \xi \to 0}} \mathcal{G}_{i,j,k}^{n+1}\left(\varepsilon, \phi_{i,j,k}^{n+1}+\xi, \{\phi_{l,j,k}^{n+1}+\xi\}_{l \neq i}, \{\phi_{\hat{i},j,k}^{n}+\xi\}\right) \leq F^{*}\left(D^2\phi(\hat{\mathbf{x}}), D\phi(\hat{\mathbf{x}}), I\phi(\hat{\mathbf{x}}), \phi(\hat{\mathbf{x}}), \hat{\mathbf{x}}\right), \tag{70}$$

and

$$\liminf_{\substack{\varepsilon \to 0 \\ \mathbf{x} \to \hat{\mathbf{x}} \\ \xi \to 0}} \mathcal{G}_{i,j,k}^{n+1}\left(\varepsilon, \phi_{i,j,k}^{n+1}+\xi, \{\phi_{l,j,k}^{n+1}+\xi\}_{l \neq i}, \{\phi_{\hat{i},j,k}^{n}+\xi\}\right) \geq F_{*}\left(D^2\phi(\hat{\mathbf{x}}), D\phi(\hat{\mathbf{x}}), I\phi(\hat{\mathbf{x}}), \phi(\hat{\mathbf{x}}), \hat{\mathbf{x}}\right), \tag{71}$$

where F^* and F_* in (70-71) are upper semi-continuous (usc) and lower semi-continuous (lsc) envelopes of the function F defined in (69), respectively. See, for example, [7] for the definition of usc and lsc envelopes.

We first show the following properties of the schemes (49) and (56).

Lemma 4.4. *If the function $G(c, h)$ satisfies condition (10) and $\Delta\tau$ satisfies the assumption (39), then by taking ε sufficiently small, (48) becomes*

$$h_{\hat{k}}^{n} = h_k + \frac{f - c_j}{a} G(c_j, h_k)\Delta\tau. \tag{72}$$

If the control $\zeta_{i,j,k}^{n+1}$ satisfies condition (15-16) and $\Delta\tau$ satisfies the assumption (39), then by taking ε sufficiently small, we have

$$Z_j = [z_{\min}, z_{\max}]. \tag{73}$$

Proof. The proof follows from the steps in [8, Lemma B.1] by showing that (72) follows if $\Delta\tau < Const.$ and $\Delta\tau = o(\varepsilon^{1-\nu})$, and that (73) holds if $\Delta\tau < Const.$ and $\Delta\tau = o(\varepsilon^{1-\theta})$, where $Const.$ represents a positive constant. Since $\nu > 0$, $\theta > 0$, these conditions are weaker than the assumption $\Delta\tau = C_4\varepsilon$ in (39) and will be satisfied if ε is sufficiently small. □

Lemma 4.5 (Consistency). *Suppose that the mesh size and timestep size satisfy assumption (39). Then the discretizations (50) and (60) are consistent as defined in Definition 4.3. In particular, if the value function is sufficiently smooth, and assuming that linear interpolation is used to compute $V_{\hat{i},\hat{j},\hat{k}}^{n}$, then the global discretization errors of both schemes are $O(\varepsilon)$.*

Proof. The proof follows the lines in [7, 8] and the results in Lemma 4.4. We omit the details here. □

4.3. Monotonicity

The following result shows that schemes (50) and (60) are monotone according to the definition in [2, 5].

Lemma 4.6 (Monotonicity). *If discretizations (50) and (60) satisfy the positive coefficient condition (41) and conditions (43-44) and if linear interpolation is used to compute $V^n_{i,\hat{j},\hat{k}}$, then discretizations (50) and (60) are monotone according to the definition in [2, 5], i.e.,*

$$\begin{aligned} &\mathcal{G}^{n+1}_{i,j,k}\left(\varepsilon, V^{n+1}_{i,j,k}, \{X^{n+1}_{l,j,k}\}_{l\neq i}, \{X^n_{i,j,k}\}\right) \\ &\le \mathcal{G}^{n+1}_{i,j,k}\left(\varepsilon, V^{n+1}_{i,j,k}, \{Y^{n+1}_{l,j,k}\}_{l\neq i}, \{Y^n_{i,j,k}\}\right); \text{ for all } X^n_{i,j,k} \ge Y^n_{i,j,k}, \forall i, j, k, n. \end{aligned} \tag{74}$$

Proof. Inequality (74) can be easily verified for all mesh nodes $(P_i, c_j, h_k, \tau^{n+1})$ for schemes (50) and (60) (e.g. see [16]). □

4.4. Convergence

In order to prove the convergence of our schemes using the results in [2, 5, 6], we need to assume the following strong comparison result, as defined in [2, 5], for equations (21) and (28).

Assumption 4.7. *For either the localized pricing problem (21), (29), (30), (33) or the localized pricing problem (28), (29), (35-38), if u and v are an upper semi-continuous (usc) subsolution and a lower semi-continuous (lsc) supersolution of the problem, respectively, then*

$$u \le v \quad \text{on } \Omega_{in} = (0, P_{\max}) \times (c_{\min}, c_{\max}) \times (h_{\min}, h_{\max}) \times (0, T]. \tag{75}$$

The strong comparison result is proved for other similar (but not identical) bounded stochastic control problems in [3, 4] and impulse control problems in [1, 20, 22, 26]. From Lemmas 4.2, 4.5 and 4.6 and Assumption 4.7, using the results in [2, 5, 6], we can obtain the following convergence result:

Theorem 4.8 (Convergence to the viscosity solution). *Assuming that discretizations (50) and (60) satisfy all conditions required for Lemmas 4.2, 4.5 and 4.6, and that Assumption 4.7 is satisfied, then schemes (50) and (60) converge to the viscosity solutions of the pricing problem (21), (29), (30), (33) and the pricing problem (28), (29), (35-38), respectively.*

5. Numerical Results

In this section, we conduct numerical experiments for the hydroelectric power plant valuation problem. We use "dollars per megawatt-hour" ($/MWhr), "meter" ($m$), "hour" ($hr$), "cubic meters per second" (m^3/s), and "cubic meters per second per hour" ($m^3/s - hr$) as the default units for electricity spot price, water head, time, inflow/outflow rate, and ramping rate, respectively. We will value the cash flows over a one week cycle ($T = 1$ week), similar to the approach in [24]. We will use the terminal condition (29), as in [24].

Table 1. Values of electricity spot price parameters in (17-19). The values of α, K_0, β, t_0, σ, P_0, $\bar{\lambda}_2$ are chosen from [24].

Parameter	Value	Parameter	Value
α	0.4	$\bar{\lambda}_2$	0.85 hr^{-1}
K_0	27 \$/MWhr	μ^1	0.3
β	15	ψ_0^1	0
t_0	7.7π	ψ_1^1	3.2
σ	0.2 $hr^{-1/2}$	μ^2	0.4
λ_1	0.01 hr^{-1}	ψ_0^2	-3.6
P_0	100 \$/MWhr	ψ_1^2	0

Prior to illustrating results, we list the parameter values in our experiments. Following [17], we assume the logarithm of the random jump sizes J_1 and J_2 in (17) satisfies a truncated version of an exponential distribution with parameters μ, ψ_0 and ψ_1, $\psi_0 < \psi_1$. The corresponding probability density function is given by

$$p(x;\mu,\psi_0,\psi_1) = \begin{cases} \frac{\mu\exp(-\mu x)}{\exp(-\mu\psi_0)-\exp(-\mu\psi_1)} & \text{if } x \in [\psi_0,\psi_1], \\ 0 & \text{otherwise.} \end{cases} \tag{76}$$

Function (76) reveals that the value of x is bounded between ψ_0 and ψ_1. We use $p(\log J_1;\mu^1,\psi_0^1,\psi_1^1)$ and $p(\log J_2;\mu^2,\psi_0^2,\psi_1^2)$ to represent the probability density function for $\log J_1$ and $\log J_2$, respectively, where $\psi_0^1 = \psi_1^2 = 0$, $\psi_1^1 > 0$ and $\psi_0^2 < 0$. These parameter values are given in Table 1, where the values of μ^1,ψ_0^1,ψ_1^1 are chosen similar to those calibrated from the time series of the electricity spot price in [17] and the values of μ^2,ψ_0^2,ψ_1^2 are calculated such that $E[J_1]\cdot E[J_2] = 1$. This approximately reflects the idea that after an up-jump immediately followed by a down-jump (this approximates a typical price spike), on average the spot price will return to its starting value before the spike occurs. The values of ψ_0^1, ψ_1^1, ψ_0^2, ψ_0^2 indicate that $J_1 \geq 1$ and $0 < J_2 \leq 1$.

Table 1 also provides values of other electricity spot price parameters in equations (17-19).

The other parameters for our experiments are given in Table 2. We set $c_{\min} = 40\ m^3/s$ if the minimum flow rate constraint is imposed; otherwise we set $c_{\min} = 0$. According to (2), the values of g, ρ, $c_{\max}$, $h_{\max}$ imply that the theoretical maximum electricity power is $H_m(c_{\max}, h_{\max}) = 138.18 \times 10^6$ Watt $= 138.18$ MW. The values of $c_{\min}$, $c_{\max}$, $z_{\max}$ in the table imply that it takes $\frac{c_{\max}-c_{\min}}{z_{\max}} \approx 18$ hours to move the outflow rate from the minimum to the maximum when the minimum flow rate and ramping constraints are imposed. The values of a, f, $c_{\max}$, $h_{\min}$ and $h_{\max}$ indicate that reducing the head from its highest value to the lowest value will take at least $\frac{a(h_{\max}-h_{\min})}{3600(c_{\max}-f)} \approx 22$ hours. We choose small fixed costs $d_- = d_+ = 10^{-8}$ for the impulse control problem (28), which implies that the operator can switch between $z = -\infty$ and $z = +\infty$ almost freely due to the negligible costs associated with the operation.

After presenting the parameters, we first carry out a convergence analysis for the hydroelectric power plant valuation with/without operational constraints. Table 3 shows

Table 2. Other input parameters used to price the value of the hydroelectric power plant. Parameters g, ρ, k_1, k_2, a, f are shown in (2-5). r is the annual riskless interest rate and T is the operational time interval. $c_{\min}$ is the minimum outflow rate in case the minimum flow rate constraint is imposed; otherwise $c_{\min} = 0$. Parameters d_- and d_+ are the fixed costs in equation (28).

Parameter	Value	Parameter	Value
g	9.8 m/s^2	r	0.05 annually
ρ	1000 kg/m^3	T	1 week
k_1	85%	$h_{\min}$	90 m
k_2	120 MW	$h_{\max}$	94 m
a	1.8×10^6 m^2	$c_{\min}$	40 m^3/s
f	60 m^3/s	$c_{\max}$	150 m^3/s
d_-	10^{-8}	$z_{\min}$	-6 $m^3/s - hr$
d_+	10^{-8}	$z_{\max}$	6 $m^3/s - hr$

the convergence results with respect to different mesh size/timestep parameters when $P = K_0 = 27$ \$/MWhr (the average of daily electricity spot price), $c = 100$ m^3/s, $h = 92$ m and $t = 0$. The convergence ratio in the table is defined as the ratio of successive changes in the solution, as the timestep and mesh size are reduced by a factor of two. A ratio of two indicates first order convergence. As shown in Table 3, our schemes are able to achieve a first-order convergence for both the pricing equation (21) with operating constraints incorporated and the impulse control equation (28) without incorporating operating constraints, as the convergence ratios are approximately two. Table 3 also implies that imposing the operational constraints reduces power plant value by more than 37%.

Recall that the computational domain has been localized in the P direction to $[0, P_{\max}]$. Initially, we set $P_{\max} \approx 7 \times 10^5$ \$/MWhr. We repeated the computations with $P_{\max} = 7 \times 10^6$ \$/MWhr. All the numerical results at $t = 0$, $P = K_0 = 27$ \$/MWhr, $c = 100$ m^3/s, $h = 92$ m are identical (to the number of digits shown) to those in the Table 3 for all refinement levels. As a result, all the subsequent results will be reported using $P_{\max} \approx 7 \times 10^5$ \$/MWhr, since the corresponding solution error incurred by the domain localization is negligible.

Figure 1 plots the optimal operational strategies for the hydroelectric power plant as a function of outflow rate c and electricity spot price P at the current time $t = 0$ when $h = 92$ m, where the minimum flow rate and ramping rate constraints are imposed. The figure shows that the control strategy is of the bang-bang type: the optimal ramping rate is either $z_{\min}$, 0 or $z_{\max}$. According to Remark 2.1, $z = z_{\max}$ implies $V_c > 0$ in pricing equation (21), while $z = z_{\min}$ implies $V_c < 0$ in (21). The region where $z = 0$ corresponds to the following three cases:

- if $c \neq c_{\min}$ and $c \neq c_{\max}$ in this region, then $z = 0$ indicates $V_c = 0$ in the region. As shown in Remark 2.1, when $V_c = 0$, the optimal z can be any value residing in $Z(c)$. Our numerical algorithm will choose $z = 0$ in this case, since there is no advantage in changing the flow rate. This choice will be consistent with the impulse control

Table 3. Convergence study for the value of the hydroelectric power plant at $t = 0$, $P = K_0 = 27$ \$/MWhr, $c = 100\ m^3/s$, $h = 92\ m$ with/without incorporating operational constraints. Input parameter data are given in Tables 1-2. Due to CPU time consideration, we do not report the value with respect to the finest mesh/timestep size for the case without incorporating the operational constraints.

P nodes	c nodes	h nodes	Timesteps	Value	Ratio
With minimum flow rate and ramping constraints					
66	12	5	337	187207	n.a.
131	23	9	673	194484	n.a.
261	45	17	1345	199182	1.55
521	89	33	2689	201848	1.76
1041	177	65	5377	203393	1.73
Without minimum flow rate or ramping constraint					
66	16	5	337	288084	n.a.
131	31	9	673	306584	n.a.
261	61	17	1345	317034	1.77
521	121	33	2689	322391	1.95

formulation with an infinitesimal fixed cost.

- if $c = c_{\max}$ in the region, then $z = 0$ represents $V_c \geq 0$. This is because if $V_c > 0$, then Remark 2.1 implies that it is optimal to choose $z = \max\{Z(c)\}$, and $\max\{Z(c)\} = 0$ in this region due to condition (14).

- if $c = c_{\min}$ in the region, then $z = 0$ indicates $V_c \leq 0$ since condition (13) implies that $\min\{Z(c)\} = 0$ in this region.

From the figure, we can observe that for most outflow rates, it is optimal to ramp up at the maximum rate when electricity price is high so that more electricity will be generated and sold to the spot market; similarly, it is optimal to ramp down at the maximum rate when electricity price is low in order to store water in the reservoir for producing electricity in the future. Meanwhile, the ramp-up region expands when the outflow rate decreases, since a lower outflow rate gives more incentive for the ramp-up operation. Similarly, the ramp-down region expands when the outflow rate increases.

Figure 2 plots the optimal control strategy that evolves over time as a function of electricity price when $c = 100\ m^3/s$ and $h = 92\ m$. In order to see the patterns more clearly, we use $T = 4$ days for this plot. The figure clearly reveals the daily trend of the electricity price: the control pattern is repeated every day. The plot also shows that the optimal control z forms three regions in which $z = z_{\max}$, $z = 0$ and $z = z_{\min}$, respectively. When $t \to T$, it is optimal to keep ramping up (i.e., to increase the outflow rate c) in order to produce as much profit as possible. This is, of course, an artifact of the terminal condition (29). Another possibility would be to impose a penalty unless the hydro plant was returned to its original state at the beginning of the weekly cycle. In fact, such penalties are common in

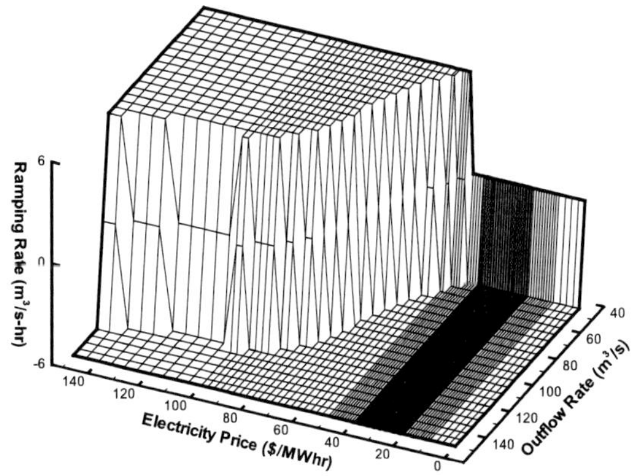

Figure 1. Optimal operational strategy as a function of outflow rate c and electricity spot price P when $t = 0$, $h = 92\ m$. The minimum flow rate and ramping constraints are imposed. Input parameter data are given in Tables 1-2.

leasing gas storage facilities [8]. However, to keep things simple we will impose condition (29).

Finally, we study the implication of the operational restrictions on the value of the hydroelectric power plant. For both cases with and without the minimum flow rate constraint, Table 4 lists values of the hydroelectric power plant with respect to different ramping rate constraints. The table shows that imposing the ramping constraint at $|z_{\min}| = |z_{\max}| = 6\ m^3/s-hr$ reduces the value by 20-32%, while imposing the minimum flow rate constraint reduces the value by 9-22%. Therefore, we conclude that it is important to take the operational constraints into consideration in order to accurately price the power plant cash flows.

Table 4 also indicates that as the maximum ramp-up and ramp-down rates $|z_{\max}|$, $|z_{\min}|$ increase, the value of the power plant converges to the value resulting from eliminating the ramping rate constraint (i.e., setting $|z_{\max}| = |z_{\min}| = \infty$). The convergence appears to be rapid. In fact, under the minimum flow rate constraint, as $|z_{\max}| = |z_{\min}| = 96\ m^3/s - hr$ (i.e., switching between the minimum outflow rate $c_{\min}$ and maximum outflow rate $c_{\max}$ can be finished in about 1.5 hours), the value of the power plant is very close to that obtained without any ramping constraint.

Conclusion

In this chapter, we determine the risk neutral value of the cash flows to a hydroelectric power plant over a one week cycle under a stochastic control framework. We take into consideration operational constraints such as ramping and minimum flow rate constraints required for environmental protection.

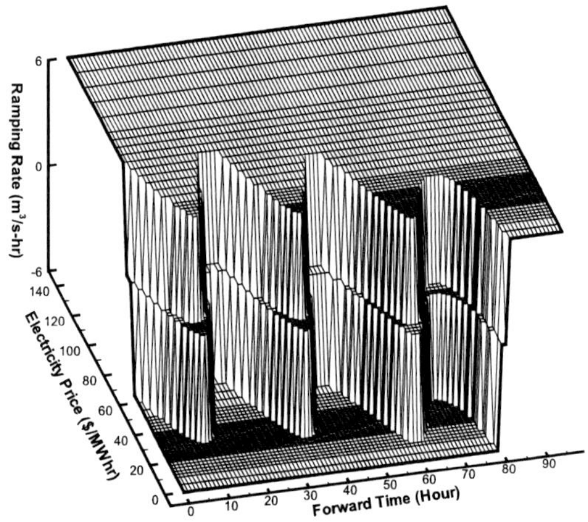

Figure 2. Optimal operational strategy as a function of forward time t and electricity spot price P when $c = 100\ m^3/s$ and $h = 92\ m$. The minimum flow rate and ramping constraints are imposed. In order to be able to see some of the daily patterns, we use $T = 4$ days in this case. Other input parameters are given in Tables 1-2.

Table 4. Comparison on values of the hydroelectric power plant with respect to different values of ramp-up and ramp-down rates. The ∞ corresponds to the case without imposing the ramping rate constraint. The column "With MFR" represents the power plant values under the minimum flow rate constraint (i.e., setting $c_{\min} = 40\ m^3/s$); the column "Without MFR" represents the values without the minimum flow rate constraint (i.e., setting $c_{\min} = 0$). The values in the table are accurate up to the first two digits. Other input parameters are given in Tables 1-2.

Maximum Ramp-up/-down Rates $\lvert z_{\max}\rvert, \lvert z_{\min}\rvert$	Value	
	With MFR	Without MFR
$6\ m^3/s - hr$	2.0×10^5	2.2×10^5
$12\ m^3/s - hr$	2.2×10^5	2.5×10^5
$24\ m^3/s - hr$	2.3×10^5	2.8×10^5
$48\ m^3/s - hr$	2.4×10^5	3.0×10^5
$96\ m^3/s - hr$	2.5×10^5	3.1×10^5
$\infty\ m^3/s - hr$	2.5×10^5	3.2×10^5

We formulate the power plant valuation problem under a ramping constraint as a bounded stochastic control problem, resulting in an HJB PIDE. We also formulate the valuation problem without the ramping restriction as an impulse control problem, resulting in an HJB variational inequality.

We develop a consistent numerical scheme for solving both the HJB PIDE for the bounded control problem and the HJB variational inequality for the impulse control problem. Our discretization scheme is essentially the same for either the bounded control case or the impulse control case. This makes software implementation very straightforward, and ensures that we obtain the correct limits as the ramping rate becomes unbounded. We prove the convergence of the numerical scheme to the viscosity solution of each pricing problem, provided a strong comparison result holds. Numerical results indicate that our scheme can achieve first order convergence.

We also study the implication of the operational restrictions on the value of a hydroelectric power plant. We observe through an example that both the ramping and the minimum flow rate constraints can considerably affect the value of the power plant, where imposing either constraint can reduce the value by 9-32%, while imposing both constraints will reduce the value by more than 37%. Therefore, we conclude that it is important to take the operational constraints into consideration in order to accurately price the power plant cash flows.

Our numerical experiments also show that as the maximum ramp-up and ramp-down rates increase, the value of the hydroelectric power plant rapidly converges to the value obtained without imposing any constraints.

Acknowledgements

This work was supported by the Natural Sciences and Engineering Research Council of Canada, and by a Morgan Stanley Equity Market Microstructure Research Grant. The views expressed herein are solely those of the authors, and not those of any other person or entity, including Morgan Stanley.

References

[1] A.L. Amadori, Quasi-variational inequalities with Dirichlet boundary condition related to exit time problems for impulse control, *SIAM J. Control Optim.* **43** (2004), 570–589.

[2] G. Barles, *Convergence of numerical schemes for degenerate parabolic equations arising in finance*, In: L.C.G. Rogers and D. Talay, editors, *Numerical methods in finance*, pages 1–21. Cambridge University Press, Cambridge, 1997.

[3] G. Barles and J. Burdeau, The Dirichlet problem for semilinear second-order degenerate elliptic equations and applications to stochastic exit time control problems, *Communications in Partial Differential Equations* **20** (1995), 129–178.

[4] G. Barles and E. Rouy, A strong comparison result for the Bellman equation arising in stochastic exit time control problems and its applications, *Communications in Partial Differential Equations* **23** (1998), 1945–2033.

[5] G. Barles and P.E. Souganidis, Convergence of approximation schemes for fully nonlinear equations, *Asymptotic Analysis* **4** (1991), 271–283.

[6] M. Briani, C. L. Chioma and R. Natalini, Convergence of numerical schemes for viscosity solutions to integro-differential degenerate parabolic problems arising in financial theory, *Numerische Mathematik* **98** (2004), 607–646.

[7] Z. Chen and P.A. Forsyth, *A numerical scheme for the impulse control formulation for pricing variable annuities with a guaranteed minimum withdrawal benefit (GMWB)*, Working paper, University of Waterloo, Submitted to: Numerische Mathematik, July 2007.

[8] Z. Chen and P.A. Forsyth, A semi-Lagrangian approach for natural gas storage valuation and optimal operation, To appear in: *SIAM Journal on Scientific Computing*, 2007.

[9] C. de Jong and R. Huisman *Option formulas for mean-reverting power prices with spikes*, Working paper, Rotterdam School of Management at Erasmus University, 2002.

[10] S. Deng and S.S. Oren. Incorporating operational characteristics and start-up costs in option-based valuation of power generation capacity, *Probability in the Engineering and Informational Sciences* **17** (2003), 151–181.

[11] Y. d'Halluin, P. A. Forsyth and G. Labahn, A penalty method for American options with jump diffusion processes, *Numerische Mathematik* **97** (2004), 321–352.

[12] Y. D'Halluin, P. A. Forsyth and G. Labahn. A Semi-Lagrangian approach for American asian options under jump diffusion, *SIAM Journal on Scientific Computing* **27** (2005), 315–345.

[13] Y. d'Halluin, P.A. Forsyth and K.R. Vetzal, Robust numerical methods for contingent claims under jump diffusion processes, *IMA Journal of Numerical Analysis* **25** (2005), 65–92.

[14] B.K. Edwards, S.J. Flaim and R.E. Howitt, Optimal provision of hydroelectric power under environmental and regulatory constraints, *Land Economics* **75** (1999), 267–283.

[15] W.H. Fleming and H.M. Soner, *Controlled Markov processes and viscosity solutions,*. Springer, 2006.

[16] P.A. Forsyth and G. Labahn, Numerical methods for controlled Hamilton-Jacobi-Bellman PDEs in finance, *Journal of Computational Finance* **11** (2007), 1–41.

[17] H. Geman and A. Roncoroni, Understanding the fine structure of electricity prices, *Journal of Business* **79** (2006), 1225–1261.

[18] X. Guan, A. Svoboda and C. Li, Scheduling hydro power systems with restricted operating zones and discharge ramping constraints, *IEEE Transactions on Power Systems* **14** (1999), 126–131.

[19] X.S. Han and H.B. Gooi, Optimal dynamic dispatch in short-term hydrothermal generation systems, *Power Engineering Society Winter Meeting* **2** (2000), 1231–1236.

[20] K. Ishii, Viscosity solutions of nonlinear second order elliptic PDEs associated with impulse control problems II, *Funkcialaj Ekvacioj* **38** (1995), 297–328.

[21] J. Keppo and E. Nasakkala, Hydropower production planning and hedging under inflow and forward uncertainty, *Working paper*, University of Michigan, 2004.

[22] B. Øksendal and A. Sulem, Optimal consumption and portfolio with both fixed and proportional transaction costs, *SIAM Journal on Control and Optimization* **40** (2002), 1765–1790.

[23] H. Pham, On some recent aspects of stochastic control and their applications, *Probability Surveys* **2** (2005), 506–549.

[24] M. Thompson, M. Davison and H. Rasmussen, Valuation and optimal operation of electric power plants in competitive markets, *Operations Research* **52** (2004), 546–562.

[25] C. Tseng and G. Barz, Short-term generation asset valuation: a real options approach, *Operations Research* **50** (2002), 297–310.

[26] V.L. Vath, M. Mnif and H. Pham, A model of optimal portfolio selection under liquidity risk and price impact, *Finance and Stochastics* **11** (2007), 51–90.

[27] H. Windcliff, P.A. Forsyth and K.R. Vetzal, Analysis of the stability of the linear boundary condition for the Black-Scholes equation, *Journal of Computational Finance* **8** (2004), 65–92.

[28] W. Zhu, *Thermal generation asset valuation problems in a competitive market*, PhD thesis, University of Maryland, College Park, 2004.

In: Nonlinear Models in Mathematical Finance...
Editor: Matthias Ehrhardt, pp. 303-318
ISBN 978-1-60456-931-5

Chapter 10

NUMERICAL SOLUTIONS OF CERTAIN NONLINEAR MODELS IN EUROPEAN OPTIONS ON A DISTRIBUTED COMPUTING ENVIRONMENT

Choi-Hong Lai
School of Computing and Mathematical Sciences
University of Greenwich
Park Row, Greenwich, London SE10 9LS, UK

Abstract

Financial modelling in the area of option pricing involves the understanding of the correlations between asset and movements of buy/sell in order to reduce risk in investment. Such activities depend on financial analysis tools being available to the trader with which he can make rapid and systematic evaluation of buy/sell contracts. In turn, analysis tools rely on fast numerical algorithms for the solution of financial mathematical models. There are many different financial activities apart from shares buy/sell activities. However it is not the intention of this chapter to discuss various financial activities. The main aim of this chapter is to discuss a distributed algorithm for the numerical solution of a European option. Both linear and non-linear cases are considered.

The algorithm is based on the concept of the Laplace transform and its numerical inverse. The scalability of the algorithm is examined. Numerical tests are used to demonstrate the effectiveness of the algorithm for financial analysis. Time dependent functions for volatility and interest rates are also discussed. Applications of the algorithm to non-linear Black-Scholes equation where the volatility and the interest rate are functions of the option value are included. Some qualitative results of the convergence behaviour of the algorithm is examined. This chapter also examines the various computational issues of the Laplace transformation method in terms of distributed computing. The idea of using a two-level temporal mesh in order to achieve distributed computation along the temporal axis is introduced. Finally, the chapter ends with some conclusions.

1. Introduction

Financial modelling in the area of option pricing involves the understanding of the correlations between assets and movements of buy/sell in order to reduce various investment risks. Full simulation of the entire market is almost impossible at this stage due to the limitations of computing power and numerical models. However analysts would like to see the best inclusion of possible factors and number of underlying assets. Such market analysing activities depends on financial analysis tools being available to the trader with which he can make rapid and systematic evaluation of buy/sell contracts. In turn, analysis tools rely on fast numerical algorithms for the solution of financial mathematical models. There are many different financial activities apart from shares buy/sell activities. It is not the intention to discuss various financial activities within the scope of this chapter. The main aim here is to discuss how distributed computational properties may be built into numerical algorithms for the solutions of European options. Both linear and non-linear cases are considered and detailed results are presented for analysis and comparison.

The organisation of this chapter is as below. First a brief introduction is given of the Black-Scholes equation for option pricing in Section 2.. The purpose of this section is to discuss the deterministic partial differential equations rather than giving an account on the background stochastic theory. Section 3. discusses some frequently used temporal integration methods for time dependent partial differential equations. Motivation of using parallel and/or distributed computing environment is discussed. The reasons of using the Laplace transformation method is explained and applied to the linear Black-Scholes equation. The concept of induced distribution properties due to the transformation is also discussed. The scalability of the algorithm has been studied theoretically in [3] and [4] for linear heat conduction problems. The work in this chapter provides numerical tests to demonstrate the effectiveness of the algorithm for financial analysis. Time dependent functions for volatility and interest rates are also discussed. Section 4. introduces the nonlinear volatility model to be examined in this chapter. In Section 5., an extension is given of the algorithm for non-linear Black-Scholes equations where the volatility is a function of the option value. Linearization of the non-linear Black-Scholes equation is discussed and combination of the linearisation with the Laplace transform method is explained. Some qualitative results of the convergence behaviour of the algorithm are provided. The algorithm also relies on fast computation of the numerical inverse Laplace transform. The main goal of this chapter is to demonstrate the feasibility and effectiveness of using the Laplace transformation method and its numerical inverse for linear and nonlinear Black-Scholes equations. Various computational issues of numerical inverse Laplace in terms of distributed computing are also discussed. Finally, Section 6. gives an early investigation on the introduction of parallel time domain computation using the concept of a two-level temporal mesh.

2. The Black-Scholes Equation

Let $v(S, t)$ denote the value of an option, where S is the current value of the underlying asset and t is the time. The value of the option relates to the current value of the underlying asset via two stochastic parameters, namely the volatility σ and the interest rate r, of the

Black-Scholes equation:

$$\frac{\partial v}{\partial t} + \frac{1}{2}\sigma^2 S^2 \frac{\partial^2 v}{\partial S^2} + rS\frac{\partial v}{\partial S} - rv = 0 \in \Omega^+ \times [T, 0), \tag{1}$$

where $\Omega^+ = \{S : S \geq 0\}$. The stochastic background of the equation is not discussed in this chapter, and readers who are interested should consult reference [12].

In this chapter, attention is paid only to European options for the purpose of demonstrating the numerical algorithms. The definition of a European option is that the holder of the option may exccute at expiry a prescribed asset, known as the underlying asset, for a prescribed amount, known as the strike price. There are two different types of options, namely the call option and the put option. At expiry, the holder of the call option has the right to buy the underlying asset and the holder of the put option has the right to sell the underlying asset. For an European put option with strike price K and expiry date T, suitable boundary conditions are needed:

$$v(0,t) = Ke^{-r(T-t)}, \quad v(L,t) = 0, \tag{2}$$

where L is usually chosen as a large value. At expiry, if $S < K$ then one should exercise the call option, i.e. handing over an amount to obtain an asset with S. However, if $S > K$ at expiry, then one should not exercise the option because of the loss $K - S$. Therefore instead of providing an initial condition, the final condition:

$$v(S,T) = \max\{K - S, 0\} \tag{3}$$

needs to be imposed. The above problem requires the solution of v for $t < T$. The financial implication of the above boundary value problem is as follows: the difference between the return on an option portfolio, which involves the first two terms of equation 1, and the return on a bank deposit, which involves the last two terms of eqn 1, should be zero for a European option. Note that within a given short period, it is possible to assume the interest rate to be a constant rather than a stochastic parameter.

Note that equation (1) is a backward equation in time. It can be transformed to a forward equation by using $\tau = T - t$ as below.

$$\frac{\partial V}{\partial \tau} = \frac{1}{2}\sigma^2 S^2 \frac{\partial^2 V}{\partial S^2} + rs\frac{\partial V}{\partial S} - rV \in \Omega^+ \times (0, T] \tag{4}$$

subject to initial condition:

$$V(S,0) = \max\{K - S, 0\} \tag{5}$$

and boundary conditions:

$$V(0,\tau) = Ke^{-r\tau}, \; V(L,\tau) = 0. \tag{6}$$

3. Numerical Solutions of the Black-Scholes Equation with Linear Volatility

An analytic solution may be derived if a change of variable is made where the Black-Scholes model is converted to a time-dependent heat conduction equation with constant coefficients [12]. However a field method, such as the finite difference method, is of more interest. One reason being that real problems of multi-factor model or many underlying assets contain time dependent coefficients and/or nonlinear coefficients are not necessarily reducible to constant coefficient. As a result not all problems have an analytic form of solutions. Therefore numerical methods should be employed in this respect. This section provides several techniques behind using numerical methods.

3.1. Temporal Marching Schemes

Typical temporal marching schemes include forward difference and backward difference in the temporal axis. Let $V^{(n)}(S)$ denotes the numerical approximation of $V(S, n\delta\tau)$ where $\delta\tau$ is the time step length in a temporal marching scheme. A brief description is given of two one-step methods.

First, apply forward difference in the temporal axis to eqn (4) leads to the following semi-discretised problem.

$$\frac{V^{(n+1)} - V^{(n)}}{\delta\tau} = \frac{1}{2}\sigma^2 S^2 \frac{\partial^2 V^{(n)}}{\partial S^2} + rS\frac{\partial V^{(n)}}{\partial S} - rV \in \Omega^+ \tag{7}$$

subject to boundary conditions:

$$V^{(n+1)}(0) = Ke^{-r(n+1)\delta\tau}, \quad V^{(n+1)}(L) = 0 \tag{8}$$

One can use a finite difference approximation to replace the spatial derivatives in eqn (7). This scheme is in essence Euler's method with first order accuracy along the temporal axis. Note that there is a stability criterion relating $\delta\tau$ and the choice of the mesh size in the spatial discretisation imposes sometimes very small step sizes along the temporal axis. This method is particularly easy to implement because it is an explicit scheme.

Second, apply backward difference in the temporal axis to eqn (4) leads to the following semi-discretised problem.

$$\frac{V^{(n+1)} - V^{(n)}}{\delta\tau} = \frac{1}{2}\sigma^2 S^2 \frac{\partial^2 V^{(n+1)}}{\partial S^2} + rS\frac{\partial V^{(n+1)}}{\partial S} - rV \in \Omega^+ \tag{9}$$

subject to boundary conditions:

$$V^{(n+1)}(0) = Ke^{-r(n+1)\delta\tau}, \quad V^{(n+1)}(L) = 0 \tag{10}$$

Again one can use a finite difference approximation to replace the spatial derivatives in eqn (9). This method leads to an implicit scheme which requires a linear solver to compute the solution of the resulting linear system of equations. Although there is no stability criterion governing the step size δt and the spatial discretisation, it is important to note that it is impossible to use an unreasonably large δt in eqn (9). In essence it is impossible to compute numerical data of the final step simply based on the initial data without going through any intermediate steps.

In many financial modelling problems, typical one spatial dimension is not of interests. Almost all realistic problems usually involve many underlying assets to be considered within a basket. Therefore numerical solutions using a field method become important in the sense that error control is far more easier than other stochastic simulation methods, such as Monte Carlo method. On the other hand, large number of simulations and solutions at intermediate time steps are usually not of interest. Efficiency of the numerical algorithm is very important in order to make evaluation and decision before the agreement of a contract is reached. Ideally one would like to use an algorithm which can be completely distributed onto a number of processors with only minimal communications between processors.

It can easily see that for both methods of temporal integration described above, concurrent computation of V at many different times is not possible using a temporal marching scheme. Attempts have been made by several researchers in this direction. For examples Miranker and Liniger proposed a family of parallel Runge-Kutta methods [8]; Lions, Maday and Turinici proposed the 'parareal' algorithm in time discretisation [7]; and more recently Gander and Vandewalle provided an anlysis of the parareal algorithm for parallel temporal integration [5]. These concepts largely based on partitioning the temporal axis into a number of non-overlapped sections known as time subdomains and the same time-dependent problem is considered in each of these time subdomains. The partitioning of the temporal axis resembles very similar to a multiple shooting method [6]. Methods developed within this context have been further enhanced by the use of a coarser temporal partitioning in conjunction with a finer temporal partitioning based on the concept of a multigrid correction technique.

3.2. The Laplace Transformation Method

In this section a class of transformation method based on the concept of the Laplace transform and its numerical inverse is discussed. Since the solutions at intermediate steps of eqn (4) are usually not of interest, the best way is to induce distribution properties into the financial model leading to a set of mutually independet boundary value problems that can be solved on a distributed or parallel computing environment. The Laplace transform [11] provides such induced distribution property.

Let

$$\mathcal{L}(V) \equiv \int_0^\infty e^{-\lambda\tau} V(S,\tau)d\tau = U(\lambda; S) \tag{11}$$

be the Laplace transform of the function $V(S,\tau)$, then the Laplace transform of eqn (4) leads to

$$\frac{1}{2}\sigma^2 S^2 \frac{d^2U}{dS^2} + rs\frac{dU}{dS} - (r+\lambda)U = -V(S,0) \in \Omega^+ \tag{12}$$

subject to boundary conditions:

$$U(\lambda;0) = \frac{K}{\lambda + r}, \quad U(\lambda; L) = 0 \tag{13}$$

Note that the dependence on time is now converted to the dependence on the parameter $\lambda \in \{\lambda_j = j\frac{ln2}{T} : j = 1, 2, ..., m\}$, which is a finite set of transformation parameters. Each element of the set of parametric equations $\{U(x; \lambda_j)$ satisfies eqn (12) $: j = 1, 2, ..., m\}$ is decoupled from each other and may be solved independently from the others by solving eqn (12). Thus an important distribution property is induced to the mathematical model leading to independent parametric equations which may be solved without the knowledge of each other. In terms of distributed computing, this independence amounts to null communication between each computer in the distributed computing environment, assuming there are as many computers as the number of parametric equations.

In order to retrieve $V(S,T)$, the approximate inverse Laplace transform due to Stehfest [10] may be computed as a weighted sum of the Laplace transformed solutions, i.e.

$$V(S,T) \approx \sum_{j=1}^{m} w_j U(\lambda_j; S) \tag{14}$$

where

$$w_j = \frac{\ln 2}{T}(-1)^{m/2+j} \sum_{k=(1+j)/2}^{\min(j,m/2)} \frac{k^{m/2}(2k)!}{(m/2-k)!k!(k-1)!(j-k)!(2k-j)!}$$

is the weighting factor.

Note that m is required to be chosen as an even number if eqn (14) is used to compute the inverse Laplace transform. This approximate inverse Laplace transform is by no means the most accurate one. Stehfest method is selected because of previous experience with the method used for linear problems [3][4][10]. One aim of this chapter is to investigate the application of the inverse method to option pricing problems. From experience, the value of m is usually a small even number not larger than 10 [3]. Numerical experiments in Section 3.3. and Section 5.3. also confirm such experience.

Note also that the weighted sum is in essence an intrinsic sequential process. There are other possible algorithms performing summation in an inefficient manner. The complexity analysis shown later has assumed no parallel process is used in the above summation process.

3.3. Numerical Tests with Constant Volatility

To demonstrate the viability of using the Laplace transformation method for computational finance, an example of European put option is examined here. The spatial domain is chosen to be $\Omega^+ = \{S : 0 \le S \le 320\}$, the strike price being $K = 100$ and the expiry date being $T = 0.25$ (i.e. three months). The two parameters σ and r of eqn (4) are chosen to be 0.4 and 0.5 respectively, throughout the simulation period. A second order finite difference method is applied along the spatial axis to each element of the set of parametric equations as given by eqn (12). The spatial mesh size is chosen to be $\delta S = 320/2^9$. Numerical tests were done on a Sun Ultra-5 workstation using an F90 program which implements the two methods described below.

A backward difference in time as given by eqn (9) is used to provide a reference solution for the model eqn (4). Time step-length is chosen as $\delta\tau = 1/365$, i.e. 1 day. the same finite difference scheme as the one applied in the Laplace transform method is applied here along the spatial axis in order to obtain the reference solution. The discretisation is an implicit scheme leads to a set of tri-diagonal system of equations at every time-step, which may then be solved by a direct method. The numerical solution of $V(S,T)$ obtained in this case is denoted by V_{TI}. The reason for choosing this method is that the computational complexity is reasonably low for the accuracy that it can achieve compared with higher order

Table 1. Linear volatility - Comparison of timings and discrepancy in solutions.

m	$\|\|V_{TI} - V_{IL}\|\|_2$	Time (V_{IL})
2	1.0767	0.006
4	0.0812	0.009
6	0.0111	0.014
8	0.0037	0.017
10	0.0032	0.018
12	0.0032	0.021
14	0.0032	0.028
16	0.0032	0.028
18	0.0032	0.035

schemes. Ignoring overheads due to various input and output facilities and assumption the computational complexity of solving one set of tri-diagonal system of equations is η. The total computational complexity by using this method is $\eta\lceil\frac{T}{\delta_T}\rceil$. For the present calculation $\lceil\frac{T}{\delta_T}\rceil = \lceil\frac{0.25}{1/365}\rceil$ which leads to 92η.

Next the set of parametric equations in the Laplace space is solved sequentially in the same computational environment, with different values of m. An approximation to $V(S, T)$ corresponding to each value of m is found by using the inverse Laplace transform as described in Section 3.2. and is denoted as V_{IL}. Discrepancies between the two numerical solutions, $||V_{TI} - V_{IL}||_2$, are recorded and shown in Table 1 for comparisons. The timing of each run observed in this example consists of two parts. First, the second order finite difference solver time and, second, overheads due to the computation of inverse Laplace transforms. The first part can easily be calculated as $m\eta$. Let the overheads due to the second part be μ. Total sequential computational complexity is $m\eta + \mu$. Dividing the timings obtained by using the Laplace transformation method gives a crude estimation of the distributed processing time if $\eta \gg \mu$. Suppose there are as many processors available as the value of m, the scalability of the algorithm can be easily observed from the sequential timings recorded in Table 1 using the above crude estimation of the corresponding distributed timings. The true parallel computational complexity is $\eta + \mu$. The discrepancy $||V_{TI} - V_{IL}||_2$ approaches an asymptotic value of 0.0032 when $m \geq 8$. Therefore it is not necessary to take m very much larger than 10. This result confirms the previous tests on a linear heat conduction problem [3], [4].

Note that these experiments involved one underlying asset. Hence the computational time is not particularly interesting. However the algorithm becomes significant when the problem is nonlinear.

Note also that for time varying $\sigma(t)$ and $r(t)$, it is possible to make a suitable coordinate transformation to the Black-Scholes equation in order to obtain a time independent-like heat equation [3]. Hence the above Laplace tranform method still applies.

4. Nonlinear Volatility

Very often, over a short period of time the interest rate, r, is fixed while the volatility, σ, is varying. The volatility may be a function of the transaction costs [1], the second derivative of the option value [9], or, in some cases, the solution of a nonlinear initial value problem [1]. In order to illustrate the nonlinear solver here, the volatility proposed in [2], i.e.

$$\sigma = \sigma_0 \sqrt{1+a} \tag{15}$$

is used, where a is the proportional transaction cost scaled by σ_0 and the transaction time. Here the authors adopted a heuristic approach in which the transaction cost is related to the option value and follows a Gaussian distribution. In order to demonstrate the inverse Laplace transform technique for nonlinear problems, a sine function is used to produce the effect of a pulse-like distribution instead of a Gaussian distribution. Therefore the proportional transaction cost, a, may be replaced by a function of the option value such as

$$a = \sin\left(\frac{V\pi}{K}\right) \tag{16}$$

where K is the strike price. This volatility is used in the subsequent numerical tests.

The forward Black-Scholes model as given in eqn (4) may be re-written as

$$\frac{\partial V}{\partial \tau} = A(V)\frac{\partial^2 V}{\partial S^2} + rS\frac{\partial V}{\partial S} - rV \in \Omega^+ \times (0,T] \tag{17}$$

subject to the same set of initial condition and boundary conditions given in eqns (5) and (6). Here $A(V) = \frac{1}{2}\sigma(V)^2 S^2$.

5. Numerical Solutions of the Black-Scholes Equation with Nonlinear Volatility

Linearization methods are needed when the Black-Scholes equation contains nonlinear volatility. Two different ways of linearisation, including an iterative coefficient and Netwon's linearisation, are described. Section 5.1. combines an iterative coefficient and a temporal marching scheme to provide a reference solution which is used for comparison. Section 5.2. combines each of the two linearisation methods to the Laplace transformation method.

5.1. Temporal Marching Schemes

An iterative coefficient technique is used in this method. The nonlinear coefficient $A(V)$ is computed by using an initial approximation $\bar{V}$, which is updated in every step of an iterative update process. Each step of the iterative update process involves a numerical solution to the equation

$$\frac{\partial V}{\partial \tau} = A(\bar{V})\frac{\partial^2 V}{\partial S^2} + rS\frac{\partial V}{\partial S} - rV \in \Omega^+ \tag{18}$$

defined in the time interval $\tau \in (t_i, t_{i+1}]$. Let $V^{(n)}(S, t_{i+1})$ and $V^{(n)}(S, t_i)$ be the numerical solutions of eqn (17) at $\tau = t_{i+1}$ and $\tau = t_i$ respectively. The superscript (n) denotes the number of iterative updates. The iterative update process to obtain the numerical solution $V^{(n)}(S, t_{i+1})$, using $V^{(n)}(S, t_i)$ as the initial approximation to $\bar{V}$, is described in the algorithm below.

Algorithm C1: Iterative coefficient - temporal itegration.
Initial approximation:- $V^{(0)}(S, t_{i+1}) := V^{(n)}(S, t_i)$;
$k := 0$;
Iterate
$k := k + 1$;
Compute $A(\bar{V})$ such that $\bar{V} := V^{(k-1)}(S, t_{i+1})$;
Find $V^{(k)}(S, t_{i+1})$ such that

$$\frac{V^{(k)}(S,t_{i+1}) - V^{(n)}(S,t_i)}{\delta\tau} = A(\bar{V})\frac{\partial^2 V^{(k)}(S,t_{i+1})}{\partial S^2} + rS\frac{\partial V^{(k)}(S,t_{i+1})}{\partial S} - rV^{(k)}(S, t_{i+1}) \in \Omega^+$$

Until $||V^{(k)}(S, t_{i+1}) - V^{(k-1)}(S, t_{i+1})|| < \epsilon$
$n := k$

Numerical solutions for (17) may be obtained by imposing an outer iteration loop with $t_i = i\delta\tau$, $i = 1, 2, ...,$ over Algorithm C1. Here $\delta\tau$ is the step length of the temporal integration. This outer iteration loop in conjunction with Algorithm C1 is used to produce a reference solution for comparison.

5.2. The Laplace Transformation Method

The Laplace transformation method may be applied within a linearisation method in order to make use of distributed computers. Here two different linearisation methods are applied to the nonlinear Black-Scholes equation.

Iterative coefficient. Take Laplace transform of eqn (18), now being defined in the time interval $\tau \in (T_i, T_{i+1}]$, and the transformed equation in its differential form is given below.

$$A(\bar{V})\frac{d^2U}{dS^2} + rS\frac{dU}{dS} - (r + \lambda)U = -V(S, T_i) \tag{19}$$

where $U = \mathcal{L}(V)$. An update process is used to renew the value of $\bar{V}$ and an inverse Laplace transform is used to obtain the numerical solution $V^{(n)}(S, T_{i+1})$, using $V^{(n)}(S, T_i)$ as the initial approximation to $\bar{V}$. The update process is described in the algorithm below.

Algorithm C2: Iterative coefficient – Laplace transformation method.
Initial approximation:- $V^{(0)}(S, T_{i+1}) := V^{(n)}(S, T_i)$;
$k := 0$;
Iterate
$k := k + 1$;
Compute $A(\bar{V})$ such that $\bar{V} := V^{(k-1)}(S, T_{i+1})$;

Parallel for $j := 1$ to $m(i)$
Find $U(\lambda_j; S)$ such that
$A(\bar{V})\frac{d^2U}{dS^2} + rS\frac{dU}{dS} - (r + \lambda_j)U = -V(S, T_i)$
End Parallel for
Compute $V^{(k)}(S, T_{i+1})$ using inverse Laplace in eqn (14)
Until $||V^{(k)}(S, T_{i+1}) - V^{(k-1)}(S, T_{i+1})|| < \epsilon$
$n := k$

Here $m(i)$ is the number of transformation parameters and $T_i = i\Delta\tau$. In order to solve eqn (19) for $U(\lambda_j; S)$, one can employ the same finite difference technique described above. In order to solve eqn (17) for $V(S, T)$, Algorithm C2 needs to be iterated through $t_i := T_1, T_2, ..., T$ by using suitable values of $m(i)$ in the form of an outer iteration. In essence the actual implementation does not require different values of $m(i)$ for many problems, and the numerical tests shown here used the same number of transformation parameters, denoted as $\bar{m}$, for different values of during the outer iteration loop. Note that in this case $\Delta\tau$ can be chosen to be much greater than $\delta\tau$ because the fine details of at each time step of a temporal integration is not required in the inverse Laplace transformation calculation. A full account on different choices of $\Delta\tau$ is not included in this chapter. In the numerical tests shown in subsection $\Delta\tau = \frac{T}{10}, \frac{T}{20}, \frac{T}{40}, \frac{T}{80}$, all normalised with respect to one year, and $t_0 = 0$.

Newton's linearisation. This is a common linearisation method. Consider a small perturbation δV applied to eqn (17), defined in the time interval $\tau \in (t_i, t_{i+1}]$, leads to

$$\left\{\frac{\partial}{\partial\tau} - \left(A'(V)\frac{\partial^2 V}{\partial S^2} + A(V)\frac{\partial^2}{\partial S^2} + rS\frac{\partial}{\partial S} - r\right)\right\}\delta V$$
$$= -\left\{\frac{\partial V}{\partial\tau} - \left(A(V)\frac{\partial^2 V}{\partial S^2} + rS\frac{\partial V}{\partial S} - rV\right)\right\} \in \Omega^+ \times (0, T] \quad (20)$$

where δV is a small incremental change of V. The initial and boundary conditions can be obtained by taking a small perturbation to eqns (5) and (6) leads to the initial condition:

$$\delta V(S, 0) = 0 \quad (21)$$

and the boundary conditions:

$$\delta V(0, \tau) = 0, \;\; \delta V(L, \tau) = 0 \quad (22)$$

A Laplace transform can be applied to eqn (20), now being defined in the time interval $\tau \in (T_i, T_{i+1}]$. The transformed equation in its differential form becomes

$$\mathcal{L}(\delta V) - \delta V(S, T_i) - \left(A'(V)\frac{d^2 V}{dS^2} + A(V)\frac{d^2}{dS^2} + rS\frac{d}{dS} - r\right)\mathcal{L}(\delta V)$$
$$= -\mathcal{L}(V) - V(S, T_i) - \left(A(V)\frac{d^2 V}{dS^2} + rS\frac{dV}{dS} - rV\right) \quad (23)$$

Using $V^{(n)}(S, T_i)$ as the initial approximation to $V^{(0)}(S, T_{i+1})$ to start the inner iterative loop in order to obtain the numerical solution $\mathcal{L}(\delta V^{(n)}(S, T_{i+1}))$. The inner iterative loop is described in the algorithm below.

Algorithm NM2: Newton's Method - Laplace transformation method.
Initial approximation:- $V^{(0)}(S, T_{i+1}) := V^{(n)}(S, T_i)$;
$k := 0$;
Iterate
    $k := k + 1$;
    Compute $\{ A(\bar{V}), A'(\bar{V}), A'(\bar{V})\frac{d^2\bar{V}}{dS^2},$
        $-\mathcal{L}(\bar{V}) - V(S, T_i) - \left(A(\bar{V})\frac{d^2\bar{V}}{dS^2} + rS\frac{d\bar{V}}{dS} - r\bar{V}\right)$
    $\}$ such that $\bar{V} := V^{(k-1)}(S, T_{i+1})$;
    Parallel for $j := 1$ to $m(i)$
        Find $\mathcal{L}(\delta V^{(k)}(S, T_{i+1}))$ such that
        $\mathcal{L}(\delta V^{(k)}(S, T_{i+1})) - \delta V(S, T_i)$
        $- \left(A'(\bar{V})\frac{d^2V}{dS^2} + A(\bar{V})\frac{d^2}{dS^2} + rS\frac{d}{dS} - r\right) \mathcal{L}(\delta V^{(k)}(S, T_{i+1}))$
        $= -\mathcal{L}(\bar{V}) - V(S, T_i) - \left(A(\bar{V})\frac{d^2\bar{V}}{dS^2} + rS\frac{d\bar{V}}{dS} - r\bar{V}\right)$
    End Parallel for
    Compute $\delta V^{(k)}(S, T_{i+1})$ using inverse Laplace in eqn (14)
    $V^{(k)}(S, T_{i+1}) := \bar{V} + \delta V^{(k)}(S, T_{i+1})$;
Until $||\delta V^{(k)}(S, T_{i+1})|| < \epsilon$
$n := k$

Here $m(i)$ is the number of transformation parameters. Similar to the spatial discretisation method described above the same finite difference technique is employed here. The Algorithm NM2 can now be iterated through $T_i := T_1, T_2, ..., T$ by choosing suitable values of $m(i)$ in the form of an outer iteration. The other discussion on the choice of $m(i)$ and $\Delta\tau$ also applies.

5.3. Numerical Tests with Nonlinear Volatility

The same mathematical model of European put option described in Section 3.3. is used to demonstrate the efficiency of the two linearization methods. In particular the computational complexity is to be examined when the algorithms are implemented on distributed computers. Here the volatility is chosen as a nonlinear function as described in section 4. and the parameters σ_0 and r are chosen to be 0.4 and 0.5 respectively. A second order finite difference method is applied to each parametric equation as given by eqn (23). The spatial mesh size is chosen to be $\frac{320}{2^9}$, which is the same as that used in Section 3.3..

Algorithm C1 is used to obtain a reference solution for comparison. It is being run sequentially on a COMPAQ laptop using a F90 program implementing a backward finite difference method along the temporal axis with $\delta\tau = 1/365$, i.e. 1 day, leading to an implicit scheme in which a second order finite difference methods along the spatial axis S can be implemented. The stopping criterion used in the linearization step is chosen as $\epsilon = 10^{-5}$. The numerical solution $V(S, T)$ obtained by this temporal integration is denoted as V_{C1}.

Table 2. Noninear volatility - Computational work unit

$\bar{m}$	Algorithm C2				Algorithm NM2			
	$\Delta\tau$				$\Delta\tau$			
	1/40	1/80	1/160	1/320	1/40	1/80	1/160	1/320
4	58	103	177	326	43	83	134	249
6	58	103	177	326	43	70	126	245
8	58	103	177	326	43	71	126	220
10	58	103	177	326	43	71	126	214
12	180	123	186	327	43	71	126	213
Sequential work unit required by using Algorithm C1 is 246								

Each of the two linearisation methods, Algorithms C2 and NM2, leads to a set of transformed equations in the Laplace space. The set of equations is solved, sequentially in the same computational environment, by taking different values of m. An approximation to $V(S,T)$ corresponds to each value of m is found by using the inverse Laplace transform as described in section 3.2.. The approximations obtained by means of the iterative coefficient and Newton's methods are denoted as V_{C2} and V_{NM2} respectively. By using the choices of $\Delta\tau = \frac{T}{10}, \frac{T}{20}, \frac{T}{40}, \frac{T}{80}$, the number of outer iterations required for Algorithms C2 and NM2 is 10, 20, 40, and 80 respectively.

As the problems set up are the same in the linear and nonlinear cases of volatilities, the computational complexity of solving a set of tri-diagonal system of equations with a fixed spatial mesh size is the same as before, i.e. η. For simplicity it is possible to call the computational complexity η as one work unit. For the purpose of comparison the work unit required to obtain V_{C1} is 246. Table 2 shows the distributed work unit when the solutions of the set of parametric equations are done independently. The work unit involved in Algorithm NM2 is less than that of Algorithm C2 in general. On the other hand there is no sudden increase of work when $m = 12$. The total sequential work unit may be obtained by multiplying the distributed work unit to m. The total parallel complexity is simply the complexity for solving the set of equations plus inverse Laplace transform and communication overheads, i.e.

$$W_m\eta + \mu + c$$

Here W_m denotes the work unit when there are m computers and m parametric equations, and c denotes communication. In general the dominant contribution to the total work is the solutions process of the set of parametric equations and contribution from c is almost negligible.

For comparison purposes the reference solution V_{C1} is used in comparing results using the Laplace transformation method. Discrepancies in solutions, i.e. $\|V_{C1} - V_{C2}\|_2$ and $\|V_{C1} - V_{NM2}\|_2$ using various $\Delta\tau$, are recorded in Figures 1 and 2. In general the discrepancy levels off when $m = 8$, which suggests that the use of more terms in the inverse Laplace transform at a fixed value of $\Delta\tau$ has no effect on the accuracy. On the other hand smaller $\Delta\tau$ produces small discrepancy at the expense of requiring more work unit as recorded in Table 2 for comparisons.

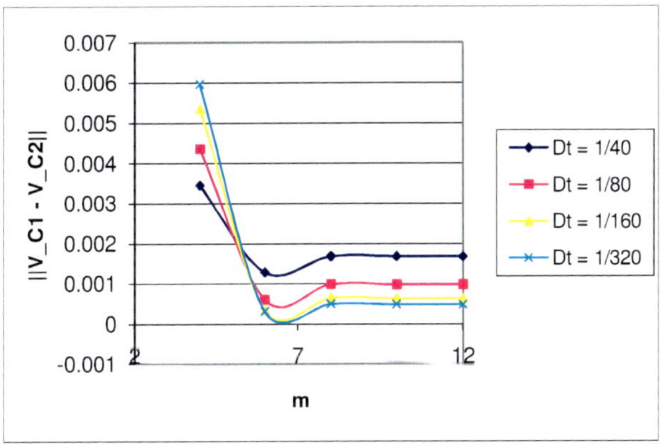

Figure 1. Discrepancies of solutions: $\|V_{C1} - V_{C2}\|_2$.

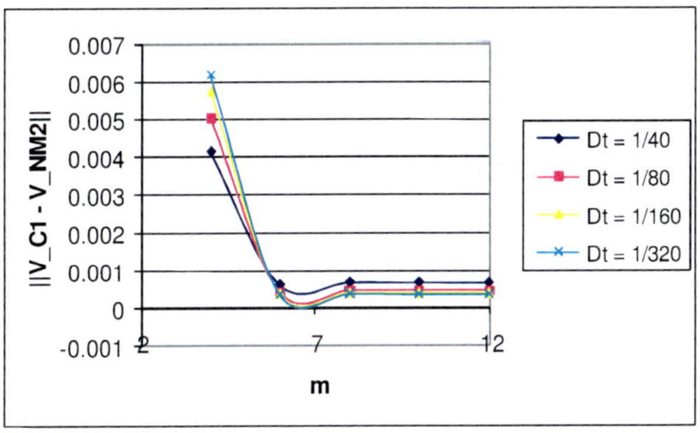

Figure 2. Discrepancies of solutions: $\|V_{C1} - V_{NM2}\|_2$.

The other feature of the two linearization methods is that the use of the iterative coefficient method has no advantage over Newton's method. The pointwise discrepancies of the solutions, $|V_{C1}(S,T) - V_{C2}(S,T)|$ and $|V_{C1}(S,T) - V_{NM2}(S,T)|$, as compared with a time-stepping method applied to the nonlinear problem with $\bar{m} = 8$ and $\Delta\tau = 100$ for the case $h = 320/2^9$ is recorded in Figure 3. Such comparison is not the best way of comparing results, but it gives an idea of the deviation from the numerical solution obtained by an implicit temporal integration based on backward difference. It shows that the largest discrepancies occur nearby the strike price. Therefore it is the resolution problem of the spatial finite difference scheme near the strike price. On the other hand the order of discrepancies is of the order of 10-3 which is in consistent with the results for linear problems. Finally the numerical approximations to $V(S,T)$ using various methods and the initial condition $V(S,0)$ are plotted in Figure 4.

Note that the linearization techniques described above may be extended to handle American Options and Asian Options with little difficulties. A complimentary problem can always be formed for American options, in which the free boundary is embedded into the formulation. A Laplace transform can then be applied to the complimentary problem [12] followed with an inner iteration loop to drive the inequality to convergence.

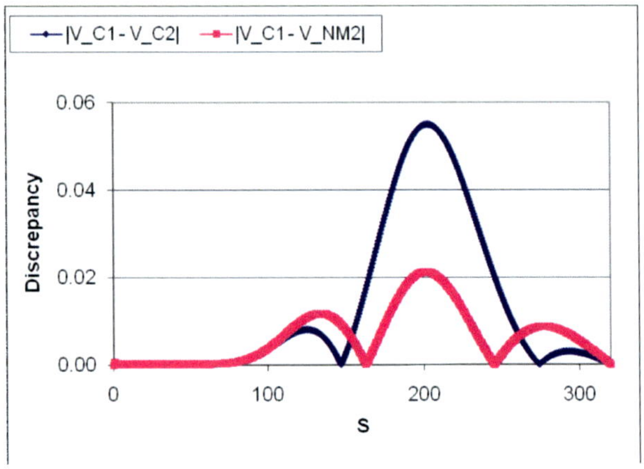

Figure 3. Pointwise comparison of numerical solutions for nonlinear volatility.

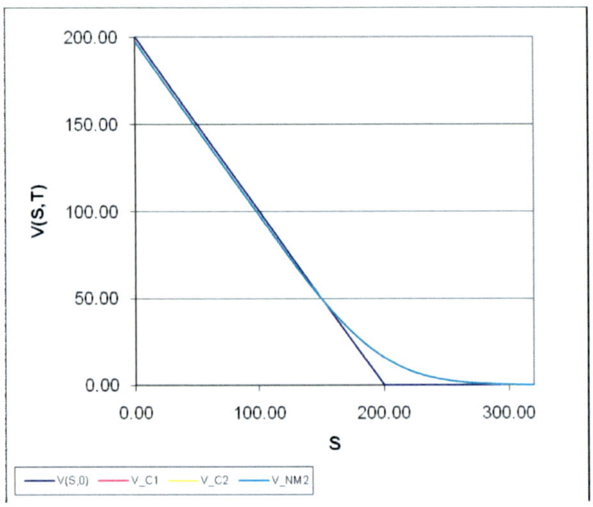

Figure 4. Numerical solution at $t = T$.

6. The Two Level Time-Domain Algorithm

The Laplace transformation method discussed in Section 5.2. is defined on a much larger time step size $\Delta\tau$ when it is compared with the time step size $\delta\tau$ used in the temporal integration method. Algorithms C2 and NM2 may be considered as a temporal integration technique with a larger temporal step $\Delta\tau$ such that distributed computation may be induced at each temporal step. Hence one obtains numerical solutions on a coarse temporal mesh. The accuracy of the numerical results obviously depends on the accuracy of the inverse Laplace transform method used and is out of the scope of this chapter. The objective of this section is to examine the possibility of using two different temporal meshes in order to achieve an intrinsic parallel time-domain algorithm.

Let the time domain $(0, T]$ be divided into n temporal steps, i.e. $\delta\tau = T/n$. The result of this leads to a fine time level where temporal integration may take place. This is essentially applying Algorithm C1 to eqn (18) and an outer-loop to run through each time interval $(i\delta\tau, (i+1)\delta\tau], i = 0, 1, \cdots, n-1$. Note that it is no reason why the temporal step length needs to be restricted to a constant value, and the algorithm may be easily generalised to variable temporal step length.

Suppose the n temporal steps are equally divided into N parts each represents a larger time interval, i.e. $(j\frac{n}{N}\delta\tau, (j+1)\frac{n}{N}\delta\tau]$, $j = 0, 1, \cdots, N-1$. This corresponds to a coarser time mesh with time step $\Delta\tau = T/N$. If solutions are obtained at $j\Delta\tau$, where $j = 0, 1, \cdots, N$, which may be considered as initial conditions for each of the time intervals $(j\frac{n}{N}\delta\tau, (j+1)\frac{n}{N}\delta\tau], j = 0, 1, \cdots, N-1$, the temporal integration in these intervals can then be performed in indepdently.

Clearly it is not sensible to employ the same temporal integration method to obtain solutions at the coarse time mesh for two reasons. First the accuracy of these solutions is not of the same order of those obtained by using a finer time step. Second distributed computation cannot be introduced into the temporal integration method. However using the Laplace transformation method described in above the solution of the nonlinear problem at each time interval of the coarse temporal mesh may be obtained by means of a parallel computer. Sequential applications of such parallel algorithm allows stepping through all of the time steps in the coarse level problem. Finally it should be noted also that the solution accuracy at time steps $j\Delta\tau$, $T = 1, \cdots, N$ are not degraded and can be used as the initial values for the fine level problem. The two-level time domain parallel algorithm is given as below.

```
Algorithm: Two-level time domain parallel algorithm
    do j := 0 to N − 1
        Coarse level (Δτ)- Apply Algorithm C2 or NM2:-
        Inverse Laplace Method for τ ∈ (T_j to T_{j+1}]
    enddo
    Distribute j := 0 to N − 1
        Fine level (δτ = (N/n)Δτ) - Apply Algorithm C1:-
        Temporal integration for τ ∈ (T_j to T_{j+1}].
    end Distribute
```

7. Conclusion

This chapter investigates the concept of using the Laplace transformation method in order to induce distributed computational properties into an otherwise sequential algorithm. The idea is applied to the numerical solutions of some nonlinear Black-Scholes equation in financial analysis of option pricing. Two linearisation processes are examined and combined with the Laplace transformation method resulting to methods which require the use of only a coarser temporal mesh compared with those used in temporal integration methods. Distributed properties exist on the coarse temporal mesh with the set of parametric equations in the Laplace space. Solutions on the coarse temporal mesh may be used to obtain fine details of the pricing at finer temporal mesh concurrently. The algorithm presented in this chapter can be easily extended to other financial analysis problems.

References

[1] G. Barles and H.M. Soner, Option pricing with transaction costs and a nonlinear Black-Scholes equation, *Finance Stochast.* **2** (1998), 369–397.

[2] P. Boyle and T. Vorst, Option replication in discrete time with transaction costs, *J. Finance* **47** (1973), 271–293.

[3] D. Crann, The Laplace transform: Numerical inversion for computational methods, *Technical Report* No. **21**, University of Hertfordshire, 1996.

[4] D. Crann, A.J. Davies, C.-H. Lai, and S.W. Leong, Time domain decomposition for European options in financial modelling, In: *Proceedings of the 10th International Conference on Domain Decomposition Methods*, 1997, Colorado. J. Mandel, C. Farhat, and X.-C. Cai, (eds.), *Domain Decomposition Methods* **10**, American Mathematical Society, 1998.

[5] M.J. Gander and S. Vandewalle, Analysis of the parareal time-parallel time-integration method, *SIAM J. Sci. Comput.* **29** (2007), 558–578.

[6] M.J. Gander and E. Hairer, Nonlinear convergence analysis for the parareal algorithm, *Lecture Notes in Computational Science* **60** (2008), 45–56.

[7] J.-L. Lions, Y. Maday and G. Turinici, A parareal in time discretisation of PDE's, *C.R. Acad. Sci. Paris, Serie I* **332** (2001), 661–668.

[8] W.L. Miranker and W. Liniger, Parallel methods for the numerical integration of ordinary differential equations, *Math. Comp.* **91** (1967), 303–320.

[9] A. Parás and M. Avellaneda, Dynamic hedging portfolios for derivative securities in the presence of large transaction costs, *Appl. Math. Finance* **1** (1994), 165–193.

[10] H. Stehfest, Numerical inversion of Laplace transforms, *Comm. ACM* **13** (1970), 47–49.

[11] D.V. Widder, *The Laplace Transform*, Princeton University Press, Princeton, 1946.

[12] P. Wilmott, S. Howison and J. Dewynne, *The Mathematics of Financial Derivatives*, Press Syndicate of the University of Cambridge, New York, 1993.

Part IV.
Parameter Identification (Inverse Problems)

In: Nonlinear Models in Mathematical Finance...
Editor: Matthias Ehrhardt, pp. 321-350
ISBN 978-1-60456-931-5

Chapter 11

CALIBRATION PROBLEMS IN OPTION PRICING

B. Düring
Institut für Analysis und Scientific Computing,
Technische Universität Wien, 1040 Wien, Austria

Abstract

We review different approaches to calibrate volatility functions in Black–Scholes type models to market data. We focus on an optimal control approach, where a regularized cost functional is minimized over a suitable set of admissible volatilities. The cost functional measures the deviations of option prices obtained from a pricing model to the given market data. We discuss the Black–Scholes case as well as the extension to pricing models in markets with frictions, e.g. models for option pricing in the presence of transaction costs.

1. Introduction

Financial derivatives, in particular options, can be used, to hedge assets and portfolios in order to control the risk due to movements in the underlying share prices. We recall that a European Call (Put) option provides the right to buy (to sell) a fixed number of assets at the fixed exercise price E at the expiry time T. In an idealized financial market the price of a European option $V(t, S)$ on an underlying asset S at time t can be obtained as the solution of the celebrated Black–Scholes equation

$$V_t(t, S) + \frac{1}{2}\sigma^2 S^2 V_{SS}(t, S) + rSV_S(t, S) - rV(t, S) = 0, \quad t \in (0, T),\ S > 0, \qquad \text{(1a)}$$

where $r > 0$ is the riskless interest rate and $T > 0$ the time of maturity, with the final condition

$$V(T, S) = P(S), \quad S > 0, \qquad \text{(1b)}$$

with given pay–off $P(S)$ and appropriate boundary conditions. Note that the parabolic equation (1a) is solved for the price $S > 0$ of the underlying asset and time $T \geq t \geq 0$, i.e. backward in time.

The Black–Scholes equation has been derived under several assumptions, in particular the asset price $S(t)$ is supposed to follow a stochastic process

$$dS(t) = \mu S(t)\,dt + \sigma S(t)\,dW(t), \tag{2}$$

where $\mu \in \mathbb{R}$, $\sigma > 0$ are the constant drift and constant volatility of the underlying asset, respectively, and $W(t)$ denotes a Brownian motion. The drift and the volatility are not directly observable. The drift is removed from the model by a hedging argument and does not enter explicitly in the Black–Scholes equation. The unrealistic assumptions that are made in the derivation of this model prove to be too restrictive in practice. Therefore, in recent years different models have been proposed to weaken one or more of these assumptions. This gives rise to a whole class of Black–Scholes type equations. We illustrate this by recalling several examples.

In the original Black–Scholes model the influence of transaction costs was neglected and it was possible to construct a riskless portfolio that perfectly replicates the option pay–off. If transaction costs are taken into account perfect replication of the contingent claim is no longer possible, and it has been shown in [42] that further restrictions have to be imposed in the model. One possibility is to restrict trading to discrete points in time. Leland [36] derived an option price equation that takes into account transaction costs and that is equal to a Black–Scholes equation but with a modified volatility of the form

$$\hat{\sigma} = \sigma(1 + cA)^{1/2}, \quad A = \frac{\alpha}{\sigma\sqrt{\triangle t}}, \tag{3}$$

with a constant $c = (2/\pi)^{1/2}$. Herein, α is the proportional transaction cost, $\triangle t$ the (fixed) transaction period, and σ is the original constant volatility of the underlying. Boyle and Vorst [8] obtained a modified volatility of the form (3) with $c = 1$ from a binomial model. Kusuoka [33] then showed that the 'optimal' c depends on the risk structure of the market. Parás and Avellaneda [39] derived the modified volatility

$$\hat{\sigma} = \sigma\big(1 + A\,\mathrm{sign}(V_{SS})\big)^{1/2} \tag{4}$$

from a binomial model. Here, V_{SS} denotes the second derivative of V with respect to S (the 'Gamma'). In particular, the option price does not need to be convex.

Another popular approach is to introduce preferences by assuming that the investor's behavior is characterized by a given utility function. In [27] it has been shown that the option price can be obtained as the initial cash increment which offsets the difference between the maximum utility of terminal wealth when there is no option liability and when there is such a liability. Davis et al. [12] extended this approach to markets with transaction costs (see also [44]). It has the disadvantage that the option price depends on the special choice of the utility function but Constantinides and Zariphopoulou [9] obtained universal bounds independent of the utility function. Using this utility maximization approach, the following model has been proposed by Barles and Soner [3].

Consider a portfolio of a bond and a stock, which evolves according to (2) with constant mean return rate $\mu \in \mathbb{R}$ and constant volatility $\sigma > 0$. For the sake of brevity, set the riskless interest rate r to zero for the moment. Let $\alpha \in (0, 1)$ be a proportional transaction cost. With initial values x, y, the corresponding portfolio of stock

$X(s) = X(s; t, x, y, L(\cdot), M(\cdot))$ and bond $Y(s) = Y(s; t, x, y, L(\cdot), M(\cdot))$ is assumed to evolve according to

$$X(s) = x - \int_t^s S(\tau)(1+\alpha)\,dL(\tau) + \int_t^s S(\tau)(1-\alpha)\,dM(\tau), \tag{5}$$

$$Y(s) = y + L(s) - M(s), \tag{6}$$

for $s \in [t, T]$, with a trading strategy $(L(s), M(s))$ as a pair of nondecreasing processes with $L(t) = M(t) = 0$, which are interpreted as the cumulative transfers, measured in shares of stock, from bond market to stock and vice versa. In (5) the first integral represents buying of shares of stock at a price increased by the proportional transaction cost, the second integral represents selling stock where the stock price is reduced by the proportional transaction cost.

Following the utility maximization approach of Hodges and Neuberger [27], Barles and Soner consider two optimization problems. In the first optimization problem the investor has no option liability and in the second she has sold N European call options. They choose the exponential utility function $U(\xi) = 1 - e^{-\gamma\xi}$, $\xi \in \mathbb{R}$, with risk aversion factor $\gamma > 0$. According to [27] the price of an option equals the increment of initial capital at time t that is needed to cope with the option liabilities arising at T. Hence, the option price is a function of the initial data $x, y, S(t), t$ and γ, N. Due to the linearity of the underlying equations, selling N options with risk aversion factor of γ yields the same price as selling one option with risk aversion factor of γN. This scaling argument leads to performing an asymptotic analysis as γN tends to infinity. Hence, they set $\varepsilon = 1/(\gamma N)$ and consider $U_\varepsilon(\xi) = 1 - e^{-\xi/\varepsilon}$, $\xi \in \mathbb{R}$. The value functions of the optimization problems are solutions of an associated dynamic programming equation. The coefficients in this equation and the terminal data are independent of the variable x and hence the option price is independent of x as well [12]. Supposing that $\alpha = a\sqrt{\varepsilon}$ for some constant $a > 0$, Barles and Soner now perform an asymptotic analysis and prove that as $\alpha, \varepsilon \to 0$ with $a = \frac{\alpha}{\sqrt{\varepsilon}}$ constant, the option price V is the unique (viscosity) solution of the fully nonlinear Black–Scholes type equation

$$V_t + \frac{1}{2}\hat{\sigma}^2(t, S, V_{SS})S^2 V_{SS} = 0, \quad t \in (0, T),\ S > 0.$$

In the case of non–zero interest rate r the equation reads

$$V_t + \frac{1}{2}\hat{\sigma}^2(t, S, V_{SS})S^2 V_{SS} + rSV_S - rV = 0, \quad t \in (0, T),\ S > 0, \tag{7}$$

where the nonlinear volatility $\hat{\sigma}(t, S, V_{SS})$ is given by

$$\hat{\sigma}(t, S, V_{SS}) = \sigma\Big(1 + \Psi\big[\exp\big(r(T-t)\big)a^2 S^2 V_{SS}\big]\Big)^{1/2}. \tag{8}$$

Herein, σ is the constant volatility of the underlying. The function Ψ is given as the solution of the nonlinear initial–value problem

$$\Psi'(A) = \frac{\Psi(A) + 1}{2\sqrt{A\Psi(A)} - A}, \quad A \neq 0, \tag{9}$$

$$\Psi(0) = 0.$$

For the case of no transaction costs occurring, i.e. $a = 0$, we recover the standard Black–Scholes equation (1a). In [3] the existence of a unique viscosity solution to (7), (8) is proved and some numerical results using an explicit finite difference approximation are given. A high-order compact finite difference scheme for the numerical solution was developed in [16,17]. The numerical results indicate an economically significant price difference between the standard Black–Scholes prices and the prices with transaction costs.

Let us consider another example. To obtain parameter values for σ often the so–called *implied volatility* is computed from market option prices by inverting the closed–form solution to (1), the so–called Black–Scholes formula. A widely observed phenomenon is that these computed volatilities are not constant. The pattern of implied volatilities for different exercise prices sometimes forms a *smile* shape, i.e., implied volatilities of in-the-money and out-of-the-money options are generally higher than that of at-the-money options. In equity option markets, typically, one observes a so–called volatility *skew*, i.e. the implied volatility for in-the-money calls is significantly higher than the implied volatility of at-the-money calls and out-of-the-money calls. Additionally, often variation with respect to time to maturity is present as well. This is usually referred to as the *volatility term structure*.

These observations lead to a natural generalization of the Black–Scholes model, where the constant volatility σ is replaced by a (deterministic) *local volatility function* $\hat{\sigma}(T, E)$, where T denotes the time to maturity and E the exercise price. Then, the option price $V = V(T, E)$ as a function of the exercise time T and the exercise price E satisfies the (forward) differential equation [13]

$$V_T(T,E) - \frac{1}{2}\hat{\sigma}^2(T,E)E^2 V_{EE}(T,E) + rEV_E(T,E) = 0, \quad T > 0,\ E > 0 \tag{10}$$

with initial condition and boundary conditions. It is derived from a Fokker–Planck equation integrated twice with respect to the space variable E and using the (formal) identity $(S_0 - E)^+_{EE} = \delta_{S_0}(E)$, where δ_{S_0} denotes the Dirac mass at $S_0 = S(0)$. This so-called local volatility function model was extended in [14, 32] by defining the local variance as the expectation of the future instantaneous variance conditional on a given asset price. In this way, the approach can be made consistent with (univariate diffusion) stochastic volatility models.

We have given several examples of *Black–Scholes type equations* which arise in option pricing models which try to weaken one or more of the assumptions present in the derivation of the original Black–Scholes equation. All of the above models result in linear or (fully) nonlinear parabolic diffusion–convection equations for $u = u(t, x)$ of the form

$$u_t + \frac{1}{2}\hat{\sigma}^2(t, x, u_x, u_{xx})x^2 u_{xx} + rxu_x - ru = 0, \quad t \in (0, T),\ x > 0, \tag{11}$$

with a nonlinear volatility function $\hat{\sigma}$, supplemented by final and boundary conditions. In the case of multiple underlying assets the partial derivatives have to be interpreted as the gradient and the Hessian, respectively. Equations of this form arise quite frequently in mathematical finance, not only in option pricing in the presence of transaction costs as in the models above, but also e.g. in modeling feedback effects due to large traders [22,41], option prices under generalized pricing kernels [19] and inverse problems (implied volatility) [4]. Often the source of such nonlinearities is market incompleteness due to trading restrictions, additional randomness in the model (e.g. stochastic volatility) or the presence of transaction costs.

Frequently, such advanced Black–Scholes type models will allow only for a numerical solution. As the error of numerical solutions can be controlled efficiently with modern discretization techniques (at least in the situation of a single or a small number of underlyings), there is a growing interest in model uncertainty. In particular, when considering hedged portfolios as in the (super–)replication of option prices, the hedging error is proportional to deviations in the volatility and the average Gamma over the lifetime of the option. Therefore, it arises the question of how to determine the volatility function $\hat{\sigma}$ in Black–Scholes type models of the form (11) from market data, such that the model under consideration replicates the market data. This parameter identification problem is often referred to as the *calibration problem*, since the model is calibrated to the market.

The outline of this chapter is as follows: In the next section we review the two main approaches to address the calibration problem. In Section 3. we focus on its formulation as an optimal control problem and prove the existence of local optimal solutions. Moreover, any optimal solution is characterized by an optimality system involving an adjoint equation for the Lagrange multiplier. The numerical optimization method is proposed in Section 4.. We apply a globalized sequential quadratic programming (SQP) method with a modified Hessian matrix to ensure that every SQP step is a descent direction and implement a line search strategy. In each level of the SQP method a linear–quadratic optimal control problem with box constraints is solved by a primal–dual active set strategy. In Section 5. numerical examples are presented and discussed.

2. The Calibration Problem

The problem of determining the volatility in (11) from market data is an ill–posed optimization problem in the sense of the lack of continuous dependence of the minimizers with respect to perturbations of the problem. In the mathematical literature, there are two main approaches to address the calibration problem. The first is to formally solve in equation (11) and apply the obtained formula on interpolated data sets of market data. This approach can be unstable and depends largely on the interpolation method but it is computationally cheap. The second approach is to use regularization techniques, e.g. based on entropic or Tikhonov regularization [21]. These methods address the problem of ill–posedness of the inverse parameter identification problem. They are based on a sound mathematical framework, their practical application may however result in increased effort for the numerical solution. In the following subsections we give an (incomplete) overview over the different approaches. For a complete review of the literature we refer to [10].

2.1. Direct Approach

The first idea to solve the calibration problem is to formally solve in equation (11) for the volatility. Then one can apply the obtained formula and compute, in principle, the volatility from given interpolated data sets of market data. Solving, for example, in (10) for the volatility [13] leads to

$$\hat{\sigma}(E,T) = \left(\frac{2[V_T(T,E) + rEV_E(T,E)]}{E^2 V_{EE}(T,E)} \right)^{1/2}. \qquad (12)$$

The terms on the right-hand side could then be evaluated by some numerical differentiation procedure. Typical theoretical option prices are strictly convex in E which implies positivity of the denominator, and applying the square root is possible. In practice, however, only discrete data are available that have to be interpolated. Depending on the interpolation method used, positivity of the denominator may not be guaranteed and the radicand changes sign. Even if such problems can be avoided, volatilities obtained from this approach feature rough oscillations which are not consistent with economical theory.

In [23] a stabilized method is proposed. After a data-preprocessing step with smoothing cubic splines which introduces additional regularity, the formula is rewritten as a linear system of equations for the unknown volatilities. Then, Tikhonov regularization is applied to obtain a smooth solution.

Although the direct approach is computationally cheap, it can be unstable if the ill–posedness of the problem is not addressed. Furthermore, it depends largely on the interpolation method that is used and prevalent data noise may induce violations of arbitrage bounds.

2.2. Optimal Control Approach

As an alternative approach, one can use optimization methods for an objective functional. To address the ill–posedness regularization techniques are applied where the regularization depends on a suitable norm of the volatility. Typically, the norm is evaluated relative to some given prior volatility, which allows e.g. to incorporate subjective estimates of a trader. These methods are more challenging from the computational point of view. Various forms for the objective functional have been suggested.

In [2], the problem is reformulated as a constrained stochastic optimal control problem and a so–called entropic regularization is performed. Using a probabalistic representation of the solution u, a dual form using Lagrange multipliers is derived. Associated with it is a Bellmann partial differential equation, that is solved by an iterative numerical procedure.

Another approach is to use a Tikhonov regularized cost functional of tracking type in a (deterministic) inverse problem [35]. For a survey on Tikhonov regularization see [21]. This approach has been adopted in many works, see, e.g., in [1,30]. However, most of the references mentioned above focus on the numerical results obtained by standard methods without analyzing in–depth the employed algorithms. A theoretical foundation of the approach with Tikhonov regularized cost functional is given in [10]. In this work a trinomial tree method using Tikhonov regularization is proposed. A probabilistic interpretation of the cost function's gradient is analyzed and numerical results are shown. Convergence rates for Tikhonov regularization under interpretable conditions have been derived in [20].

We only note that one can also follow a *discretize-then-optimize* approach. In this case one can apply methods of large scale optimization to solve a discrete optimization problem. Here, we pursue the *optimize-then-discretize* strategy where we formulate optimality conditions in the continuous form and then discretize by a suitable numerical method.

In particular, we follow an optimal control approach using a Lagrangian framework. The algorithm we employ is based on a sequential quadratic programming method (SQP) and on a primal–dual active set strategy that guarantees pointwise bilateral constraints for the volatility. SQP methods have been widely applied to optimization problems of the form

$$\text{minimize } J(x) \text{ subject to } e(x) = 0,$$

where the cost functional $J : X \to \mathbb{R}$ and the constraint $e : X \to Y$ are sufficiently smooth functions and X, Y are real Hilbert spaces. SQP methods for such problems occur frequently in constrained optimal control of systems described by partial differential equations and have been studied widely. The basic idea of SQP methods is to minimize at each iteration a quadratic approximation of the Lagrangian associated with the cost functional over an affine subspace of solutions of the linearized constraint. For a general survey on SQP methods we refer to [6] and the references therein. To base the numerical method on a solid foundation, we carry out a thorough analysis of first– and second–order optimality conditions.

3. Analysis of the Optimal Control Problem

Our goal is to identify from given market data the volatility function in (11). In our analysis, we focus on volatility functions which depend on time and space only like in (10), i.e. $\hat{\sigma} = \hat{\sigma}(t, x)$. Our numerical algorithm can be carried over without too many difficulties to the case of volatilities of the form $\hat{\sigma} = \hat{\sigma}(t, x, u_x)$. Indeed, the sequential quadratic programming algorithm can be viewed as a Newton method applied to a system composed of first–order necessary conditions for the Lagrangian associated with the optimization problem and the equality constraint given by (11). Hence, if a (local) solution is a *regular point* [38], it can be characterized by a standard constraint qualification and the algorithm presented in Section 4. can be modified accordingly.

For fully nonlinear models like (7) the situation is more involved. Concerning the analysis, the suitable notion of solutions in this case are viscosity solutions and our mathematical framework which is based on a Sobolev space setting is not applicable. Moreover, since this notion of solution provides only very weak regularity, it is not clear how to develop an optimal control framework. However, away from the initial conditions numerical evidence suggests that solutions of these equations have higher regularity (see, e.g. [16]). Hence, one could be tempted to apply the numerical solution method in a similar fashion. When formally computing the partial derivatives of the associated Lagrangian e.g. in the case of the model (7), the derivative of the nonlinear function Ψ appears. Since the right hand side of (9) is not bounded one needs to study an approximative solution with bounded derivative. Confer also [17], where similar problems appeared in proving convergence of a high-order compact finite difference scheme for (7). Alternatively, one can first identify the volatility function as a function $\hat{\sigma}(t, x)$ depending on time and space only and afterwards employ a postprocessing step to fit $\hat{\sigma}(t, x, u_x, u_{xx})$.

We will consider market data that are to be matched at only one maturity. Our results also extend naturally to the case of multiple constraints at different maturities. However, market data for different maturities are typically very scarcely distributed in time, e.g. four times per year. Hence, it will be difficult in practice to successfully identify the time dependence of the volatility from this information.

3.1. Formulation of the Optimal Control Problem

In this section we formulate the calibration problem as a constrained optimal control problem. We prove the existence of at least one optimal solution. Furthermore, we investigate first–order necessary and sufficient second–order optimality conditions.

We consider in the following the situation, where we have a time–dependent Dirichlet boundary condition on one boundary, and a homogeneous Dirichlet boundary on the other. It is straightforward to extend the following analysis to other situations, e.g. with Dirichlet data on both boundaries. Homogeneous Dirichlet conditions can be incorporated into the definition of the function spaces in which we consider the problem. Concerning the error inflicted by introducing artificial boundary conditions we refer to [31].

Let us start by introducing some notation. For $0 < a < x < b$ and $T > 0$ we define $\Omega = (a, b)$ as the one–dimensional spatial domain and $Q = (0, T) \times \Omega$ the time–spatial domain. We define the Hilbert space

$$V = \{\varphi \in H^1(\Omega) : \varphi(b) = 0\}$$

endowed with the inner product

$$\langle \varphi, \psi \rangle_V = \int_\Omega \varphi_x \psi_x \, \mathrm{d}x \quad \text{for all } \varphi, \psi \in V.$$

By $L^2(0, T; V)$ we denote the space of (equivalence classes) of measurable functions $\varphi : [0, T] \to V$, which are square integrable, i.e.,

$$\int_0^T \|\varphi(t)\|_V^2 \, \mathrm{d}t < \infty.$$

Other spaces used below are defined analogously. Moreover, we make use of the space

$$W(0, T) = \{\varphi \in L^2(0, T; V) : \varphi_t \in L^2(0, T; V')\},$$

which is a Hilbert space endowed with the common inner product; see [11, p. 473]. Let us recall the Hilbert space

$$\begin{aligned} H^{2,1}(Q) &= H^1(0, T; L^2(\Omega)) \cap L^2(0, T; H^2(\Omega)) \\ &= \{\varphi : Q \to \mathbb{R} \mid \varphi, \varphi_t, \varphi_x, \varphi_{xx} \in L^2(Q)\}, \end{aligned}$$

supplied with the inner product

$$\langle \varphi, \psi \rangle_{H^{2,1}(Q)} = \int_0^T \int_\Omega \varphi_t \psi_t + \varphi_{xx} \psi_{xx} + \varphi_x \psi_x + \varphi \psi \, \mathrm{d}x \mathrm{d}t \quad \text{for } \varphi, \psi \in H^{2,1}(Q)$$

and the induced norm $\|\cdot\|_{H^{2,1}(Q)} = \langle \cdot, \cdot \rangle_{H^{2,1}(Q)}^{1/2}$. Recall that from $\Omega \subset \mathbb{R}$ it follows that $H^{2,1}(Q)$ is continuously embedded into $L^\infty(Q)$; see, e.g. [37, p. 24]. When t is fixed, the expression $\varphi(t)$ stands for the function $\varphi(t, \cdot)$ considered as a function in Ω only.

Next we specify the set of admissible coefficient functions. Suppose that q_{min} and q_{max} are given functions in $H^{2,1}(Q) \cap L^\infty(0,T;H^2(\Omega))$ satisfying $\overline{q}_{\mathsf{min}} \le q_{\mathsf{min}} < q_{\mathsf{max}}$ in Q almost everywhere (a.e.) with $\overline{q}_{\mathsf{min}} = \operatorname{essinf}\{q(t,x) : (t,x) \in Q\} > 0$. In particular, there exists $C_{\mathsf{ad}} > 0$ such that

$$\max\left\{||q_{\mathsf{min}}||_{L^\infty(0,T;H^2(\Omega))}, ||q_{\mathsf{max}}||_{L^\infty(0,T;H^2(\Omega))}\right\} \le C_{\mathsf{ad}}.$$

We introduce the set for the admissible coefficient functions by

$$\mathcal{Q}_{\mathsf{ad}} = \left\{q \in H^{2,1}(Q) : ||q||_{L^\infty(0,T;H^2(\Omega))} \le C_{\mathsf{ad}},\ q_{\mathsf{min}} \le q \le q_{\mathsf{max}} \text{ in } Q \text{ a.e.}\right\}, \tag{13}$$

which is a closed, bounded and convex set in $H^{2,1}(Q)$. Note, that the bound $C_{\mathsf{ad}} > 0$ is purely technical, and can be chosen arbitrarily large. The functions q_{min} and q_{max} reflect some knowledge about volatilities. One could set them to historical extremal values, or, if a statistical distribution of volatilities is known, one could set them as some lower and upper quantile of this distribution.

The goal of the parameter identification is to determine the volatility function $\hat{\sigma}(t,x)$ in (11). For streamlining the presentation we restrict ourselves to the case $r = 0$ of zero interest rate. Hence, we need to determine the coefficient function $q = q(t,x)$ in the parabolic problem

$$u_t(t,x) - q(t,x)u_{xx}(t,x) = 0 \qquad \text{for all } (t,x) \in Q, \tag{14a}$$

$$u(t,a) = u_D(t) \qquad \text{for all } t \in (0,T), \tag{14b}$$

$$u(t,b) = 0 \qquad \text{for all } t \in (0,T), \tag{14c}$$

$$u(0,x) = u_0(x) \qquad \text{for all } x \in \Omega \tag{14d}$$

from given, observed market data $u_T \in L^2(\Omega)$ for the solution u of (14) at the final time T.

Definition 3.1. *For given $q \in \mathcal{Q}_{\mathsf{ad}}$, $u_D \in H^1(0,T)$ and $u_0 \in L^2(\Omega)$ a function u is called a* weak solution to (14) *if $u \in W(0,T)$, $u(\cdot,a) = u_D$ in $L^2(0,T)$, $u(0) = u_0$ in $L^2(\Omega)$ and*

$$\int_0^T \langle u_t, \varphi\rangle_{H^{-1},H_0^1} + \Big(\int_\Omega q u_x\varphi_x + q_x u_x \varphi \,\mathrm{d}x\Big)\mathrm{d}t = 0, \tag{15}$$

for all $\varphi \in L^2(0,T;H_0^1(\Omega))$. In (15) $\langle\cdot,\cdot\rangle_{H^{-1},H_0^1}$ denotes the duality pairing between $H_0^1(\Omega)$ and its dual space $H^{-1}(\Omega)$.

The following theorem ensures existence of a weak solution to (14) for positive coefficient functions. Its proof follows from standard arguments [34].

Theorem 3.2. *Suppose that $u_0 \in L^2(\Omega)$ and $u_D \in H^1(0,T)$. Then, for every $q \in \mathcal{Q}_{\mathsf{ad}}$ there exists a unique weak solution u to (14) and a constant $C > 0$ such that*

$$\|u\|_{W(0,T)} \le C\left(\|u_0\|_{L^2(\Omega)} + \|u_D\|_{H^1(0,T)}\right). \tag{16}$$

If the initial condition u_0 is more regular, we have the following corollary. Its proof is omitted, because it is standard.

Corollary 3.3. *If $u_0 \in V$ holds with the compatibility condition $u_0(a) = u_D(0)$, it follows that $u \in H^{2,1}(Q)$ and there exists a constant $C > 0$ such that*

$$\|u\|_{H^{2,1}(Q)} \leq C\left(\|u_0\|_V + \|u_D\|_{H^1(0,T)}\right). \tag{17}$$

To write the state equations (14) in an abstract form we define the two Hilbert spaces

$$X = H^{2,1}(Q) \times W(0,T) \quad \text{and} \quad Y = L^2(0,T;H_0^1(\Omega)) \times L^2(0,T) \times L^2(\Omega)$$

endowed with their product topologies. Moreover, let

$$K_{\mathsf{ad}} = \mathcal{Q}_{\mathsf{ad}} \times W(0,T)$$

which is closed and convex. In the sequel we identify the dual Y' of Y with the product space $L^2(0,T;H^{-1}(\Omega)) \times L^2(0,T) \times L^2(\Omega)$.

Next we introduce the bilinear operator $e = (e_1, e_2, e_3) : X \to Y'$ by

$$e_1(\omega) = u_t - qu_{xx}, \tag{18a}$$
$$e_2(\omega) = u(\cdot\,, a) - u_D, \tag{18b}$$
$$e_3(\omega) = u(0) - u_0, \tag{18c}$$

where $\omega = (q, u)$ holds and the identity $e_1(\omega) = u_t - qu_{xx}$ in $L^2(0,T;H^{-1}(\Omega))$ stands for

$$\langle e_1(\omega), \varphi\rangle_{L^2(0,T;H^{-1}(\Omega)),L^2(0,T;H_0^1(\Omega))}$$
$$= \int_0^T \langle u_t, \varphi\rangle_{H^{-1},H_0^1}\,\mathrm{d}t + \int_0^T\int_\Omega qu_x\varphi_x + q_x u_x\varphi\,\mathrm{d}x\mathrm{d}t \quad \text{for } \varphi \in L^2(0,T;H_0^1(\Omega)).$$

The following proposition summarizes properties of the operator e. In particular, e is Fréchet differentiable and its linearization $e'(\omega)$ is surjective at any point $\omega \in K_{\mathsf{ad}}$. The surjectivity guarantees a constraint qualification, so that there exists a (unique) Lagrange multiplier λ^* satisfying the first–order necessary optimality condition (see Theorem 3.7). The Fréchet derivatives with respect to ω are denoted by primes, where subscripts denote as usual the associated partial Fréchet derivative.

Proposition 3.4 (cf. [18]). *The bilinear operator $e : X \to Y'$ is twice continuously Fréchet differentiable and the mapping $\omega \mapsto e''(\omega)$ is Lipschitz continuous on X. Moreover, its linearization $e'(\omega) : X \to Y'$ at any point $\omega = (q,u) \in K_{\mathsf{ad}}$ is surjective. Furthermore, we have*

$$\|\delta u\|_{W(0,T)} \leq C_1 \|\delta q\|_{H^{2,1}(Q)} \quad \textit{for all } \delta\omega = (\delta q, \delta u) \in N(e'(\omega)), \tag{19}$$

where $N(e'(\omega)) \subset X$ denotes the null space of $e'(\omega)$. For arbitrary directions $\delta\omega = (\delta q, \delta u)$, $\widetilde{\delta\omega} = (\widetilde{\delta q}, \widetilde{\delta u}) \in X$ the directional derivatives are given by

$$e'(\omega)\delta\omega = \begin{pmatrix} \delta u_t - q\delta u_{xx} - \delta q u_{xx} \\ \delta u(\cdot\,, a) \\ \delta u(0) \end{pmatrix} \tag{20}$$

and

$$e''(\omega)(\delta\omega, \widetilde{\delta\omega}) = \begin{pmatrix} -\widetilde{\delta q}\delta u_{xx} - \delta q \widetilde{\delta u}_{xx} \\ 0 \\ 0 \end{pmatrix}. \tag{21}$$

At any point $\omega \in K_{\text{ad}}$ the operator $e_u : W(0,T) \to Y'$ is even bijective.

To measure deviations of the solution of the model to observed market data we introduce the cost functional $J : X \to [0, \infty)$ of tracking type by

$$J(\omega) = \frac{1}{2} \int_\Omega |u(T) - u_T|^2 \, \mathrm{d}x + \frac{\beta}{2} \|q\|^2_{H^{2,1}(Q)} \quad \text{for } \omega = (q, u) \in X, \tag{22}$$

where u_T are give observed market data at the end–time T, and $\beta > 0$ is a regularization parameter.

Proposition 3.5 (cf. [18]). *The cost functional $J : X \to [0, \infty)$ is twice Fréchet differentiable and its Fréchet derivatives are given by*

$$J'(\omega)\delta\omega = \int_\Omega \big(u(T) - u_T\big)\delta u(T) \, \mathrm{d}x + \beta \, \langle q, \delta q \rangle_{H^{2,1}(Q)} \tag{23}$$

and

$$J''(\omega)(\delta\omega, \widetilde{\delta\omega}) = \int_\Omega \delta u(T)\widetilde{\delta u}(T) \, \mathrm{d}x + \beta \, \langle \delta q, \widetilde{\delta q} \rangle_{H^{2,1}(Q)} \tag{24}$$

for arbitrary directions $\delta\omega = (\delta q, \delta u), \widetilde{\delta\omega} = (\widetilde{\delta q}, \widetilde{\delta u}) \in X$. In particular, the mapping $\omega \mapsto J''(\omega)$ is Lipschitz–continuous on X.

The calibration problem then is given by a constrained optimal control problem

$$\min J(\omega) \quad \text{s.t.} \quad \omega \in K_{\text{ad}} \text{ and } e(\omega) = 0. \tag{\textbf{P}}$$

In this formulation, the state variable u and the coefficient q are considered as independent variables while the realization of (14) is an explicit constraint. Alternatively, one could use the equality constraint to treat $u = u(q)$ as a variable depending on the unknown coefficient q and solve the nonlinear least–squares problem by the Gauss–Newton method. Then, in each iteration one would need to solve the possibly nonlinear state equation.

Here, we choose an SQP approach with independent variables. SQP methods can be viewed as a natural extension of Newton methods to the present situation. Also they are expected to inherit the fast local convergence property. Indeed, the iterates of the SQP method are identical to those generated by Newton's method when applied to the system composed of the first–order necessary conditions for the Lagrangian associated with (**P**) that we will present in Section 3.3. and the equality constraint $e(\omega) = 0$. Note that SQP methods are not feasible–point methods, i.e. its iterates need not be points satisfying the constraints.

3.2. Existence of Optimal Solutions

We introduce the admissible set

$$E = \{\omega = (q, u) \in X : e(\omega) = 0 \text{ in } Y' \text{ and } \omega \in K_{\mathsf{ad}}\}. \tag{25}$$

The next theorem guarantees that **(P)** possesses an optimal solution.

Theorem 3.6. *Problem* **(P)** *has at least one (global) solution* $\omega^* = (q^*, u^*) \in K_{\mathsf{ad}}$.

Proof. In view of Theorem 3.2, the set E is non–empty, since $q_{\mathsf{min}} \in \mathcal{Q}_{\mathsf{ad}}$ implies that $(q_{\mathsf{min}}, u(q_{min})) \in E$. Moreover, by definition $J(\omega) \geq 0$ holds for all $\omega \in E$. Thus there exists a $\zeta \geq 0$ such that

$$\zeta = \inf\{J(\omega) : \omega \in E\}. \tag{26}$$

We infer that there exists a minimizing sequence $(\omega^n)_{n\in\mathbb{N}} \subset E$, $\omega^n = (q^n, u^n)$, with

$$\lim_{n\to\infty} J(\omega^n) = \zeta.$$

Due to (16) and

$$J(\omega^n) \geq \frac{\beta}{2} \|q^n\|^2_{H^{2,1}(Q)} \quad \text{for all } n,$$

we infer that the sequence $(\omega^n)_{n\in\mathbb{N}}$ is bounded in X. Thus, there exist subsequences, again denoted by $(\omega^n)_{n\in\mathbb{N}}$, and a pair $\omega^* = (q^*, u^*) \in X$ satisfying

$$\begin{aligned} q^n &\rightharpoonup q^* \text{ in } H^{2,1}(Q) \text{ as } n \to \infty, \\ u^n &\rightharpoonup u^* \text{ in } W(0,T) \text{ as } n \to \infty. \end{aligned} \tag{27}$$

Furthermore, since $q^n \in \mathcal{Q}_{\mathsf{ad}}$ and it holds $L^\infty(0,T;H^2(\Omega)) \cap H^1(0,T;L^2(\Omega)) \hookrightarrow C([0,T];H^1(\Omega))$ compactly due to Aubin's lemma [40], we obtain

$$q^n \to q^* \text{ in } C([0,T];H^1(\Omega)) \text{ as } n \to \infty. \tag{28}$$

In view of (27) and (28) it holds

$$\int_0^T \int_\Omega q^n u^n_x \varphi_x + q^n_x u^n_x \varphi \, dx dt \to \int_0^T \int_\Omega q^* u^*_x \varphi_x + q^*_x u^*_x \varphi \, dx dt$$

as $n \to \infty$ for every $\varphi \in L^2(0,T;H^1_0(\Omega))$. Therefore,

$$\lim_{n\to\infty} e_1(\omega^n) = e_1(\omega^*) \quad \text{in } L^2(0,T;H^{-1}(\Omega)).$$

From $e_1(\omega^n) = 0$ for all $n \in \mathbb{N}$ we conclude that $e_1(\omega^*) = 0$. Since the operators e_2 and e_3 are linear, we find $e(\omega^*) = 0$. Since J is convex and continuous, and therefore weakly lower semi–continuous, we obtain $J(\omega^*) \leq \lim_{n\to\infty} J(\omega^n) = \zeta$. Finally, since $\mathcal{Q}_{\mathsf{ad}}$ is convex and closed in $H^{2,1}(Q)$, and therefore weakly closed, we have $q^* \in \mathcal{Q}_{\mathsf{ad}}$, and the claim follows. □

3.3. First–Order Necessary Optimality Conditions

In this section we introduce the Lagrange functional associated with **(P)**. We derive first–order necessary optimality conditions and prove the existence of a unique Lagrange multiplier associated with the inequality constraints for the optimal coefficient q^*.

Since problem **(P)** is a non–convex programming problem, uniqueness of a minimizer is not guaranteed. Different local minima may occur and a numerical method will reveal a local minimum close to its starting value. Hence, we do not restrict our investigations to global solutions of **(P)**. Rather we assume that a fixed reference solution $\omega^* = (q^*, u^*) \in K_{\text{ad}}$ is given satisfying certain first– and second–order optimality conditions, which ensure local optimality of the solution.

To formulate the optimality conditions we introduce the Lagrange functional $L : X \times Y \to \mathbb{R}$ associated with **(P)** by

$$\begin{aligned} L(\omega, p) &= J(\omega) + \langle e(\omega), (\lambda, \mu, \nu) \rangle_{Y',Y} \\ &= \frac{1}{2} \|u(T) - u_T\|^2_{L^2(\Omega)} + \frac{\beta}{2} \|q\|^2_{H^{2,1}(Q)} + \int_\Omega \big(u(0) - u_0\big)\nu \, \mathrm{d}x \\ &\quad + \int_0^T \langle u_t, \lambda \rangle_{H^{-1},H^1_0} \, \mathrm{d}t + \int_0^T \int_\Omega (q\lambda)_x u_x \, \mathrm{d}x\mathrm{d}t + \int_0^T \big(u(\cdot, a) - u_D\big)\mu \, \mathrm{d}t, \end{aligned}$$

with $\omega = (q, u) \in X$ and $p = (\lambda, \mu, \nu) \in Y$. Due to Proposition 3.4 and Proposition 3.5 the Lagrangian is twice continuously Fréchet differentiable with respect to $\omega \in X$ for each fixed $p \in Y$ and its second Fréchet derivative is Lipschitz–continuous.

An optimal solution to **(P)** can be characterized by first–order necessary optimality conditions. In particular, the following theorem holds.

Theorem 3.7. *Let $B_\rho(\omega)$ denote the open ball in X with radius $\rho > 0$ and mid point $\omega \in X$. Suppose that $\omega^* = (q^*, u^*) \in K_{\text{ad}}$ is a local solution to* **(P)***, i.e., $\omega^* \in E$ and there exists a constant $\rho > 0$ such that*

$$J(\omega^*) \le J(\omega) \quad \textit{for all } \omega \in E \cap B_\rho(\omega^*).$$

Then there is a unique Lagrange multiplier $p^* = (\lambda^*, \mu^*, \nu^*) \in Y$ satisfying the adjoint equations

$$-\lambda^*_t - (q^*\lambda^*)_{xx} = 0 \qquad \textit{in } Q, \tag{29a}$$
$$\lambda^*(\cdot, a) = \lambda^*(\cdot, b) = 0 \qquad \textit{in } (0, T), \tag{29b}$$
$$\lambda^*(T) = -(u^*(T) - u_T) \qquad \textit{in } \Omega \tag{29c}$$

in the weak sense and the identities

$$\mu^* = (q^*\lambda^*)_x(\cdot, a) \qquad \textit{in } L^2(0, T), \tag{30}$$
$$\nu^* = \lambda^*(0) \qquad \textit{in } L^2(\Omega) \tag{31}$$

hold. Moreover, the variational inequality

$$\langle \beta q^* - \mathcal{R}(\lambda^* u^*_{xx}), q - q^* \rangle_{H^{2,1}(Q)} \ge 0 \quad \textit{for all } q \in \mathcal{Q}_{\text{ad}} \tag{32}$$

holds, where $\mathcal{R} : (H^{2,1}(Q))' \to H^{2,1}(Q)$ denotes the Riesz isomorphism, i.e., $q = \mathcal{R}(f) \in H^{2,1}(Q)$ solves

$$\int_0^T \int_\Omega q_t\varphi_t + q_{xx}\varphi_{xx} + q_x\varphi_x + q\varphi \,\mathrm{d}x\mathrm{d}t = \langle f, \varphi\rangle_{(H^{2,1}(Q))',H^{2,1}(Q)} \text{ for all } \varphi \in H^{2,1}(Q)$$

with $f \in (H^{2,1}(Q))'$. Here, $\langle\cdot\,,\cdot\rangle_{(H^{2,1}(Q))',H^{2,1}(Q)}$ denotes the duality pairing between $H^{2,1}(Q)$ and its dual.

Proof. We infer from Proposition 3.4 that a standard constraint qualification holds at (q^*, u^*) [38]. Therefore, there exists a unique Lagrange multiplier $p^* = (\lambda^*, \mu^*, \nu^*) \in Y$ such that

$$L_q(\omega^*, p^*)(q - q^*) \geq 0 \qquad \text{for all } q \in \mathcal{Q}_{\mathrm{ad}}, \tag{33}$$
$$L_u(\omega^*, p^*)u = 0 \qquad \text{for all } u \in W(0,T), \tag{34}$$
$$L_p(\omega^*, p^*)p = 0 \qquad \text{for all } p \in Y. \tag{35}$$

Equation (35) is equivalent to the equality constraint $e(\omega^*) = 0$ and is fulfilled since ω^* solves **(P)**. Next we turn to (34), which is equivalent to

$$\begin{aligned} 0 \;=\; & \int_\Omega (u^*(T) - u_T)u(T)\,\mathrm{d}x + \int_0^T \langle u_t, \lambda^*\rangle_{H^{-1},H_0^1(\Omega)}\,\mathrm{d}t \\ & + \int_0^T\int_\Omega (q^*\lambda^*)_x u_x \,\mathrm{d}x\mathrm{d}t + \int_0^T u(\cdot,a)\mu^*\,\mathrm{d}t + \int_\Omega u(0)\nu^*\,\mathrm{d}x \end{aligned} \tag{36}$$

for all $u \in W(0,T)$. In particular, (36) holds for all $u(t,x) = \chi(t)\psi(x)$ with $\chi \in C_0^1(0,T)$ and $\psi \in H_0^1(\Omega) \subset V$. Consequently,

$$\int_0^T \int_\Omega \chi_t\psi\lambda^* + (q^*\lambda^*)_x\chi\psi' \,\mathrm{d}x\mathrm{d}t = 0 \tag{37}$$

for all $\chi \in C_0^1(0,T)$ and $\psi \in H_0^1(\Omega)$. Notice that

$$\int_0^T\int_\Omega \chi_t\psi\lambda^*\,\mathrm{d}x\mathrm{d}t = \int_\Omega \Big(\int_0^T \chi_t\lambda^*\,\mathrm{d}t\Big)\psi\,\mathrm{d}x = \Big\langle -\int_0^T \lambda_t^*\chi\,\mathrm{d}t, \psi\Big\rangle_{H^{-1},H_0^1}, \tag{38}$$

where λ_t^* denotes the distributional derivative of λ^* with respect to t. The remaining term in (37) leads to

$$\int_0^T\int_\Omega (q^*\lambda^*)_x\psi'\chi\,\mathrm{d}x\mathrm{d}t = \Big\langle -\int_0^T (q^*\lambda^*)_{xx}\chi\,\mathrm{d}t, \psi\Big\rangle_{H^{-1},H_0^1}. \tag{39}$$

Inserting (38) and (39) into (37) we get

$$\Big\langle \int_0^T \big(-\lambda_t^* - (q^*\lambda^*)_{xx}\big)\chi\,\mathrm{d}t, \psi\Big\rangle_{H^{-1},H_0^1} = 0, \tag{40}$$

for all $\chi \in C_0^1(0,T)$ and $\psi \in H_0^1(\Omega)$. Notice that $q^* \in \mathcal{Q}_{\text{ad}}$ implies $q^* \in L^\infty(Q)$ as well as $q_x^* \in L^\infty(0,T;L^2(\Omega))$. Therefore, it follows $(q^*\lambda^*)_x \in L^2(Q)$ and, consequently, $(q^*\lambda^*)_{xx} \in L^2(0,T;H^{-1}(\Omega))$. The set

$$\{\varphi \in L^2(0,T;H_0^1(\Omega)) : \varphi(t,x) = \chi(t)\psi(x) \text{ with } \chi \in C_0^1(0,T) \text{ and } \psi \in H_0^1(\Omega)\}$$

is dense in $L^2(0,T;H_0^1(\Omega))$ so that $\lambda_t^* \in L^2(0,T;H^{-1}(\Omega))$ and (29a) hold. Moreover,

$$\begin{aligned} \int_0^T \frac{\mathrm{d}}{\mathrm{d}t} \langle \lambda^*, u\rangle_{L^2(\Omega)} \,\mathrm{d}t &= \langle \lambda_t^*, u\rangle_{L^2(0,T;H^{-1}(\Omega)),L^2(0,T;H_0^1(\Omega))} \\ &\quad + \langle u_t, \lambda^*\rangle_{L^2(0,T;H^{-1}(\Omega)),L^2(0,T;H_0^1(\Omega))} \end{aligned} \tag{41}$$

for $u \in W(0,T)$. Hence, we may apply (36), (40), and (41) to obtain

$$\begin{aligned} 0 &= \int_\Omega (u^*(T) - u_T)u(T)\,\mathrm{d}x + \int_0^T \frac{\mathrm{d}}{\mathrm{d}t}\langle \lambda^*, u\rangle_{L^2(\Omega)}\,\mathrm{d}t \\ &\quad + \int_0^T \langle -\lambda_t^* - (q^*\lambda^*)_{xx}, u\rangle_{H^{-1},H_0^1(\Omega)} + \int_0^T (q^*\lambda^*)_x u\Big|_{x=a}^{x=b}\,\mathrm{d}t \\ &\quad + \int_0^T \mu^* u(\cdot,a)\,\mathrm{d}t + \int_\Omega \nu^* u(0)\,\mathrm{d}x \\ &= \langle (u^*(T) - u_T + \lambda^*(T), u(T)\rangle_{L^2(\Omega)} + \langle \nu^* - \lambda^*(0), u(0)\rangle_{L^2(\Omega)} \\ &\quad + \langle \mu^* - (q^*(\cdot,a)\lambda^*(\cdot,a))_x, u(\cdot,a)\rangle_{L^2(0,T)}. \end{aligned}$$

Choosing appropriate test functions in $W(0,T)$, we find (29c), (30), and (31). Finally, we consider (33). We compute

$$L_q(q^*,u^*,p^*)q = \int_0^T \int_\Omega \beta\big(q_t^* q_t + q^* q + q_x^* q_x + q_{xx}^* q_{xx}\big) + (q\lambda^*)_x u_x^*\,\mathrm{d}x\mathrm{d}t \tag{42}$$

for all $q \in \mathcal{Q}_{\text{ad}}$. For $\lambda^* \in L^2(0,T;H_0^1(\Omega))$ and $u_x^* \in L^2(Q)$ the integral

$$\int_0^T \int_\Omega (q\lambda^*)_x u_x^*\,\mathrm{d}x\mathrm{d}t$$

is bounded for all $q \in \mathcal{Q}_{\text{ad}}$. Moreover, $(\lambda^* u_x^*)(\cdot,a) = (\lambda^* u_x^*)(\cdot,b) = 0$ holds. Thus, the function $g = -\lambda^* u_{xx}^*$ can be identified with an element in $(H^{2,1}(Q))'$ and we derive from (42)

$$L_q(q^*,u^*,p^*)q = \beta\langle q^*, q\rangle_{H^{2,1}(Q)} + \langle g, q\rangle_{(H^{2,1}(Q))',H^{2,1}(Q)} \tag{43}$$

for all $q \in \mathcal{Q}_{\text{ad}}$. Employing the Riesz isomorphism $\mathcal{R}$, inserting (43) into (33) we find

$$\langle \beta q^* - \mathcal{R}(\lambda^* u_{xx}^*), q - q^*\rangle_{H^{2,1}(Q)} \geq 0 \quad \text{for all } q \in \mathcal{Q}_{\text{ad}},$$

which is the variational inequality (32). □

Remark 3.8. The usage of the Riesz operator $\mathcal{R} : (H^{2,1}(Q))' \to H^{2,1}(Q)$ in (32) requires to solve a problem of the form

$$-u_{tt} + u_{xxxx} - u_{xx} + u = f \quad \text{in } Q,$$

including initial and boundary conditions. Hence, in our numeric realization we will employ the 'weaker' norm in $L^2(0,T;H^1(\Omega))$, see Section 4.. Then $\mathcal{R}$ can be replaced by the Riesz operator $\widetilde{\mathcal{R}} : (H^1(\Omega))' \to H^1(\Omega)$, that requires only the solution of the Neumann problem

$$-u(t)_{xx} + u(t) = f(t) \quad \text{in } \Omega, \quad u(t)_x|_{\delta\Omega} = 0,$$

for a.e. $t \in (0,T)$.

Utilizing variational techniques we can prove the following error estimate for the adjoint variable λ^*.

Corollary 3.9. *Let all hypotheses of Theorem* 3.7 *hold. Then there exists a constant* $C > 0$ *depending on* $\|q^*\|_{L^\infty(0,T;L^4(\Omega))}$ *and* $\overline{q}_{\mathsf{min}}$ *such that*

$$\|\lambda^*\|_{L^2(0,T;H^1_0(\Omega))} \le C\,\|u^*(T) - u_T\|_{L^2(\Omega)}$$

Hence, if the residual $\|u^*(T) - u_T\|_{L^2(\Omega)}$ becomes small the norm of the Lagrange multiplier λ^* is small. We will make use of this estimate in the next section.

From Theorem 3.7 we infer the existence of a Lagrange multiplier associated with the constraint $q^* \in \mathcal{Q}_{\mathsf{ad}}$. To formulate the result we introduce the following sets.

Definition 3.10. *Let* K *be a convex subset of a (real) Banach space* Z *and* $z^* \in K$. *The* cone of feasible directions R_K at the point z^*, *the* tangent cone T_K at the point z^* *and the* normal cone N_K at the point z^* *are defined by*

$$\begin{aligned} R_K(z^*) &= \{z \in Z : \exists\sigma > 0 : z^* + \sigma z \in K\},\\ T_K(z^*) &= \{z \in Z : \exists z^*(\sigma) = z^* + \sigma z + o(\sigma) \in K, \sigma \ge 0\},\\ N_K(z^*) &= \{z \in Z' : \langle z, \tilde{z} - z^*\rangle_{Z',Z} \le 0 \textit{ for all } \tilde{z} \in K\}. \end{aligned}$$

In case of $z^* \notin K$ *the normal cone* $N_K(z^*)$ *is set equal to the empty set.*

In the following we choose $Z = H^{2,1}(Q)$, $K = \mathcal{Q}_{\mathsf{ad}}$ and $z^* = q^*$.

Corollary 3.11. *Let all hypotheses of Theorem* 3.7 *be satisfied. Then there exists a Lagrange multiplier* $\tilde{\xi}^* \in N_{\mathcal{Q}_{\mathsf{ad}}}(q^*)$ *associated with the inequality constraints such that*

$$L_q(\omega^*, p^*) + \tilde{\xi}^* = 0 \quad \textit{in } (H^{2,1}(Q))'. \tag{44}$$

Proof. Defining $\tilde{\xi}^* = -\beta q^* + \lambda^* u^*_{xx} \in (H^{2,1}(Q))'$ and using (32) we obtain $\tilde{\xi}^* \in N_{\mathcal{Q}_{\mathsf{ad}}}(q^*)$. In particular, (44) follows. □

Remark 3.12. Using the Riesz isomorphism $\mathcal{R}$ introduced in Theorem 3.7 we can identify $\tilde{\xi}^* \in (H^{2,1}(Q))'$ with an element in the Hilbert space $H^{2,1}(Q)$ by setting $\xi^* = -\beta q^* + \mathcal{R}(\lambda^* u^*_{xx})$.

Let $\omega^* = (q^*, u^*) \in K_{\mathsf{ad}}$ denote a local solution to **(P)**. If the solution $q^* \in \mathcal{Q}_{\mathsf{ad}}$ is inactive with respect to the norm constraint, i.e., $\|q^*\|_{L^\infty(0,T;H^2(\Omega))} < C_{\mathsf{ad}}$, then **(P)** is locally equivalent to

$$\min J(\omega) \quad \text{s.t.} \quad \omega \in \hat{K}_{\mathsf{ad}} \text{ and } e(\omega) = 0, \tag{$\hat{\mathbf{P}}$}$$

where $\hat{K}_{\mathsf{ad}} = \hat{Q}_{\mathsf{ad}} \times W(0,T)$ and

$$\hat{Q}_{\mathsf{ad}} = \big\{ q \in H^{2,1}(Q) \ : \ q_{\mathsf{min}} \le q \le q_{\mathsf{max}} \text{ in } Q \text{ a.e.} \big\},$$

which is a closed, convex and bounded subset in $L^2(Q)$. We define by

$$\hat{E} = \{ \omega \in \hat{K}_{\mathsf{ad}} : e(\omega) = 0 \}$$

the admissible set of $(\hat{\mathbf{P}})$. Monitoring the sequence $\|q^n\|_{L^\infty(0,T;H^2(\Omega))}$ we solve $(\hat{\mathbf{P}})$ in our numerical experiments, see Section 5. below. For that reason we focus on $(\hat{\mathbf{P}})$ in the remainder of this section.

The Lagrange multiplier ξ^* associated with the inequality constraints for the optimal coefficient q^* is characterized by the following corollary.

Corollary 3.13 (cf. [18]). *Let all hypotheses of Theorem 3.7 be satisfied. Suppose that $\|q^*\|_{L^\infty(0,T;H^2(\Omega))} < C_{\mathsf{ad}}$. Then ξ^* satisfies*

$$\xi^*\big|_{A^*_-} \le 0, \quad \xi^*\big|_{A^*_+} \ge 0, \quad \xi^*\big|_{I^*} = 0, \tag{45}$$

where

$$\begin{aligned} A^*_- &= \big\{ (t,x) \in \overline{Q} : q^*(t,x) = q_{\mathsf{min}}(t,x) \big\}, \\ A^*_+ &= \big\{ (t,x) \in \overline{Q} : q^*(t,x) = q_{\mathsf{max}}(t,x) \big\}, \\ I^* &= \big\{ (t,x) \in \overline{Q} : q_{\mathsf{min}}(t,x) < q^*(t,x) < q_{\mathsf{max}}(t,x) \big\} \end{aligned}$$

are the active and inactive sets for the optimal coefficient q^*.

The primal–dual active set algorithm used below employs a result from convex analysis [24, 29] to show that one can replace $q^* \in \hat{Q}_{\mathsf{ad}}$ and condition (45) by

$$q^* = \mathcal{P}_{\mathsf{ad}}\left(q^* + \frac{\xi^*}{c} \right) \quad \text{for every } c > 0, \tag{46}$$

where

$$\mathcal{P}_{\mathsf{ad}} : L^2(Q) \to \{ q \in L^2(Q) \mid q_{\mathsf{min}} \le q \le q_{\mathsf{max}} \text{ in } Q \text{ a.e.} \}$$

is defined by

$$\mathcal{P}_{\mathsf{ad}}(q)(t,x) = \begin{cases} q_{\mathsf{min}}(t,x) & \text{if } q(t,x) < q_{\mathsf{min}}(t,x), \\ q(t,x) & \text{if } q_{\mathsf{min}}(t,x) \le q(t,x) \le q_{\mathsf{max}}(t,x), \\ q_{\mathsf{max}}(t,x) & \text{if } q(t,x) > q_{\mathsf{max}}(t,x) \end{cases}$$

for almost all $(t,x) \in Q$. The primal–dual active set method uses the identification (46) as a prediction strategy, see Section 4.2..

3.4. Second–Order Optimality

In Section 3.3. we have investigated first–order necessary optimality conditions for $(\hat{\mathbf{P}})$. To ensure that a solution (ω^*, p^*) satisfying $\omega^* = (q^*, u^*) \in \hat{E}$, $q^* \in \hat{\mathcal{Q}}_{\text{ad}}$, (29) and (45) indeed solves $(\hat{\mathbf{P}})$, we have to guarantee second–order sufficient optimality. This is the focus of this section. We review different second–order optimality conditions and set them into relation. Moreover, it can be shown that the second–order sufficient optimality condition holds, provided the residual $\|u^*(T) - u_T\|_{L^2(\Omega)}$ is sufficiently small.

For any directions $\delta\omega = (\delta q, \delta u), \widetilde{\delta\omega} = (\widetilde{\delta q}, \widetilde{\delta u}) \in X$ the second Fréchet derivative of the Lagrangian is given by

$$\begin{aligned} L''(\omega, p)(\delta\omega, \widetilde{\delta\omega}) = \beta \int_0^T \int_\Omega \delta q_t \widetilde{\delta q}_t + \delta q \widetilde{\delta q} + \delta q_x \widetilde{\delta q}_x + \delta q_{xx} \widetilde{\delta q}_{xx} \, \mathrm{d}x\mathrm{d}t \\ + \int_\Omega \delta u(T) \widetilde{\delta u}(T) \, \mathrm{d}x + \int_0^T \int_\Omega (\delta q \lambda)_x \widetilde{\delta u}_x + (\widetilde{\delta q} \lambda)_x \delta u_x \, \mathrm{d}x\mathrm{d}t \end{aligned}$$

with $\omega = (q, u) \in X$ and $p = (\lambda, \mu, \nu) \in Y$.

We introduce the *second–order necessary optimality condition*

$$L''(\omega^*, p^*)(\delta\omega, \delta\omega) \geq 0 \quad \text{for all } \delta\omega \in C(\omega^*), \tag{47}$$

where

$$C(\omega^*) = \left\{ \delta\omega \in \left(T_{\hat{\mathcal{Q}}_{\text{ad}}}(q^*) \cap \{\xi^*\}^\perp \right) \times W(0,T) : \delta\omega \in N(e'(\omega^*)) \right\}$$

denotes the *critical cone at* ω^*, $^\perp$ denotes the orthogonal complement in $H^{2,1}(Q)$ and $T_{\hat{\mathcal{Q}}_{\text{ad}}}(q^*)$ is the tangential cone at q^* (introduced in Def. 3.10). The critical cone $C(\omega^*)$ is the set of directions that are tangent to the feasible set.

The naming of (47) calls for justification. To this end let us recall the concept of polyhedricity.

Definition 3.14. *Let K be a closed convex subset of the Hilbert space Z, $z \in Z$ and $v \in N_K(z)$. Then K is called* polyhedric at z for the normal direction v *provided*

$$T_K(z) \cap \{v\}^\perp = \overline{R_K(z) \cap \{v\}^\perp}.$$

If K is polyhedric at each $z \in K$ for all directions $v \in N_K(z)$, we call K polyhedric.

In the present situation, $Z = H^{2,1}(Q)$, $K = \hat{\mathcal{Q}}_{\text{ad}}$ and $z = q$ and the following proposition follows directly from [7, Prop. 4.3].

Proposition 3.15. *The closed convex set $\hat{\mathcal{Q}}_{\text{ad}}$ is polyhedric.*

The polyhedricity of the set of admissible coefficients allows to prove the following theorem which justifies the name of (47).

Theorem 3.16 (cf. [7]). *Let ω^* be a local solution to $(\hat{\mathbf{P}})$. Then ω^* satisfies* (47).

Next, we turn to *sufficient* conditions. Let us recall the following definition, see [7].

Definition 3.17. *Let $\omega^* = (q^*, u^*) \in \hat{E}$.*

a) *The point ω^* is a local solution to* $(\hat{\mathbf{P}})$ *satisfying the* quadratic growth condition *if there exists a $\rho > 0$ satisfying*

$$J(\omega) \geq J(\omega^*) + \rho \, \|\omega - \omega^*\|_X^2 + o(\|\omega - \omega^*\|_X^2) \quad \text{for all } \omega \in \hat{E}. \tag{48}$$

b) *Suppose that ω^* satisfies the first–order necessary optimality conditions with associated unique Lagrange multipliers $p^* \in Y$ and $\xi^* \in N_{\hat{\mathcal{Q}}_{\text{ad}}}(q^*)$. At (ω^*, p^*) the* second–order sufficient optimality condition *holds if there exists a constant $\kappa > 0$ such that*

$$L''(\omega^*, p^*)(\delta\omega, \delta\omega) \geq \kappa \, \|\delta\omega\|_X^2 \quad \text{for all } \delta\omega \in C(\omega^*). \tag{49}$$

It turns out that (48) and (49) are related to the weaker condition

$$L''(\omega^*, p^*)(\delta\omega, \delta\omega) > 0 \quad \text{for all } \delta\omega \in C(\omega^*) \setminus \{0\}, \tag{50}$$

which is very close to the necessary optimality condition. In particular, the following theorem holds.

Theorem 3.18 (cf. [18]). *The quadratic growth condition* (48)*, the second–order sufficient optimality condition* (49)*, and condition* (50) *are equivalent.*

Hence, to prove that a solution (ω^*, p^*) satisfying $\omega^* = (q^*, u^*) \in \hat{E}$, $q^* \in \hat{\mathcal{Q}}_{\text{ad}}$, (29) and (45) truly solves $(\hat{\mathbf{P}})$, one needs to guarantee second–order sufficient optimality by ensuring that one of these sufficient second–order conditions is fulfilled. Indeed, using the estimate (3.9) for the Lagrange multiplier, it is possible to show the following theorem. It presents a sufficient condition for the second–order sufficient optimality condition (49).

Theorem 3.19 (cf. [18]). *Let all hypotheses of Theorem* 3.7 *be satisfied. Then the second–order sufficient optimality condition* (49) *holds provided the residual $\|u^*(T) - u_T\|_{L^2(\Omega)}$ is sufficiently small.*

As we have characterized a (local) solution by its optimality conditions, we now turn to the numerical algorithm to determine a solution numerically.

4. A Sequential Quadratic Programming Algorithm

In this section we turn to the optimization algorithm used to solve the parameter identification problem $(\hat{\mathbf{P}})$. We employ a globalized sequential quadratic programming (SQP) method that was originally proposed in [18]. The method consists of an outer and an inner loop. In each SQP iteration of the outer loop a linear–quadratic optimal control problem with box constraints is solved by a primal–dual active set strategy in an inner loop. The globalization is implemented by a modification of the Hessian matrix to ensure that every SQP step is a descent direction and by a line search strategy.

The globalized SQP method is addressed in Section 4.1.. Section 4.2. is concerned with the primal–dual active set method. The line search strategy is discussed in Section 4.3.. In Section 4.4. we propose a variation of the algorithm which leads to a semi–smooth Newton method.

4.1. The Globalized SQP Method

In the numerical realization we initialize our SQP method with a function $u^0 \in W(0,T)$, which satisfies the boundary condition $u^0(\cdot, a) = u_D$ in $(0,T)$ and the initial condition $u^0(0) = u_0$ in Ω. Hence, the iterates $u^{n+1} = u^n + \delta u^n$, $n \geq 0$, can be determined by looking for δu^n in the linear space

$$\mathfrak{U} = \left\{ \delta u \in W(0,T) \,:\, \delta u(t,\cdot) \in H_0^1(\Omega) \text{ for } t \in (0,T) \text{ and } \delta u(0) = 0 \text{ in } \Omega \right\},$$

which is a Hilbert space endowed with the topology of $W(0,T)$. Consequently, the constraints $e_2(\omega^n) = 0$ and $e_3(\omega^n) = 0$ are satisfied by construction for any $n \geq 0$ and the only constraint left is $e_1(\omega) = 0$ in $L^2(0,T;H^{-1}(\Omega))$. Set $Y_1 = L^2(0,T;H_0^1(\Omega))$ and identify its dual space Y_1' with $L^2(0,T;H^{-1}(\Omega))$.

In our numerical implementation we make use of the the 'weaker' norm in $L^2(0,T;H^1(\Omega))$, relative to a prior q_d, instead of using the $H^{2,1}(Q)$–norm in the regularization term in (22). Hence, as already noted in Remark 3.8 we only have to solve a Neumann problem for q in each time step. Therefore, let $X = X_1 \times X_2 = L^2(0,T;H^1(\Omega)) \times W(0,T)$ in the following.

The linear–quadratic minimization problems that have to be solved in each step of the SQP method are well–defined provided $L''(\omega^n, \lambda^n)$ is coercive on $N(e'(\omega^n))$ and $e'(\omega^n)$ on $N(e'(\omega^n))$ is surjective for every iterate. Often these requirements hold only locally so we consider in the following a globalization strategy using a modified Hessian (cf. [24]). For $\gamma \in [0,1]$ let us define the function ${}^\gamma\mathsf{L} : X \times Y_1 \to \mathbb{R}$ by (see [25])

$${}^\gamma\mathsf{L}(\omega, \lambda) = J(\omega) + \gamma \left\langle e_1(\omega), \lambda \right\rangle_{Y_1', Y_1} \quad \text{for } (\omega, \lambda) \in X \times Y_1.$$

Notice that for $\gamma = 1$ the function ${}^\gamma\mathsf{L}$ is the usual Lagrangian associated with the single constraint $e_1(\omega) = 0$ in Y_1'. Therefore, we set $\mathsf{L} = {}^\gamma\mathsf{L}$ for $\gamma = 1$.

Algorithm 1 (Globalized SQP method).

1) Choose $\omega^0 = (q^0, u^0) \in X = X_1 \times X_2$ *that satisfies* $u^0(\cdot, a) = u_D$ *in* $(0,T)$ *and* $u^0(0) = u_0$ *in* Ω, $\lambda^0 \in Y_1$ *with* $\lambda^0(T) = u_T - u^0(T)$ *in* Ω. *Fix relative and absolute stopping tolerances* $1 > \varepsilon_{\mathsf{rel}} \geq \varepsilon_{\mathsf{abs}} > 0$. *Choose the maximal number of SQP iterations* $n_{\mathsf{sqp}} \in \mathbb{N}$. *Set* $n := 0$ *and* $\bar{\kappa} \in (0,1]$.

2) For $\omega^n = (q^n, u^n)$ *evaluate* $e_1(\omega^n)$, $e_1'(\omega^n)$, $J(\omega^n)$, $J'(\omega^n)$. *If*

$$\|\nabla \mathsf{L}(\omega^n, \lambda^n)\|^2_{X' \times Y_1'} = \|\mathsf{L}'(\omega^n, \lambda^n)\|^2_{X'} + \|e_1(\omega^n)\|^2_{Y_1'} < \varepsilon_{\mathsf{abs}}$$

or

$$\|\nabla \mathsf{L}(\omega^n, \lambda^n)\|^2_{X' \times Y_1'} < \varepsilon_{\mathsf{rel}} \, \|\nabla \mathsf{L}(\omega^0, \lambda^0)\|^2_{X' \times Y_1'}$$

or $n = n_{\mathsf{sqp}}$, *then STOP. Otherwise continue with step* 3).

3) Set $\gamma = 1$.

4) *Solve for* $\delta\omega^n = (\delta q^n, \delta u^n)$ *the following linear–quadratic optimal control problem*

$$\min \mathsf{L}(\omega^n, \lambda^n) + \mathsf{L}'(\omega^n, \lambda^n)\delta\omega + \frac{1}{2}{}^{\gamma}\mathsf{L}''(\omega^n, \lambda^n)(\delta\omega^n, \delta\omega^n)$$

$$\text{s.t.} \begin{cases} e_1'(\omega^n)\delta\omega^n + e_1(\omega^n) = 0 & \text{in } Y_1', \\ q_{\mathsf{min}} \le q^n + \delta q^n \le q_{\mathsf{max}} & \text{in } X_1. \end{cases} \qquad \textbf{(QP)}$$

5) *If* $\|\delta\omega\|_X > 0$ *evaluate the quotient*

$$\kappa^n = \frac{{}^{\gamma}\mathsf{L}''(\omega^n, \lambda^n)(\delta\omega^n, \delta\omega^n)}{\|\delta q^n\|_{X_1}^2 + \|\delta u^n\|_{L^2(0,T;L^2(\Omega))}^2}.$$

where

$${}^{\gamma}\mathsf{L}''(\omega^n, \lambda^n)(\delta\omega^n, \delta\omega^n) = J''(\omega^n)(\delta\omega^n, \delta\omega^n) + \gamma \left\langle e_1''(\omega^n)(\delta\omega^n, \delta\omega^n), \lambda^n \right\rangle_{Y_1', Y_1}$$

holds. If $\kappa^n \le 0$ *then set* $\gamma = 0$ *and go back to step* 4). *If* $\kappa^n \in (0, \bar\kappa)$ *then set* $\bar\kappa = \kappa^n$.

6) *Determine a step size parameter* $\alpha_n \in (0, 1]$ *by a backtracking line search (see Section 4.3.).*

7) *Set* $q^{n+1} = q^n + \alpha_n \delta q^n$, $u^{n+1} = u^n + \alpha_n \delta u^n$, $\lambda^{n+1} = \lambda^n + \alpha_n \delta\lambda^n$, *set* $n := n + 1$, *and go back to step* 2).

Remark 4.1. In Algorithm 1, the positive scalar $\bar\kappa$ serves as an estimate for the coercivity constant in (49). Instead of setting $\gamma = 0$ directly if $\kappa^n \le 0$, γ can be adjusted by using an iterative strategy, see [18] for details.

4.2. The Primal–Dual Active Set Method

In each iteration of Algorithm 1 we have to solve the linear–quadratic optimal control problem (**QP**) in step 4). To this end we apply the primal–dual active set method [25]. In practice, this algorithm behaves like an infeasible one, since its iterates violate the constraints up to the last–but–one iterate. It stops at a feasible and optimal solution. Based on the identification (46) for the inactive and active sets, the algorithm exhibits a low number of iterations to find the optimal solution and is very robust [5,25]. Note also that the algorithm uses only one Lagrange multiplier to realize the two inequality constraints. This reduces the number of variables and hence the amount of memory needed by its implementation.

Suppose that we are at level n of the SQP method. Thus, we have iterates (q^n, u^n, λ^n) and start our primal–dual active set strategy to determine the directions $(\delta q^n, \delta u^n, \delta\lambda^n)$. Let us define $q^n_{\mathsf{min}} = q_{\mathsf{min}} - q^n$ and $q^n_{\mathsf{max}} = q_{\mathsf{max}} - q^n$. Then we are looking for a step δq^n satisfying the inequality constraints $q^n_{\mathsf{min}} \le \delta q^n \le q^n_{\mathsf{max}}$ in Q. The method uses the identification (46) as a prediction strategy, i.e. given a current primal–dual iteration pair $(\delta q_k, \xi_k)$ the next inactive and active sets are given by (51)–(53) (see below). The method stops at a feasible and optimal solution as soon as two consecutive tuples of active and inactive sets are equal. This solution is then used in Algorithm 1.

Instead of solving problem (**QP**) with inequality constraints we solve

$$\min \mathsf{L}(\omega^n,\lambda^n)+\mathsf{L}'(\omega^n,\lambda^n)\delta\omega+\frac{1}{2}\gamma\mathsf{L}''(\omega^n,\lambda^n)(\delta\omega^n,\delta\omega^n)$$

$$\text{s.t.}\begin{cases} e_1'(\omega^n)\delta\omega^n+e_1(\omega^n)=0 & \text{in } Y_1',\\ \delta q_k=q^n_{\mathsf{min}} & \text{in } \mathcal{A}^k_-,\\ \delta q_k=q^n_{\mathsf{max}} & \text{in } \mathcal{A}^k_+,\\ q^n_{\mathsf{min}}\le q^n+\delta q_k\le q^n_{\mathsf{max}} & \text{in } X_1.\end{cases} \qquad (\mathbf{QP}_{\mathcal{I}_k})$$

Hence, it is solved for δq_k only on the inactive set $\mathcal{I}_k$, but δq_k is fixed on the active set $\mathcal{A}^k=\mathcal{A}^k_-\cup\mathcal{A}^k_+$. We have the following algorithm.

Algorithm 2 (Primal–dual active set method).

1) *Choose initial values* $(\delta q_0,\xi_0)\in X_1\times X_1$*, the parameter* $c>0$ *and set* $k:=0$.

2) *Determine the active and inactive sets*

$$\mathcal{A}^k_-=\{(t,x)\in Q\mid \delta q_k+\frac{\xi_k}{c}<q^n_{\mathsf{min}}\}, \tag{51}$$

$$\mathcal{A}^k_+=\{(t,x)\in Q\mid \delta q_k+\frac{\xi_k}{c}>q^n_{\mathsf{max}}\}, \tag{52}$$

$$\mathcal{I}^k=Q\setminus\mathcal{A}^k. \tag{53}$$

with $\mathcal{A}^k=\mathcal{A}^k_-\cup\mathcal{A}^k_+$*, and set*

$$\widetilde{\delta q_k}=\begin{cases} q^n_{\mathsf{min}} & \textit{in } \mathcal{A}^k_-,\\ q^n_{\mathsf{max}} & \textit{in } \mathcal{A}^k_+,\\ \delta q_k & \textit{in } \mathcal{I}^k.\end{cases}$$

3) *If* $k>0$ *and* $\mathcal{A}^k=\mathcal{A}^{k-1}$ *holds then STOP.*

4) *Solve for* $(\delta q_k,\delta u_k,\delta\lambda_k)\in X_1\times\mathfrak{U}\times Y_1$ *the linearized state equations*

$$(\delta u_k)_t-q^n(\delta u_k)_{xx}-\widetilde{\delta q_k}u^n_{xx}=-(u^n_t-q^nu^n_{xx}) \quad \textit{in } Q, \tag{54a}$$

$$\delta u_k(\cdot,a)=\delta u_k(\cdot,b)=0 \quad \textit{in } (0,T), \tag{54b}$$

$$\delta u_k(0)=0 \quad \textit{in } \Omega, \tag{54c}$$

the linearized adjoint system

$$-(\delta\lambda_k)_t-(q^n\delta\lambda_k)_{xx}-\gamma(\widetilde{\delta q_k}\lambda^n)_{xx}=-(-\lambda^n_t-(q^n\lambda^n)_{xx}) \quad \textit{in } Q, \tag{54d}$$

$$\delta\lambda_k(\cdot,a)=\delta\lambda_k(\cdot,b)=0 \quad \textit{in } (0,T), \tag{54e}$$

$$\delta\lambda_k(T)+\delta u_k(T)=u_T-u^n(T)-\lambda^n(T) \quad \textit{in } \Omega, \tag{54f}$$

and the linearized optimality condition on the inactive set

$$\begin{aligned}&\beta(-(\delta q_k)_{xx}+\delta q_k)-\delta\lambda_k u^n_{xx}-\gamma\lambda^n(\delta u_k)_{xx}\\&\quad=-\Big(\beta\big(-(q^n-q_d)_{xx}+(q^n-q_d)\big)-\widetilde{\mathcal{R}}(\lambda^n u^n_{xx})\Big)\quad \text{in } \mathcal{I}^k,\end{aligned}\tag{54g}$$

where (q^n, u^n, λ^n) *denotes the current SQP iterate and* $\widetilde{\mathcal{R}} = (-\Delta + \mathrm{Id})^{-1}$ *is the Riesz operator between* $(H^1(\Omega))'$ *and* $H^1(\Omega)$.

5) *Set* $q^n_k = q^n + \delta q_k$, $u^n_k = u^n + \delta u_k$, $\lambda^n_k = \lambda^n + \delta\lambda_k$ *and*

$$\xi_k = -\beta(-(q^n_k - q_d)_{xx} + q^n_k - q_d) + \widetilde{\mathcal{R}}(\lambda^n_k (u^n_k)_{xx}) \quad \text{in } Q.$$

From (54g) *we infer that* $\xi_k = 0$ *on* $\mathcal{I}_k$. *Set k:=k+1 and go back to step* 2).

Recall that we have chosen λ^0 in such a way that $\lambda^0(T) = u_T - u^0(T)$ holds in Ω. Thus, by induction we have $\lambda^n(T) = u_T - u^n(T)$ in Ω and (54f) reads

$$\delta\lambda_k(T) + \delta u_k(T) = 0 \quad \text{in } \Omega.$$

4.3. The Line Search Strategy

The line search strategy employed in step 6) of Algorithm 1 is the subject of this subsection. Algorithm 2 ensures feasibility of each solution to **(QP)** with respect of the solution to the inequality constraints. Thus, the goal is to find a compromise between the descent of the cost functional J and the reduction of the violation of the equality constraint $e(\omega) = 0$. This is realized by a line search method utilizing a suitable chosen merit function. We use the exact penalty functional

$$\varphi^n(\alpha^n) = J(\omega^n + \alpha^n\delta\omega^n) + \mu\,\|e_1(\omega^n + \alpha^n\delta\omega^n)\|_{Y_1'},\tag{55}$$

where $\alpha^n \in (0, 1]$ is the step size parameter, which has to be determined, and $\mu > 0$ is a penalty parameter for violations of the constraint $e_1(\omega) = 0$. This choice does not introduce an additional nonlinearity into our problem. In a first–order variation we use the approximation

$$\begin{aligned}\bar\varphi^n(\alpha^n) &= J(\omega^n) + \alpha^n J'(\omega^n)\delta\omega^n + \mu\,\|e_1(\omega^n) + \alpha^n e_1'(\omega^n)\delta\omega^n\|_{Y_1'}\\&= J(\omega^n) + \alpha^n J'(\omega^n)\delta\omega^n + \mu(1-\alpha^n)\,\|e_1(\omega^n)\|_{Y_1'},\end{aligned}$$

where we have used that $e_1(\omega^n) + e_1'(\omega^n)\delta\omega^n = 0$ holds for the SQP step $\delta\omega^n$. Notice that $\bar\varphi^n(1)$ does not depend on μ. For appropriately chosen penalty parameter $\mu > 0$, our line search strategy is based on the well–known Armijo rule

$$\varphi^n(\alpha^n) - \varphi^n(0) \le c\alpha^n\left(\bar\varphi^n(1) - \bar\varphi^n(0)\right) \quad \text{for } c \in \left(0, \frac{1}{2}\right).\tag{56}$$

There exists a sufficiently large penalty parameter $\bar\mu > 0$ [24] such that

$$\bar\varphi^n(1) - \bar\varphi^n(0) < 0 \quad \text{for all } \mu \ge \bar\mu.\tag{57}$$

To find an appropriate value for μ we check whether (57) holds and increase μ if not and iterate. If (57) is fulfilled we determine α^n from (56) by a backtracking strategy starting with $\alpha^n = 1$. If (56) is violated we decrease α^n by setting $\alpha^n := \zeta\alpha^n$ with $\zeta \in (0, 1)$ and iterate until (56) holds or α^n falls below a minimal step size parameter.

4.4. Variant: Semi–smooth Newton Method

One can also consider the following variant of the algorithm proposed above (cf. [15]). In each SQP iteration, in contrast to the above, one does not iterate until convergence, but only performs one step of the primal–dual active set method. This yields a semi–smooth Newton method [43], that converges locally superlinearly. The relationship between the primal dual active set strategy and semi–smooth Newton methods has been investigated in [26]. We keep the implemented globalization strategy that is realized by the modification of the Hessian and by the line search strategy to ensure that every SQP step is a descent direction. Since we do not iterate the *inner loop* until convergence now the inequality constraints are not necessarily fulfilled in each SQP iteration. Therefore, we need to modify the line search strategy presented above. To determine the step size parameter α^n we here use a first–order variation of the merit function

$$\varphi^n(\alpha^n) = J(\omega^n + \alpha^n \delta\omega^n) + \mu \, \|e_1(\omega^n + \alpha^n \delta\omega^n)\|_{Y_1'} \\ + \zeta \, \|\max\{0, q^n - q_{\mathsf{max}}\}\|_{L^2(Q)} + \eta \, \|\max\{0, q_{\mathsf{min}} - q^n\}\|_{L^2(Q)}.$$

For appropriately chosen penalty parameters, our line search strategy is then based on the Armijo rule as above.

5. Numerical Experiments

In this section we present some numerical results. The algorithm proposed in Section 4. has to be discretized for the numerical realization. Here, we make use of a finite element method with piecewise linear finite elements for the spatial discretization. To solve the discrete version of (54) we apply a preconditioned GMRES method.

Some words are in order concerning the market data u_T. As index option prices are concerned, many data are freely available over the internet. To ensure that these price quotes really contain information of market participants' expectations of future market developments one should try to use prices, at which options were actually traded. Moreover, it can be useful to exclude data which violate well–known arbitrage bounds. To carry out the above analysis and numerics, we implicitly assumed, that there exists a continuum of market data, that can be used. In practice, one will obtain discrete data which need to be interpolated to be used. The data can be interpolated using a cubic spline. Simple, linear interpolation is not advisable, since it may introduce violations of arbitrage bounds. Note also that meaningful empirical data are usually only available for strike prices E in a small region around the actual spot price S_0, typically $S_0 \pm 10\%$ or even only $S_0 \pm 5\%$. Restricting the computational domain to this small region is not advisable, hence one needs to extrapolate the data outside of this region. For example, one can compute the implied volatilities of the options with the largest and the smallest strike price available by inverting the Black–Scholes formula, and use prices computed by the Black–Scholes formula with these volatilities in those regions in which no market data are available.

As a rule of thumb, the regularization parameter β should be chosen of the same order as prevalent data noise. In market prices, a the spread between bid and ask quotes can be used as a proxy. A typical number would be a noise level of $\delta = 0.1\%$. In [20], the following

Table 1. True option price and reconstructed option prices computed for different strikes E and different prior volatilities q_d. The true values were computed using the Black–Scholes formula with constant volatility $\sigma = 0.15$.

Strike E	95	97.5	100	102.5	105
True value	5.24433	3.23921	1.72734	0.77577	0.28866
Good guess	5.24429	3.23921	1.72734	0.77576	0.28861
Good guess & noise	5.24430	3.23921	1.72734	0.77576	0.28861
Bad guess	5.24435	3.23922	1.72733	0.77578	0.28866
Bad guess & noise	5.24435	3.23922	1.72733	0.77578	0.28866
Good guess, fine grid	5.24433	3.23921	1.72734	0.77577	0.28866

strategy was proposed, that we apply here. Define a decreasing sequence of admissible regularization parameters by

$$(\beta_1, \ldots, \beta_6) = (10^2, 10^1, \ldots, 10^{-3})\delta = (10^{-1}, 10^{-2}, \ldots, 10^{-6}).$$

Start the minimization procedure with the highest value $\beta = \beta_1$ and subsequently decrease the regularization parameter, restarting the method at the minimizers obtained in the previous step.

As an example we apply our method to an artificial data set of Black–Scholes European call option prices, i.e. prices computed with the Black–Scholes formula, with $S_0 = 100$, $r = 0$, one month to maturity and constant volatility $\sigma = 0.15$. Our goal is to identify the volatility in equation (10), that is supplemented with the initial condition

$$V(0, E) = V_0(E) = \max(S_0 - E, 0), \quad E > 0,$$

and boundary conditions

$$V(T, 0) = S_0, \quad \lim_{E \to \infty} V(T, E) = 0, \quad T > 0.$$

For the SQP method we choose stopping tolerances $\varepsilon_{\mathsf{abs}} = 10^{-6}$ and $\varepsilon_{\mathsf{rel}} = 10^{-3}$ and a maximal number of iterations $n_{\mathsf{sqp}} = 20$. We use a non–uniform grid with 140 nodes locally refined around $x = S_0$ for the spatial discretization. In time, we employ a fixed, non–equidistant grid consisting of 35 points with small time steps close to $t = 0$.

We consider four different cases with a *prior* volatility $q_d = \frac{1}{2}\sigma_d^2 x^2$. In our first simulation we use the 'good' guess $\sigma_d = 0.16$, in a second one we add 0.1% uniformly distributed noise. We compare these results to those from a third and fourth run using a 'bad' guess $\sigma_d = 0.1$, where in the fourth run again we added 0.1% uniformly distributed noise. The resulting option prices are given in Table 1 and the corresponding volatilities are shown in Table 2. The error–free true values for the option prices were computed using the Black–Scholes formula. Table 3 displays the residuals $\|u(T) - u_T\|_{L^2(\Omega)}$ remaining after reconstruction. In all four tests, the identified option prices correspond very well to the true

Table 2. Reconstructed volatilities for different strikes E and different prior volatilities q_d.

Strike E	95	97.5	100	102.5	105
True value	0.1500	0.1500	0.1500	0.1500	0.1500
Good guess	0.1454	0.1500	0.1517	0.1506	0.1470
Good guess & noise	0.1457	0.1500	0.1517	0.1506	0.1470
Bad guess	0.1458	0.1500	0.1517	0.1506	0.1472
Bad guess & noise	0.1460	0.1500	0.1517	0.1506	0.1472
Good guess, fine grid	0.1488	0.1500	0.1509	0.1503	0.1494

Table 3. Residuals remaining after reconstruction.

	$\|u(T) - u_T\|_{L^2(\Omega)}$
Good guess	6.76×10^{-3}
Good guess & noise	3.56×10^{-2}
Bad guess	6.50×10^{-3}
Bad guess & noise	3.47×10^{-2}
Good guess, fine grid	1.51×10^{-3}

values, the difference is neglectably small. The corresponding volatilities are well identified, with small differences remaining due to discretization errors, which can be reduced by using a finer grid. This can be seen from the results of a fifth run which was executed on a grid with halved mesh size in space and time. Figure 1 displays the error in reconstructed option prices (left figure) and the relative error in the associated volatilities (right figure) for different values of the regularization parameter β. Decreasing β further is not advisable since for very small values of β, the reconstructed σ shows oscillations, since we do not control the H^1 norm (Remark 3.8), while the residual still decreases.

Conclusion

To conclude, the globalized SQP algorithm provides a stable method to identify option prices and associated volatilities. It takes very few iterations, typically one to three, to meet the prescribed stopping tolerances. Moreover, the algorithm showed very small dependence regarding to the chosen *prior* volatility and is robust with respect to additional data noise.

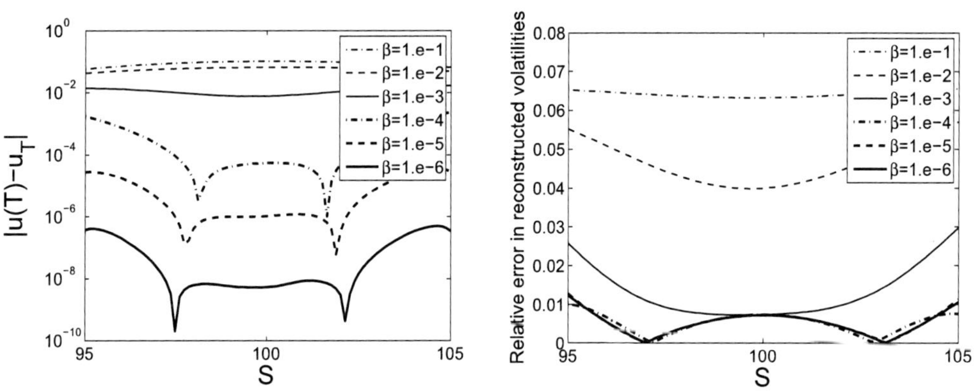

Figure 1. Error in reconstructed option prices (left figure) and relative error in the associated volatilities (right figure) for different values of the regularization parameter β.

Acknowledgements

The author acknowledges support from the Deutsche Forschungsgemeinschaft, grant JU 359/6 (Forschergruppe 518).

References

[1] Y. Achdou and O. Pironneau, Volatility smile by multilevel least square, *Int. J. Theor. Appl. Finance* **5** (2002), 619–643.

[2] M. Avellaneda, C. Friedman, R. Holmes and D. Samperi, Calibrating volatility surfaces via relative-entropy minimization, *Appl. Math. Finance* **4** (1997), 37–64.

[3] G. Barles and H.M. Soner, Option Pricing with transaction costs and a nonlinear Black–Scholes equation, *Finance Stochast.* **2** (1998), 369–397.

[4] H. Berestycki, J. Busca and I. Florent, Asymptotics and calibration of local volatility models, *Quant. Finance* **2** (2002), 61–69.

[5] M. Bergounioux, K. Ito and K. Kunisch, Primal–dual strategy for constrained optimal control problems, *SIAM J. Control Optim.* **35** (1997), 1524–1543.

[6] P.T. Boggs and J.W. Tolle, Sequential quadratic programming, In: *Acta Numerica* 1995, A. Iserles (ed.), Cambridge University Press, 1995, pp. 1–51.

[7] J.F. Bonnans, Second-order analysis for control constrained optimal control problems of semilinear elliptic systems, *Appl. Math. Optim.* **38** (1998), 303–325.

[8] P. Boyle and T. Vorst, Option replication in discrete time with transaction costs, *J. Finance* **47** (1992), 271–293.

[9] G.M. Constantinides and T. Zariphopoulou, Bounds on process of contingent claims in an intertemporal economy with proportional transaction costs and general preferences, *Finance Stoch.* **3** (1999), 345–369.

[10] S. Crépey, Calibration of the local volatility in a trinomial tree using Tikhonov regularization, *Inverse Probl.* **19** (2003), 91–127.

[11] R. Dautray and J.-P. Lions, Mathematical Analysis and Numerical *Methods for Science and Technology* **5**: Evolution Problems, Springer-Verlag, Berlin, 1992.

[12] M. Davis, V. Panis and T. Zariphopoulou, European option pricing with transaction fees, *SIAM J. Contr. Optim.* **31** (1993), 470–493.

[13] B. Dupire, Pricing with a smile, *Risk* **7** (1994), 18–20.

[14] B. Dupire, A unified theory of volatility, In: *Derivatives Pricing: The Classic Collection*, P. Carr (ed.), Risk Books, 2004.

[15] B. Düring, A semi-smooth Newton method for an inverse problem in option pricing, To appear in: *Proc. Appl. Math. Mech. (PAMM), Proceedings of ICIAM07*, 2008.

[16] B. Düring, M. Fournié and A. Jüngel, High order compact finite difference schemes for a nonlinear Black–Scholes equation, *Int. J. Theor. Appl. Finance* **6** (2003), 767–789.

[17] B. Düring, M. Fournié and A. Jüngel, Convergence of a high–order compact finite difference scheme for a nonlinear Black–Scholes equation, *Math. Mod. Num. Anal.* **38** (2004), 359–369.

[18] B. Düring, A. Jüngel and S. Volkwein, Sequential Quadratic Programming method for volatility estimation in option pricing, *J. Optim. Theory Appl.* **38**, 2008.

[19] B. Düring and E. Lüders, Option prices under generalized pricing kernels, *Rev. Der. Research* **8** (2005), 97–123.

[20] H. Egger and H.W. Engl. Tikhonov regularization applied to the inverse problem of option pricing: convergence analysis and rates, *Inverse Probl.* **21** (2005), 1027–1045.

[21] H.W. Engl, M. Hanke and A. Neubauer, *Regularization of Inverse Problems*, Kluwer Academic Publishers, Dordrecht, 1996.

[22] R. Frey, Perfect option hedging for a large trader, *Finance Stochast.* **2** (1998), 115–141.

[23] M. Hanke and E. Rösler, Computation of local volatilities from regularized Dupire equations, *Int. J. Theor. Appl. Finance* **8** (2005), 207–221.

[24] M. Hintermüller. On a globalized augmented Lagrangian–SQP algorithm for nonlinear optimal control problems with box constraints, in K.-H. Hoffmann, R. H. W. Hoppe, V. Schulz, editors, *Fast Solution of Discretized Optimization Problems, International Series of Numerical Mathematics* **138** (2001), 139–153.

[25] M. Hintermüller, A primal–dual active set algorithm for bilaterally control constrained optimal control problems, *Quarterly Appl. Math.* **61** (2003), 131–161.

[26] M. Hintermüller, K. Ito and K. Kunisch, The primal–dual active set strategy as a semi–smooth Newton method. *SIAM J. Optim.* **13** (2003), 865–888.

[27] S.D. Hodges and A. Neuberger, Optimal replication of contingent claims under transaction costs, *Rev. Future Markets* **8** (1989), 222–239.

[28] K. Ito and K. Kunisch, The primal–dual active set method for nonlinear problems with bilateral constraints, *SIAM J. Control Optim.* **43** (2004), 357–376.

[29] K. Ito and K. Kunisch, Augmented Lagrangian methods for nonsmooth convex optimization in Hilbert spaces, *Nonlin. Anal. TMA* **41** (2000), 591–616.

[30] N. Jackson, E. Sli and S. Howison, Computation of deterministic volatility surfaces, *J. Comp. Finance* **2** (1999), 5–32.

[31] P. Kangro and R. Nicolaides, Far field boundary conditions for Black–Scholes equations, *SIAM J. Numer. Anal.* **38** (2000), 1357–1368.

[32] I. Kani, E. Derman and M. Kamal, Trading and hedging local volatility, *J. Financ. Engin.* **6** (1997), 1233–1268.

[33] S. Kusuoka, Option replication cost with transaction costs, In: *Nonlinear and convex analysis in economic theory,* T. Maruyama (ed.), *Proceedings of the T.I. Tech/ K.E.S. conference on nonlinear and convex analysis in economic theory,* Keio University, Japan, July 2-4, 1993. *Lect. Notes Econ. Math. Syst.* **419**, 1995, pp. 197–202.

[34] O.A. Ladyzenskaja, V.A. Solonnikov and N.N. Ural'ceva, Linear and Quasilinear Equations of Parabolic Type, *AMS Transl.*, Vol. 23, Providence, 1968.

[35] R. Lagnado and S. Osher, A technique for calibrating derivative security pricing models: numerical solution of an inverse problem, *J. Comput. Finance* **1** (1997), 13–25.

[36] H. Leland, Option pricing and replication with transaction costs, *J. Finance* **40** (1985), 1283–1301.

[37] J.-L. Lions, *Control of Distributed Singular Systems*, Gauthier–Villars, Paris, 1983.

[38] D.G. Luenberger, *Optimization by Vector Space Methods*, John Wiley & Sons, New York, 1969.

[39] A. Parás and M. Avellaneda, Dynamic hedging portfolios for derivative securities in the presence of large transaction costs, *Appl. Math. Finance* **1** (1994), 165–193.

[40] J. Simon, Compact sets in the space $L^p(0, T; B)$, *Ann. Mat. Pura Appl., IV. Ser.* **146** (1987), 65–96.

[41] K.R. Sircar and G. Papanicolaou, General Black–Scholes models accounting for increased market volatility from hedging strategies, *Appl. Math. Finance* **5** (1998), 45–82.

[42] H.M. Soner, S.E. Shreve and J. Cvitanic, There is no nontrivial hedging portfolio for option pricing with transaction costs, *Ann. Appl. Probab.* **5** (1995), 327–355.

[43] M. Ulbrich, Semismooth Newton methods for operator equations in function spaces, *SIAM J. Optim.* **13** (2003), 805–842.

[44] A. Whalley and P. Wilmott, An asymptotic analysis of an optimal hedging model for option pricing with transaction costs, *Math. Finance* **7** (1997), 307–324.

INDEX

A

B

C

D

M

N

O

P

Q

R

U

V

W

Y